Alternative Sweeteners

FOOD SCIENCE AND TECHNOLOGY

A Series of Monographs, Textbooks, and Reference Books

1. Flavor Research: Principles and Techniques, *R. Teranishi, I. Hornstein, P. Issenberg, and E. L. Wick*
2. Principles of Enzymology for the Food Sciences, *John R. Whitaker*
3. Low-Temperature Preservation of Foods and Living Matter, *Owen R. Fennema, William D. Powrie, and Elmer H. Marth*
4. Principles of Food Science
   Part I: Food Chemistry, *edited by Owen R. Fennema*
   Part II: Physical Methods of Food Preservation, *Marcus Karel, Owen R. Fennema, and Daryl B. Lund*
5. Food Emulsions, *edited by Stig E. Friberg*
6. Nutritional and Safety Aspects of Food Processing, *edited by Steven R. Tannenbaum*
7. Flavor Research: Recent Advances, *edited by R. Teranishi, Robert A. Flath, and Hiroshi Sugisawa*
8. Computer-Aided Techniques in Food Technology, *edited by Israel Saguy*
9. Handbook of Tropical Foods, *edited by Harvey T. Chan*
10. Antimicrobials in Foods, *edited by Alfred Larry Branen and P. Michael Davidson*

11. Food Constituents and Food Residues: Their Chromatographic Determination, *edited by James F. Lawrence*
12. Aspartame: Physiology and Biochemistry, *edited by Lewis D. Stegink and L. J. Filer, Jr.*
13. Handbook of Vitamins: Nutritional, Biochemical, and Clinical Aspects, *edited by Lawrence J. Machlin*
14. Starch Conversion Technology, *edited by G. M. A. van Beynum and J. A. Roels*
15. Food Chemistry: Second Edition, Revised and Expanded, *edited by Owen R. Fennema*
16. Sensory Evaluation of Food: Statistical Methods and Procedures, *Michael O'Mahony*
17. Alternative Sweeteners, *edited by Lyn O'Brien Nabors and Robert C. Gelardi*
18. Citrus Fruits and Their Products: Analysis and Technology, *S. V. Ting and Russell L. Rouseff*
19. Engineering Properties of Foods, *edited by M. A. Rao and S. S. H. Rizvi*
20. Umami: A Basic Taste, *edited by Yojiro Kawamura and Morley R. Kare*
21. Food Biotechnology, *edited by Dietrich Knorr*
22. Food Texture: Instrumental and Sensory Measurement, *edited by Howard R. Moskowitz*
23. Seafoods and Fish Oils in Human Health and Disease, *John E. Kinsella*
24. Postharvest Physiology of Vegetables, *edited by J. Weichmann*
25. Handbook of Dietary Fiber: An Applied Approach, *Mark L. Dreher*
26. Food Toxicology, Parts A and B, *Jose M. Concon*
27. Modern Carbohydrate Chemistry, *Roger W. Binkley*
28. Trace Minerals in Foods, *edited by Kenneth T. Smith*
29. Protein Quality and the Effects of Processing, *edited by R. Dixon Phillips and John W. Finley*
30. Adulteration of Fruit Juice Beverages, *edited by Steven Nagy, John A. Attaway, and Martha E. Rhodes*
31. Foodborne Bacterial Pathogens, *edited by Michael P. Doyle*
32. Legumes: Chemistry, Technology, and Human Nutrition, *edited by Ruth H. Matthews*
33. Industrialization of Indigenous Fermented Foods, *edited by Keith H. Steinkraus*
34. International Food Regulation Handbook: Policy • Science • Law, *edited by Roger D. Middlekauff and Philippe Shubik*
35. Food Additives, *edited by A. Larry Branen, P. Michael Davidson, and Seppo Salminen*
36. Safety of Irradiated Foods, *J. F. Diehl*
37. Omega-3 Fatty Acids in Health and Disease, *edited by Robert S. Lees and Marcus Karel*
38. Food Emulsions: Second Edition, Revised and Expanded, *edited by Kåre Larsson and Stig E. Friberg*
39. Seafood: Effects of Technology on Nutrition, *George M. Pigott and Barbee W. Tucker*
40. Handbook of Vitamins: Second Edition, Revised and Expanded, *edited by Lawrence J. Machlin*

41. Handbook of Cereal Science and Technology, *Klaus J. Lorenz and Karel Kulp*
42. Food Processing Operations and Scale-Up, *Kenneth J. Valentas, Leon Levine, and J. Peter Clark*
43. Fish Quality Control by Computer Vision, *edited by L. F. Pau and R. Olafsson*
44. Volatile Compounds in Foods and Beverages, *edited by Henk Maarse*
45. Instrumental Methods for Quality Assurance in Foods, *edited by Daniel Y. C. Fung and Richard F. Matthews*
46. *Listeria*, Listeriosis, and Food Safety, *Elliot T. Ryser and Elmer H. Marth*
47. Acesulfame-K, *edited by D. G. Mayer and F. H. Kemper*
48. Alternative Sweeteners: Second Edition, Revised and Expanded, *edited by Lyn O'Brien Nabors and Robert C. Gelardi*
49. Food Extrusion Science and Technology, *edited by Jozef L. Kokini, Chi-Tang Ho, and Mukund V. Karwe*
50. Surimi Technology, *edited by Tyre C. Lanier and Chong M. Lee*
51. Handbook of Food Engineering, *edited by Dennis R. Heldman and Daryl B. Lund*
52. Food Analysis by HPLC, *edited by Leo M. L. Nollet*
53. Fatty Acids in Foods and Their Health Implications, *edited by Ching Kuang Chow*
54. *Clostridium botulinum*: Ecology and Control in Foods, *edited by Andreas H. W. Hauschild and Karen L. Dodds*
55. Cereals in Breadmaking: A Molecular Colloidal Approach, *Ann-Charlotte Eliasson and Kåre Larsson*
56. Low-Calorie Foods Handbook, *edited by Aaron M. Altschul*
57. Antimicrobials in Foods: Second Edition, Revised and Expanded, *edited by P. Michael Davidson and Alfred Larry Branen*
58. Lactic Acid Bacteria, *edited by Seppo Salminen and Atte von Wright*
59. Rice Science and Technology, *edited by Wayne E. Marshall and James I. Wadsworth*
60. Food Biosensor Analysis, *edited by Gabriele Wagner and George G. Guilbault*
61. Principles of Enzymology for the Food Sciences: Second Edition, *John R. Whitaker*
62. Carbohydrate Polyesters as Fat Substitutes, *edited by Casimir C. Akoh and Barry G. Swanson*
63. Engineering Properties of Foods: Second Edition, Revised and Expanded, *edited by M. A. Rao and S. S. H. Rizvi*
64. Handbook of Brewing, *edited by William A. Hardwick*
65. Analyzing Food for Nutrition Labeling and Hazardous Contaminants, *edited by Ike J. Jeon and William G. Ikins*
66. Ingredient Interactions: Effects on Food Quality, *edited by Anilkumar G. Gaonkar*
67. Food Polysaccharides and Their Applications, *edited by Alistair M. Stephen*
68. Safety of Irradiated Foods: Second Edition, Revised and Expanded, *J. F. Diehl*
69. Nutrition Labeling Handbook, *edited by Ralph Shapiro*

70. Handbook of Fruit Science and Technology: Production, Composition, Storage, and Processing, *edited by D. K. Salunkhe and S. S. Kadam*
71. Food Antioxidants: Technological, Toxicological, and Health Perspectives, *edited by D. L. Madhavi, S. S. Deshpande, and D. K. Salunkhe*
72. Freezing Effects on Food Quality, *edited by Lester E. Jeremiah*
73. Handbook of Indigenous Fermented Foods: Second Edition, Revised and Expanded, *edited by Keith H. Steinkraus*
74. Carbohydrates in Food, *edited by Ann-Charlotte Eliasson*
75. Baked Goods Freshness: Technology, Evaluation, and Inhibition of Staling, *edited by Ronald E. Hebeda and Henry F. Zobel*
76. Food Chemistry: Third Edition, *edited by Owen R. Fennema*
77. Handbook of Food Analysis: Volumes 1 and 2, *edited by Leo M. L. Nollet*
78. Computerized Control Systems in the Food Industry, *edited by Gauri S. Mittal*
79. Techniques for Analyzing Food Aroma, *edited by Ray Marsili*
80. Food Proteins and Their Applications, *edited by Srinivasan Damodaran and Alain Paraf*
81. Food Emulsions: Third Edition, Revised and Expanded, *edited by Stig E. Friberg and Kåre Larsson*
82. Nonthermal Preservation of Foods, *Gustavo V. Barbosa-Cánovas, Usha R. Pothakamury, Enrique Palou, and Barry G. Swanson*
83. Milk and Dairy Product Technology, *Edgar Spreer*
84. Applied Dairy Microbiology, *edited by Elmer H. Marth and James L. Steele*
85. Lactic Acid Bacteria: Microbiology and Functional Aspects: Second Edition, Revised and Expanded, *edited by Seppo Salminen and Atte von Wright*
86. Handbook of Vegetable Science and Technology: Production, Composition, Storage, and Processing, *edited by D. K. Salunkhe and S. S. Kadam*
87. Polysaccharide Association Structures in Food, *edited by Reginald H. Walter*
88. Food Lipids: Chemistry, Nutrition, and Biotechnology, *edited by Casimir C. Akoh and David B. Min*
89. Spice Science and Technology, *Kenji Hirasa and Mitsuo Takemasa*
90. Dairy Technology: Principles of Milk Properties and Processes, *P. Walstra, T. J. Geurts, A. Noomen, A. Jellema, and M. A. J. S. van Boekel*
91. Coloring of Food, Drugs, and Cosmetics, *Gisbert Otterstätter*
92. *Listeria*, Listeriosis, and Food Safety: Second Edition, Revised and Expanded, *edited by Elliot T. Ryser and Elmer H. Marth*
93. Complex Carbohydrates in Foods, *edited by Susan Sungsoo Cho, Leon Prosky, and Mark Dreher*
94. Handbook of Food Preservation, *edited by M. Shafiur Rahman*
95. International Food Safety Handbook: Science, International Regulation, and Control, *edited by Kees van der Heijden, Maged Younes, Lawrence Fishbein, and Sanford Miller*
96. Fatty Acids in Foods and Their Health Implications: Second Edition, Revised and Expanded, *edited by Ching Kuang Chow*

97. Seafood Enzymes: Utilization and Influence on Postharvest Seafood Quality, *edited by Norman F. Haard and Benjamin K. Simpson*
98. Safe Handling of Foods, edited by Jeffrey M. Farber and Ewen C. D. Todd
99. Handbook of Cereal Science and Technology: Second Edition, Revised and Expanded, *edited by Karel Kulp and Joseph G. Ponte, Jr.*
100. Food Analysis by HPLC: Second Edition, Revised and Expanded, *edited by Leo M. L. Nollet*
101. Surimi and Surimi Seafood, *edited by Jae W. Park*
102. Drug Residues in Foods: Pharmacology, Food Safety, and Analysis, *Nickos A. Botsoglou and Dimitrios J. Fletouris*
103. Seafood and Freshwater Toxins: Pharmacology, Physiology, and Detection, *edited by Luis M. Botana*
104. Handbook of Nutrition and Diet, *Babasaheb B. Desai*
105. Nondestructive Food Evaluation: Techniques to Analyze Properties and Quality, *edited by Sundaram Gunasekaran*
106. Green Tea: Health Benefits and Applications, *Yukihiko Hara*
107. Food Processing Operations Modeling: Design and Analysis, *edited by Joseph Irudayaraj*
108. Wine Microbiology: Science and Technology, *Claudio Delfini and Joseph V. Formica*
109. Handbook of Microwave Technology for Food Applications, *edited by Ashim K. Datta and Ramaswamy C. Anantheswaran*
110. Applied Dairy Microbiology: Second Edition, Revised and Expanded, *edited by Elmer H. Marth and James L. Steele*
111. Transport Properties of Foods, *George D. Saravacos and Zacharias B. Maroulis*
112. Alternative Sweeteners: Third Edition, Revised and Expanded, *edited by Lyn O'Brien Nabors*

Additional Volumes in Preparation

Food Additives: Second Edition, Revised and Expanded, *edited by Alfred Larry Branen, P. Michael Davidson, Seppo Salminen, and John H. Thorngate*

Control of Foodborne Microorganisms, *edited by Vijay K. Juneja and John N. Sofos*

Handbook of Dietary Fiber, *edited by Susan Sungsoo Cho and Mark L. Dreher*

Flavor, Fragrance, and Odor Analysis, *edited by Ray Marsili*

Alternative Sweeteners

Third Edition, Revised and Expanded

edited by

Lyn O'Brien Nabors

Calorie Control Council
Atlanta, Georgia

MARCEL DEKKER, INC. NEW YORK • BASEL

ISBN: 0-8247-0437-1

This book is printed on acid-free paper.

Headquarters
Marcel Dekker, Inc.
270 Madison Avenue, New York, NY 10016
tel: 212-696-9000; fax: 212-685-4540

Eastern Hemisphere Distribution
Marcel Dekker AG
Hutgasse 4, Postfach 812, CH-4001 Basel, Switzerland
tel: 41-61-261-8482; fax: 41-61-261-8896

World Wide Web
http://www.dekker.com

The publisher offers discounts on this book when ordered in bulk quantities. For more information, write to Special Sales/Professional Marketing at the headquarters address above.

Current printing (last digit):
10 9 8 7 6 5 4 3 2 1

PRINTED IN THE UNITED STATES OF AMERICA

Preface

Alternative sweeteners, both as a group and in some cases individually, are among the most studied food ingredients. Controversy surrounding them dates back almost a century. Consumers are probably more aware of sweeteners than any other category of food additive. The industry continues to develop new sweeteners, each declared better than the alternatives preceding it and duplicative of the taste of sugar, the gold standard for alternative sweeteners. In truth, no sweetener is perfect—not even sugar. Combination use is often the best alternative.

While new developments in alternative sweeteners continue to abound, their history remains fascinating. Saccharin and cyclamates, among the earliest of the low-calorie sweeteners, have served as scientific test cases. They have been used to ''test the test.'' For example, saccharin's fate has rested on adverse findings in second-generation rat studies. Only in the late 1990s did scientific technology become sophisticated enough to understand these findings and demonstrate conclusively that man is not a big rat. The International Agency for Research on Cancer (IARC) and the U.S. National Toxicology Program (NTP) now incorporate mechanistic data into their determinations, and saccharin was the first substance to be evaluated on the basis of these new criteria by NTP, and among the first for IARC.

The numerous sweetener developments throughout the 1990s have facilitated combination use. With the availability of numerous low-calorie and reduced-calorie sweeteners and improved technology, higher-quality products can be produced, and in greater quantity. In some parts of the world, foods and beverages are available that contain as many as three or more alternative sweeteners. Regulatory authorities recognize the reduced caloric value of polyols, and these sweeteners are being used increasingly. Researchers have developed a sweetener-sweetener salt in which aspartame and acesulfame are combined at the molecular level. Petitions are pending in various countries for new sweeteners, such as neotame and D-tagatose.

This book provides the latest information on numerous alternative sweeteners and their combination use. Some are currently approved and used in various countries, some are expected to be available in the future, and some are presented as a matter of scientific interest. All chapters that appeared in the second edition of *Alternative Sweeteners* have been updated, and chapters on neotame, D-tagatose, trehalose, erythritol, and the aspartame-acesulfame salt have been added. As in the earlier editions, a chapter on fat replacers is also included. More and more low-calorie foods are expected to contain not only sweetener combinations but also sweeteners plus fat replacers. A concerted effort has been made to provide the reader with comprehensive, current information on a wide variety of alternative sweeteners and substantial references for those who wish to learn more.

Lyn O'Brien Nabors

Contents

Contributors

Sue E. Andress The NutraSweet Company, Mount Prospect, Illinois

Michael H. Auerbach Danisco Cultor America, Ardsley, New York

Abraham I. Bakal ABIC International Consultants, Inc., Fairfield, New Jersey

Hans Bertelsen Arla Foods Ingredients amba, Videbaek, Denmark

Barbara A. Bopp TAP Pharmaceutical Products, Inc., Deerfield, Illinois

Francisco Borrego Zoster, S. A. (Grupo Ferrer), Murcia, Spain

Saskia Brokx PURAC biochem bv., Gorinchem, The Netherlands

Allan W. Buck Archer Daniels Midland Company, Decatur, Illinois

Harriett H. Butchko The NutraSweet Company, Mount Prospect, Illinois

C. Phil Comer The NutraSweet Company, Mount Prospect, Illinois

César M. Compadre University of Arkansas, Little Rock, Arkansas

Ronald C. Deis Product and Process Development, SPI Polyols, Inc., New Castle, Delaware

Lee B. Dexter Lee B. Dexter & Associates, Austin, Texas

Laura Eberhardt Calorie Control Council, Atlanta, Georgia

Milda E. Embuscado Cerestar USA, Inc., Hammond, Indiana

Kristian Eriknauer Arla Foods Ingredients amba, Viby, Denmark

John C. Fry Connect Consulting, Horsham, Sussex, United Kingdom

Leslie A. Goldsmith McNeil Specialty Products Company, New Brunswick, New Jersey

Lisa Y. Hanger Nutrinova, Inc., Somerset, New Jersey

Søren Juhl Hansen Arla Foods Ingredients amba, Videbaek, Denmark

Michael E. Hendrick† Pfizer Inc., Groton, Connecticut

Annet C. Hoek Business Support Center, Holland Sweetener Company, Geleen, The Netherlands

William E. Irwin* Palatinit Süßungsmittel GmbH, Elkhart, Indiana

Kazuaki Kato Towa Chemical Industry Co., Ltd., Tokyo, Japan

A. Douglas Kinghorn Department of Medicinal Chemistry and Pharmacognosy, College of Pharmacy, University of Illinois at Chicago, Chicago, Illinois

Rene Soegaard Laursen Arla Foods Ingredients amba, Brabrand, Denmark

Anh S. Le SPI Polyols, Inc., New Castle, Delaware

Graeme Locke‡ Cultor Food Science, Redhill, Surrey, United Kingdom

Dale A. Mayhew The NutraSweet Company, Mount Prospect, Illinois

† Deceased.
* Retired.
‡ Cultor Food Science is now Danisco Sweeteners.

Carolyn M. Merkel McNeil Specialty Products Company, New Brunswick, New Jersey

Paul H. J. Mesters PURAC biochem bv., Gorinchem, The Netherlands

Helen Mitchell Danisco Sweeteners, Redhill, Surrey, United Kingdom

Helena Montijano Zoster, S. A. (Grupo Ferrer), Murcia, Spain

Frances K. Moppett Pfizer, Inc., Groton, Connecticut

Alan H. Moskowitz Operations Development, Adams Division of Pfizer, Parsippany, New Jersey

Kathleen Bowe Mulderrig SPI Pharma Group, SPI Polyols, Inc., New Castle, Delaware

Lyn O'Brien Nabors Calorie Control Council, Atlanta, Georgia

Philip M. Olinger Polyol Innovations, Inc., Reno, Nevada

Thomas F. Osberger Food Industry Consultant, Upland, California

Sakharam K. Patil Cerestar USA, Inc., Hammond, Indiana

Ronald L. Pearson PMC Specialties Group, Inc., Cincinnati, Ohio

Tammy Pepper Danisco Sweeteners, Redhill, Surrey, United Kingdom

Paul Price Price International, Libertyville, Illinois

Alan B. Richards Hayashibara International Inc., Westminster, Colorado

James Saunders Biospherics Incorporated, Beltsville, Maryland

Djaja Djendoel Soejarto Department of Medicinal Chemistry and Pharmacognosy, College of Pharmacy, University of Illinois at Chicago, Chicago, Illinois

W. Wayne Stargel The NutraSweet Company, Mount Prospect, Illinois

Peter Jozef Sträter Palatinit Süßungsmittel GmbH, Mannheim, Germany

John A. van Velthuijsen PURAC biochem bv., Gorinchem, The Netherlands

Gert-Wolfhard von Rymon Lipinski Scientific and Regulatory Affairs, Nutrinova Nutrition Specialties and Food Ingredients GmbH, Frankfurt/Main, Germany

John S. White White Technical Research Group, Argenta, Illinois

Marie-Christel Wijers* Palatinit of America, Inc., Morris Plains, New Jersey

Christine D. Wu College of Dentistry, University of Illinois at Chicago, Chicago, Illinois

* Retired.

1

Alternative Sweeteners: An Overview

Lyn O'Brien Nabors
Calorie Control Council, Atlanta, Georgia

The use of low-calorie sugar-free products tripled in the final two decades of the 20th century. In the United States alone, more than 150 million people use these products regularly. Even though hundreds of good-tasting low-calorie, sugar-free products are now available, most light product consumers say they would like to have additional low-calorie sugar-free products available. Of particular interest are baked goods and desserts (1).

A number of events that occurred in the late 1990s are expected to facilitate providing additional good-tasting, low-calorie, sugar-free products. The approval of acesulfame potassium for soft drinks and aspartame and sucralose as general purpose sweeteners in the United States and recognition by regulatory agencies around the world that polyols have reduced caloric values compared with sucrose are examples. (A general purpose sweetener may be used in accordance with good manufacturing practices to sweeten any food when a standard of identity does not preclude its use.) These events should expand the use of sweeteners alone and in combination as well.

On the scientific front, after more than 100 years of use, scientists around the world are publicly acknowledging that saccharin is safe for humans. For example, in 1997, a special International Agency for Research on Cancer (IARC) panel determined the bladder tumors in male rats resulting from the ingestion of high doses of sodium saccharin are not relevant to man. And, in late 1998, IARC downgraded saccharin from a Category 2B substance, possible carcinogenic to humans, to Category 3, not classifiable as to its carcinogenicity to humans. Agents for which the evidence of carcinogenicity is inadequate in humans but sufficient in experimental animals may be placed in Group 3 when strong evidence exists that the mechanism of carcinogenicity in experimental animals does not operate in humans. This is the first time IARC has considered mechanistic data, and the IARC panel voted unanimously that saccharin could cause tumors in rats but that this is not predictive of human cancer (2).

Unfortunately, not all events surrounding sweeteners have been positive. Although some new technologies have provided a means of supporting the safety of sweeteners, the Internet has made disseminating negative information without accountability a new art form.

No adverse health effects related to aspartame have been confirmed, but this has not stopped its critics. An extremely negative, inaccurate article making absurd claims about aspartame began circulating on the Internet in late 1998. The article asserts that aspartame is responsible for any number of ailments, without supporting data and creating new challenges for industry. Fortunately, many of the negative claims from this article and other antiaspartame fanatics are so absurd that the sources are not considered credible by many.

I. THE IDEAL SWEETENER

The search for the perfect sweetener continues, but it has long been recognized that the ideal sweetener does not exist. Even sucrose, the "gold standard," is not perfect and is unsuitable for some pharmaceuticals and chewing gums.

Alternative sweeteners (a) provide and expand food and beverage choices to control caloric, carbohydrate, or specific sugar intake; (b) assist in weight maintenance or reduction; (c) aid in the management of diabetes; (d) assist in the control of dental caries; (e) enhance the usability of pharmaceuticals and cosmetics; (f) provide sweetness in times of sugar shortage; and (g) assist in the cost-effective use of limited resources.

The ideal sweetener should be at least as sweet as sucrose, colorless, odorless, and noncariogenic. It should have a clean, pleasant taste with immediate onset without lingering. The more a sweetener tastes and functions like sucrose the greater the consumer acceptability. If it can be processed much like sucrose with existing equipment, the more desirable it is to industry.

The ideal sweetener should be water soluble and stable in both acidic and basic conditions and over a wide range of temperatures. Length of stability and consequently the shelf-life of the final product are also important. The final food product should taste much like the traditional one. A sweetener must be compatible with a wide range of food ingredients because sweetness is but one component of complex flavor systems.

Safety is essential. The sweetener must be nontoxic and metabolized normally or excreted unchanged, and studies verifying its safety should be in the public domain.

To be successful, a sweetener should be competitively priced with sucrose and other comparable sweeteners. It should be easily produced, stored, and transported.

II. RELATIVE SWEETNESS

Perceived sweetness is subjective and depends on or can be modified by a number of factors. The chemical and physical composition of the medium in which the sweetener is dispersed has an impact on the taste and intensity. The concentration of the sweetener, the temperature at which the product is consumed, pH, other ingredients in the product, and the sensitivity of the taster are all important. Again, sucrose is the usual standard. Intensity of the sweetness of a given substance in relation to sucrose is made on a weight basis. Table 1 provides the approximate relative sweetness of many of the alternative sweeteners discussed in this book.

Table 1 Relative Sweetness of Alternatives to Sucrose

	Approximate sweetness (sucrose = 1)
Lactitol	0.4
Hydrogenated starch hydrolysates	0.4–0.9
Trehalose	0.45
Isomalt	0.45–0.65
Isomaltulose	0.48
Sorbitol	0.6
Erythritol	0.7
Mannitol	0.7
Maltitol	0.9
D-Tagatose	0.9
Xylitol	1.0
High fructose corn syrup, 55%	1.0
High fructose corn syrup, 90%	1.0
Crystalline fructose	1.2–1.7
Cyclamate	30
Glycyrrihizin	50–100
Aspartame	180
Acesulfame potassium	200
Saccharin	300
Stevioside	300
Sucralose	600
Hernandulcin	1000
Monellin	1500–2000
Neohesperidine dihydrochalcone	1800
Alitame	2000
Thaumatin	2000–3000
Neotame	8000

III. THE MULTIPLE INGREDIENT APPROACH

The advantages of the multiple sweetener approach have long been known. A variety of approved sweeteners are essential because no sweetener is perfect for all uses. With several available, each sweetener can be used in the application(s) for which it is best suited. Manufacturers also can overcome limitations of individual sweeteners by using them in blends.

During the 1960s, cyclamate and saccharin were blended together in a variety of popular diet soft drinks and other products. This was really the first practical application of the multiple sweetener approach. The primary advantage of this sweetener blend was that saccharin (300 times sweeter than sucrose) boosted the sweetening power of cyclamate (30 times sweeter), whereas cyclamate masked the aftertaste that some people associate with saccharin.

The two sweeteners when combined have a synergistic effect—that is the sweetness of the combination is greater than the sum of the individual parts. This is true for most sweetener blends. Cyclamate was the major factor in launching the diet segment of the carbonated beverage industry. By the time it was banned in the United States in 1970, the products and trademarks had been well established. Such a large market for diet beverages provided a tremendous incentive to develop new sweeteners.

After cyclamate was taken off the market in 1970, saccharin was the only available low-calorie alternative to sugar available in the United States for more than a decade. But now with the availability of acesulfame potassium, aspartame, sucralose, and saccharin, the multiple sweetener approach is a visible reality in the United States. Fountain soft drinks generally contain a combination of saccharin and aspartame and bottled drinks are available with combinations of aspartame and acesulfame K, as well as sucralose and aspartame. Triple blends, such as acesulfame potassium, aspartame, and saccharin and aspartame, cyclamate, and acesulfame potassium are being used in some parts of the world.

The polyols also are important adjuncts to sugar-free product development. These sweeteners provide the bulk of sugar but are generally less sweet than sucrose. The polyols, which are reduced in calories, combine well (e.g., they are synergistic) with low-calorie sweeteners, resulting in good-tasting reduced-calorie products that are similar to their traditional counterparts.

With the availability of fat replacers and low-calorie bulking agents (e.g., polydextrose), not just a multiple sweetener approach but a multiple ingredient approach to calorie control is being used. In addition to the evidence that humans have an innate desire for sweets (3), research indicates that the obese and those who once were obese may have a greater preference than others for fatty liquids mixed with sugar (4). Replacing the fat and the sugar is therefore important in the development of products to assist in calorie control.

IV. INTERNATIONAL REGULATORY GROUPS

Food ingredients are evaluated and/or regulated by numerous national and international bodies. International groups include the Joint Food and Agriculture Organization/World Health Organization Expert Committee on Food Additives (JECFA), the Codex Alimentarius Commission, and the Scientific Committee for Food of the Commission of the European Union.

The objective of the joint Food and Agriculture Organization/World Health Organization (FAO/WHO) program on food additives is to make systematic evaluations of food additives and provide advice to member states of FAO and WHO on the control of additives and related health aspects. The two groups responsible for implementing the program are JECFA and the Committee on Food Additives of the Joint FAO/WHO Codex Alimentarius Commission (5).

JECFA is made up of an international group of experts who serve without remuneration in their personal capacities rather than as representatives of their governments or other bodies. Members are selected primarily for their ability and technical experience, with consideration given to adequate geographical distribution. Their reports contain the collective views of the group and do not necessarily represent the decision or the stated policy of the WHO or FAO. The experts convene to give advice on technical and scientific matters, establishing specifications for identity and purity for food additives, evaluating the toxicological data, and recommending, where appropriate, acceptable daily intakes for humans. The Expert Committee also acts in an advisory capacity for the Codex Committee on Food Additives and Contaminants (5).

The *Codex Alimentarius* Commission was established in 1962 to implement the Joint FAO/WHO Food Standards Program. Membership is made up of those member nations and associate members of FAO and WHO that have notified the Director-General of FAO or WHO of their wish to be members. The stated purpose of the program is:

> To protect the health of consumers and to ensure fair practices of the food-trade; to promote coordination of all food standards work undertaken by international governmental and non-governmental organizations; to determine priorities and initiate and guide the preparation of draft standards through and with the aid of appropriate organizations; to finalize standards and after acceptance by governments publish them in a *Codex Alimentarius* either as regional or worldwide standards.

The *Codex Alimentarius* is intended to guide and promote the elaboration and establishment of definitions and requirements for foods, including food additives, to assist in their harmonization and, thereby, facilitate trade. The *Codex*

Committee for Food Additives and Contaminants (CCFAC) is the body that deals with food additives (6).

The World Trade Organization (WTO) encourages countries to harmonize food standards on the basis of *Codex* standards and uses its decisions to settle trade disputes. In addition, the WTO recognizes JECFA specifications for food additives in international trade, increasing the importance of both *Codex* and JECFA.

The Scientific Committee for Food (SCF) of the Commission of the European Union was established by the Commission in 1974. The Committee advises the Commission "on any problem relating to the protection of the health and safety of persons arising from the consumption of food, and in particular the composition of food, processes which are liable to modify food, the use of food additives and other processing aids as well as the presence of contaminants." Committee members are independent persons qualified in medicine, nutrition, toxicology, biology, chemistry, or other similar disciplines. Committee opinions are submitted to the Commission (7).

The European Union has agreed to harmonize its member states' laws on food additives, including sweeteners. The Sweeteners Directive, adopted in 1994, provides for the use of acesulfame K, aspartame, cyclamate, saccharin, thaumatin, and neohesperidine DC, as well as sorbitol, sorbitol syrup, mannitol, isomalt, maltitol, maltitol syrup, lactitol, and xylitol (8). This Directive is under review and expected to be revised in 2001–2002.

The Committee on Food Chemicals Codex (FCC), a full committee of the Food and Nutrition Board, Institute of Medicine, National Research Council of the U.S. National Academy of Sciences, provides information on the quality and purity of food-grade substances. Specifications and test methods are included in almost 1000 FCC monographs on substances that are added to or come in contact with foods. The members of the Committee on FCC are chosen for their special competencies and with regard to appropriate balance.

Food Chemicals Codex is recognized internationally. FCC specifications are cited, by reference, in the U.S. Code of Federal Regulations as the reference for specifications to define specific safe ingredients. In Canada, FCC and its supplements are officially recognized in the Canadian Food and Drug Regulations as the reference for specifications for food additives. Under New Zealand food regulations, a food additive is defined as being of appropriate quality "if it complies with the monograph for that food additive (if any) in the current edition of the *Food Chemicals Codex* published by the National Academy of Sciences and the National Research Council of the United States of America in Washington, D.C." Similarly, the national food authority of Australia frequently refers to the *Food Chemicals Codex* specifications to define food additives (9).

V. U.S. REGULATION OF SWEETENERS

Food additives were first subjected to regulation in the United States under the Food and Drug Act of 1906. Section 402(a)(1) of the Act states that a food shall be deemed adulterated

> if it bears or contains any poisonous or deleterious substance which may render it injurious to health; but in case the substance is not an added substance, such food shall not be considered adulterated under this clause if the quantity of such substance in such food does not ordinarily render it injurious to health (10, p. 33).

''Added'' is not defined but is generally understood to mean a substance not present in a food in its natural state. The intent of this section is to prohibit any level of added food substance inconsistent with public health. The Federal Food, Drug, and Cosmetic Act of 1938 contains food safety provisions similar to those in the 1906 Act.

The basic Food, Drug, and Cosmetic Act was last updated in 1958. Section 201(s) defines a ''food additive'' as

> any substance the intended use of which results or may reasonably be expected to result, directly or indirectly, in its becoming a component or otherwise affecting the characteristics of any food (including any substance intended for use in producing, manufacturing, packing, processing, preparing, treating, packaging, transporting, or holding food; and including any source of radiation intended for any such use) (10, p. 5).

Congress passed the Food Additives Amendment, Section 409 of the Food, Drug, and Cosmetic Act, in 1958 as well. This amendment exempts two important groups of substances from the food additive definition. Those exempted are (a) substances generally recognized as safe (GRAS) among experts qualified by scientific training and experience to evaluate safety, and (b) substances that either the FDA or the U.S. Department of Agriculture (USDA) had sanctioned for use in food before 1958 (so-called ''prior sanction'' substances). The amendment does not pertain to pesticide chemicals in or on raw agricultural commodities.

The ''Delaney Clause'' is part of the 1958 Foods Additives Amendment. The clause states that ''no additive shall be deemed to be safe if it is found to induce cancer when ingested by man or animal, or if it is found after tests which are appropriate for the evaluation of the safety of food additives, to induce cancer in man or animal'' (10, p. 55). The Delaney Clause is often referred to but rarely used. Debate often centers on the undefined phrases ''induce cancer'' and ''tests which are appropriate.''

The 1958 Food Additives Amendment forbids the use of any food additive not approved by the FDA, and the agency may only approve additives shown to be "safe." Section 409 of the Act outlines the requirements for requesting approval for a food additive (i.e., "Petition to establish safety") and details the action to be taken by the FDA in dealing with such a petition.

A petitioner, requesting the issuance of a food additive regulation, must provide, in addition to any explanatory or supporting data:

> (2)(A) the name and all pertinent information concerning such food additive, including, where available, its chemical identity and composition; (B) a statement of the conditions of proposed use of such additive, including all directions, recommendations, and suggestions proposed for the use of such additive, and including samples of its proposed labeling; (C) all relevant data bearing on the physical or other technical effect such additive is intended to produce, and the quantity of the additive required to produce such effect; (D) a description of practicable methods of determining the quantity of such additive in or on food, and any substance formed in or on food, because of its use; and (E) full reports of investigations made with respect to the safety of such additive including full information as to the methods and controls used in conducting such investigations (10, p. 54).

The Federal Food, Drug, and Cosmetic Act does not describe the safety investigations to be conducted on the proposed food additive. The FDA, therefore, issued a document entitled "Toxicological Principles for the Safety Assessment of Direct Food Additives and Color Additives Used in Food" (referred to as the "Redbook") in 1982 (11).

Redbook II was issued in 1993 and is intended to provide guidance on criteria used for the safety assessment of direct food additives and color additives used in food and to assist petitioners in developing and submitting toxicological safety data for FDA review. Although the Redbook is not legally binding and FDA states that a petitioner may follow the guidelines and protocols in Redbook II or choose to use alternative procedures, the agency notes that alternative procedures should be discussed informally with the agency "to prevent expenditure of money and effort on activities that may later be determined to be unacceptable to the FDA" (12).

The cost in both time and money for the approval of a new food additive is almost prohibitive, resulting in an increased number of GRAS applications. General recognition of safety (i.e., a GRAS determination), as noted previously, must reflect the views of experts qualified by scientific training and experience to evaluate the safety of substances directly or indirectly added to food. Those expert views must be based on "scientific procedures," supplemented in the case of substances used in food before 1958 by experience based on common use in food. The term "safe" means that there is reasonable certainty in the minds of competent scientists that a substance is not harmful under intended conditions of use.

General recognition of safety based on scientific procedures calls for the same quantity and quality of scientific evidence as would be required to obtain a food additive regulation for a substance. Scientific procedures include human, animal, analytical, and other studies whether published or unpublished, appropriate to establish the safety of a substance. General recognition of safety through scientific procedures ordinarily must be based on published studies, although those studies can be bolstered by unpublished evidence.

Common use means a substantial history of consumption by a significant number of people before January 1, 1958. General recognition of safety through experience based on common use in food should be based on generally available data (usually published data) and information and can include well-documented use in foreign countries.

Any interested person may make a determination that a substance is GRAS for a particular use(s) (i.e., GRAS self-determination) as outlined previously. Current regulations acknowledge that current GRAS lists do not include all GRAS substances and are not expected to include pre-1958 natural, nutritional substances. The regulations do invite any interested party to petition FDA to affirm that a substance is GRAS (13).

A determination of GRAS, with or without a petition to FDA, does not require FDA action. If a petition is submitted requesting GRAS affirmation, once the petition is accepted for filing, the substance may be legally used for the petitioned uses while awaiting FDA affirmation.

In April of 1997, FDA proposed a replacement for the system under which manufacturers may get affirmation from FDA that a food substance is generally recognized as safe (GRAS) (14). Under this "GRAS notification system," manufacturers may still make a self-determined GRAS declaration, claiming exemption from the premarket or food additive approval requirements. Instead of petitioning FDA for affirmation, manufacturers notify FDA of their GRAS determination and provide evidence supporting their decision. After evaluating the notification, FDA is to respond to the manufacturer, conveying the agency's disposition within 90 days. A response that does not identify a problem is *not* equivalent to an affirmation of GRAS status. The proposal allows for a notification to be revisited if new information indicates a reason for concern.

The notification procedure is designed to inform FDA of GRAS actions without the need for rulemaking. Under the proposal, any GRAS affirmation petition pending when the notification rule is finalized would be converted or dropped from review. Because the substantive requirements of an acceptable GRAS notification differ from those for a GRAS affirmation petition, any pending petition would have to be amended. For example, a GRAS exemption claim, signed by the notifiee, and explicitly accepting full responsibility for the GRAS determination, would be necessary. In lieu of this, the GRAS affirmation petitioner could petition for food additive approval, cross-referencing information in the GRAS petition, or submit a complete GRAS notification.

In notifying FDA of their GRAS determination, manufacturers must provide evidence supporting their decisions. Such data include generally available and accepted scientific data, information, methods, or principles. Under certain circumstances other scientific data, as well as analytical methods, methods of manufacture and/or accepted scientific principles could be relied on as part of the technical information. FDA notes that the quantity and quality of scientific evidence required to demonstrate the safety may vary depending on the estimated dietary exposure and the chemical, physical, and physiological properties of the substance. The notice summary must consider the totality of the publicly available information and evidence about the safety of the substance for its intended use, including favorable and potentially unfavorable information.

Although the proposed procedure facilitates a prompt response to self-GRAS affirmations, FDA is not required to provide its affirmation. FDA would affirm GRAS status only when the agency of its own volition wants to so affirm a substance.

Until the GRAS notification proposal is finalized, FDA has invited interested parties to use the proposed GRAS notification procedure. This invitation includes those who wish to convert pending GRAS affirmation petitions to GRAS notification. FDA will acknowledge receipt and make a "Good Faith" effort to meet the 90-day time frame but is not required to do so. As of December 2000, more than 30 GRAS notifications have been reviewed and so acted on by the agency.

Both the new and old procedures of GRAS self-affirmation allow ingredients to be used in the U.S. food supply without a published FDA GRAS or food additive regulation.

VI. ACCEPTABLE DAILY INTAKE

As part of the evaluation of a food additive, many regulatory bodies establish an acceptable daily intake (ADI) level. The ADI "for man, expressed on a body weight basis, is the amount of a food additive that can be taken daily in the diet, even over a lifetime, without risk" (5). The ADI may be used as a benchmark to evaluate the actual intake of a substance and as an aid in reviewing possible additional uses for a food ingredient. The ADI is expressed in milligrams per kilogram of body weight.

The ADI is a conservative estimate that incorporates a considerable safety factor. It is established from toxicological testing in animals, and sometimes humans, and is usually estimated by applying an intentionally conservative safe factor (generally a 100-fold safety factor). Animal tests are used to determine the maximum dietary level of an additive demonstrating no toxic effects, a ''no observable effect level'' or NOEL. The NOEL is then used to determine the ADI. For example, if safety evaluation studies of a given substance demonstrate a NOEL of 1000 mg/kg, using a 100-fold safety factor the ADI would be 10 mg/kg body weight per day for humans.

The ADI does not represent a maximum allowable daily intake level. It should not be regarded as a specific point at which safety ends and possible health concerns begin. In fact, the U.S. FDA has said it is not concerned that consumption occasionally may exceed the ADI. The agency has stressed that because the ADI has a built-in safety margin and is based on a chronic lifetime exposure, occasional consumption in amounts greater than the ADI ''would not cause adverse effects'' (15).

VII. CONCLUSION

Low-calorie products are in demand and consequently so are the ingredients that make them possible. With recent approvals more and more good-tasting products will become available. In the near future, additional products that are reduced in fat and calories, incorporating both fat replacers and low-calorie and reduced-calorie sweeteners are expected. With the increase in obesity in many parts of the world and as consumers become increasingly aware that ''calories still count,'' the number of successful light products should soar.

REFERENCES

1. Light Products Usage and Weight Control Habits Survey. Conducted by Booth Research Services, Inc. for the Calorie Control Council, March, 1998.
2. J Rice, et al. Rodent tumors of urinary bladder, renal cortex, and thyroid gland in IARC monographs evaluations of carcinogenic risk to humans. Toxicol Sci 49:166–171, 1999.
3. JA Desor, O Maller, LS Green. Preference for sweet in humans. In: JM Weiffenbach, ed. Infants, Children and Adults in Taste and Development: The Genesis of Sweet Preference. Washington, DC: U.S. Department of Health, Education, and Welfare, 1977, p 171.
4. A Drewnowski. Sweetness and obesity. In: J Dobbing, ed. Sweetness. Berlin: Springer-Verlag, 1987.
5. Joint FAO/WHO Expert Committee on Food Additives. Toxicological Evaluation of Certain Food Additives with a Review of General Principles and Specifications. Geneva: World Health Organization, 1974.
6. Codex Alimentarius Commission. Procedural Manual. Rome: 5th ed. Food and Agricultural Organization of the United Nations, 1981, p 21.

7. Commission Decision 74/234/EEC of 16 April 1974 relating to the institution of a Scientific Committee for Food. Official Journal of the European Communities No. L 136/1, 20 May 1974.
8. European Parliament and Council Directive 94/35/EC of 30 June 1994 on sweeteners for use in foodstuffs. Official Journal of the European Communities No. L 237/3, 10 September 1994.
9. Food Chemicals Codex. 4th ed. Washington, DC: National Academy Press, 1996.
10. Federal Food, Drug, and Cosmetic Act, As Amended. Washington, DC: U.S. Government Printing Office, July 1993.
11. Federal Register [Docket No. 80N-0446], October 15, 1982, p 46141.
12. Toxicological principles for the safety assessment of direct food additives and color additives used in food (draft). "Redbook II." U.S. Food and Drug Administration, Center for Food Safety and Applied Nutrition, 1993.
13. U.S. Code of Federal Regulations, Title 21, Sections 170.30. 170.35. Washington, DC: U.S. Government Printing Office, 1999.
14. Federal Register. Substances Generally Recognized As Safe [Docket No. 97N-0103], April 17, 1997, pp 18937–18964.
15. Federal Register. Food additives permitted for direct addition to food for human consumption; Aspartame; Denial of Requests for Hearing; Final Rule [Docket 75F-0355 and 82F-0305], February 22, 1984, pp 6672–6682.

2
Acesulfame K

Gert-Wolfhard von Rymon Lipinski
Nutrinova Nutrition Specialties and Food Ingredients GmbH, Frankfurt/Main, Germany

Lisa Y. Hanger
Nutrinova, Inc., Somerset, New Jersey

I. INTRODUCTION

Clauss and Jensen (1) in 1967 incidentally discovered a sweet-tasting compound, 5,6 dimethyl-1,2,3-oxathiazin-4(3H)-one 2,2-dioxide, having a new ring system that had not been previously synthesized. Systematic research on dihydrooxathiazinone dioxides revealed quite a number of sweet-tasting compounds in this group of substances. Variations of substituents in positions 5 and 6 of the ring system showed noticeable influence on the intensity and purity of the sweetness. All synthesized dihydrooxathiazinone dioxides, however, exhibited some sweetness, even the ring system without substituents. The maximum sweetness was found in compounds with short-chain alkyl groups.

Sensory evaluations of the different dihydrooxathiazinone dioxides showed that the substitutions on the ring system not only influenced the intensity but also the purity of the sweetness (Fig. 1).

In addition to variations of substituents on the dihydrooxathiazinone dioxide ring system, structurally similar compounds were synthesized to investigate whether variations within the ring system would influence the sweet taste. These evaluations did not reveal any new sweet-tasting compounds; even methylation on the nitrogen in the ring furnishes a compound without sweetness (Fig. 2). An evaluation of the different compounds clearly demonstrated that 6-methyl-1,2,3-oxathiazine-4(3H)-one 2,2-dioxide exhibited the most favorable taste properties. Because production of this compound seemed to be less difficult than that of

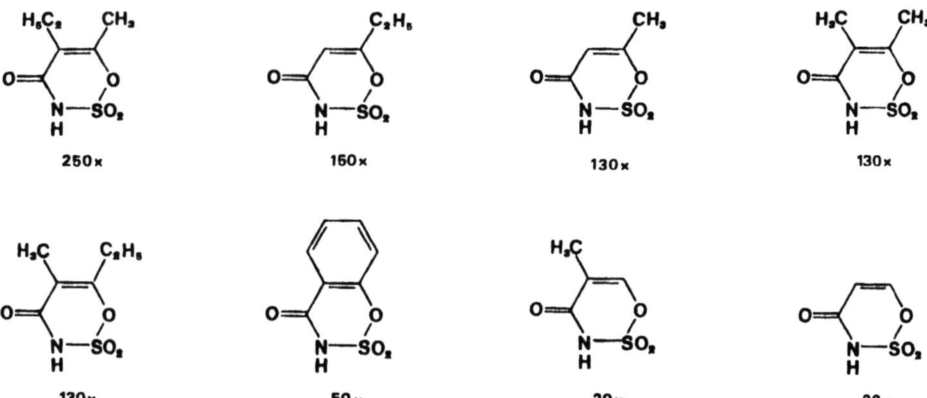

Figure 1 Dihydrooxathiazinone dioxides with different substituents and sweetness intensities of their sodium salts compared with a 4% sucrose solution [tentatively measured by Clauss and Jensen (1)].

other dihydrooxathiazinone dioxides, it was chosen for systematic evaluation of its suitability as an intense sweetener for use in foods and beverages (Fig. 3).

In 1978, the World Health Organization registered acesulfame potassium salt as the generic name for this compound. This is often abbreviated to acesulfame K.

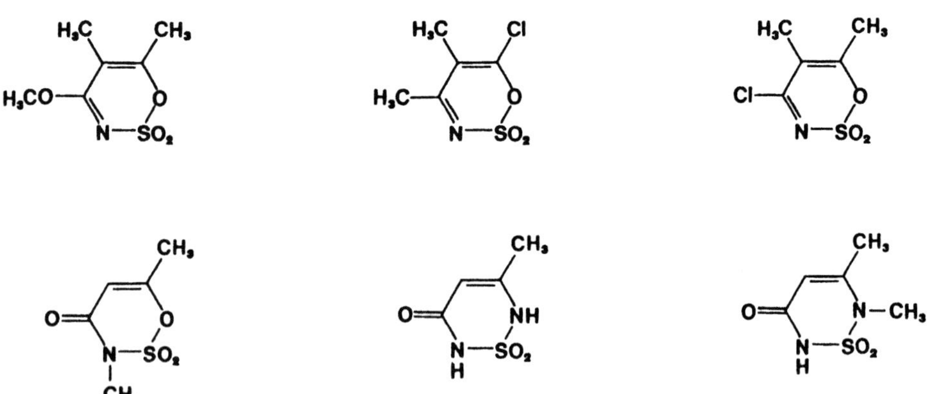

Figure 2 Heterocyclic compounds without sweet taste.

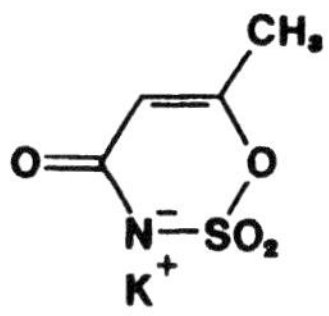

Figure 3 Acesulfame K.

II. SYNTHESIS

Dihydrooxathiazinone dioxides basically can be synthesized from different raw materials using different production routes.

Suitable starting materials are ketones, β-diketones, derivatives of β-oxocarbonic acids, and alkynes that may be reacted with halogen sulfonyl isocyanates. The compounds formed from such reactions are transformed into *N*-halogen sulfonyl acetoacetic acid amide. In the presence of potassium hydroxide, this compound cyclizes to the dihydrooxathiazinone dioxide ring system by separating out the corresponding potassium salts. Because dihydrooxathiazinone dioxides are highly acidic compounds, salts of the ring system are formed. The production of acesulfame potassium salt requires KOH; however, NaOH or $Ca(OH)_2$ can also be used (1–4).

Acetoacetamide-*N*-sulfonic acid is another suitable starting material. In the presence of sulfur trioxide this compound cyclizes to form the dihydrooxathiazinone dioxide ring system, which may be reacted with KOH to yield acesulfame potassium salt. Again, the production of other salts than the potassium salt seems basically possible (5).

Continuous production of acesulfame K is possible using this route of synthesis. This allows large-scale production.

Table 1 Solubilities of Acesulfame K (2)

Solvent	Temperature (°C)	g/100 ml solvent
Water	0	15
Water	20	27
Water	100	ca. 130
Ethanol	20	0.1
Glacial acetic acid	20	13
Dimethyl sulfoxide	20	30

III. PROPERTIES

Acesulfame K is a white, crystalline powder. The crystals are monoclinic, of the P 21/c order. X-ray diffraction demonstrated that the ring system is almost plane, whereas the distances between the single atoms are less than those calculated from the theoretical values. The specific gravity of acesulfame K is 1.83 g/cm^3 (6).

The shelf-life of pure solid acesulfame K seems to be almost unlimited at room temperature. Samples kept at room temperature for more than 6 years and either exposed to light or protected from it did not show any signs of decomposition or differences in analytical data compared with freshly produced material (4).

Acesulfame K does not show a definitive melting point. When the product is heated under conditions used for melting point conditions, decomposition is normally observed at temperatures well above 200°C. The decomposition limit seems to depend on the heating rate. No decomposition of acesulfame K has been observed under conditions of temperature exposure normally found for food additives. In contrast to acesulfame K, acesulfame acid has a sharp and definitive melting point at 123°C (2).

Even at room temperature, acesulfame K dissolves readily in water. The solubility at 20°C is about 270 g/l water. With rising temperatures, the solubility increases sharply to more than 1000 g/l at 100°C. In alcohols, however, acesulfame K is much less soluble. At 20°C, only about 1 g/l dissolves in anhydrous ethanol. In mixtures of alcohol and water, the solubility increases with rising water content. In 50% ethanol (v/v), about 100 g/l can be dissolved (2). Aqueous solutions of acesulfame K are almost neutral (See Table 1).

In view of the temperature/solubility ratio of acesulfame K solutions in water, the product can be easily purified by recrystallization. High-purity acesulfame K can thus be produced on a technical scale while meeting the purity requirements for food additives.

A. Sensory Properties

Acesulfame K exhibits about 200 times the sweetness of a 3% sucrose solution, although sometimes slightly higher values have been reported (7). The sweetness intensity depends on the concentrations of the sucrose solution to which it is compared. At the threshold level, the intensity is much greater and decreases with increasing sucrose concentrations to values from 130 to 100 times the sucrose value (Fig. 4). Normally, acesulfame K can be considered to be about half as sweet as sodium saccharin, similarly sweet as aspartame, and four to five times sweeter than sodium cyclamate. In acid foods and beverages with the same

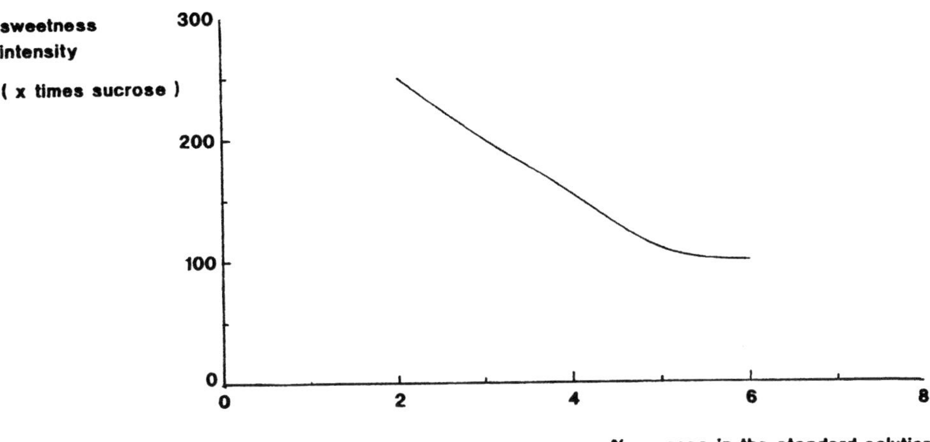

Figure 4 Sweetness intensity of acesulfame K in aqueous solution.

concentrations, a slightly greater sweetness may be perceived compared with neutral solutions.

The sweet taste of acesulfame K is perceived quickly and without unpleasant delay. Compared with other intense sweeteners (aspartame, alitame), acesulfame K has a faster sweetness onset (8). The sweet taste is not lingering and does not persist longer than the intrinsic taste of foods. In aqueous solutions with high concentrations of acesulfame K, a bitter taste can sometimes be detected (9). In foodstuffs with lower concentrations of acesulfame K, this effect is not of great importance. As for all intense sweeteners, however, assessments of different taste characteristics depend on the product in which the sweeteners are used (9). It was further observed that the sweetness of acesulfame K solutions does not decrease with rising temperatures to the extent of other intense sweeteners (10).

A strong synergistic taste enhancement was noted in mixtures of acesulfame K and aspartame or sodium cyclamate, whereas only slight taste enhancement was perceived in mixtures of acesulfame K and saccharin (11–13). Acesulfame and aspartame both taste approximately or even less than 100 times sweeter than a sucrose solution containing 100 g/l (9). A blend of both sweeteners, however, tastes more than 300 times sweeter than the same sucrose solution. This effect apparently cannot be explained from the individual sweetness intensity curves and should therefore be attributable to synergism (14). Although strongest taste enhancement was found in 1:1 (w/w) blends, pronounced synergism was observed for other blend ratios also. Acesulfame K exhibits synergism when combined with many of the other intense or nutritive sweeteners, including sucralose, high-fructose corn syrup, alitame, thaumatin, and fructose (15).

Although acesulfame K can be used as an intense sweetener by itself and does not show particular taste problems when used in appropriate concentrations, mixtures with other sweetening agents are of high practical interest (16–19). Studies showed that, in particular, mixtures of acesulfame with various other intense sweeteners had especially desirable sensory properties. This qualitative improvement of taste seems to be caused by an addition of the time-intensity profiles of the individual sweeteners. Thus, combinations balance the characteristics of the sweeteners used in such blends (20). Favorably viewed were mixtures of acesulfame K and aspartame (ca. 1:1 by weight), acesulfame K and sodium cyclamate (ca. 1:5 by weight), and sometimes other mixtures, in which both sweetening agents almost equally contributed to the sweetness of the mixture. When comparing single and mixtures of intense sweeteners to sucrose, mixtures of sweeteners containing acesulfame K are found to be closer in taste to sucrose than single sweeteners (18). In particular, acesulfame K/ aspartame and acesulfame K/aspartame/saccharin/cyclamate mixtures are not different from sucrose for taste attributes known to cause differences among sweeteners (sweet or bitter side tastes or aftertastes). The lingering sweetness of aspartame and sucralose was substantially reduced when blending either of the sweeteners with acesulfame K (18). In addition, blends of acesulfame K and fructose are found to be not different from sucrose in sweetened beverages (20). Blending acesulfame K with other intense or bulk sweeteners can help maximize the flavor profile of a food or beverage because the sweetener blend tastes more like sucrose.

Favorable taste reports were given for mixtures of acesulfame K with sugar alcohols or sugars. Mixtures of acesulfame K with sugar alcohols have a full and well-balanced sweetness. Therefore, these mixtures are particularly suitable for sugarless confectionery, fruit preparations, and other foods that require a bulking agent. The rounding effect seems attributable to the fast onset of acesulfame's sweetness. When acesulfame K is combined with bulk sweeteners (e.g., maltitol), the mixture provides a taste profile closer to sugar because nonsweet tastes and aftertastes versus sucrose are significantly reduced (19). Suitable blend ratios of acesulfame K with bulk sweeteners are approximately 1:100 in a mixture with xylitol, 1:150 in a mixture with maltitol, 1:150–200 in a mixture with sorbitol, and 1:250 in a mixture with isomalt (21).

B. Toxicology

Because toxicological evaluations of intense sweeteners are of crucial importance for approval and subsequent use, a full range of toxicological studies was carried out with acesulfame K.

The acute oral toxicity of acesulfame K is so low that it can be regarded as practically nontoxic. The LD_{50} was orally determined at 6.9–8.0 g/kg body weight. The intraperitoneal LD_{50} is 2.2 g/kg body weight (22). Subchronic toxic effects were investigated in a 90-day study with rats. The animals were fed concentrations from 0–10% acesulfame K in the diet. Potential carcinogenicity and chronic toxicity were studied in rats that were fed up to 3% acesulfame K in the diet. A carcinogenicity study was conducted in mice fed with concentrations up to 3% acesulfame K. In addition, chronic toxicity effects of acesulfame K were studied in beagle dogs for 2 years (22).

No mutagenicity was found in various respective studies. Among such studies, there were a dominant lethal test; a micronucleus test; bone marrow investigations in hamsters; tests for malignant transformation, DNA binding, and others (22).

In reproduction studies with rats fed acesulfame K, no deviation was seen in fertility, the number of young animals per litter, body weight, growth, or mortality (22).

The toxicological studies on acesulfame K demonstrated that the compound would be safe for use as an intense sweetener. This view was confirmed by the Joint Expert Committee on Food Additives of the FAO and WHO, according to which the data showed that acesulfame K did not exhibit mutagenicity or carcinogenicity. Therefore, an acceptable daily intake (ADI) of 0–9 mg/kg of body weight was allocated, which was later increased to 0–15 mg/kg (23). The Scientific Committee for Foods of the EU published an assessment stating that long-term studies did not show any dose-related increase in specific tumors nor any treatment-related pathological changes of significance. Therefore, an acceptable daily intake of 0–9 mg/kg of body weight was allocated (24). On the basis of a detailed evaluation of the available animal studies, the Food and Drug Administration in the United States also allocated an acceptable daily intake of up to 15 mg/kg of body weight (25).

C. Metabolism and Physiological Characteristics

Acesulfame K is not metabolized by the human body. To investigate possible metabolic transformations, ^{14}C-labeled acesulfame K was used. Studies were carried out in rats, dogs, and pigs. Because animal studies did not show any metabolism, human volunteers were also given labeled acesulfame K. The different animal species, as well as the human volunteers, excreted the original compound. No activity attributable to metabolites was found.

Because acesulfame K is excreted completely unmetabolized, it does not have any caloric value.

In conjunction with the metabolic studies, the pharmacokinetics of acesulfame K were also investigated. These studies were carried out with rats, dogs, pigs, and later with human volunteers. All animal species, as well as humans, quickly absorbed acesulfame K, but there was rapid excretion of the compound, mainly in the urine. A multiple-dose study showed no accumulation in tissues (22). Serum determination of acesulfame K can be carried out by high-performance liquid chromatographic (HPLC) analysis (26). No activity attributable to metabolites was found. After prolonged exposure to acesulfame K, animals did not show any sign of induced metabolism. Again, after administration of ^{14}C-labeled acesulfame K, only the original substance was found in the excreta (22).

Acesulfame K is not metabolized by bacteria. This applies, too, for *Streptococcus mutans* and other microorganisms that may contribute to the formation of caries. Acesulfame K was tested in several studies. Although an inhibition of dental plaque microorganisms or *S. mutans* was not always reported, in other test systems a clear inhibition was demonstrated (27–29). Synergism in the inhibition of bacteria was observed in mixtures of intense sweeteners (30) or mixtures of acesulfame K, saccharin, and fluoride (31). In all studies, however, the concentrations necessary for an inhibitory effect were well above the concentrations used for customary sweetness levels.

D. Stability and Reactions in Foods

Long-term and heat stability are important factors for the use of intense sweeteners in many food products and in beverages. In this regard, various conditions have to be met. In foods and beverages, pH levels vary from neutral to the acid range and may, in extreme cases like certain soft drinks, go down to values around and even less than 3. In this wide pH range, even after prolonged storage, no decrease of sweetness intensity is detected.

In aqueous media, acesulfame K is distinguished by very good stability. After several months of storage at room temperature, virtually no change in acesulfame K concentration was found in the pH range common for beverages. Prolonged continuous exposure to 30°C, conditions that will hardly be found in practice, does not cause losses exceeding 10%, the threshold for recognition of sweetness differences (7). Even at temperatures of 40°C, the threshold for detection of sweetness differences is exceeded after several months only for products having pH 3.0 or less (32).

Extensive studies were performed with buffered aqueous solutions. Results for pH levels and storage conditions commonly found for soft drinks are given in Table 2. After 10 years' storage of a solution buffered to pH 7.5 at room temperature, no significant loss of acesulfame K was detected.

Acesulfame K-containing beverages can be pasteurized under normal pasteurization conditions without loss of sweetness. Pasteurizing for longer periods at lower temperatures is possible, as is short-term pasteurization for a few seconds at high temperatures. Sterilization is possible without losses under the normal

Table 2 Stability of Acesulfame K in Buffered Aqueous Solution

20°C Storage time (weeks)	pH 3.0 recovery (%)	pH 3.5 recovery (%)
16	98	98
30	98	99
50	98	99
100	95	98
30°C **Storage time (weeks)**	**pH 3.0 recovery (%)**	**pH 3.5 recovery (%)**
16	97	100
30	95	97
50	91	96

conditions (i.e., temperatures around 100°C for products having lower pH levels and 121°C for products around and greater than 4). In a solution of pH 4.0, which was heated to 120°C for 1 h, no loss of acesulfame could be measured. Half-life values determined at 100°C demonstrate that the common treatments of foods and beverages should not cause any substantial decomposition of acesulfame K (33).

In baking studies, no indication of decomposition of acesulfame K was found even when biscuits with a low water content were baked at high oven temperatures for short periods. This corresponds to the observation that acesulfame K decomposes at temperatures well above 200°C.

Potential decomposition products of acesulfame K can, therefore, be found only under extreme conditions. Under such conditions, compounds of hydrolytic decomposition are mainly acetone, CO_2, ammonium salts, sulfate, and amidosulfonate. In the hydrolytic decomposition, the ring system is initially opened, which quickly yields the end products of hydrolysis. Only occasionally and then in very acidic media can traces of derivatives of acetoacetic acid be detected (3).

IV. APPLICATIONS

Acesulfame K can be used as a sweetening agent in a wide range of products, for instance in low-calorie products, diabetic foods, sugarless products, oral hygiene preparations, pharmaceuticals, and animal feeds. The product groups acesulfame K has been used in are shown in Table 3.

Table 3 Present Applications of Acesulfame K

Tabletop sweetener	Cookies
Carbonated soft drinks	Jams
Noncarbonated soft drinks	Fruit purees
Squashes and dilutables	Hard candy
Fruit nectars	Soft candy
Powdered beverages	Gum confections
Instant tea	Chocolate confections
Cider	Marzipan
Flavored milk	Chewing gum
Cocoa	Pickled vegetables
Yogurt	Marinated fish
Flavored whey	Tobacco
Desserts	Mouthwashes
Rice pudding	Toothpaste
Ice cream	Drugs
Slimming diets	Feedstuffs
Cake	

Acesulfame K is suitable for low-calorie and diet beverages because of its good stability in aqueous solutions even at low pH typical of diet soft drinks. If used by itself, acesulfame K can impart sweetness comparable to 8–10% sucrose, but mixtures of acesulfame K with other intense sweeteners are more predominantly used because of the sucroselike taste these blends provide. In countries where mixtures of intense sweeteners with other sweetening agents are permitted, mixtures of acesulfame K with fructose, glucose, high-fructose corn syrup, or sucrose can be used. Generally, beverages containing such mixtures of acesulfame K with bulk sweeteners are rated to be fuller bodied because of the slightly higher viscosity and the different taste profiles of sugars and acesulfame K. A substantial number of low-calorie beverages, however, are sweetened with mixtures of acesulfame K and aspartame or other intense sweeteners. These beverages benefit from the synergism and the improved taste characteristics provided by such blends. For example, a sweetness level equivalent to approximately 10% of sucrose is imparted by concentrations in a range of 500–600 mg/l of acesulfame K or aspartame, whereas the same sweetness level can be achieved by using a blend of only 160 mg/l of each of the sweeteners. Blend ratios between acesulfame and aspartame, in particular, are different in beverages having different flavors to match flavor and sweetness profiles (25). Of particular importance in beverages is sweetener stability, and acesulfame K's stability at lower pH increases sweetness retention in beverages versus sweeteners that have less stability (e.g., aspartame). In sensory testing, acesulfame K/aspartame–sweetened colas were preferred over aspartame-sweetened colas throughout an 18-week storage period (34). Because the cola with the sweetener blend exhibited greater sweetness retention as a result of the stability of acesulfame, the preference was attributed to this.

When blending acesulfame K with other intense sweeteners for beverage applications, the blend ratio may depend on different factors, including the flavor or flavor type. Blends of acesulfame K and aspartame (40:60) in orange-flavored beverages have been noted to have time intensity curves of sweetness and fruitiness similar to sucrose-sweetened beverages (35). In raspberry-flavored beverages containing natural flavors, 40/60 to 25/75 (acesulfame K/aspartame) blend ratios are considered optimum, whereas beverages with artificial raspberry flavors were considered optimum with blend ratios of 50/50 to 20/80 (36). Optimum taste profiles when using intense sweeteners are considered to have high fruit flavor and minimum side tastes or aftertastes, and sweetener mixtures containing acesulfame K exhibit these properties (18). When only single sweeteners are substituted for sucrose in beverages, flavor problems are encountered, and it has been noted that mixtures of sweeteners can minimize these flavor problems (37). Most notable are mixtures including acesulfame K, which produce taste profiles similar to sucrose (18). Optimum blend ratios complement the taste profile of the specific flavor of beverage and are best determined by sensory testing various blends.

Emerging trends in beverages include replacement of sugar in fully sugared beverages with intense sweeteners like acesulfame K. In regions of the world where sugar is expensive or has inconsistent quality, intense sweeteners are used in combination with sucrose to sweeten many different beverages. Stability in warm climates for longer storage conditions and stability at low pH are necessary for sweeteners that can replace sugar. Acesulfame K exhibits excellent properties for use in this type of application.

Producing jams and marmalades with intense sweeteners only is difficult. Bulking agents are important to improve the texture and shelf-life of such products. In trials for the production of jams with acesulfame K, highly acceptable preparations were prepared with sorbitol as bulking material. Other sugar alcohols and polydextrose proved to be suitable, too. In combinations with acesulfame K, the concentration of sorbitol or other bulking agents can be reduced, yielding products with a noticeable reduction in caloric values compared with to sucrose-containing products. These jams and marmalades are less protected from microbial spoilage compared with sucrose-containing products. In view of the low concentration of osmotically active compounds, the addition of preservatives can help to avoid such microbial spoilage. The flavor stability of jams and marmalades containing low levels of dry solids, however, is normally lower than the storage stability of standard products (20).

Confectionery items can be made with acesulfame K if suitable bulking ingredients or bulk sweeteners are added to give the necessary volume. Because the sweetness intensity of most sugar alcohols and bulking ingredients is lower than the sweetness of sucrose, acesulfame K imparts the desired sweetness, whereas sugar alcohols or low-calorie bulking ingredients provide the necessary bulk and texture. Combinations of acesulfame K with sorbitol can have a good taste pattern. Gum confections containing sorbitol instead of sucrose, and, in addition, 1000 mg/kg acesulfame K were produced without difficulties. They showed a good, fruity sweet taste. Hard-boiled candies can be manufactured using acesulfame as the intense sweetener and isomalt, maltitol or lactitol as the bulking ingredient. Acesulfame rounds the sweetness of these sugar alcohols and brings the taste close to standard, sugar-containing products. Because of the good temperature stability of acesulfame K, it can be added before cooking. Alternatively, it can be added together with acids, flavorings and colorants after cooking. In soft candies, acesulfame shows a good shelf stability. Again, suitable blends of sugar alcohols and acesulfame allow the production of sugarless soft candy coming close to standard products in taste and texture. Sugarless marzipan having good taste, texture, and shelf stability can be manufactured using acesulfame as the intense sweetener and isomalt as a bulking ingredient. In chocolate and related products, acesulfame can be added at the beginning of the production process (e.g., before rolling). It withstands all treatments including conching without detectable decomposition (38). Starch-based confectionery items also benefit from the addition of acesulfame K because it withstands the extrusion temperatures while adding sweetness because these types of products are limited in the amount of sugar that can be incorporated.

Bulking agents with a sweet taste like sorbitol are generally used for the production of sugarless chewing gums. The low sweetness of sugar alcohols is normally enhanced by artificial sweeteners. The fast onset of sweetness of acesulfame K is beneficial in forming the initial taste. Therefore, acesulfame K–containing chewing gum has a pleasant sweet taste from the beginning. Because of its good solubility, acesulfame K may be dissolved fairly quickly by the saliva. Prolonged sweetness may be achieved by encapsulation of some of the acesulfame K or acesulfame in combinations with other sweetening ingredients. More recently, use of acesulfame K to enhance and extend sweetness in fully sugared chewing gums has been seen.

In reduced-calorie baked goods, bulking agents like polydextrose substitute for sugar and flour and may help to reduce the level of fats. Acesulfame combines well with suitable bulking ingredients and bulk sweeteners and therefore allows production of sweet-tasting baked goods having fewer calories. In diabetic products, combinations of acesulfame K and sugar alcohols like isomalt, lactitol, maltitol, or sorbitol can provide volume and sweetness. Texture and the sweetness intensity can be similar to sucrose-containing products. Although lower sweetness for baked compared with unbaked goods was reported in one study (39), extensive work on acesulfame K in baked goods did not give any indication for losses during baking. Even after prolonged exposure to elevated temperatures or elevated oven temperatures, no statistically significant deviation from the added concentrations was analyzed. Sensory studies showed good and pleasant sweetness for baked goods containing acesulfame alone or in combination with bulking ingredients or bulk sweeteners (40). Reduced calorie baked goods (muffins and frostings) sweetened partially with acesulfame K were found to be equal to full-calorie products in sensory testing (41), and sweetener combinations of acesulfame K (with aspartame/saccharin or aspartame alone) in shortbread cookies were found to have sweetness time intensity profiles similar to sucrose (42).

Acesulfame can easily be used in fruit-flavored dairy products. It withstands pasteurization of fruit preparations and pasteurization of the product itself. For nonfermented products like flavored milk, cocoa beverages, and similar products, the use of acesulfame K as the single sweetener is advisable because good heat and shelf-life stability is required. In certain countries where cyclamate is approved, blends of acesulfame K and cyclamate are used in this application, too. Whenever temperature stability is not very important, particularly well-rounded taste profiles are obtained from blends of acesulfame and aspartame, for example, in fruit flavored yogurt and yogurt analog products. Acesulfame K is apparently not attacked by lactic acid bacteria and other microorganisms used in fermented milk products. In ice cream and analog products, the modification of freezing and crystallization of water has to be achieved by the use of appropriate levels of sugar substitutes. Again, acesulfame can round and enhance the sweetness because it combines particularly well with the taste of sugar alcohols (43).

In addition to the listed food categories, many other products can be sweetened with acesulfame K; examples are shown in Table 3.

The production of tabletop preparations containing acesulfame K is similar to those containing other intense sweeteners. Effervescent tablets, granules, powder, and solutions have been produced. Because of the good solubility of acesulfame K in water, highly concentrated solutions suitable for household use can be manufactured. No shelf-life problems have to be anticipated for solutions under normal storage conditions. Similarly, no problems have been reported in the dissolution of tablets or powders. Solid tabletop preparations are generally used in hot beverages, and the solubility of acesulfame K is extremely good at such elevated temperatures.

Outside the food sector, acesulfame K can be used to sweeten oral hygiene products, pharmaceuticals, tobacco products, and animal feedstuffs. In cosmetics, such as toothpaste and mouthwash preparations, and in similar products, some ingredients (e.g., surfactants) tend to impart a bitter taste. Ordinarily, special flavors and sweeteners are needed to mask such taste and to produce an initial pleasant flavor. Because the sweetness of acesulfame K is perceived quickly, it is particularly suitable for these oral hygiene products. It is compatible with commonly used flavoring agents in toothpastes and mouthwash preparations.

If the flavor concentration is not crucial for the taste character of the cosmetic product, the fast onset of the sweetness of acesulfame K requires low concentrations of flavoring ingredients. As mentioned earlier, acesulfame K is highly compatible with sorbitol, which is often used in toothpaste. If glycerol is used as a humectant and brightener, no solubility problems are incurred. Problems have also not been encountered when acesulfame K is dissolved in mouthwash formulations containing alcohol, despite the low solubility of acesulfame K in pure ethanol. In mixtures of ethanol and water, the solubility of acesulfame K normally is greater than concentrations that are used in mouthwash preparations (44, 45).

Pharmaceuticals sometimes have unpleasant flavor or taste characteristics; again, acesulfame K can mask such bitter taste patterns.

V. ANALYTICAL METHODS

The thin-layer chromatographic detection of acesulfame K is simple. Quantitative determinations can be carried out by liquid chromatographic methods and isotachophoresis. Because the volatility of acesulfame K is low and methylation produces differing ratios of methyl derivatives, the quantitative determination by gas chromatography is impossible. Apart from the direct spectroscopic determination of acesulfame K at 227 nm, no spectroscopic methods for the determination are available.

The thin-layer chromatographic detection of acesulfame K can be carried out on polyamide layers. With fluorescent compounds in the layer, the detection can be carried out by extinction of fluorescence. In visible light, detection is possible after spraying with fluorescein and exposing to vapors of bromine. The limit of detection is about 2 μg (46).

Liquid chromatographic determination of acesulfame K in foods is simple, because beverages or aqueous extracts from foods often can be injected onto the columns immediately after filtration. The methods described are basically reverse-phase separations with detection in ultraviolet (UV) light (47, 48). The same liquid chromatographic procedure can be used in assaying feeds which, in contrast to foods, should be clarified by adding zinc sulfate and potassium hexacyanoferrate (49).

Ion chromatography on anion separator columns may be used alternatively. Acesulfame and other sweeteners including cyclamate can be detected by a conductivity detector (50).

Because conductivity detectors are used in isotachophoresis, this method can also be used for separation of acesulfame from other food ingredients and intense sweeteners. Quantitative determination of acesulfame in the presence of other sweeteners has been described (48, 51).

All methods described allow the separation of acesulfame K from other intense sweeteners that may be used in combination.

VI. REGULATORY SITUATION

The completion of a comprehensive safety evaluation program paved the way for the approval of acesulfame K as a food additive.

The evaluation of the data by the Joint Expert Committee for Food Additives of the WHO and FAO resulted in approval for food use with the allocation of an ADI of 0–15 mg/kg of body weight (23). In addition, specifications were published that were revised later (52). The Scientific Committee for Foods of the EEC performed another evaluation of acesulfame's safety. Again, acesulfame was approved for food use, and an ADI of 0–9 mg/kg of body weight was allocated (24).

In the United Kingdom, the first country approving Acesulfame, the Food Additives and Contaminants Committee published a report on sweetening agents for foods, listing acesulfame K among "substances that the available evidence suggests are acceptable for use in food" (53). In the United States, the approval of acesulfame for a number of applications was granted in 1988 (25) followed by approvals for many other product categories including soft drinks. All approvals allow use of acesulfame K under conditions of good manufacturing practice (i.e., without numerical limitation) (54).

After the favorable assessments published by the WHO and FAO and the Scientific Committee for Foods of the EU, more than 100 countries have approved the use of acesulfame K in at least some products. They include all member states of the European Union (because acesulfame K is listed for many applications in the Directive on Sweeteners for Use in Foodstuffs), Canada, Japan, Australia, and New Zealand. In several countries, approvals according to good manufacturing practice have been granted.

Because the ADI values allocated for acesulfame K may not be easy to compare with use levels, a conversion into the corresponding sucrose levels offers a simple possibility to get an idea of the potential applications. Using the sweetness intensity factor of 200 times sucrose, the ADI allocated by the WHO and FAO corresponds to 180 g of sucrose for the 60-kg adult. These values are well above the average daily sucrose consumption. Therefore, it can be concluded that the intake of acesulfame K will generally be well below the ADI. This view has been confirmed by intake studies and calculations.

REFERENCES

1. K Clauss, H Jensen. Oxathiazinon dioxides—a new group of sweetening agents. Angew Chem Engl 85:965, 1973; Angew Chem (Int Ed), 12:869, 1973.

2. K Clauss, E Lück, G-W von Rymon Lipinski. Acetosulfam, ein neuer Süssstoff. Z Lebensm Unters Forsch 162:37, 1976.
3. H-J Arpe. Acesulfame K, a new noncaloric sweetener. In: B Guggenheim, ed. Health and Sugar Substitutes. Proceeding of the ERGOB Conference, Geneva 1978. Basel: Karger, 1978, p 178, and personal communications.
4. G-W von Rymon Lipinski, BE Huddart. Acesulfame K Chem Ind 1983: 427.
5. European Patent 1 59 516 (22.03.1984) to Hoechst AG.
6. EF Paulus. 6-Methyl-1,2,3-oxathiazin-4(3H)-on 2,2-dioxide. Acta Cryst (B) 31: 1191, 1975.
7. K Hoppe, B Gassmann. Vergleichstabellen zur Süsseintensität von 16 Süßungsmitteln. Lebensmittelindustrie 32:227, 1985.
8. DB Ott, CL Edwards, SJ Palmer. Perceived taste intensity and duration of nutritive and non-nutritive sweeteners in water using time-intensity (T-I) evaluations. J Food Sci 56:535, 1991.
9. K Paulus, M Braun. Süsskraft und Geschmacksprofil von Süssstoffen. Ernährungs-Umschau 35:384, 1988.
10. K Hoppe, B Gassmann. Neue Aspekte der Süsskraftbeurteilung und des Einsatzes von Süssungsmitteln. Nahrung 24:423, 1980.
11. U.S. Patent 4 158 068 (12.06.1979) to Hoechst AG.
12. RA Frank, SJ Mize, R Carter. An assessment of binary mixture interactions for nine sweeteners. Chem Senses 14:621, 1989.
13. N Ayya, HT Lawless. Quantitative and qualitative evaluation of high-intensity sweeteners and sweetener mixtures. Chem Senses, 17:245, 1992.
14. G-W von Rymon Lipinski, E Klein. Synergistic effects in blends of Acesulfame K and Aspartame. ACS Conference ''Sweeteners—Carbohydrate and Low Calorie.'' Los Angles, 1988, No. 27.
15. LY Hanger, G Ritter, R Baron. unpublished results.
16. AI Bakal. Functionality of combined sweeteners in several food applications. Chem Ind 18:700, 1983.
17. RC Gelardi. The multiple sweetener approach and new sweeteners on the horizon. Food Technol 41(1):123, 1987.
18. LY Hanger, A Lotz, S Lepeniotis. Descriptive profiles of selected high intensity sweeteners (HIS), HIS blends and sucrose. J Food Sci 61:456, 1996.
19. MO Portman, D Kilcast. Descriptive profiles of synergistic mixtures of bulk and intense sweeteners. Food Quality Preference, 9:221, 1998.
20. PV Tournot, J Pelgroms, JV DerMeeren. Sweetness evaluation of mixtures of fructose with saccharin, aspartame or acesulfame K. J Food Sci 50:469, 1985.
21. G-W von Rymon Lipinski. The new intense sweetener Acesulfame K. Food Chem 16:259, 1985.
22. Toxicological Evaluation of Certain Food Additives. WHO Food Additives Series, 16:11, 1980.
23. Evaluàtion of Certain Food Additives and Contaminants. Twenty-seventh Report of the Joint FAO/WHO Expert Committee on Food Additives. WHO Technical Report Series No. 696. Geneva, 1983, p 21; Evaluation of Certain Food Additives and Contaminants. Thirty-seventh Report of the Joint FAO/WHO Expert Committee on Food Additives. WHO Technical Report Series No. 806. Geneva 1991, p 20.

24. Report of the Scientific Committee for Food No. 16—Sweeteners. Commission of the European Communities, Luxembourg 1985, p
25. Federal Register 53(145):28379, 1988.
26. M Uihlein. Bestimmung von Acesulfam, einem neuen Süssstoff, in Serum mittels Hochdruckflüssigkeitschromatographie. Dtsch Lebensm Rdsch 74:157, 1977.
27. TH Grenby, MG Saldanha. Studies of the inhibitory action of intense sweeteners on oral microorganisms relating to dental health. Caries Res 20:17, 1986.
28. HA Linke. Adherence of *Streptococcus mutans* to smooth surface in the presence of artificial sweeteners. Microbios 36:41, 1983.
29. SC Ziesenitz, G Siebert. Nonnutritive sweeteners as inhibitors of acid formation by oral microorganisms. Caries Res 20:498, 1986.
30. G Siebert, SC Ziesenitz, J Lotter. Marked caries inhibition in the sucrose-challenged rat by a mixture of nonnutritive sweeteners. Caries Res 21:141, 1987.
31. AT Brown, GM Best. Apparent synergism between the interaction of saccharin, Acesulfame K and fluoride with hexitol metabolism by *Streptococcus mutans*. Caries Res 22:2, 1988.
32. G-W von Rymon Lipinski. Einsatz von Sunett® in Erfrischungsgetränken. Swiss Food 10(5):25, 1988.
33. K Clauss. Unpublished results.
34. LY Hanger. Unpublished results.
35. NL Matysiak, AC Nobel. Comparison of temporal perception of fruitiness in model systems sweetened with aspartame, an aspartame+acesulfame K blend, or sucrose. J Food Sci 56:823, 1991.
36. R Baron, LY Hanger. Using acid level, acesulfame potassium/aspartame blend ratio and flavor type to determine optimum flavor profiles of fruit flavored beverages. J Sens Studies 13:269, 1998.
37. DF Nahon, JP Roozen, C De Graaf. Sweetness flavor interactions in soft drinks. Food Chemistry 56:283, 1996.
38. G-W von Rymon Lipinski, E Klein. Sunett®—Anwendungen in Süsswaren. Süsswaren 32:336, 1988.
39. PE Redlinger, CS Setser. Sensory quality of selected sweeteners: Unbaked and baked flour doughs. J Food Sci 52:1391, 1987.
40. G-W von Rymon Lipinski, Ch Klug, E Gorstelle. Unpublished results.
41. LY Hanger. Unpublished results.
42. H Lim, CS Setser, SS Kim. Sensory studies of high potency multiple sweetener systems for shortbread cookies with and without polydextrose. J Food Sci 54:625, 1989.
43. G-W von Rymon Lipinski. Einsatz von Sunett® in Joghurt und anderen Milcherzeugnissen. Swiss Food 10(6):29, 1988.
44. G-W von Raymon Lipinski, E Lück, F-J Dany. Acetosulfam—ein neuer Süssstoff für die Verwendung in Mundkosmetika. Seifen Öle Fette Wachse 102:243, 1976.
45. G-W von Rymon Lipinski, E Lück. Acesulfame K: A new sweetener for oral cosmetics. Manufacturing Chemist 52:37, 1981.
46. G-W von Rymon Lipinski, H-Ch Brixius. Dünnschichtchromatographischer Nachweis von Acesulfam, Saccharin und Cyclamat. Z Lebensm Unters Forsch 168:212, 1979.

47. U Hagenauer-Hener, C Frank, U Hener, A Mosandl. Bestimmung von Aspartame, Acesulfame K, Saccharin, Coffein, Sorbinsaure und Benzoesaure in Lebensmitteln mittels HPLC. Dtsch Lebensm Rdsch 86:348, 1986.
48. U Zache, H Gründing. Bestimmung von Acesulfam-K, Aspartam, Cyclamat und Saccharin in fruchtsafthaltigen Erfrischungsgetränken. Z Lebensm Unters Forsch 184:530, 1987.
49. A Görner, H Rückemann. Qualitative Bestimmung von Acesulfam-K in Mischfuttermitteln mittels Hochdruckflüssigkeitschromatographie. Landwirtsch Forschung 33: 213, 1980.
50. Th A Biemer. Analysis of saccharin, acesulfame K and sodium cyclamate by high-performance ion chromatography. J Chromatog. 463:463, 1989.
51. H Klein, W Stoya. Isotachophoretische Bestimmung von Acesulfam-K in Lebensmitteln. Ernährung/Nutrition 11:322, 1987.
52. Specifications for Identity and Purity, FAO Food and Nutrition Paper No. 28, Rome, 1983, p 3; Compendium of Food Additive Specifications Addendum 4, FAO Food and Nutrition Paper No. 52 Add. 4, Rome, 1996, p 1.
53. Food Additives and Contaminants Committee, Report on the Review of Sweeteners in Food, London, 1982.
54. Federal Register 57(236):57960, 1992; Federal Register 59(230):61538, and 61543, 1994; Federal Register 60(85):21700, 1995; Federal Register 63(128):36344, 1998.

3
Alitame

Michael H. Auerbach
Danisco Cultor America, Ardsley, New York

Graeme Locke
Cultor Food Science, Redhill, Surrey, United Kingdom

Michael E. Hendrick†
Pfizer Inc., Groton, Connecticut

I. INTRODUCTION

The field of high-potency sweeteners has long been an active area of research. Stimulated by the accidental discovery of the potent sweet taste of aspartame in 1965 (1,2), dipeptides became an area of increasing interest for chemists and food technologists alike. Starting in the early 1970s an intensive, systematic program to develop high-potency sweeteners was carried out at Pfizer Central Research. This program, which involved the synthesis of a large number of dipeptides of diverse structural types, culminated in the synthesis of alitame, which is the subject of this review. Alitame and structurally related peptide sweeteners are the subject of a U.S. patent (3).

II. STRUCTURE

The structure of alitame is shown in Fig. 1, which emphasizes the component parts of the molecule. Alitame is formed from the amino acids L-aspartic acid and D-alanine, with a novel C-terminal amide moeity. It is this novel amide moeity (formed from 2,2,4,4-tetramethylthienanylamine) that is the key to the very high sweetness potency of alitame. The structure of alitame was developed by following leads from a number of synthesized model compounds. Within the

†Deceased.

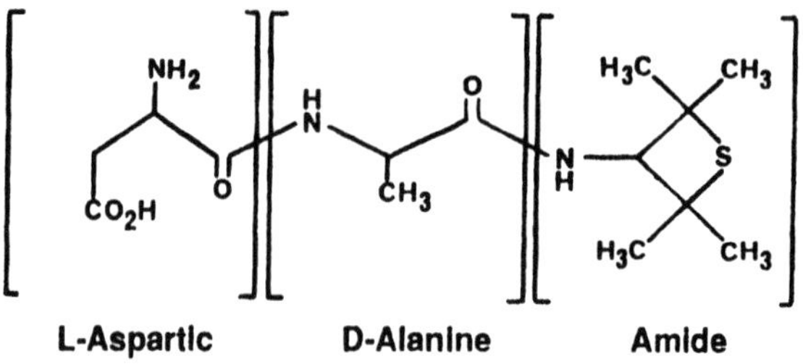

L-Aspartic D-Alanine Amide

DEVELOPMENT:

Discovered: 1979 Pfizer Central Research
Patented: 1983 (U.S. 4,411,925)
FDA Filing: 1986 Food Additive Petition

Figure 1 Alitame: structure and development.

series of L-aspartyl-D-alanine amides those structural features that were found to be most conducive to high sweetness potency include small to moderate ring size, presence of small-chain branching α to the amine-bearing carbon, and the introduction of the sulfur atom into the carbocyclic ring.

III. ORGANOLEPTIC PROPERTIES

Alitame is a crystalline, odorless, nonhygroscopic powder. Its sweetness potency, determined by comparison of the sweetness intensity of alitame solutions with concentrations in the range of 50 μg/ml to a 10% solution of sucrose, is approximately 2000 times that of sucrose. Compared with threshold concentrations of sucrose (typically 2–3%), the potency of alitame increases to about 2900 times that of sucrose. This phenomenon is typical of high-potency sweeteners. Use of higher concentrations of alitame allows preparation of solutions with a sweetness intensity equivalent to sucrose solutions of 40% or greater.

The sweetness of alitame is of a high quality, sucroselike, without accompanying bitter or metallic notes often typical of high-potency sweeteners. The sweetness of alitame develops rapidly in the mouth and lingers slightly, in a manner similar to that of aspartame.

Alitame has been found to exhibit synergy when combined with both acesulfame-K and cyclamate. High-quality blends may be obtained with these and other sweeteners, including saccharin.

Table 1 Solubility of Alitame in Various Solvents at 25°C

Solvent	Solubility (%w/v)
Water	13.1
Methanol	41.9
Ethanol	61.0
Propylene glycol	53.7
Chloroform	0.02
N-Heptane	0.001

IV. SOLUBILITY

At the isoelectric pH, alitame is very soluble in water. Excellent solubility is also found in other polar solvents, as is shown in Table 1. As expected from the molecule's polar structure, alitame is virtually insoluble in lipophilic solvents. In aqueous solutions the solubility rapidly increases with temperature and as the pH deviates from the isoelectric pH. This effect is shown in Table 2.

V. DECOMPOSITION PATHWAYS

The principal pathways for the reaction of alitame with water are relatively simple compared with aspartame (Fig. 2). The major pathway involves hydrolysis of the aspartylalanine dipeptide bond to give aspartic acid and alanyl-2,2,4,4-tetramethylthietane amide (''alanine amide''). The α,β-aspartic rearrangement, common

Table 2 Solubility[a] of Alitame as a Function of pH and Temperature

pH	5°C	20°C	30°C	40°C	50°C
2.0	41.7	48.7	56.4	50.3	54.0
3.0	32.2	39.2	46.5	50.9	53.9
4.0	12.9	13.9	17.3	20.4	37.6
5.0	11.7	12.8	14.9	16.8	29.2
6.0	11.6	13.2	14.9	19.5	32.8
7.0	11.8	14.3	17.6	29.5	51.8
8.0	14.8	24.9	46.8	56.2	52.1

[a]Water solubility (%w/v).

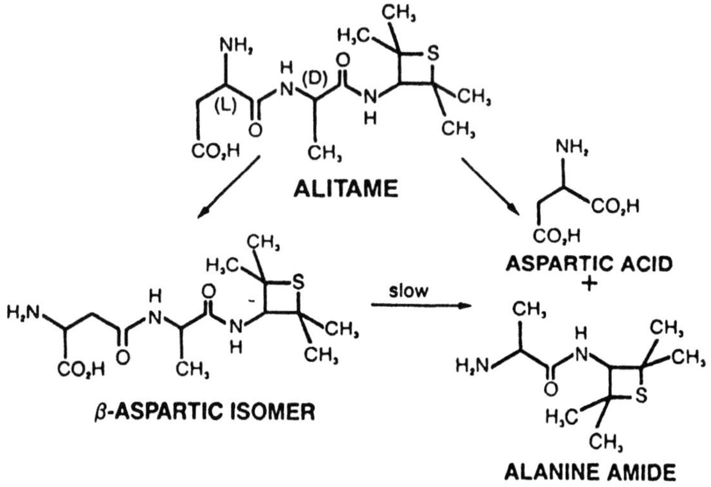

Figure 2 Main degradation pathways of alitame.

to all peptides bearing N-terminal aspartic acid (4), also occurs to give the β-aspartic isomer of alitame. This rearranged dipeptide hydrolyzes at a slower rate than alitame to give the same products as those arising from the parent compound (i.e., aspartic acid and alanine amide). No cyclization to diketopiperazine or hydrolysis of the alanine amide bond is detectable in solutions of alitame that have undergone up to 90% decomposition. All three major decomposition products are completely tasteless at levels that are possible in foods.

VI. STABILITY

In Fig. 3 the half-lives for alitame and aspartame (5) in buffer solutions of various pH are compared. As can be seen, the solution stability of alitame approaches the optimum for aspartic acid dipeptides. At acid pH (2–4), alitame solution half-lives are more than twice those of aspartame. As the pH increases, this stability advantage increases dramatically. In particular in the neutral pH range (5–8) alitame is completely stable for more than 1 year at room temperature.

Alitame is sufficiently stable for use in hard and soft candies, heat-pasteurized foods, and in neutral pH foods processed at high temperatures, such as sweet baked goods. The dramatic heat stability advantage of alitame over aspartame under simulated baking conditions is illustrated in Fig. 4. Actual half-life data for the two sweeteners under these conditions may be found in Table 3. This

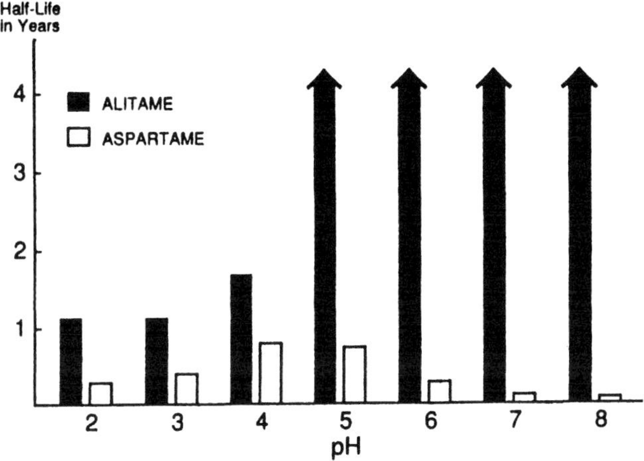

Figure 3 Alitame and aspartame stability in buffer solutions at 23°C.

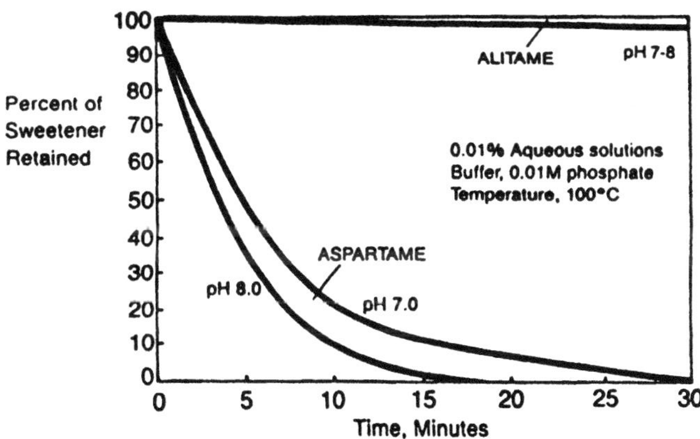

Figure 4 Thermal stability of alitame and aspartame.

Table 3 Stability of Alitame and Aspartame at Elevated Temperature in Aqueous Solution

Sweetener	Temp (°C)	Half-life (hr) pH 6.0	Half-life (hr) pH 7.0	Half-life (hr) pH 8.0
Alitame	100	13.5	13.4	12.6
Alitame	115	2.1	2.1	2.1
Aspartame	100	0.4	0.1	0.04
Aspartame	115	0.1	0.03	0.02

allows alitame to survive the thermal and pH conditions of the baking process with insignificant decomposition.

VII. COMPATIBILITY

Alitame exhibits excellent functionality and is compatible with a wide variety of freshly prepared foods. In accordance with its chemical structure, it can undergo chemical reaction with certain food components. In particular, high levels of reducing sugars, such as glucose and lactose, may react with alitame in heated liquid or semiliquid systems, such as baked goods, to form Maillard reaction products. High levels of flavor aldehydes can behave similarly. Such reactions have been reported for aspartame and other aspartyl dipeptides (6–9). Therefore, the compatibility of alitame with a given recipe will depend on the actual ingredients present and the thermal and pH exposure involved in the manufacturing process.

Prolonged storage of alitame in a few standard acidic liquid beverages recipes may result in an incompatibility as measured organoleptically (off-flavors). This is not reflected in storage stability as measured by chemical assay for alitame and its degradation products. Levels of off-flavorant(s) are below the limits of modern analytical detection. Substances, which may produce off-flavors on storage with alitame in liquid products, are hydrogen peroxide, sodium bisulfite, ascorbic acid, and some types of caramel color at pHs less than 4.0. However, a research program is underway to find appropriate solutions to overcome these problems.

VIII. METABOLISM

After oral administration to the mouse, rat, dog, or man, alitame is well absorbed. Most of an oral dose (77–96%) is excreted in urine as a mixture of metabolites (Fig. 5). The remainder (7–22%) is excreted in the feces, primarily as unchanged alitame. Radiochemical balances of 97–105% were obtained in the four species.

In all four species the metabolism of alitame is characterized by the loss of aspartic acid followed by conjugation and/or oxidation at the sulfur atom of the alanine amide fragment yielding the corresponding sulfoxide isomers and the sulfone. Further hydrolysis of the alanine amide fragments does not take place. In the rat and dog the alanine amide is partially acetylated, and in man it is partially conjugated with glucuronic acid. No observable cleavage of the alanine amide bond or rupture of the thietane ring takes place.

Because the aspartic acid portion of the molecule is available for normal amino acid metabolism, alitame is partially caloric. The maximum caloric contri-

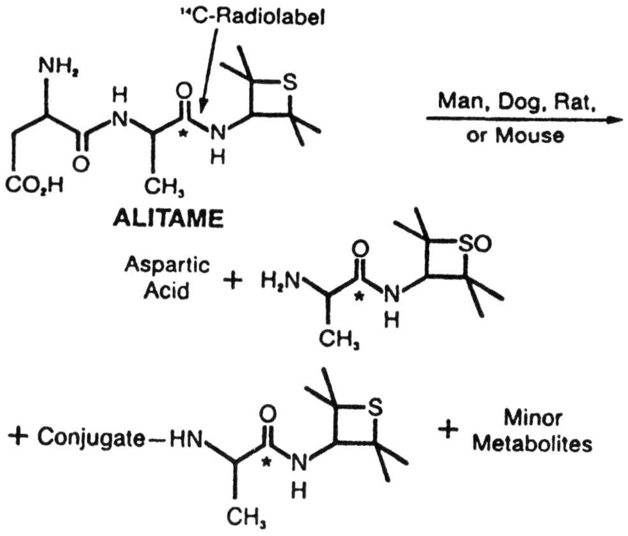

Figure 5 Metabolism of alitame.

bution of 1.4 calories per gram is clearly insignificant at the use levels in the diet.

IX. SAFETY

Alitame has been evaluated in the extensive series of studies designed to establish its safety as a sweetener in the human diet. A list of the major studies completed is shown in Table 4. On the basis of substantial safety factors determined in these studies, it is concluded that alitame is safe for its intended use as a component of the diet of man.

The appropriateness of the animal studies is supported by the fact that the major metabolites of the compound are common to laboratory animals and man. The cumulative data amassed during the safety evaluation demonstrate that this sweetener has an extremely low order of toxicity that supports its estimated mean chronic daily human dietary intake of 0.34 mg/kg body weight, assuming that alitame is the sole sweetener in all food categories requested in the Food Additive Petition. In all the studies the no-effect level (NOEL) was consistently greater than 100 mg/kg, or >300 times the mean chronic human exposure.

The primary effect of alitame in animals treated with levels greater than the NOEL (>100 mg/kg) for periods ranging from 5 days to 2 years was the dose-

Table 4 Major Safety Studies Completed

Genetic toxicology
Three-month dog (0.5, 1, 2% of the diet)
Three-month rat (0.5, 1, 2% of the diet)
Rat teratology (100, 300, 1000 mg/kg/day)
Rabbit teratology (100, 300, 1000 mg/kg/day)
One-month mouse (1, 2, 5% of the diet)
Twelve-month rat (0.01, 0.03, 0.1, 0.3, 1% of the diet)
Eighteen-month dog (10, 30, 100, 500 mg/kg/day)
Rat reproduction (0.1, 0.3, 1% of the diet)
Twenty-four-month mouse oncogenicity (0.1, 0.3, 0.7% of the diet)
Twenty-four-month rat oncogenicity (0.1, 0.3, 1% of the diet; two species)
Rat, neurobehavioral (3 studies)
Man, metabolism
Man, no effects (15 mg/kg/day, 14 days)
Man, no effects (10 mg/kg/day, 90 days)
Diabetic men and women, no effects (10 mg/kg/day, 90 days; two studies)

related increase in liver weight secondary to the induction of hepatic microsomal metabolizing enzymes, a common adaptive response of the liver to xenobiotics. This high-dose effect abated during chronic studies and after cessation of treatment. At a dose in man of 15 mg/kg/day (44 times the mean chronic intake estimate and equivalent to consumption of 18 liters of alitame-sweetened carbonated beverage per day by a 60-kg individual), no enzyme induction was observed over a 14-day period.

Furthermore, neither enzyme induction nor any other effects were observed over a 90-day period in a double-blind toleration study of 130 subjects of both sexes of three races (Caucasian, African-American, and Asian) receiving 10 mg alitame/kg/day, a dose equivalent to consumption of 12 liters carbonated beverage per day.

No evidence of carcinogenic potential was noted in rats and mice treated for 2 years at doses as high as 564 and 1055 mg/kg/day, respectively. Alitame, in doses as high as 1000 mg/kg/day during organogenesis, was devoid of embryotoxic or teratogenic potential in rats and rabbits. In two-generation reproduction studies there were no compound effects on mating behavior, pregnancy rate, the course of gestation, litter size, or maternal or offspring survival.

Mutagenicity assays, both in vitro and in vivo, revealed no genotoxicity at gene or chromosomal levels in microbial and mammalian test systems. Alitame had no detectable activity in a battery of pharmacological systems used to assess autonomic, gastrointestinal, renal, and central nervous system functions. No effects were noted on fasting blood glucose or on the disposition of an oral glucose load.

X. REGULATORY STATUS

In 1986 a Food Additive Petition was submitted to the U.S. Food and Drug Administration requesting broad clearance for alitame (10). The petition requests approval of alitame as a sweetener and flavoring in specified foods in amounts necessary to achieve the intended effect and in accordance with good manufacturing practice. Table 5 presents the 16 food product categories requested in this petition.

In addition to the United States, permission for the use of alitame in food has been requested from a number of other countries and regulatory agencies. Alitame was approved for use in Australia in December 1993; in Mexico in May 1994; in New Zealand in October 1994; in People's Republic of China in November 1994; in Indonesia in October 1995; in Colombia in April 1996, and in Chile in June 1997. Alitame was reviewed by Joint FAO/WHO Expert Committee on

Table 5 Food Product Categories Requested in 1986 Food Additive Petition to U.S. FDA

Baked goods and baking mixes (restricted to fruit-, custard- and pudding-filled pies; cakes; cookies, and similar baked products)
Presweetened, ready-to-eat breakfast cereals
Milk products (restricted to flavored milk- or dairy-based beverages and mixes for their preparations, yogurt, and dietetic milk products)
Frozen desserts and mixes
Fruit and water ices and mixes
Fruit drinks, ades, and mixes, including diluted juice beverages and concentrates (frozen and non-frozen) for dilute juice beverages
Confections and frosting
Jams, jellies, preserves, sweet spreads
Sweet sauces, toppings, and mixes
Gelatins, puddings, custards, fillings, and mixes
Beverages, nonalcoholic and mixes
Dairy product analogs (restricted to toppings and topping mixes)
Sugar substitutes
Sweetened coffee and tea beverages, including mixes and concentrates
Candy (including soft and hard candies and cough drops)
Chewing gum

Food Additives (JEFCA) in 1995 (11) and 1996 (12) and allocated an acceptable daily intake (ADI) of 1 mg/kg of body weight (12). A WHO/FAO specification monograph has been published (13). Alitame INS No. 956 is also listed in the draft codex Alimentarius General Standard for Food Additives (14).

REFERENCES

1. A Ripper, BE Homler, GA Miller. Aspartame. In: LO Nabors, RC Gelardi, eds. Alternative Sweeteners. New York: Marcel Dekker, Inc., 1986, pp 43–70.
2. RH Mazur, JM Schlatter, AH Goldkamp. J Am Chem Soc 91:2684, 1969.
3. TM Brennan, ME Hendrick. U.S. Patent No. 4,411,925 (October 25, 1983).
4. M Bodanszky, J Martinez. Side reactions in peptide synthesis. In: E Gross, J Meienhofer, eds. The Peptides, Vol. 5. New York: Academic Press, Inc., 1983, pp 143–148.
5. BM Homler. Food Technology (July): 50–55, 1984.
6. AS Cha, C-T Ho. J Food Science, 53: 562, 1988.
7. Y Ariyoshi, N Sato. Bull Chem Soc Japan 45:2015, 1972.
8. MM Hussein, RP D'Amelia, AL Manz, H Jacin, W-TC Chen. J Food Science 49: 520, 1984.
9. JA Stamp, TP Labuza. J Food Science 48:543, 1983.
10. Federal Register 51(188):34503, 1986.
11. PJ Abbott. Alitame. In: Toxicological Evaluation of Certain Food Additives and Contaminants. 44th JECFA, WHO Food Additives Series 35, pp 209–254, 1996.
12. PJ Abbott. Alitame. In: Toxicological Evaluation of Certain Food Additives. 46th JECFA, WHO Food Additives Series 37, pp 9–11, 1996.
13. Alitame. In: Compendium of Food Additive Specifications. 44th JECFA, FAO Food and Nutrition Paper 52 Add. 3, pp 9–13, 1995.
14. Proposed Draft Schedule 1 on Sweeteners—Additives with Numerical ADIs. In: Report of the 29th Session of the Codex Committee on Food Additives and Contaminants, Alinorm 97/12A, March 1997, pp 156–157 lines 67–78.

4

Aspartame

Harriett H. Butchko, W. Wayne Stargel, C. Phil Comer, Dale A. Mayhew, and Sue E. Andress
The NutraSweet Company, Mount Prospect, Illinois

I. INTRODUCTION AND HISTORY

Over three decades have elapsed since aspartame was discovered accidentally in 1965 by G. D. Searle and Co. chemist James Schlatter (1, 2). At present, it is estimated that aspartame is used in approximately 6000 different products worldwide.

The safety of aspartame has been tested extensively in animal and human studies. It is undoubtedly the most thoroughly studied of the high-intensity sweeteners. The safety of aspartame has been affirmed by numerous scientific bodies and regulatory agencies, including the Joint FAO/WHO Expert Committee on Food Additives (JECFA) of the Codex Alimentarius (Food and Agriculture Organization/World Health Organization) (3), the Scientific Committee for Food of the Commission of European Communities (4), the U.S. Food and Drug Administration (FDA) (5, 6), and the regulatory agencies of more than 100 other countries around the world.

II. PHYSICAL CHARACTERISTICS AND CHEMISTRY

A. Structure

Aspartame is a dipeptide composed of two amino acids, L-aspartic acid and the methyl ester of L-phenylalanine. The chemical structure of aspartame is depicted in Fig. 1. Aspartame sold for commercial use meets all requirements of the Food Chemical Codex (7).

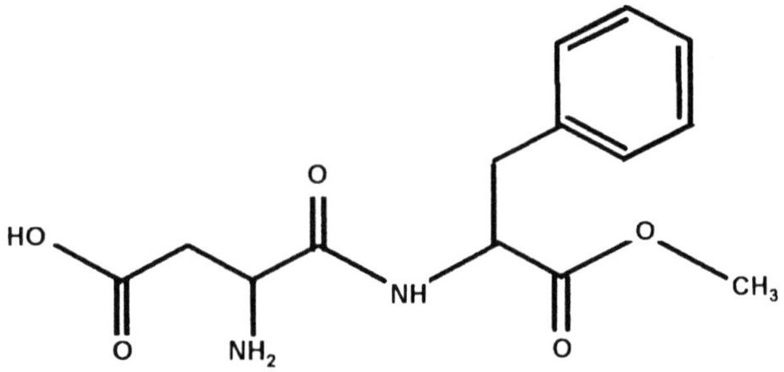

L-aspartyl-L-phenylalanine methyl ester

Figure 1 Chemical structure of aspartame.

B. Stability

In liquids and under certain conditions of moisture, temperature, and pH, the ester bond is hydrolyzed, forming the dipeptide, aspartylphenylalanine, and methanol. Ultimately, aspartylphenylalanine can be hydrolyzed to its individual amino acids—aspartate and phenylalanine (8, 9). Alternatively, methanol may also be hydrolyzed by the cyclization of aspartame to form its diketopiperazine (DKP) (Fig. 2).

The decomposition of aspartame is indicative of first-order kinetics, and stability is determined by time, moisture, temperature, and pH. Under dry conditions, the stability of aspartame is excellent; it is, however, affected by extremely high temperatures that are not typical for the production of dry food products (10, 11). The combined effect of time, temperature, and pH on the stability of aspartame in solutions is shown in Fig. 3.

At 25°C, the maximum stability is observed at pH ~4.3. Aspartame functions very well over a broad range of pH conditions but is most stable in the weak acidic range in which most foods exist—between pH 3 and 5. A frozen dairy dessert may have a pH ranging from 6.5 to more than 7.0, but, because of the frozen state, the rate of reaction is dramatically reduced. And, because of the lower free moisture, the shelf-life stability of aspartame exceeds the predicted shelf-life stability of these products.

C. Solubility in Food and Beverage Applications

Aspartame is slightly soluble in water (about 1.0% at 25°C) and is sparingly soluble in alcohol (10, 11). It is not soluble in fats or oils. An important consideration, especially for many products that are prepared by mixing aspartame-

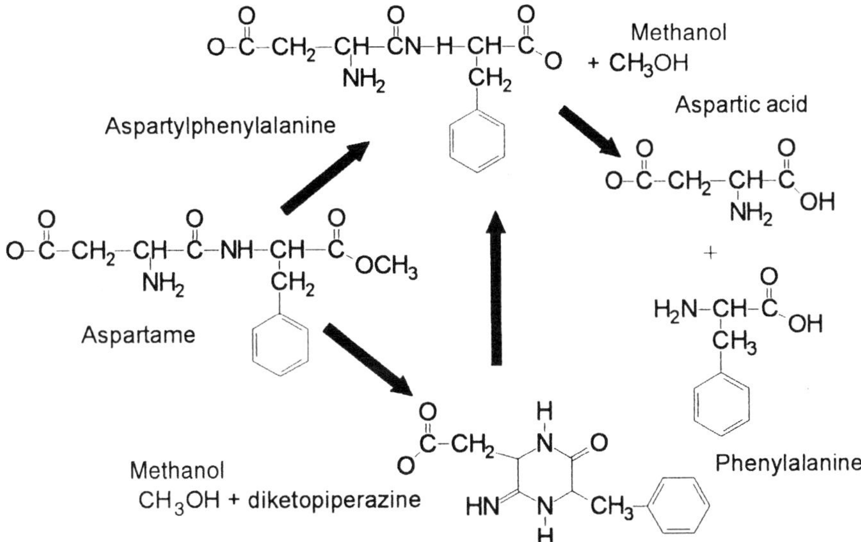

Figure 2 Principal conversion products of aspartame.

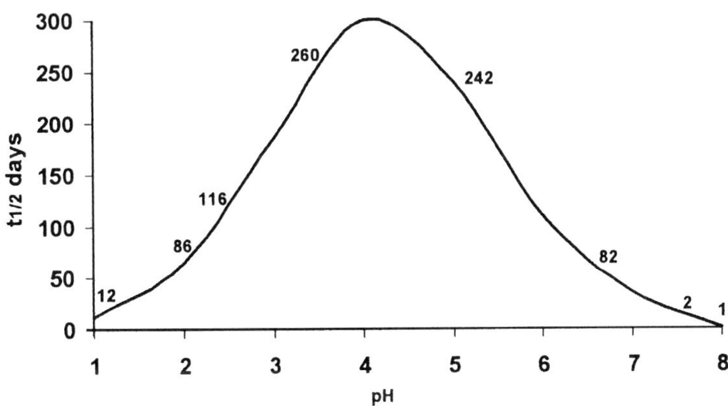

Figure 3 Stability of aspartame in aqueous buffers at 25°C.

sweetened products with water or with milk at different temperatures, is the rate of dry mix dissolution. Solubility is a function of both temperature and pH.

III. TASTE

The taste of aspartame is described as clean and sweet like sugar but without the bitter chemical or metallic aftertaste often associated with some other high-intensity sweeteners. Comparisons of the sweetness of aspartame and sucrose that use quantitative descriptive analyses reveal that the taste profile for aspartame closely resembles that of sucrose. Food industry and university-based studies have documented this sugarlike taste of aspartame (8, 12).

A. Flavor-Enhancing Property

Various food and beverage flavors are enhanced or extended by aspartame, especially acid fruit flavors. Such flavor enhancements or flavor extensions are particularly evident with naturally derived flavors (13). This flavor-enhancing property, as evident in chewing gum, may extend flavor up to four times longer (14). Such a characteristic is important in many food applications.

B. Sweetness Intensity

Depending on the food or beverage system, the intensity (i.e., potency) of aspartame has been determined to be 160 to 220 times the sweetness of sucrose (10, 11). Generally, an inverse relationship exists between the intensity of aspartame and the concentrations of sucrose being replaced. Overall, the relative sweetness of aspartame may vary depending on the flavor system, the pH, and the amount of sucrose or other sugars being replaced (10, 11, 15).

IV. FOOD AND BEVERAGE APPLICATIONS

Aspartame is approved for general use in foods (16), including carbonated soft drinks, powdered soft drinks, yogurt, hard candy, and confectionery. The stability of aspartame is excellent in dry-product applications (e.g., tabletop sweetener, powdered drinks, dessert mixes). Aspartame can withstand the heat processing used for dairy products and juices, aseptic processing and other processes in which high-temperature short-time and ultra-high temperature conditions are used. The potential for aspartame to hydrolyze or cyclize may, under some conditions of excessive heat, limit some applications of aspartame.

A wide range exists over which aspartame sweetness levels are acceptable. The loss of aspartame because of certain combinations of pH, moisture, and temperature can lead to a gradual loss of perceived sweetness, with no development of off-flavors, because conversion products of aspartame are tasteless (10, 11).

Blends or combinations of sweeteners are often used to achieve the desired level of sweetness in food and beverage products that traditionally have been sweetened with single sweeteners. Aspartame works well in admixture with other sweeteners, including sugar. The flavor-enhancement quality of aspartame masks bitter flavors even at subsweetening levels and makes aspartame a desirable choice in blends with those sweeteners that possess potentially undesirable or more complex taste profiles.

V. THE ESTABLISHED SAFETY OF ASPARTAME AND ITS COMPONENTS

Aspartame has been proven to be a remarkably safe sweetener, with more than 200 scientific studies in animals and humans confirming its safety. Vigilant postmarketing surveillance of anecdotal complaints from consumers revealed no consistent pattern of symptoms related to consumption of aspartame, and an extensive postmarketing research program to evaluate these allegations in controlled, scientific studies further confirmed that aspartame is not associated with adverse health effects.

A. Acceptable Daily Intake vs. Actual Intake

On the basis of the results of the comprehensive safety studies in animals (17, 18), an Acceptable Daily Intake (ADI) of 40 mg/kg/day for aspartame was set by JECFA (3). On the basis of both animal and human data, the US FDA set an ADI for aspartame of 50 mg/kg body weight/day (19).

The ADI represents the amount of a food additive that can be consumed daily for a lifetime with no ill effect (20–22). It is not a maximum amount that can be safely consumed on a given day. A person may occasionally consume a food additive in quantities exceeding the ADI without adverse effects. Aspartame is about 200 times sweeter than sugar, so an ADI of 50 mg/kg body weight is the sweetness equivalent of approximately 600 g (1.3 pounds) of sucrose consumed daily by a 60-kg person over a lifetime. If aspartame replaced all the sucrose in our diet, consumption would be well below the ADI at approximately 8.3 mg/kg/day (5).

Actual consumption levels of aspartame were monitored from 1984 to 1992 through dietary surveys in the United States (23–26). Average daily aspartame consumption at the 90th percentile (''eaters'' only) in the general population ranged from about 2 to 3 mg/kg body weight. Consumption by 2- to 5-year-old children in these surveys ranged from about 2.5 to 5 mg/kg/day. Aspartame consumption has also been estimated in several other countries around the world. Although survey methods differed among these evaluations, aspartame consumption is remarkably consistent and is only a fraction of the ADI (27–37).

B. Extensive Safety Studies with Aspartame in Humans

In addition to the comprehensive battery of toxicology studies in animals, the safety of aspartame and its metabolic constituents has been exhaustively assessed in humans. In addition to healthy men and women, several human subgroups were studied, including infants, children, adolescents, obese individuals, diabetic individuals, individuals with renal disease, individuals with liver disease, lactating women, and individuals heterozygous for the genetic disease, phenylketonuria (PKU), who have a somewhat decreased ability to metabolize phenylalanine. These and longer term studies showed no untoward health consequences from aspartame (38–59). The results of the human studies, along with the animal research, have provided convincing evidence that aspartame is safe for the general population, including pregnant women and children.

C. Safety of Dietary Components of Aspartame

After consumption, aspartame is rapidly metabolized into its components: aspartate, phenylalanine, and methanol. These components have sometimes been the objects of speculation regarding their potential for adverse effects. As discussed later, the components of aspartame are normal constituents of the diet, and no evidence exists for adverse effects associated with consumption of aspartame.

1. Aspartate

Plasma concentrations of aspartate are not altered by even enormous amounts of aspartame (38–42, 44, 46). Oral administration at a dosage of 34 mg/kg/day, more than 10 times the estimated 90th percentile consumption, does not change plasma aspartate concentrations in humans (38). Chronic administration of even higher doses of aspartame (75 mg/kg/day for 24 weeks) did not change mean fasting plasma concentrations of aspartate (60). The addition of glutamate and aspartame (each at 34 mg/kg) to a meal providing 1 g protein/kg did not increase plasma concentrations of either glutamate or aspartate beyond the changes caused by the meal itself (61, 62). As Fig. 4 illustrates, at current levels of consumption only a small fraction of the daily dietary intake of aspartate is derived from aspartame in adults or children (26).

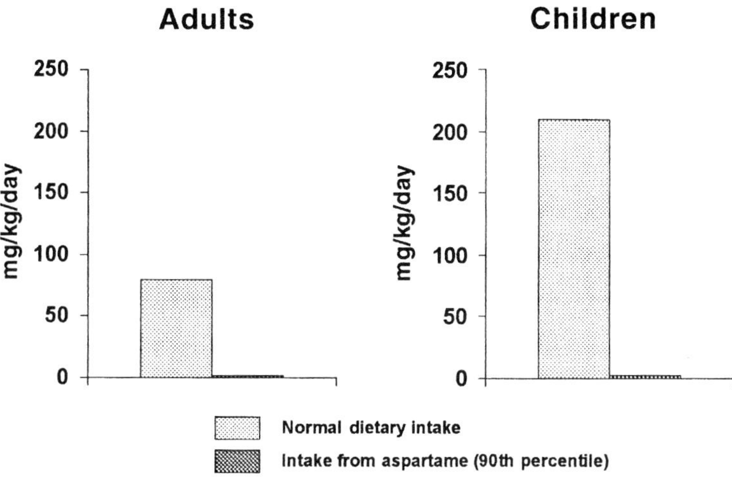

Figure 4 Dietary intake of aspartic acid vs. intake from aspartame in adults and children.

2. Phenylalanine

Consumption of aspartame-sweetened foods does not increase plasma phenylalanine concentrations beyond those that normally occur postprandially (38, 39, 41). Doses of aspartame at approximately 30 mg/kg/day did not increase plasma phenylalanine concentrations above those observed after eating a protein-containing meal in normal adults, phenylketonuric heterozygotes, or non-insulin-dependent diabetic populations (63). As Fig. 5 illustrates, at current levels of consumption only a small fraction of daily dietary intake of the essential amino acid phenylalanine is derived from aspartame in adults or children (26). The only individuals who must be concerned regarding aspartame's phenylalanine content are those with phenylketonuria, a rare genetic disease in which the body cannot properly metabolize phenylalanine. These individuals must severely restrict phenylalanine intake from all dietary sources, including aspartame.

3. Methanol

Aspartame yields approximately 10% methanol by weight. The amount of methanol released from aspartame is well below normal dietary exposures to methanol from fruits, vegetables, and juices (Fig. 6) (5, 23). Aspartame consumption, even in amounts many times those consumed in products, does not alter baseline blood concentrations of methanol or formate (43). Whereas methanol exposure at the 90th percentile of chronic aspartame consumption is 0.3 mg/kg/day, the FDA has established acceptable levels of exposure to methanol at 7.1 to 8.4 mg/kg/

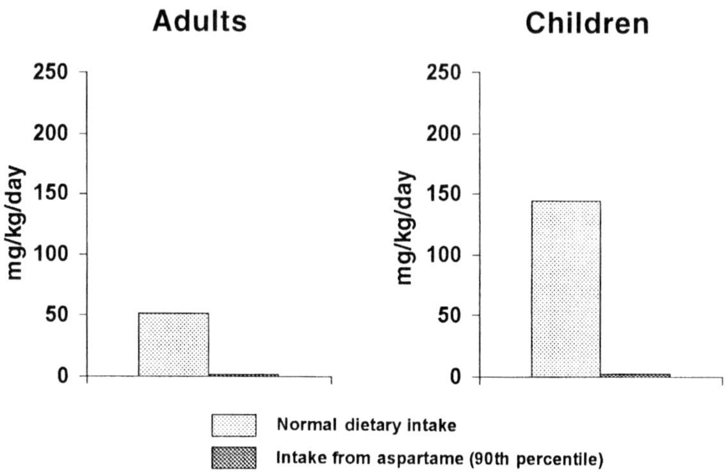

Figure 5 Dietary intake of phenylalanine vs. intake from aspartame in adults and children.

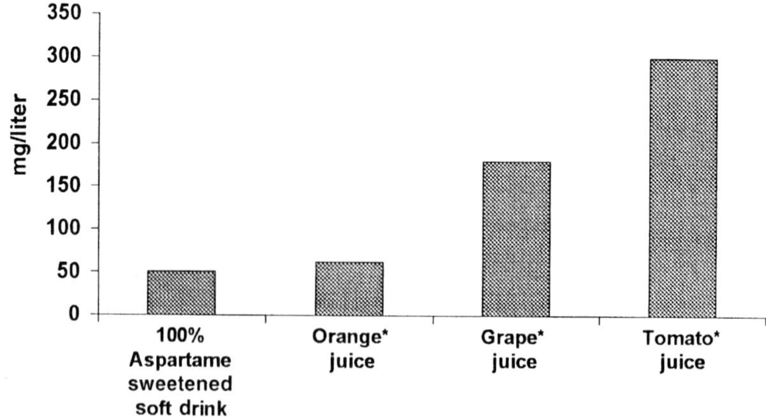

Figure 6 Methanol content of beverage sweetened with aspartame vs. fruit juices. (*From Wucherpfennig et al., Flussuges Obst., 348–354, 1983.)

day for 60-kg adults (64). Thus, acceptable dietary exposures to methanol are approximately 25 times potential exposures to methanol after 90th percentile consumption of aspartame.

Recently, Trocho et al. (65) concluded from a study in rats that aspartame may be hazardous because formaldehyde adducts in tissue proteins and nucleic acids from aspartame may accumulate. However, according to Tephly (66), the doses of aspartame used in the study do not even yield blood methanol concentrations outside control values. Furthermore, the amounts of aspartame equal to that in about 75 cans of beverage as a single bolus result in no detectable increase in blood formate concentrations in humans, whereas increased urinary formate excretion shows that the body is easily able to handle even excessive amounts of aspartame. In addition, there is no accumulation of blood or urinary methanol or formate with long-term exposure to aspartame. Thus, Tephly concluded that "the normal flux of one-carbon moieties whether derived from pectin, aspartame, or fruit juices is a physiologic phenomenon and not a toxic event" (66).

D. Postmarketing Surveillance of Anecdotal Health Reports

Aspartame has been the subject of extensive postmarketing surveillance. Shortly after its widespread marketing, a number of anecdotal reports of health effects appeared which some consumers related to their consumption of aspartame-containing products. Not unexpectedly, negative media stories influenced the numbers and types of these reports. The NutraSweet Company took the unprecedented step of instituting a postmarketing surveillance system to document and evaluate these anecdotal reports (24, 25, 67). Data from this system were evaluated by the company and also shared with the US FDA, as discussed later.

After the approval of aspartame in carbonated beverages in 1983, an increase in the reporting of adverse health events allegedly associated with the consumption of aspartame-containing products led the FDA to request the Centers for Disease Control and Prevention (CDC) to evaluate these reports (68, 69). The CDC was charged with describing the types of complaints and determining whether there were any specific clusters or types of complaints that would indicate the need for further study. The CDC concluded, "Despite great variety overall, the majority of frequently reported symptoms were mild and are symptoms that are common in the general populace" (68). The CDC could not identify any specific group of symptoms that was clearly related to aspartame but believed that focused clinical studies would be the best mechanism to address the issues raised by these reports.

In the mid-1980s, the FDA Center for Food Safety and Applied Nutrition (CFSAN) established its own passive surveillance system, the Adverse Reaction Monitoring System (ARMS), to monitor and evaluate anecdotal reports of adverse health effects thought to be related to foods, food and color additives, and vitamin/mineral supplements (70, 71). Through this system, spontaneous reports of food-associated adverse health events received from consumers, physicians, and industry are documented, investigated, and evaluated.

On the basis of reviews of anecdotal complaints for aspartame, the FDA concluded that there is no "reasonable evidence of possible public health harm" and "no consistent or unique patterns of symptoms reported with respect to aspartame that can be causally linked to its use" (70, 71). In a recent evaluation, they further concluded that there was a gradual decrease in the number of reports regarding aspartame received over time and that the reports remained comparable to previous ones in terms of demographics, severity, strength of association, and symptoms (72).

In considering the anecdotal reports, it is important to keep in mind that, because about 100 million people consume aspartame in the United States alone, it is not unlikely that a consumer may experience a medical event or ailment near the same time aspartame is consumed. The error of inclinations to associate causality with coincidence is perhaps best stated by one scientist who indicated, "As aspartame is estimated to be consumed by about half the U.S. population, one need not be an epidemiologist to grasp the problem of establishing a cause-and-effect relationship. Half the headaches in America would be expected to occur in aspartame users, as would half the seizures and half the purchases of Chevrolets." (73).

E. Research to Investigate Issues Raised in the Postmarketing Period

To expand the knowledge base about aspartame and further address the anecdotal reports and other scientific issues raised during the postmarketing period, a number of studies, including "focused" studies in humans as suggested by the CDC, were done. For example, the results of a long-term clinical study with high doses of aspartame (75 mg/kg/day for 24 weeks, or about 25 to 30 times current consumption levels at the 90th percentile) provided additional confirmatory evidence of aspartame's safety (60). Even after these enormous daily doses of aspartame, there were no changes in clinical or biochemical parameters or increased adverse experiences compared with a placebo. Clinical studies to evaluate whether aspartame causes headache, seizures, or allergic-type reactions were done (74–87). In some of these studies, individuals identified through the company's medical postmarketing surveillance system, who were convinced that aspartame caused their symptoms, were studied (77, 83–85). In addition, studies were done to evaluate whether aspartame had any effect on brain neurotransmitter concentrations or neurotransmission or on indicators of brain function, such as memory, learning, mood, and behavior (88–111). The weight of evidence from the results of the research done with aspartame clearly demonstrates that, even in amounts many times what people typically consume, aspartame is not associated with adverse health effects (112–125).

F. False Allegations of Brain Tumors

In 1996, a group led by long-time aspartame critic John Olney contended that the reported increase in the rate of brain tumors in the United States was related to the marketing of aspartame (126). Olney and colleagues described what they termed a "surge in brain tumors in the mid 1980s" on the basis of selective analysis of the US Surveillance Epidemiology and End Results (SEER) tumor database.

The arguments of Olney et al. implicitly require two biologically indefensible assumptions: first, that a certain factor (aspartame) could cause an observed increase in the incidence of brain cancer in less than 4 years and, second, that even more widespread exposure to this factor would cause no further increase in the incidence of that cancer in subsequent years. However, the trend of increased brain tumor rates started well before aspartame was approved, and overall brain tumor rates have actually been decelerating in recent years (127).

Furthermore, the pattern of increased brain tumor rates has been noted primarily in the very elderly (128–131), not the typical age group of aspartame consumers. In addition, it is widely thought that apparent increases in brain tumor rates in the mid-1980s may not reflect genuine increases in brain tumors but rather enhanced detection, largely resulting from the availability of sophisticated noninvasive diagnostic technology, such as computed tomography and magnetic resonance imaging (128–135).

Epidemiologists have criticized Olney and coworkers' attempted association between the introduction of aspartame and occurrence of brain tumors (136, 137). For example, Ross (137) stated, "From an epidemiologic perspective, the conclusion of the report may well represent a classic example of 'ecologic fallacy.' . . . There is no information available regarding whether the individuals who developed brain tumors consumed aspartame. For example, one might also invoke (a) cellular phone, home computer, and VCR usage; (b) depletion of the ozone layer; or (c) increased use of stereo headphones as potentially causative agents . . . some or all of these possibilities may or may not have any biological plausibility to the observed associations." Seife (138) humorously chided that Olney and coworkers had neglected to consider the close statistical correlation that exists between increased brain tumor incidence and the rise of the national debt driven by supply-side economics in the mid-1980s!

Furthermore, a case-controlled study specifically evaluating aspartame consumption and the risk of childhood brain tumors was published by Gurney et al. (139). The results of the study showed that children with brain tumors were no more likely to have consumed aspartame than control children nor was there any elevated risk from maternal consumption of aspartame during pregnancy. Gurney and coworkers concluded, ''. . . it appears unlikely that any carcinogenic effect of aspartame ingestion could have accounted for the recent brain tumor trends as Olney et al. contend'' (139).

Nonetheless, the allegations regarding aspartame and brain tumors have been carefully evaluated by scientists at regulatory agencies in the United States, Australia/New Zealand, the United Kingdom, and the European Union with the unanimous conclusion that aspartame does not cause cancer (140–143).

G. Internet Misinformation

Although the Internet is a revolutionary source of information, anyone can post information on the Internet without being held accountable for accuracy. The Internet has unfortunately been exploited and provided an unwitting forum for dissemination of calculated misinformation, scientifically unfounded allegations, and speculations by a handful of individuals. Society has unfortunately not yet developed the critical evaluation skills necessary for distinguishing Internet fact from Internet fiction. For example, a number of websites use pseudoscience to attribute any number of maladies to aspartame. Virtually all the information on these websites is distorted, anecdotal, from anonymous sources, or is scientifically implausible. In a number of cases, responsible medical organizations have been compelled to respond to bogus Internet allegations. For example, the senior medical advisor of one organization so targeted, the Multiple Sclerosis Foundation, stated, ''This campaign by the 'aspartame activists' is not innocent drum banging,'' because they have created a danger in that individuals who should seek appropriate medical treatment instead blame aspartame for their medical conditions. He further stated that whatever the ultimate agenda of ''aspartame activists,'' ''it is not public health.'' (144, 145)

Misinformation on the Internet has caught the attention of various regulatory agencies, such as the US Food and Drug Administration (146), the Ministry of Health in Brazil (147), and the Ministry of Health in the UK (148). These agencies have evaluated the allegations, including those regarding a number of serious diseases such as multiple sclerosis, lupus erythematosus, Gulf War syndrome, brain tumors, and a variety of other diseases. They have concluded that these allegations are anecdotal and that there is no reliable scientific evidence that aspartame is responsible for any of these conditions, thus reaffirming the safety of aspartame. Furthermore, scientific organizations all around the globe have also rebutted these attacks on the Internet and issued statements of support for aspartame's safety (149).

VI. WORLDWIDE REGULATORY STATUS

In addition to the United States, aspartame has been approved for food and beverage and/or tabletop sweetener use in more than 100 countries.

VII. CONCLUSION

The availability of aspartame to food manufacturers worldwide has been one of the major factors responsible for the growth of the "light" and "low-calorie" segments of the food industry. Aspartame provides many opportunities for formulating new products while lowering or limiting calories and sugar consumption. Aspartame's clean, sugarlike taste and unique flavor-enhancing properties, combined with the exhaustive documentation of its safety, have contributed to acceptance by consumers, the food industry, and health professionals worldwide.

REFERENCES

1. RH Mazur. Aspartame—a sweet surprise. J Toxicol Environ Health 2:243–249, 1976.
2. RH Mazur, A Ripper. Peptide-based sweeteners. In: CAM Hough, KJ Parker, AJ Vlitos, eds. Developments in Sweeteners. London: Applied Science, 1979, pp 125–134.
3. JECFA. Aspartame. Toxicological evaluation of certain food additives. WHO Tech Rep Ser. 653. Rome: World Health Organization, 1980, pp 1–86.
4. Scientific Committee for Food. Aspartame. Food Science and Techniques. Reports of the Scientific Committee for Food. 21st series. Commission of the European Communities, 1989, pp 22–23.
5. FDA. Aspartame: commissioner's final decision. Federal Register 46. 1981, pp 38285–38308.
6. FDA. Food additives permitted for direct addition to food for human consumption; aspartame. Federal Register 48. 1983, pp 31376–31382.
7. Aspartame. Food Chemicals Codex. Washington: National Academy Press, 1996, pp 35–36.
8. RH Mazur, J Schlatter, AH Goldkamp. Structure-taste relationships of some dipeptides. J Am Chem Soc 91(10):2684–2691, 1969.
9. RH Mazur. Aspartic acid-based sweeteners. In: GE Inglett, ed. ACS Sweetener Symposium. Westport, CT: AVI Publishing, 1974.
10. CI Beck. Sweetness, character and applications of aspartic acid-based sweeteners. In: G Inglett, ed. ACS Sweetener Symposium. Westport, CT: AVI Publishing, 1974, pp 164–181.
11. CI Beck. Application potential for aspartame in low calorie and dietetic foods. In:

BK Dwivedi, ed. Low Calorie and Special Dietary Foods. West Palm Beach, FL: CRC Press, 1978, pp 59–114.

12. SS Schiffman. Comparison of taste properties of aspartame with other sweeteners. In: LD Stegink, LJ Filer, eds. Aspartame: Physiology and Biochemistry. New York: Marcel Dekker, 1984, pp 207–246.
13. RE Baldwin, BM Korschgen. Intensification of fruit-flavors by aspartame. J Food Sci 44:938–939, 1979.
14. BJ Bahoshy, RE Klose, HA Nordstrom. Chewing gums of longer lasting sweetness and flavor. General Foods Corp. US 3,943,258. March 9, 1976.
15. BE Homler, RC Deis, WH Shazer. Aspartame. In: LO Nabors, RC Gelardi, eds. Alternative Sweeteners. 2nd ed. New York: Marcel Dekker, 1991, pp 39–69.
16. FDA. Food Additives Permitted for Direct Addition to Food for Human Consumption; Aspartame. Federal Register 61. 1996, pp 33654–33656.
17. SV Molinary. Preclinical studies of aspartame in nonprimate animals. In: LD Stegink, LJ Filer Jr, eds. Aspartame Physiology and Biochemistry. New York: Marcel Dekker, 1984, pp 289–306.
18. FN Kotsonis, JJ Hjelle. The safety assessment of aspartame: scientific and regulatory considerations. In: C. Tschanz, HH Butchko, WW Stargel, FN Kotsonis, eds. The Clinical Evaluation of a Food Additive: Assessment of Aspartame. Boca Raton, FL: CRC Press, 1996, pp 23–41.
19. FDA. Food additives permitted for direct addition to food for human consumption; aspartame. Federal Register 49. 1984, pp 6672–6682.
20. FC Lu. Acceptable daily intake: inception, evolution, and application. Regul Toxicol Pharmacol 8:45–60, 1988.
21. AG Renwick. Acceptable daily intake and the regulation of intense sweeteners. Food Addit Contam 7:463–475, 1990.
22. AG Renwick. Safety factors and establishment of acceptable daily intakes. Food Addit Contam 8:135–149, 1991.
23. HH Butchko, FN Kotsonis. Acceptable daily intake vs. actual intake: the aspartame example. J Am Coll Nutr 10:258–266, 1991.
24. HH Butchko, C Tschanz, FN Kotsonis. Postmarketing surveillance of food additives. Reg Toxicol Pharmacol 20:105–118, 1994.
25. HH Butchko, FN Kotsonis. Postmarketing surveillance in the food industry: the aspartame case study. In: FN Kotsonis, M Mackey, J Hjelle, eds. Nutritional Toxicology. New York: Raven Press, 1994, pp 235–250.
26. HH Butchko, FN Kotsonis. Acceptable daily intake and estimation of consumption. In: C Tschanz, HH Butchko, WW Stargel, FN Kotsonis, eds. The Clinical Evaluation of A Food Additive: Assessment of Aspartame. New York: CRC Press, 1996, pp 43–53.
27. SM Virtanen, L Rasanen, A Paganus, P Varo, HK Akerblom. Intake of sugars and artificial sweeteners by adolescent diabetics. Nutr Rep Int 38(6):1211–1218, 1988.
28. JP Heybach, C Ross. Aspartame consumption in a representative sample of Canadians. J Can Diet Assoc 50:166–170, 1989.
29. AL Hinson, WM Nicol. Monitoring sweetener consumption in Great Britain. Food Addit Contam 9(6):669–681, 1992.

30. A Bar, C Biermann. Intake of intense sweeteners in Germany. Z Ernahrungswiss 31(1):25–39, 1992.
31. C Bergsten. Intake of acesulfame K, aspartame, cyclamate and saccharin in Norway. SNT Norwegian Food Control Authority; Report 3:1–56, 1993.
32. A Collerie de Borely, C Renault. Intake of intense sweeteners in France (1989–1992). Paris: CREDOC, Departement Prospective de la Consommation; 1994.
33. Ministry of Agriculture, Fisheries and Food (UK). Survey of the intake of sweeteners by diabetics. Food Surveillance Info Sheet. London: 1995.
34. KFAM Hulshof, M Bouman. Use of various types of sweeteners in different population groups: 1992 Dutch National Food Consumption Survey. TNO Nutrition and Food Research Institute, Netherlands: 1–49, 1995.
35. MCF Toldeo, SH Ioshi. Potential intake of intense sweeteners in Brazil. Food Addit and Contam 12;799–808, 1995.
36. National Food Authority, Australia. Survey of intense sweetener consumption in Australia: final report: 1–91, 1995.
37. C Leclercq, D Berardi, MR Sorbillo, J Lambe. Intake of saccharin, aspartame, acesulfame K and cyclamate in Italian teenagers: present levels and projections. Food Addit Contam 16(3):99–109, 1999.
38. LD Stegink, LJ Filer, GL Baker. Effect of aspartame and aspartate loading upon plasma and erythrocyte free amino acid levels in normal adult volunteers. J Nutr 107:1837–1845, 1977.
39. LD Stegink, LJ Filer, GL Baker, J McDonnell. Effect of aspartame loading upon plasma and erythrocyte amino acid levels in phenylketonuric heterozygotes and normal adult subjects. J Nutr 109:708 -717, 1979.
40. LD Stegink, LJ Filer Jr, GL Baker. Plasma, erythrocyte and human milk levels of free amino acids in lactating women administered aspartame or lactose. J Nutr 109: 2173–2181, 1979.
41. LD Stegink, LJ Filer Jr, GL Baker, JE McDonnell. Effect of an abuse dose of aspartame upon plasma and erythrocyte levels of amino acids in phenylketonuric heterozygous and normal adults. J Nutr 110:2216–2224, 1980.
42. LD Stegink, LJ Filer Jr, GL Baker. Plasma and erythrocyte concentrations of free amino acids in adult humans administered abuse doses of aspartame. J Toxicol Environ Health 7:291–305, 1981.
43. LD Stegink, MC Brummel, K McMartin, G Martin-Amat, LJ Filer, GL Baker, TR Tephly. Blood methanol concentrations in normal adult subjects administered abuse doses of aspartame. J Toxicol Environ Health 7:281–290, 1981.
44. LJ Filer Jr, GL Baker, LD Stegink. Effect of aspartame loading on plasma and erythrocyte free amino acid concentrations in one-year-old infants. J Nutr 113: 1591–1599, 1983.
45. LD Stegink, MC Brummel, LJ Filer Jr, GL Baker. Blood methanol concentrations in one-year-old infants administered graded doses of aspartame. J Nutr 13:1600–1606, 1983.
46. LD Stegink, LJ Filer, GL Baker. Repeated ingestion of aspartame-sweetened beverage: effect on plasma amino acid concentrations in normal adults. Metabolism 37(3):246–251, 1988.
47. B Caballero, BE Mahon, FJ Rohr, HL Levy, RJ Wurtman. Plasma amino acid levels

after single-dose aspartame consumption in phenylketonuria, mild hyperphenylalaninemia, and heterozygous state for phenylketonuria. J Pediatr 109:668–671, 1986.

48. R Koch, KNF Shaw, M Williamson, M Haber. Use of aspartame in phenylketonuric heterozygous adults. J Toxicol Environ Health 2:453–457, 1976.
49. GH Frey. Use of aspartame by apparently healthy children and adolescents. J Toxicol Environ Health 2:401–415, 1976.
50. RH Knopp, K Brandt, RA Arky. Effects of aspartame in young persons during weight reduction. J Toxicol Environ Health 2:417–428, 1976.
51. SB Stern, SJ Bleicher, A Flores, G Gombos, D Recitas, J Shu. Administration of aspartame in non-insulin-dependent diabetics. J Toxicol Environ Health 2:429–439, 1976.
52. K Langlois. Short term tolerance of aspartame by normal adults. GD Searle & Co. Research Report (E-23), Submitted to the FDA November 30, 1972.
53. R Hoffman. Long term tolerance of aspartame by obese adults. GD Searle & Co. Research Report (E-64), Submitted to FDA June 14, 1973.
54. R Hoffman. Short term tolerance of aspartame by obese adults. GD Searle & Co. Research Report (E-24), Submitted to FDA November 30, 1972.
55. GH Frey. Long term tolerance of aspartame by normal adults. GD Searle & Co. Research Report (E-60), Submitted to FDA February 9, 1973.
56. V Gupta, C Cochran, TF Parker, DL Long, J Ashby, MA Gorman, GU Liepa. Effect of aspartame on plasma amino acid profiles of diabetic patients with chronic renal failure. Am J Clin Nutr 49:1302–1306, 1989.
57. ZI Hertelendy, CI Mendenhall, SD Rouster, L Marshall, R Weesner. Biochemical and clinical effects of aspartame in patients with chronic, stable alcoholic liver disease. Am J Gastroenterol 88(5):737–743, 1993.
58. JK Nerhling, P Kobe, MP McLane, RE Olson, S Kamath, DL Horwitz. Aspartame use by persons with diabetes. Diabetes Care 8(5):415–417, 1985.
59. DL Horwitz, M McLane, P Kobe. Response to single dose of aspartame or saccharin by NIDDM patients. Diabetes Care 11(3):230–234, 1988.
60. AS Leon, DB Hunninghake, C Bell, DK Rassin, TR Tephly. Safety of long-term large doses of aspartame. Arch Intern Med 149:2318–2324, 1989.
61. LD Stegink, LJ Filer, GL Baker. Effect of aspartame plus monosodium l-glutamate ingestion on plasma and erythrocyte amino acid levels in normal adult subjects fed a high protein meal. Am J Clin Nutr 36:1145–1152, 1982.
62. LD Stegink, LJ Filer, GL Baker. Plasma amino acid concentrations in normal adults ingesting aspartame and monosodium L-glutamate as part of a soup/beverage meal. Metabolism 36:1073–1079, 1987.
63. LJ Filer, LD Stegink. Aspartame metabolism in normal adults, phenylketonuric heterozygotes, and diabetic subjects. Diabetes Care 12(1s1):67–74, 1989.
64. FDA. Food additives permitted for direct addition to food for human consumption; dimethyl dicarbonate. Federal Register 61. 1996, pp 26786–26788.
65. C Trocho, R Pardo, I Rafecas, et al. Formaldehyde derived from dietary aspartame binds to tissue components in vivo. Life Sci 63:337–349, 1998.
66. TR Tephly. Comments on the purported generation of formaldehyde and adduct formation from the sweetener aspartame. Life Sci 65:157–160, 1999.

67. HH Butchko, C Tschanz, FN Kotsonis. Postmarketing surveillance of anecdotal medical complaints. In: C Tschanz, HH Butchko, WW Stargel, FN Kotsonis, eds. The clinical evaluation of a food additive: assessment of aspartame. Boca Raton, FL: CRC Press; 1996, pp 183–193.
68. Evaluation of consumer complaints related to aspartame use. MMWR 33:605–607, 1984. Atlanta, GA, Centers for Disease Control.
69. MK Bradstock, MK Serdula, JS Marks, RJ Barnard, NT Crane, L Remington, FL Trowbridge. Evaluation of reactions to food additives: the aspartame experience. Am J Clin Nutr 43:464–469, 1986.
70. L Tollefson. Monitoring adverse reactions to food additives in the U.S. Food and Drug Administration. Regul Toxicol Pharmacol 8:438–446, 1988.
71. L Tollefson, RJ Barnard, Wh Glinsmann. Monitoring of adverse reactions to aspartame reported to the U.S. Food and Drug Administration. In: RJ Wurtman, E Ritter-Walker, eds. Dietary Phenylalanine and Brain Function. Boston: Birkhauser, 1988, pp 317–337.
72. TG Wilcox, DM Gray. Summary of adverse reactions attributed to aspartame. FDA Health Hazard Evaluation Board, 1995.
73. A Raines. Another side to aspartame (letter). Wash Post, June 2, 1987.
74. SM Koehler, A Glaros. The effect of aspartame on migraine headache. Headache 28(1):10–14, 1988.
75. SS Schiffman. Aspartame and headache. Headache 28:370, 1988.
76. WK Amery. More on aspartame and headache. Headache. 28(9):624, 1988.
77. SS Schiffman, CE Buckley, HA Sampson, EW Massey, JN Baraniuk, J Follett, ZS Warwick. Aspartame and susceptibility to headache. N Engl J Med 317(19):1181–1185, 1987.
78. SK Van den Eeden, TD Koepsell, G van Belle, WT Longstreth, B McKnight. A randomized crossover trial of aspartame and headaches. Am J Epidemiol 134(7): 789, 1991.
79. S Schiffman. Letter to editor. Neurology 45:1632, 1995.
80. PS Levy, D Hedeker, PG Sanders. Letter to editor. Neurology 45:1631–1632, 1995.
81. A Kulczycki. Aspartame-induced urticaria. Ann Intern Med 104:207–208, 1986.
82. M Garriga, C Berkebile, D Metcalfe. A combined single-blind, double-blind, placebo-controlled study to determine the reproducibility of hypersensitivity reactions to aspartame. J Allergy Clin Immunol 87(4):821–827, 1991.
83. R Geha, CE Buckley, P Greenberger, R Patterson, S Polmar, A Saxon, A Rohr, W Yang, M Drouin. Aspartame is no more likely than placebo to cause urticaria/angioedema: results of a multicenter, randomized, double-blind, placebo-controlled, crossover study. J Allergy Clin Immunol 92:513–520, 1993.
84. JA Rowan, BA Shaywitz, L Tuchman, JA French, D Luciano, CM Sullivan. Aspartame and seizure susceptibility: results of a clinical study in reportedly sensitive individuals. Epilepsia 36(3):270–275, 1995.
85. BA Shaywitz, GM Anderson, EJ Novotny, J Ebersole, CM Sullivan, SM Gillespie. Aspartame has no effect on seizures or epileptiform discharges in epileptic children. Ann Neurol 35(1):98–103, 1994.
86. PR Camfield, CS Camfield, J Dooley, K Gordon, S Jollymore, DF Weaver. Aspar-

tame exacerbates EEG spike-wave discharge in children with generalized absence epilepsy: a double-blind controlled study. Neurology 42:1000–1003, 1992.

87. BA Shaywitz, EJ Novotny. Letter to editor. Neurology 43:630, 1993.
88. RJ Wurtman. Neurochemical changes following high-dose aspartame with dietary carbohydrates. N Engl J Med 309:429–430, 1983.
89. S Garattini, S Caccia, M Romano, et al. Studies on the susceptibility to convulsions in animals receiving abuse doses of aspartame. In: RJ Wurtman, WE Ritter, eds. Dietary Phenylalanine and Brain Function. Boston: Birkhauser, 1988, pp 131–143.
90. C Perego, MG DeSimoni, F Fodritto, L Raimondi, L Diomede, M Salmona, S Algeri, S Garattini. Aspartame and the rat brain monoaminergic system. Toxicol Lett 44:331–339, 1988.
91. MA Reilly, EA Debler, A Fleischer, A Lajtha. Lack of effect of chronic aspartame ingestion on aminergic receptors in rat brain. Biochem Pharmacol 38:4339–4341, 1989.
92. MA Reilly, EA Debler, A Lajtha. Perinatal exposure to aspartame does not alter aminergic neurotransmitter systems in weanling rat brain. Res Commun Psychol Psychiatry Behav 15:141–159, 1990.
93. A Lajtha, MA Reilly, DS Dunlop. Aspartame consumption: lack of effects on neural function. J Nutr Biochem 5:266–283, 1994.
94. M Wolraich, R Milich, P Stumbo, F Schultz. Effects of sucrose ingestion on the behavior of hyperactive boys. J Pediatr 106:675–682, 1985.
95. HB Ferguson, C Stoddart, JG Simeon. Double-blind challenge studies of behavioral and cognitive effects of sucrose-aspartame ingestion in normal children. Nutr Rev 44:144–150, 1986.
96. R Milich, WE Pelham. Effects of sugar ingestion on the classroom and playgroup behavior of attention deficit disordered boys. J Consult Clin Psychol 54:714–718, 1986.
97. JA Goldman, RH Lerman, JH Contois, JN Udall. Behavioral effects of sucrose on preschool children. J Abnorm Child Psychol 14:565–577, 1986.
98. M Ryan-Harshman, LA Leiter, GH Anderson. Phenylalanine and aspartame fail to alter feeding behavior, mood and arousal in men. Physiol Behav 39:247–253, 1987.
99. MJP Kruesi, JL Rapoport, EM Cummings, CJ Berg, DR Ismond, M Flament, M Yarrow, C Zahn-Waxler. Effects of sugar and aspartame on aggression and activity in children. Am J Psychiatry 144:1487–1490, 1987.
100. HR Lieberman, B Caballero, GG Emde, JG Bernstein. The effects of aspartame on human mood, performance, and plasma amino acid levels. In: RJ Wurtman, E Ritter-Walker, eds. Dietary Phenylalanine and Brain Function. Boston: Birkhauser, 1988, pp 196–200.
101. KA Lapierre, DJ Greenblatt, JE Goddard, JS Harmatz, RI Shader. The neuropsychiatric effects of aspartame in normal volunteers. J Clin Pharmacol 30:454–460, 1990.
102. S Saravis, R Schachar, S Zlotkin, LA Leiter, GH Anderson. Aspartame: effects on learning, behavior, and mood. Pediatrics 86:75–83, 1990.
103. RE Dodge, D Warner, S Sangal, RD O'Donnell. The effect of single and multiple doses of aspartame on higher cognitive performance in humans. Aerospace Medical Association 61st Annual Scientific Meeting, 1990;A30.

104. AF Stokes, A Belger, MT Banich, H Taylor. Effects of acute aspartame and acute alcohol ingestion upon the cognitive performance of pilots. Aviat Space Environ Med 62(7):648–653, 1991.
105. RG Walton, R Hudak, RJ Green-Waite. Adverse reactions to aspartame: double-blind challenge in patients from a vulnerable population. Biol Psychiatry 34(1-2): 13–17, 1993.
106. HH Butchko. Adverse reactions to aspartame: double-blind challenge in patients from a vulnerable population by Walton et al. Biol Psychiatry 36:206–207, 1994.
107. F Trefz, L de Sonneville, P Matthis, C Benninger, Lanz-Englert B, Bickel H. Neuropsychological and biochemical investigations in heterozygotes for phenylketonuria during ingestion of high dose aspartame (a sweetener containing phenylalanine). Hum Genet 93:369–374, 1994.
108. ML Wolraich, SD Lindgren, PH Stumbo, LD Stegink, MI Appelbaum, MC Kiritsy. Effects of diets high in sucrose or aspartame on the behavior and cognitive performance of children. N Engl J Med 330(5):301–307, 1994.
109. AF Stokes, A Belger, MT Banich, E Bernadine. Effects of alcohol and chronic aspartame ingestion upon performance in aviation relevant cognitive tasks. Aviat Space Environ Med 65:7–15, 1994.
110. BA Shaywitz, CM Sullivan, GM Anderson, SM Gillespie, B Sullivan, SE Shaywitz. Aspartame, behavior, and cognitive function in children with attention deficit disorder. Pediatrics 93:70–75, 1994.
111. PA Spiers, L Sabaounjian, A Reiner, DK Myers, J Wurtman, DL Schomer. Aspartame: neuropsychologic and neurophysiologic evaluation of acute and chronic effects. Am J Clin Nutr 68(3):531–537, 1998.
112. LD Stegink, LJ Filer, eds. Aspartame: Physiology and Biochemistry. New York: Marcel Dekker, 1984.
113. JAMA. American Medical Association. Council on Scientific Affairs. Aspartame: review of safety issues. JAMA 254:400–402, 1985.
114. LD Stegink. Aspartame: review of the safety issues. Food Technol 41:119–121, 1987.
115. LD Stegink. The aspartame story: a model for the clinical testing of a food additive. Am J Clin Nutr 46:204–215, 1987.
116. PJCM Janssen, CA van der Heijden. Aspartame: review of recent experimental and observational data. Toxicology 50:1–26, 1988.
117. RS London. Saccharin and aspartame. Are they safe to consume during pregnancy? J Reprod Med 33:17–21, 1988.
118. PJCM Janssen, CA van der Heijden. Aspartame: review of recent experimental and observational data. Toxicology 50:1–26, 1988.
119. HH Butchko, FN Kotsonis. Aspartame: review of recent research. Comments Toxicol 3:253–278, 1989.
120. PY Sze. Pharmacological effects of phenylalanine on seizure susceptibility: an overview. Neurochem Res 14:103–111, 1989.
121. RS Fisher. Aspartame, neurotoxicity, and seizures: a review. J Epilepsy 2:55–64, 1989.
122. JD Fernstrom. Central nervous system effects of aspartame. In: ND Kretchmer, CB

Hallenbeck, eds. Sugars and Sweeteners. Boca Raton, FL: CRC Press, 1991, pp 151–173.

123. B Meldrum. Amino-acids as dietary excitotoxins: a contribution to understanding neurodegenerative disorders. Brain Res Rev 18(3):293–314, 1993.
124. P Jobe, J Dailey. Aspartame and seizures. Amino Acids 4(3):197–235, 1993.
125. C Tschanz, HH Butchko, WW Stargel, FN Kotsonis, eds. The Clinical Evaluation of a Food Additive: Assessment of Aspartame. New York: CRC Press, 1996.
126. JW Olney, NB Farber, E Spitznagel, LN Robins. Increasing brain tumor rates: is there a link to aspartame? J Neuropathol Exp Neurol 55(11):1115–1123, 1996.
127. PS Levy, D Hedeker. Letter to the editor. J Neuropathol Exp Neurol 55:1280, 1996.
128. CS Muir, HH Storm, A Polednak. Brain and other nervous system tumours. Cancer Surv 19/20:369–392, 1994.
129. NH Greig, LG Ries, R Yancik, SI Rapoport. Increasing annual incidence of primary malignant brain tumors in elderly. J Natl Cancer Inst 82:1621–1624, 1990.
130. MH Werner, S Phuphanich, GH Lyman. The increasing incidence of malignant gliomas and primary central nervous system lymphoma in the elderly. Cancer 76: 1634–1642, 1995.
131. DL Davis, A Ahlbom, D Hoel, C Percy. Is brain cancer mortality increasing in industrial countries? Am J Ind Med 19:421–431, 1991.
132. C La Vecchia, F Lucchini, E Negri, P Boyle, P Maisonneuve, F Levi. Trends of cancer mortality in Europe, 1955–1989:IV, urinary tract, eye, brain and nerves, and thyroid. Eur J Cancer 28A:1210–128!, 1992.
133. P Boyle, P Maisonneuve, R Saracci, CS Muir. Is the increased incidence of primary malignant brain tumors in the elderly real? J Natl Cancer Inst 82:1594–1596, 1990.
134. E Marshall. Experts clash over cancer data. Science 250:900–902, 1990.
135. B Modan, DK Wagener, JJ Feldman, HM Rosenberg, M Feinleib. Increased mortality from brain tumors: A combined outcome of diagnostic technology and change of attitude toward the elderly. Am J Epidemiol 135:1349–1357, 1992.
136. SM Davies, J Ross, WG Woods. Aspartame and brain tumors: junk food science. Causes Childhood Cancer Newslett 7(6), 1996.
137. JA Ross. Brain tumors and artificial sweeteners? A lesson on not getting soured on epidemiology. Med Pediatr Oncol 30(1):7–8, 1998.
138. C Seife. Increasing brain tumor rates: is there a link to deficit spending? J Neuropathol Exp Neurol 58(4):404–405, 1999.
139. JG Gurney, JM Pogoda, EA Holly, SS Hecht, S Preston-Martin. Aspartame consumption in relation to childhood brain tumor risk: results from a case—control study. J Natl Cancer Inst 89:1072–1074, 1997.
140. FDA. FDA statement on aspartame. FDA Talk Paper. November 18, 1996, pp 1–2.
141. EC Scientific Committee for Food. Aspartame: extract of minutes of the 107th meeting of the Scientific Committee for Food held on 12–13 June 1997 in Brussels. June 30, 1997, pp 9.
142. Australia New Zealand Food Authority (ANZFA). Information paper. Aspartame: information for consumers, 1997.
143. European Commission. Scientific Committee for Food: Minutes of the 107th Meeting of the Scientific Committee for Food held on 12–13 June, 1997. Brussels: 1997.

144. D Squillacote. Aspartame (Nutrasweet): no danger. The Multiple Sclerosis Foundation. January 12, 1999. www.msfacts.org/aspartame.htm
145. D Squillacote. Aspartame: come in from the ledge. The Multiple Sclerosis Foundation; January 26, 1999. www.msfacts.org/aspart2.htm 1
146. J Henkel. Sugar substitutes: Americans opt for sweetness and lite. FDA Consumer 12–16, 1999.
147. Agencia Saude. Sao Paulo, Brazil. Forum of scientific discussion: aspartame; October 29, 1999. http://anvsl.saude.gov.br/Alimento/aspartame.html
148. Food Standards Agency. Food safety information bulletin No 117, February, 2000. Http://www.foodstandards.gov.uk/maff/archive/food/asp9911.htm
149. Calorie Control Council. www.aspartame.org.

5
Cyclamate

Barbara A. Bopp
TAP Pharmaceutical Products, Inc., Deerfield, Illinois

Paul Price
Price International, Libertyville, Illinois

I. INTRODUCTION

Cyclamate (Fig. 1) was synthesized in 1937 by a University of Illinois graduate student, Michael Sveda, who accidentally discovered its sweet taste (1, 2). The patent for cyclamate eventually became the property of Abbott Laboratories, which performed the necessary studies and submitted a New Drug Application for the sodium salt in 1950. Cyclamate was initially marketed as tablets that were recommended for use as a tabletop sweetener for diabetics and others who had to restrict their use of sugar. In 1958, after enactment of the Food Additive Amendment to the Food, Drug, and Cosmetic Act, the Food and Drug Administration (FDA) of the United States classified cyclamate as a GRAS, or Generally Recognized as Safe, sweetener. A mixture of cyclamate and saccharin, which had been found to have synergistic sweetening properties and an improved taste (3, 4), was subsequently marketed for use in special dietary foods. When soft drinks sweetened with the cyclamate–saccharin mixture became popular in the United States during the 1960s, the consumption of cyclamate increased dramatically. Prompted by the growing use of cyclamate, additional studies were initiated. It was found that cyclamate could be metabolized to cyclohexylamine (Fig. 1) by bacteria in the intestine (5). Then in 1969, the results of a chronic toxicity study with a 10:1 cyclamate–saccharin mixture were interpreted as implicating cyclamate as a bladder carcinogen in rats (6). Cyclamate was removed from GRAS status (7), its use was initially restricted, and then in 1970, it was banned

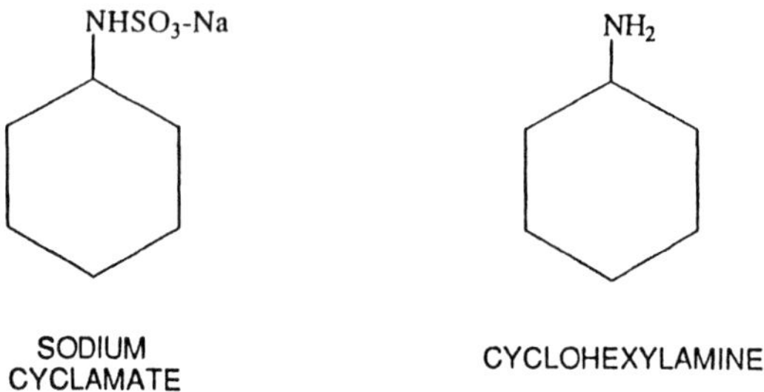

Figure 1 Structures of cyclamate and cyclohexylamine.

in the United States from use in all foods, beverages, and drugs (8). However, many foreign governments did not act as precipitously, and cyclamate continued to be used as a sweetener in these countries.

In the next few years, many additional toxicity and carcinogenicity studies were conducted with cyclamate, the 10:1 cyclamate–saccharin mixture, and cyclohexylamine, but they failed to confirm the original findings. On the basis of the results of these studies, Abbott Laboratories filed a Food Additive Petition for cyclamate in 1973 (9). After a lengthy review and an administrative hearing, cyclamate was denied food additive status in 1980 (10). Subsequently, in September 1982, the Calorie Control Council and Abbott Laboratories filed a second Food Additive Petition, containing new studies that confirmed the safety of cyclamate and assisted in determining the safe level of intake (11). Action on this petition is still pending.

II. PRODUCTION

The production of cyclamate, which is accomplished by the sulfonation of cyclohexylamine, was originally limited to Abbott Laboratories (12). However, as cyclamate became more widely used in foods and beverages, other companies, including The Pillsbury Company, Pfizer Inc., Cyclamate Corporation of America, and Miles Laboratories, entered the market. By 1970, several foreign countries, including Japan, Taiwan, and Korea, also had production capabilities. When cyclamate was banned in the United States, all domestic producers, except Abbott Laboratories, ceased production. Production was also banned in Japan, and thus the Taiwanese and Koreans became the major producers. Today, the major producers and exporters of cyclamate are located in China, Indonesia, Taiwan, and Spain.

III. PHYSICAL CHARACTERISTICS (13–16)

Cyclamic acid, or cyclohexylsulfamic acid ($C_6H_{13}NO_3S$; MW = 179.24), is a white crystalline powder, with a melting point of 169–170°C, good aqueous solubility (1 g/7.5 ml), and a lemon-sour sweetness. It is a strong acid, and the pH of a 10% aqueous solution is about 0.8–1.6. Sodium and calcium cyclamate are strong electrolytes, which are highly ionized in solution, fairly neutral in character, and have little buffering capacity. Both salts exist as white crystals or white crystalline powders. The molecular weight of the sodium salt ($C_6H_{12}NO_3S{\cdot}Na$) is 201.22 and that of the calcium salt ($C_{12}H_{24}N_2O_6S_2{\cdot}Ca$) is 396.54 (432.58 as the dihydrate). They are freely soluble in water (1 g/4–5 ml) at concentrations far in excess of those required for normal use but have limited solubility in oils and nonpolar solvents. Cyclamate solutions are stable to heat, light, and air throughout a wide pH range.

Cyclohexylamine, which is the starting material in the synthesis of cyclamate and is also a metabolite of cyclamate, has distinctly different properties. Cyclohexylamine (MW = 99.17) is a base, not an acid; it has a fishy odor and a bitter taste, not a sweet taste. It is a clear colorless liquid, which is miscible with water, alcohol, and nonpolar solvents and has a boiling point of 134.5°C. The pH of a 0.01% aqueous solution is 10.5. Cyclohexylamine has many industrial applications, including use in water treatment and rubber acceleration.

IV. RELATIVE SWEETNESS AND UTILITY

In contrast to the sweet taste from sucrose, which appears quickly and has a sharp, clean cut-off, the sweetness from cyclamate builds to its maximal level more slowly and persists for a longer time (13). Cyclamate is generally accepted as being 30 times sweeter than sucrose, but the relative sweetness of cyclamate tends to decrease at higher sweetness intensities. For example, cyclamate is about 40 times as sweet as a 2% sucrose solution but only 24 times as sweet as a 20% sucrose solution (13). This trend may be at least partially due to the increasing levels of bitterness and aftertaste that characteristically appear at very high cyclamate concentrations (3). This off-taste is, however, not a problem at normal use concentrations.

Calcium cyclamate is somewhat less sweet than sodium cyclamate, and the off-taste response starts at a lower concentration of the calcium salt than of the sodium salt. Vincent et al. (3) suggested that the differences between the two salts might be related to differences in ionization because the sweet taste is due to the cyclamate ion, whereas the off-taste may be associated with the undissociated salt. The sweetening power of cyclamate also varies with the medium, and

Table 1 Regulatory Status of Cyclamate

Country	Food[a]	Beverage	Tabletop
Angola	−	−	+
Antigua	+	+	
Argentina	+	+	+
Australia	+	+	+
Austria	+	+	+
Bahamas	+		
Belgium	+	+	+
Brazil	+	+	+
Bulgaria	+	+	+
Canada	−	−	+
Caribbean (Ind.)	+	+	+
Chile	−	+	−
China	+	+	
Denmark	+	+	+
Dominica	+	+	+
Ecuador	−	−	+
Finland	+	+	+
France	+	+	+
Germany	+	+	+
Greece	+	+	+
Guadeloupe		+	
Guatemala	−	−	+
Haiti	+	+	
Hungary	+	+	+
Iceland	−	+	+
Indonesia	+	+	+
Ireland	+	+	+
Israel	+	+	+
Italy	+	+	+
Kuwait			+
Luxembourg	+	+	+
Martinique	+	+	+
Monserrat		+	
Netherlands	+	+	+
New Zealand	+	+	+
Nicaragua	+	+	+
Norway	+	+	+
Papua New Guinea	+	+	+
Paraguay	+	+	+
Peru	+	+	−
Portugal	+	+	+
Russia	+	+	+

Table 1 Continued

Country	Food[a]	Beverage	Tabletop
Sierra Leone	+	+	+
Slovakia	+	+	+
South Africa	+	+	+
Spain	+	+	+
Srilanka		−	+
Sweden	+	+	+
Switzerland	+	+	+
Taiwan	+	+	−
Thailand		−	+
Trinidad	+	+	
Turkey	−	−	+
United Arab Emirates			+
United Kingdom	+	+	+
United States	−	−	−
Uruguay	+	+	+
Venezuela	−	−	+
Yugoslavia (former)		+	+
Zimbabwe	+	+	+

[a] May not apply to all food categories. + = Permitted; − = prohibited; blank = no information. Information current as of 1999 but may be subject to change. Information provided by the Calorie Control Council.

its actual relative sweetness should be determined in each different product (13). Most notably, the sweetening power of cyclamate is considerably enhanced in fruits. More detailed information on the relative sweetness and other properties of cyclamate can be found in articles by Beck (13, 14).

The primary use of cyclamate is as a noncaloric sweetener, generally in combination with other sweeteners, but cyclamate can also be used as a flavoring agent (i.e., to mask the unpalatable taste of drugs) (12, 17). Before being banned, cyclamate was used in the United States in tabletop sweeteners, in a variety of foods, in beverages, and in both liquid and tablet pharmaceuticals, and it is found in numerous products in more than 50 countries today (see Table 1). Many of these same applications are proposed in the current food additive petition (11) as the potential uses of cyclamate:

Tabletop sweeteners, in tablet, powder, or liquid form
Beverages, fruit juice drinks, beverage bases or mixes
Processed fruits
Chewing gum and confections
Salad dressings
Gelatin desserts, jellies, jams, and toppings

V. ADMIXTURE POTENTIAL

Cyclamate has most frequently been used in combination with saccharin. In the 1950s, it was shown that cyclamate–saccharin mixtures are sweeter than would be expected from the known sweetness of either component alone and that any off-taste is minimized with the mixtures (3, 4). About a 10–20% synergism is observed when cyclamate and saccharin are used together. For example, a combination of 5 mg saccharin and 50 mg cyclamate in a tabletop sweetener is as sweet as 125 mg cyclamate alone or 12.5 mg saccharin alone (13). Although the ratio of cyclamate to saccharin can vary considerably from product to product, the 10:1 mixture is used most frequently. With this combination, each component contributes about equally to the sweetness of the mixture because saccharin is about 10 times sweeter than cyclamate.

More recently, numerous patents have described the use of cyclamate in combination with aspartame or aspartame and saccharin (e.g., 18, 19). Applications have also been reported for cyclamate in combination with acesulfame-K and other sweeteners (20). If and when cyclamate is again approved for use in the United States, it would undoubtedly be used primarily in combination with other more potent sweeteners because no one sweetener appears to meet all the technological requirements, and mixtures of sweeteners generally appear to offer improved taste.

VI. TECHNICAL QUALITIES

Cyclamate has a number of technical qualities that make it a good alternative sweetener (13, 17, 21). It is noncaloric and noncariogenic. Although its relative sweetness is less than that of saccharin or aspartame, its sweetening power is adequate, especially when used in combination with other more intense sweeteners. Cyclamate has a favorable taste profile and does not leave an unpleasant aftertaste at normal use concentrations. It is better than sugar in masking bitterness, and it enhances fruit flavors. Cyclamate is compatible with most foods and food ingredients, as well as with natural and artificial flavoring agents, chemical preservatives, and other sweeteners. Its solubility is more than adequate for most uses, and at normal concentrations it does not change the viscosity or density of solutions. The stability of cyclamate is excellent, at both high and low temperatures, over a wide pH range, and in the presence of light, oxygen, and other food chemicals. Cyclamate is nonhygroscopic and will not support mold or bacterial growth. However, like most noncaloric sweeteners, cyclamate does not provide the bulk, texture, or body associated with sugar.

VII. UTILITY

Although most of the product development work with cyclamate was done in the 1950s and 1960s, Beck has reviewed some of the applications and presented typical formulations sweetened with cyclamate or a cyclamate–saccharin mixture (13). Perhaps the primary rule for the development of low-calorie foods and beverages is that cyclamate (or any other nonnutritive sweetener) cannot simply be substituted for the sugar; instead, the product must be reformulated (13). The two most critical aspects for a successful product are its flavor and texture. Because a flavor may not taste the same in systems sweetened with sucrose and cyclamate (or mixtures of cyclamate and other nonnutritive sweeteners), the flavoring of a product frequently has to be modified. A proper balance between the taste effects of the acid and sweetener components of a product must also be achieved. For example, if the lingering sweetness from noncaloric sweeteners results in a higher sweetness intensity than desired, some compensation can be achieved by increasing the level of acidity. On the other hand, if an aftertaste develops at the level of sweetness desired, decreasing the acidity may permit a reduction in the sweetness level, hence minimizing the aftertaste. The other major problem in the development of low-calorie products with cyclamate is that of texture or body. This largely results from the elimination of sugar solids and can frequently be solved by the addition of a suitable hydrocolloid or bulking agent. An alternative approach that has been used in soft drinks involves achieving the proper balance between sweetness and tartness, the use of special flavors, and the adjustment of the carbonation level.

Cyclamate has always been particularly useful in fruit products because it enhances fruit flavors and, even at low concentrations, can mask the natural tartness of some citrus fruits (13, 17, 21–23). The cyclamate solutions used for canned fruits have a lower specific gravity and osmotic pressure than sucrose syrups and hence do not draw water out of the fruit. Thus, fruits packed in cyclamate solutions tend to have a greater drained weight than those packed in sucrose. Cyclamate-sweetened gelatins are reasonably easy to formulate, requiring the use of high-bloom gelatins and crystalline sorbitol or mannitol as a bodying agent, dispersant, and filler (13, 22). Thickening and consistency represent the major problems with jams, jellies, and puddings sweetened with cyclamate. Low-methoxy pectin is usually used as a gelling agent in jams and jellies because it does not require sugar for gel formation (13, 21–23). However, low-methoxy pectin needs more calcium than is normally present, and hence calcium cyclamate may be preferable to the sodium salt for this application. Because of the lower concentrations of osmotically active compounds, jams and jellies containing cyclamate may require a preservative to extend their shelf-life. Body and thickening of puddings can be achieved with starches or a combination of nonnutritive gums and thickeners (13, 22). Low-calorie salad dressings require the substitution of two basic ingredients: cyclamate (or a mixture of cyclamate and another sweetener) for the sugar and a hydrocolloid or thickener for the oil (13, 22).

Baked goods are probably the most difficult foods to reformulate with noncaloric sweeteners (13, 22, 24). In addition to sweetness, sugar provides bulk and texture, has a tenderizing effect on gluten, and is important in the browning reaction. Cyclamate cannot furnish these properties, and hence the formulations must be modified to include bulking agents (e.g., modified starch or dextrins, carboxymethylcellulose) and a tenderizing agent (e.g., lecithin). Although proper browning is difficult to achieve, some success can be obtained by application of a caramel solution onto the surface. In yeast doughs, sugar also acts as an energy source for the fermentation reaction. However, satisfactory products containing cyclamate and only a small amount of sugar (≤1%) can be prepared if the salt content and the fermentation time are reduced (24). As the amount of sugar in chemically aerated products is decreased, more liquid must be added to retain the proper consistency of the batter and the eggs must be used to their best advantage for structure and aeration. Good results were obtained with a slurry technique, in which the flour and other dry ingredients were mixed with water before being added to the whipped eggs (24). However, preparation of high-quality, low-calorie baked goods still represents a major technical challenge.

In contrast to baked goods, the lack of a browning reaction with cyclamate can be an advantage in cured meats (17, 22, 25). When sugar-cured bacon or ham is fried, the sugar tends to caramelize, losing its sweet taste and giving the meats a darkened appearance. Because cyclamate has a higher melting point than sucrose, cyclamate-cured meats taste better, have an improved color, and do not scorch or stick in the frying pan.

Cyclamate has also found applications in pharmaceutical and oral hygiene products. It is particularly good at masking the bitterness and unpalatable taste of many drugs and hence is especially useful in syrups, other liquid formulations, and chewable tablets (16). Cyclamate imparts a high level of sweetness with a low solid content, thus providing suspensions that are more fluid and have fewer problems with caking (16) or tablets that disintegrate rapidly and have less bulk (26). Cyclamate is also useful as a sweetener in both film coating and compression coating of tablets, and the acid form can be used as an effervescent agent (26). Because cyclamate is noncariogenic, it is suitable for use in toothpastes and mouthwashes.

VIII. AVAILABILITY

Cyclamate is still available in the United States, as well as from foreign suppliers. The use of cyclamate as a food additive is currently prohibited in the United

States, and therefore it may not be used in foods or beverages in this country at present. It is, however, legal to manufacture cyclamate-containing foods in the United States for export to foreign countries, where the use of cyclamate as a sweetener is permitted. This must, however, be done in compliance with Section 801(d) of the Federal Food, Drug, and Cosmetic Act 21 USC 381(d).

IX. SHELF-LIFE

Samples of tablets containing a cyclamate and saccharin mixture, which were manufactured in 1969 or before, did not show any diminution in sweetening ability or any physical deterioration after at least 7 years (12). It would seem, therefore, that cyclamate in tablet form has an extremely long shelf-life. Information about the possible shelf-life of cyclamate in other applications, such as soft drinks or canned foods, was not available to the authors. However, when cyclamate was widely used in the United States, the shelf-life was more than adequate to allow the products to be sold in the ordinary course of business. There is no known instance of a recall of products because of the degradation of the sweetening content from cyclamate. The cyclamate stability data indicate that an expiration date is not needed to ensure the identity, strength, quality, and purity of either the bulk food additive or foods and beverages containing cyclamate.

X. TRANSPORT

No known problem exists with the transport of the bulk material, and in the United States cyclamate is nonregulated with respect to transport.

XI. GENERAL COST/ECONOMICS

It is anticipated that should the Food and Drug Administration once again allow the use of cyclamate as a food additive in the United States, the price would be approximately $2.00 per pound, estimated at today's production costs. The price from foreign producers has varied considerably from year to year, depending largely on product availability. As the price of sugar also fluctuates considerably, the only means of determining the economics of cyclamate use is to compare its cost at the time of production to the cost of equivalent sweetening from sugar at its prevailing price. Such comparisons generally indicate that cyclamate is one of the most economical noncaloric sweeteners.

XII. METABOLISM

Cyclamate is slowly and incompletely absorbed from the gastrointestinal tract (15). In one study involving almost 200 subjects, the absorption of cyclamate averaged only 37% (27). The volume of distribution for cyclamate in rats is approximately equal to the total body water content, and hence cyclamate does not concentrate in most tissues (15). Once absorbed, cyclamate is excreted unchanged in the urine by both glomerular filtration and active tubular secretion (15).

Although early studies had indicated that cyclamate was not metabolized to any appreciable extent, in 1966 Kojima and Ichibagase (5) detected cyclohexylamine in urine samples from humans and dogs receiving cyclamate. Subsequently, conversion was also demonstrated in mice, rats, guinea pigs, rabbits, monkeys, and pigs, as well as in dogs and humans (15). Cyclamate is, however, not metabolized to cyclohexylamine by mammalian tissues, but rather the cyclohexylamine is formed by the action of the microflora on the nonabsorbed cyclamate remaining in the intestinal tract (15).

Probably the most important feature of cyclamate metabolism is the extreme variability in cyclohexylamine formation. Not all individuals are able to convert cyclamate to cyclohexylamine, and even among converters, the extent of conversion varies greatly and often changes in the same individual over time. Retrospective analyses of studies that attempted to define the incidence of converters indicated that only about 25% of the subjects were able to metabolize cyclamate to cyclohexylamine (15, 27). The incidence of converters was slightly lower (~20%) among Europeans and North Americans who were given at least three daily doses of cyclamate but appeared to be higher among the Japanese (≥80% in studies involving about 60 subjects).

As noted previously, even among those individuals who can convert cyclamate to cyclohexylamine, there is substantial intersubject and day-to-day variability in the extent of conversion, which can range from <0.1% to >60%. The frequency distribution curve for cyclamate conversion is, however, strongly skewed, and only a few individuals are able to form large amounts of cyclohexylamine. It has been estimated that only about 3% of a population converts more than 20% of a cyclamate dose to cyclohexylamine, and ≤1% converts 60% or more (11, 28, 29). The 60% level approaches, on the average, the maximal conversion possible because only the nonabsorbed cyclamate can be metabolized, and the absorption of cyclamate averages about 40%.

The conversion of cyclamate to cyclohexylamine also appears to depend on continuous exposure to the sweetener (15). A single dose of cyclamate will frequently not be metabolized, and daily ingestion of cyclamate is usually necessary to induce and maintain the converting ability at a high but still variable level. Furthermore, if cyclamate is withdrawn from the diet for even a few days, the ability to metabolize cyclamate is diminished and gradually lost. Hence, intermittent cyclamate use (which might be a typical pattern of use by some people) would tend to limit the converting ability of an individual.

Because the no-observed effect level (NOEL) in animal toxicity studies is based on cyclohexylamine, not cyclamate, determination of the extent of conversion is critical for establishing the safe level of use for the sweetener. However, the skewed nature of the cyclamate conversion curve makes this more difficult. It is not appropriate to use the mean conversion by an entire group of converters because most of these subjects would be forming only small amounts of cyclohexylamine. Instead, the best estimate can probably be derived from the average conversion by a subgroup of high converters. The available data suggest that among those subjects who converted at least 1% of the dose to cyclohexylamine, conversion averaged slightly less than 20% (11, 15). If the group was further restricted to only those subjects converting at least 5% of the dose, the average level increased only slightly (~25%) (15, 30). A few individuals would, of course, be converting at higher levels at least some of the time. However, it is probably not necessary to use the maximal conversion rate in establishing the acceptable daily intake (ADI) because the large safety factor applied to food additives compensates for considerable intersubject variability.

In contrast to cyclamate, cyclohexylamine is rapidly and completely absorbed from the gastrointestinal tract, even from the large intestine where it is formed (15). The plasma half-life of cyclohexylamine in humans is dose-dependent, increasing from 3.5 hr with a 2.5 mg/kg oral dose to 4.8 hr with a 10 mg/kg dose (31). The apparent volume of distribution of cyclohexylamine is 2–3 liters/kg, and tissue concentrations typically exceed those in plasma (15, 31).

Cyclohexylamine is primarily excreted unchanged in the urine by both glomerular filtration and a saturable transport mechanism (31, 32). Although cyclohexylamine is not extensively metabolized, its biotransformation shows some species differences (15). The principal metabolic pathway in rats is ring hydroxylation, leading to the formation of the isomeric 3- or 4-aminocyclohexanols, which account for 5–20% of the dose (33–35). Mice and humans, however, form negligible quantities of these metabolites (33–35). Only the deamination products, cyclohexanol and *trans*-cyclohexane-1,2-diol, are found in humans given cyclohexylamine, and these metabolites represent only 1–2% of the dose (33). Both ring hydroxylation and deamination occur in guinea pigs and rabbits (33).

XIII. CARCINOGENICITY

In 1969, a 2-year chronic toxicity study with a cyclamate–saccharin mixture was nearing completion at the Food and Drug Research Laboratories (FDRL) in Maspeth, New York. In that study, rats were fed diets containing a 10:1 mixture of sodium cyclamate and sodium saccharin to provide daily doses of 500, 1120, or 2500 mg/kg/day (6, 36). The ability of some of the rats from this study to convert cyclamate to cyclohexylamine had been documented (37), and after 78 weeks, the diets of half of the animals were supplemented with cyclohexylamine at levels corresponding to about 10% conversion of the cyclamate dose (i.e., 25, 56, and 125 mg/kg/day). At the conclusion of the study, tumors were found in the urinary bladder of eight high-dose rats, with four to eight of the tumors being classified as carcinomas by the different pathologists who reviewed the slides (6). In subsequent analyses, the number of tumors increased to 12 and all were considered carcinomas (36). Calcification of the urinary tract and bladder parasites were observed in some of these rats and possibly could have affected the results of this study because both of those factors are known to contribute to the development of bladder tumors in rats. Furthermore, six of the tumors occurred in consecutively numbered and presumably consecutively housed male rats, suggesting that some extraneous environmental factor could also have been involved in the development of the tumors. Nevertheless, the results of this study were interpreted as implicating cyclamate as a bladder carcinogen in rats (6) and led to its removal from GRAS status and finally to its being banned for use in foods and beverages in the United States (7, 8).

In the years after this observation, the carcinogenic potential of cyclamate and cyclohexylamine was reevaluated in a group of well-designed and well-controlled bioassays, which were performed by independent investigators throughout the world. Cyclamate, including both the sodium and calcium salts, was tested in at least four separate studies with rats given doses up to 2.5 g/kg/day (38–43), in three separate studies with mice given doses up to 7–9 g/kg/day (43–45), and in one study with hamsters given doses up to 3 g/kg/day (46). The 10:1 cyclamate–saccharin mixture that had been used in the original FDRL study has been tested twice in rats (38, 39, 41, 42) and once in mice (45). Two studies, one in rats (40) and one in mice (45), even included in utero exposure of the animals to cyclamate. In addition to these conventional rodent bioassays, two studies have been conducted in monkeys given cyclamate, with treatment lasting 8 years in one (200 mg/kg/day) (47) and 20+ years in the other (100 or 500 mg/kg/day) (48, 49). Cyclohexylamine, as the hydrochloride or sulfate salt, was tested in three studies with rats given maximal doses of 150–300 mg/kg/day (38, 50, 51) and in two studies with mice given maximal doses of 400–600 mg/kg/day (45, 52). A 9-year study was also performed in dogs (53). None of these studies confirmed the original findings that implicated cyclamate as a bladder carcinogen. Instead, when the results of these studies are evaluated in accordance with recognized toxicological and statistical procedures, they provide strong evidence that neither cyclamate nor cyclohexylamine is carcinogenic in animals.

In 1976, the Temporary Committee of the National Cancer Institute that was charged with reviewing the cyclamate data concluded "the present evidence does not establish the carcinogenicity of cyclamate or its principal metabolite, cyclohexylamine, in experimental animals" (54). On receipt of this report, the Bureau of Foods of the U.S. FDA apparently concurred and reported to the Commissioner that "we think the issue of the carcinogenicity of cyclamate is settled" (55). In 1977, the World Health Organization's Joint Expert Committee on Food Additives (JECFA) reached a similar conclusion, stating that "it is now possible to conclude that cyclamate has been demonstrated to be non-carcinogenic in a variety of species" (56). Nevertheless, carcinogenicity was a major issue in the 1980 decision of the FDA commissioner, who stated that "cyclamate has not been shown not to cause cancer" and thus denied cyclamate food additive status (10). However, some toxicological and statistical principles used in this decision were subsequently challenged by the American Statistical Association (57) and the Task Force of the Past Presidents of the Society of Toxicology (58).

After submission of the second Food Additive Petition for cyclamate in 1982, the Cancer Assessment Committee (CAC) of the U.S. FDA's Center for Food Safety and Applied Nutrition reviewed all the cyclamate data and concluded that "there is very little credible data to implicate cyclamate as a carcinogen at any organ/tissue site to either sex of any animal species tested," that "the collective weight of the many experiments . . . indicates that cyclamate is not carcinogenic," and further that "no newly discovered toxic effects of cyclamate are likely to be revealed if additional standardized studies were performed" (59). In 1985, the National Academy of Sciences reaffirmed the CAC conclusion stating that ". . . the totality of the evidence from studies in animals does not indicate that cyclamate or its major metabolite cyclohexylamine is carcinogenic by itself" (60). The NAS report, however, raised some concern about a possible role of cyclamate as a promoter. This was largely predicated on two studies involving direct exposure of the urinary bladder to cyclamate. One involved implantation of a pellet containing cyclamate in the bladder of mice (61) and the other involved intravesicular instillation of *N*-methyl-*N*-nitrosourea into the urinary bladder of female rats, which were then given cyclamate in either their food or water (62). At the request of the FDA, the Mitre Corporation evaluated these models and concluded that "both types of direct bladder exposure studies, pellet implantation and intravesicular catherization, are considered unsuitable for predicting human carcinogenic risk" (63). Thus, once again, it would appear that the carcinogenicity issue has been settled, at least from a scientific point of view, and that it can be concluded that cyclamate and cyclohexylamine are not carcinogenic in animals.

The possible association between cancer, particularly bladder cancer, and the consumption of noncaloric sweeteners by humans has been extensively studied during the past 10–20 years. Since the widespread use of cyclamate in foods and beverages was restricted to a relatively short time span in many countries, these studies are more applicable to the assessment of the possible carcinogenicity of saccharin than cyclamate. However, the cyclamate issue was specifically addressed in some of the studies. The epidemiology studies with the noncaloric sweeteners have been reviewed by others (64–66), and it has generally been agreed that there is no conclusive evidence of an increased risk of bladder cancer associated with the use of these sweeteners.

XIV. MUTAGENICITY

In the 1960s and 1970s, the mutagenic potential of cyclamate and cyclohexylamine was evaluated in many different test systems, including the Ames test and other microbial gene mutation assays, studies in *Drosophila*, in vitro cytogenetic studies, in vivo cytogenetic studies with both somatic and germ cells, dominant lethal tests, and others. Although there were instances of conflicting or discordant results, the preponderance of evidence provided by this battery of tests suggested that neither cyclamate nor cyclohexylamine represented a significant mutagenic hazard (15, 67, 68). Nevertheless, in the 1980 decision the Commissioner of the FDA still concluded that "cyclamate has not been shown not to cause heritable genetic damage" (10).

In 1985 the NAS Committee evaluated the mutagenicity tests with cyclamate and cyclohexylamine and found little evidence to conclude that either compound was a DNA-reactive carcinogen (60). They, however, recommended that additional tests for mammalian cell DNA damage, mammalian cell gene mutation tests, and more definitive cytogenetic studies be conducted to complete and strengthen the database for the two compounds. After consultation with the FDA, both calcium cyclamate and cyclohexylamine were tested in the Chinese hamster ovary HGPRT forward mutation assay, for unscheduled DNA synthesis in rat hepatocytes, and in the *Drosophila* sex-linked recessive lethal assay (69). The results of these three tests gave no evidence of any intrinsic genotoxicity from either compound. In addition, dominant lethal and heritable translocation tests were conducted in male mice given the maximal tolerated dose of calcium cyclamate for 6 weeks because the sodium salt had been used in almost all the previous studies with mammalian germ cells (70). The results of these two tests were also negative, indicating that the calcium salt did not induce any transmissible chromosomal aberrations in the germ cells of mice. Thus, these recent studies, performed according to the currently accepted standards, confirm the previous conclusions that cyclamate and cyclohexylamine are not mutagenic and do not cause heritable genetic damage.

XV. OTHER TOXICITY

A. Cyclamate

Although a great many toxicity studies have been conducted with cyclamate and the cyclamate–saccharin mixture, they have revealed very few pathophysiological effects attributable to the sweetener (15). Perhaps the most frequently reported effects in animals and humans given cyclamate are softening of the stools and diarrhea. These effects, however, only occur at excessively high doses and are due to the osmotic activity of the nonabsorbed cyclamate remaining in the gastrointestinal tract (71).

B. Cyclohexylamine

Cyclohexylamine is considerably more toxic than cyclamate, and its toxicity limits the use of the sweetener. Because the toxicity studies with cyclohexylamine have thoroughly been reviewed (15), only the two areas of major concern—testicular atrophy and cardiovascular effects—will be discussed herein.

1. Testicular Effects of Cyclohexylamine

Numerous toxicological studies have shown that the testes of rats is the organ that is most sensitive to any adverse effects from cyclohexylamine, and this effect has been used by JECFA and others as the basis for establishing the acceptable daily intake for cyclamate. The results of three 90-day studies in rats given cyclohexylamine in the diet indicated that the testes were not adversely affected at concentrations of 0.2% (approximately 100 mg/kg/day) and below, but at concentrations of 0.6% (approximately 300 mg/kg/day) and above testicular atrophy, characterized by decreased organ weight, decreased spermatogenesis, and/or degeneration of the tubular epithelium, was observed (72–75). Because the concentration of cyclohexylamine in the food of these animals was held constant, the milligram per kilogram per day dose ingested by the animals declined over the 3-month study period as the food consumption of the rats relative to their body weight progressively decreased. To circumvent this problem and to more precisely define the no-adverse effect dose, an additional study was conducted by Dr. Horst Brune (76) in Germany. Groups of 100 young male rats were fed diets containing cyclohexylamine to provide constant daily doses of 50, 100, 200, or 300 mg/kg/day for 3 months. Ad libitum and pair-fed control groups were also included in the experimental design of this study. The results clearly demonstrated that 100 mg/kg/day was a NOEL. Slight, but statistically significant, histopathological changes were noted in the testes of the rats receiving the 200 mg/kg/day dose, and marked effects were seen at 300 mg/kg/day. Analysis of all the studies involving an effect of cyclohexylamine on the rat testes showed that the data were quite consistent and confirmed the steep nature of the dose-response curve, as demonstrated by the Brune study (15). The lack of significant histopathological changes in the testes of rats given a 150 mg/kg/day dose of cyclohexylamine for 2 years (51) and the presence of only mild, sporadic effects in rats given ~175 mg/kg/day dose for 90 days (72) are consistent with the steep dose-response curve and suggest that the no-effect level might be as high as 150 mg/kg/day (11). However, 100 mg/kg, which was clearly shown to be a NOEL in the study of Brune, was adopted as the NOEL by JECFA (77).

Early studies had indicated that mice were less sensitive to the effects of cyclohexylamine than rats because no testicular changes had been seen in mice given doses of ~300 mg/kg/day (0.3% in the diet) (52). Renwick and Roberts suggested that this species difference in testicular toxicity might be related to differences in metabolism because the ring-hydroxylated metabolites of cyclohexylamine (i.e., 3- or 4-aminocyclohexanols) account for about 15–20% of the dose in male rats but are formed in negligible amounts in mice (34–35). To test this hypothesis, the metabolism and testicular effects of a 400 mg/kg/day dose of cyclohexylamine were compared in mice, Wistar rats, and DA rats, a strain deficit in hydroxylating ability (35). Consistent with the previous findings, the mice formed only small amounts of the ring-hydroxylated metabolites and did not develop testicular toxicity. Although the Wistar rats had higher concentrations of the aminocyclohexanol metabolites in their plasma and testes than the DA rats, the DA strain appeared to be more sensitive to the cyclohexylamine-induced testicular effects than the Wistar rats. Thus, the development of testicular atrophy could not be attributed to the ring-hydroxylated metabolites. Instead, the species differences in toxicity were related to differences in the pharmacokinetics of cyclohexylamine (32). The concentrations of cyclohexylamine in the plasma and testes of rats showed a nonlinear relationship to dietary intake, with substantially elevated levels occurring at intakes greater than 200 mg/kg/day. This nonlinear accumulation of cyclohexylamine in the rat testes correlated well with the previously noted steep dose-response curve for the testicular atrophy. In contrast, both the plasma and testicular cyclohexylamine concentrations in mice increased linearly with intake, even at doses far in excess of 200 mg/kg/day. Furthermore, the dose-normalized cyclohexylamine concentrations at levels greater than 200 mg/kg/day were considerably higher in rats than in mice, suggesting that the greater exposure of the rat testes to cyclohexylamine probably contributed to the greater toxicity in that species. Because these studies clearly demonstrated that the testicular toxicity of cyclohexylamine was attributable to the parent compound per se and that the rat was more sensitive to this effect than the mouse, it is appropriate to use the NOEL in rats for determining the acceptable daily intake of cyclamate in humans.

2. Cardiovascular Effects of Cyclohexylamine

The other major question about the safety of cyclohexylamine involves its possible effects on blood pressure. Cyclohexylamine is an indirectly acting sympathomimetic agent, similar to tyramine but more than 100 times less potent than tyramine (15). Intravenous administration of cyclohexylamine to anesthetized animals causes vasoconstriction and increases both the blood pressure and heart rate. Despite these acute effects, a rise in blood pressure was not seen in subchronic or chronic toxicity studies with orally administered cyclohexylamine in rats, even at high doses (0.4–1% in the diet or approximately 200–400 mg/kg/day) (38, 72).

The cardiovascular effects of single oral doses of cyclohexylamine in humans have been thoroughly characterized by Eichelbaum et al. (31). They found that a 10 mg/kg bolus oral dose of cyclohexylamine caused a 30-mm rise in the mean arterial blood pressure of healthy human volunteers. A smaller increase was seen after a 5 mg/kg dose, but no significant change in blood pressure occurred with a 2.5 mg/kg dose. A reflex-mediated, slight decrease in heart rate accompanied the vasopressor effects of the two higher doses. The cyclohexylamine levels in the plasma of these subjects were closely correlated with the increases in blood pressure, and it was estimated that the lowest cyclohexylamine concentration to cause a significant hypertensive effect was about 0.7–0.8 μg/ml. However, the blood pressure in these subjects rapidly returned toward normal despite the presence of cyclohexylamine concentrations that were associated with a pressor effect during the absorptive phase. These observations suggested the rapid development of tolerance or tachyphylaxis to the hypertensive effects of cyclohexylamine.

Despite the potential of cyclohexylamine to increase blood pressure, there is no evidence to suggest that such effects actually occur after the administration of cyclamate. Periodic cardiovascular monitoring in some subjects participating in cyclamate metabolism studies revealed no changes in blood pressure or heart rate, even in those individuals who received high doses of cyclamate and were forming large amounts of cyclohexylamine (27, 29, 78, 79). Moreover, the cyclohexylamine blood levels in some of these subjects approached the concentrations associated with a rise in blood pressure after a bolus oral dose of cyclohexylamine, yet their blood pressure was apparently not affected (29).

Two human studies conducted in Germany during the 1970s also addressed the questions of cyclamate conversion, cyclohexylamine blood levels, and possible cardiovascular effects. In one study the high converters among a group of 44 regular users of cyclamate were initially identified on the basis of their urinary cyclohexylamine levels. After ingestion of 1.2–1.7 g of sodium cyclamate, the cyclohexylamine blood levels in three of the high converters ranged from 0.03–

0.42 μg/ml and were correlated with the urinary cyclohexylamine levels (80). In the other study, the blood pressure and heart rate of 20 subjects were monitored three times a day during a 10-day dosing period with sodium cyclamate (81). Half of these subjects did not convert cyclamate to cyclohexylamine, five were low converters, and the other five were high converters. However, no significant changes in blood pressure or heart rate were seen, even in the high converters. Furthermore, noninvasive hemodynamic studies, which were conducted in six subjects known to be good converters, failed to reveal any changes that were attributable to cyclamate or cyclohexylamine, even though the cyclohexylamine levels in serum ranged from about 0.2–0.6 μg/ml in most subjects and reached 1 μg/ml in one subject (81).

Buss et al. (82) conducted a controlled clinical study investigating the conversion of cyclamate to cyclohexylamine and its possible cardiovascular consequences in 194 diabetic patients who were given calcium cyclamate (1 g/day as cyclamic acid equivalents) for 7 days. The incidence and extent of cyclohexylamine formation in this group of subjects were similar to those reported in the earlier studies; 78% of subjects excreted <0.1% of the daily dose as cyclohexylamine in urine, but 4% (8 subjects) excreted more than 20% of the daily dose as cyclohexylamine. Similar intersubject variability was observed in the cyclohexylamine concentrations in plasma, with concentrations of 0–0.01 μg/ml in 168 subjects, 0.01–0.3 μg/ml in 18 subjects, 0.3–1.0 μg/ml in 6 subjects, and >1.0 μg/ml in 2 subjects. A significant correlation was found between the cyclohexylamine in plasma and urine. The changes in mean arterial blood pressure and heart rate in the eight subjects with plasma cyclohexylamine concentrations between 0.3 and 1.9 μg/ml were similar to those found in 150 subjects with very low (<0.01 μg/ml) cyclohexylamine concentrations. In a second study conducted 10–24 months later, 20 of the subjects were given calcium cyclamate for 2 weeks at a dose of 2 g cyclamic acid equivalents/day (0.66 g tid). Cyclohexylamine concentrations in plasma, blood pressure, and heart rate were measured every 30 minutes during the final 8-hr dosing interval. Twelve of the subjects had plasma cyclohexylamine concentrations of 0.09–2.0 μg/ml at the start of the dose interval. Transient increases or decreases in the cyclohexylamine concentrations were not observed during the 8-hr dosing interval nor were there any significant changes in blood pressure or heart rate measurements. Collectively, these results indicate that the metabolism of cyclamate (2 g/day) to cyclohexylamine does not significantly affect blood pressure or heart rate, even in those few individuals who are high converters and have relatively high concentrations of cyclohexylamine in plasma.

The lack of cardiovascular effects in the subjects who had plasma cyclohexylamine concentrations of 0.3–1.9 μg/ml in the study by Buss et al. contrasts with the observation by Eichelbaum et al. that cyclohexylamine concentrations of 0.7–0.8 μg/ml were sufficient to increase the blood pressure of subjects given a bolus oral dose of cyclohexylamine. The difference in the cyclohexylamine plasma concentration-time profiles from the two routes of delivery may offer an explanation for this discrepancy. A bolus oral dose of cyclohexylamine leads to a rapid rise and subsequent fall in the plasma concentrations. In contrast, during periods of cyclamate ingestion, the cyclohexylamine concentrations would increase gradually as metabolizing activity develops and then remain relatively constant. To further explore this possibility, Buss and Renwick (83) investigated the relationship between cyclohexylamine concentrations and blood pressure changes in rats given intravenous infusions of cyclohexylamine. The blood pressure increases in the rats were inversely related to the duration of the infusion, despite the presence of similar plasma concentrations at the end of the infusion. In addition, the plasma concentration-effect relationship showed a clockwise hysteresis, indicative of tachyphylaxis and similar to that seen in humans in the study by Eichelbaum et al. Thus, the hypertensive effects of cyclohexylamine primarily occur when the plasma concentrations are rapidly increasing. Because the kinetics of cyclamate metabolism would not lead to a rapid increase in the cyclohexylamine concentrations, hypertensive effects would not be expected to occur in those individuals who are ingesting large amounts of cyclamate and are high converters.

XVI. REGULATORY STATUS

The regulatory status of cyclamate in the United States is quite simply expressed. It is still banned from use. Whether the FDA will alter that status will not be known until their evaluation of the cyclamate Food Additive Petition is completed. Cyclamate is, however, currently available for use in more than 50 countries (84). As summarized in Table 1, some countries allow the use of cyclamate in foods, beverages, or both, whereas others only permit tabletop use. Pharmaceutical use is also permitted in some countries. In 1982, JECFA (77) increased the acceptable daily intake for cyclamate almost threefold and established an ADI of 0–11 mg/kg, expressed as cyclamic acid. However, the ADI for cyclamate may vary from country to country.

REFERENCES

1. LF Audrieth, M Sveda. Preparation and properties of some N-substituted sulfamic acids. J Org Chem 9:89–101, 1944.
2. LF Audrieth, M Sveda. U.S. Patent No. 2,275,125 (March 3, 1942).
3. HC Vincent, MJ Lynch, FM Pohley, FJ Helgren, FJ Kirchmeyer. A taste panel study of cyclamate–saccharin mixture and of its components. J Am Pharm Assoc 44:442–446, 1955.

4. U.S. Patent No. 2,803,551 (August 20, 1957).
5. S Kojima, H Ichibagase. Studies on synthetic sweetening agents. VIII. Cyclohexylamine, a metabolite of sodium cyclamate. Chem Pharm Bull 14:971–974, 1966.
6. JM Price, CG Biava, BL Oser, EE Vogin, J Steinfeld, HC Ley. Bladder tumors in rats fed cyclohexylamine or high doses of a mixture of cyclamate and saccharin. Science 167:1131–1132, 1970.
7. Federal Register. 34(202):17063 (October 21, 1969).
8. Federal Register. 35(167):13644 (August 27, 1970).
9. Abbott Laboratories. Cyclamic acid and its salts. Food Additive Petition 4A2975 (November 15, 1973).
10. Federal Register. 45:61474 (September 16, 1980).
11. The Calorie Control Council and Abbott Laboratories. Food Additive Petition for Cyclamate 2A3672 (September 22, 1982).
12. RW Kasperson, N Primack. Cyclamate. In: L O'Brien-Nabors, RC Gelardi, eds. Alternative Sweeteners. New York: Marcel Dekker, 1986, pp 71–87.
13. KM Beck. Nonnutritive sweeteners: Saccharin and cyclamate. In: TE Furia, ed. CRC Handbook of Food Additives, Vol. II, 2nd ed. Boca Raton, FL: CRC Press, 1980.
14. KM Beck. Properties of the synthetic sweetening agent, cyclamate. Food Technol 11:156–158, 1957.
15. BA Bopp, RC Sonders, JW Kesterson. Toxicological aspects of cyclamate and cyclohexylamine. CRC Critical Reviews in Toxicology 16:213–306, 1986.
16. MJ Lynch, HM Gross. Artificial sweetening of liquid pharmaceuticals. Drug and Cosmetic Industry 87:324–326, 412–413, 1960.
17. KM Beck, AS Nelson. Latest uses of synthetic sweeteners. Food Engineering 35: 96–97, 1963.
18. D Scott (GD Searle). Saccharin-dipeptide sweetening compositions. British Patent 1,256,995 (1971).
19. TP Finucane (General Foods). Sweetening compositions containing APM and other adjuncts. Canadian Patent 1,043,158 (November 28, 1978).
20. AG Hoechst. U.S. Patent 4,158,068 (June 12, 1979).
21. JA Stein. Technical aspects of non-nutritive sweeteners in dietary products. Food Technol, March 1966.
22. KM Beck, JV Ziemba. Are you using sweeteners correctly? Food Engineering 38: 71–73, 1966.
23. DK Salunkhe, RL McLaughlin, SL Day, MB Merkley. Preparation and quality evaluation of processed fruits and fruit products with sucrose and synthetic sweeteners. Food Technol 17:85–91, 1963.
24. CD Stone. Sucaryl. Calorie reduction in baked products. Bakers Digest 36:53–56, 59–64, 1962.
25. KM Beck, R Jones, LW Murphy. New sweetener for cured meats. Food Engineering 30:114, 1958.
26. CJ Endicott, HM Gross. Artificial sweetening of tablets. Drug and Cosmetic Industry 85:176–177, 254–256, 1959.
27. RC Sonders. Unpublished studies, Abbott Laboratories (1967–1968).
28. AG Renwick. The fate of cyclamate in man and rat. Third European Toxicology Forum, Geneva, Switzerland, October 1983.

29. AJ Collings. Metabolism of cyclamate and its conversion to cyclohexylamine. Diabetes Care 12(suppl. 1):50–55, 1989.
30. AG Renwick. The metabolism, distribution and elimination of non-nutritive sweeteners. In: B Guggenheim, ed. Health and Sugar Substitutes. Basel: S. Karger, 1979, pp 41–47.
31. M Eichelbaum, JH Hengstmann, HD Rost, T Brecht, HJ Dengler. Pharmacokinetics, cardiovascular and metabolic actions of cyclohexylamine in man. Arch Toxikol 31: 243–263, 1974.
32. A Roberts, AG Renwick. The pharmacokinetics and tissue concentrations of cyclohexylamine in rats and mice. Toxicol Appl Pharmacol 98:230–242, 1989.
33. AG Renwick, RT Williams. The metabolites of cyclohexylamine in man and certain animals. Biochem J 129:857–867, 1972.
34. A Roberts, AG Renwick. The metabolism of ^{14}C-cyclohexylamine in mice and two strains of rat. Xenobiotica 15:477–483, 1985.
35. A Roberts, AG Renwick, G Ford, DM Creasy, I Gaunt. The metabolism and testicular toxicity of cyclohexylamine in rats and mice during chronic dietary administration. Toxicol Appl Pharmacol 98:216–229, 1989.
36. BL Oser, S Carson, GE Cox, EE Vogin, SS Sternberg. Chronic toxicity study of cyclamate:saccharin (10:1) in rats. Toxicology 4:315–330, 1975.
37. BL Oser, S Carson, EE Vogin, RC Sonders. Conversion of cyclamate to cyclohexylamine in rats. Nature 220:178–179, 1968.
38. D Schmähl. Absence of carcinogenic activity of cyclamate, cyclohexylamine, and saccharin in rats. Arzneim-Forsch/Drug Research 23:1466–1470, 1973.
39. D Schmähl, M Habs. Investigations on the carcinogenicity of the artificial sweeteners sodium cyclamate and sodium saccharin in rats in a two-generation experiment. Arzneim-Forsch/Drug Research 34:604–606, 1984.
40. JM Taylor, MA Weinbuger, L Friedman. Chronic toxicity and carcinogenicity to the urinary bladder of sodium saccharin in the in utero exposed rat. Toxicol Appl Pharmacol 54:57–75, 1980.
41. T Furuya, K Kawamata, T Kaneko, O Uchida, S Horiuchi, and Y Ikeda. Long-term toxicity study of sodium cyclamate and sodium saccharin in rats. Japan J Pharmacol 25(suppl. 1):55P, 1975.
42. Y Ikeda, S Horiuchi, T Furuya, K Kawamata, T Kaneko, O Uchida. Long-term toxicity study of sodium cyclamate and sodium saccharin in rats. National Institute of Hygienic Sciences, Tokyo, Japan, unpublished report, 1973.
43. F Homburger. Studies on saccharin and cyclamate. Bio-Research Consultants, Inc., unpublished report, 1973.
44. PG Brantom, IF Gaunt, P Grasso. Long-term toxicity of sodium cyclamate in mice. Food Cosmet Toxicol 11:735–746, 1973.
45. R Kroes, PWJ Peters, JM Berkvens, HG Verschuuren, T deVries, and GJ Van Esch. Long-term toxicity and reproduction study (including a teratogenicity study) with cyclamate, saccharin, and cyclohexylamine. Toxicology, 8:285–300, 1977.
46. J Althoff, A Cardesa, P Pour, PA Shubik. A chronic study of artificial sweeteners in Syrian golden hamsters. Cancer Lett 1:21–24, 1975.
47. F Coulston, EW McChesney, K-F Benitz. Eight-year study of cyclamate in *Rhesus* monkeys. Toxicol Appl Pharmacol 41:164–165, 1977.

48. SM Sieber, RH Adamson. Long-term studies on the potential carcinogenicity of artificial sweeteners in non-human primates. In: B Guggenheim, ed. Health and Sugar Substitutes. Basel: S. Karger, 1979, pp 266–271.
49. S Takayama, AG Renwick, SL Johansson, UP Thorgeirsson, M Tsutsumi, DW Dalgard, SM Sieber. Long-term toxicity and carcinogenicity study of cyclamate in nonhuman primates. Toxicol Sci 53:33–39, 2000.
50. IF Gaunt, J Hardy, P Grasso, SD Gangolli, KR Butterworth. Long-term toxicity of cyclohexylamine hydrochloride in the rat. Food Cosmet Toxicol 14:255–267, 1976.
51. BL Oser, S Carson, GE Cox, EE Vogin, SS Sternberg. Long-term and multigeneration toxicity studies with cyclohexylamine hydrochloride. Toxicology 6:47–65, 1976.
52. J Hardy, IF Gaunt, J Hooson, RJ Hendy, KR Butterworth. Long-term toxicity of cyclohexylamine hydrochloride in mice. Food Cosmet Toxicol 14:269–276, 1976.
53. Industrial Bio-Test Laboratories. Chronic oral toxicity study with cyclohexylamine sulfate in beagle dogs. Unpublished report, April 21, 1981.
54. Report of the Temporary Committee for the Review of Data on Carcinogenicity of Cyclamate. DHEW Publication No. (NIH)77-1437, Department of Health, Education and Welfare, February 1976.
55. H Roberts. Actions Memorandum. Food Additive Petition for Cyclamate (FAP 4A2975). Undated and unsigned to Commissioner Schmidt.
56. World Health Organization, 21st Report of the Joint FAO/WHO Expert Committee on Food Additives. Evaluation of Certain Food Additives, April 18–27, 1977. Technical Report Series, 617, 1978.
57. American Statistical Association. Letter to the Commissioner of the Food and Drug Administration, April 7, 1981.
58. Task Force of Past Presidents of the Society of Toxicology. Animal data in hazard evaluation: Paths and pitfalls. Fundam Appl Toxicol 2:101–107, 1982.
59. Cancer Assessment Committee (CAC), Center for Food Safety and Applied Nutrition, Food and Drug Administration. Scientific review of the long-term carcinogen bioassays performed on the artificial sweetener, cyclamate, April 1984.
60. National Academy of Sciences, National Research Council, Committee on the Evaluation of Cyclamate for Carcinogenicity. Evaluation of cyclamate for carcinogenicity. Washington, DC: National Academy Press, 1985.
61. GT Byran, E Erturk. Production of mouse urinary bladder carcinomas by sodium cyclamate. Science 167:996–998, 1970.
62. RM Hicks, J St J Wakefield, J Chowaniec. Evaluation of a new model to detect bladder carcinogens or co-carcinogens: Results obtained with saccharin, cyclamate and cyclophosphamide. Chem-Biol Interactions 11:225–233, 1975.
63. JM DeSesso, JM Kelley, BB Fuller. Relevance of direct bladder exposure studies to human health concerns. The MITRE Corporation. Unpublished report, May 1987.
64. RW Morgan, O Wang. A review of epidemiological studies of artificial sweeteners and bladder cancer. Food Chem Toxicol 23:529–533, 1985.
65. OM Jensen. Artificial sweeteners and bladder cancer: Epidemiological evidence. Third European Toxicology Forum. Geneva, Switzerland, October 1983.
66. DL Arnold, D Krewski, IC Munro. Saccharin: A toxicological and historical perspective. Toxicology 27:179–256, 1983.

67. BM Cattanach. The mutagenicity of cyclamates and their metabolites. Mutat Res 39:1–28, 1976.
68. P Cooper. Resolving the cyclamate question. Food Cosmet Toxicol 15:69–70, 1977.
69. D Brusick, M Cifone, R Young, S Benson. Assessment of the genotoxicity of calcium cyclamate and cyclohexylamine. Environ Mol Mutagen 14:188–199, 1989.
70. KT Cain, CV Cornett, NLA Cacheiro, LA Hughes, JG Owens, WM Generoso. No evidence found for induction of dominant lethal mutations and heritable translocations in male mice by calcium cyclamate. Environ Mol Mutagen 11:207–213, 1988.
71. K Hwang. Mechanism of the laxative effect of sodium sulfate, sodium cyclamate and calcium cyclamate. Arch Int Pharmacodyn Ther 163:302–340, 1966.
72. AJ Collings, WW Kirkby. The toxicity of cyclohexylamine hydrochloride in the rat, 90-day feeding study, Unilever Research Laboratory. Unpublished report, 1974.
73. IF Gaunt, M Sharratt, P Grasso, ABG Lansdown, SD Gangolli. Short-term toxicity of cyclohexylamine hydrochloride in the rat. Food Cosmet Toxicol 12:609–624, 1974.
74. RF Crampton. British Industrial Biological Research Association, correspondence to H. Blumenthal, Food and Drug Administration, December 19, 1975.
75. PL Mason, GR Thompson. Testicular effects of cyclohexylamine hydrochloride in the rat. Toxicology 8:143–156, 1977.
76. H Brune, U Mohr, RP Deutsch-Wenzel. Establishment of the no-effect dosage of cyclohexylamine hydrochloride in male Sprague-Dawley rats with respect to growth and testicular atrophy. Unpublished report, 1978.
77. World Health Organization, 26th Report of the Joint FAO/WHO Expert Committee on Food Additives. Evaluation of certain food additives, April 19–28, 1982, Technical Report Series, 1982, p 683.
78. MH Litchfield, AAB Swan. Cyclohexylamine production and physiological measurements in subjects ingesting sodium cyclamate. Toxicol Appl Pharmacol 18:535–541, 1971.
79. AJ Collings. Effect of dietary level and the role of intestinal flora on the conversion of cyclamate to cyclohexylamine. Unilever Research Laboratory. Unpublished report, 1969.
80. U Schmidt. Cyclamate conversion study in humans. Supplement to Food Additive Petition 2A3672, December 14, 1987.
81. HJ Dengler, JH Hengstmann, H Lydtin. Extent and frequency of conversion of cyclamate into cyclohexylamine in a German patient group, Supplement to Food Additive Petition 2A3672, December 14, 1987.
82. NE Buss, AG Renwick, KM Donaldson, CF George. The metabolism of cyclamate to cyclohexylamine and its cardiovascular consequences in human volunteers. Toxicol Appl Pharmacol 115:199–210, 1992.
83. NE Buss, AG Renwick. Blood pressure changes and sympathetic function in rats given cyclohexylamine by intravenous infusion. Toxicol Appl Pharmacol 115:211–215, 1992.
84. Calorie Control Council, Worldwide Status of Cyclamate. Unpublished material, 1989.

6
Neohesperidin Dihydrochalcone

Francisco Borrego and Helena Montijano
Zoster, S. A. (Grupo Ferrer), Murcia, Spain

I. INTRODUCTION

The sweet taste of neohesperidin dihydrochalcone (hereinafter referred to as neohesperidine DC) was discovered by Horowitz and Gentili in 1963 while studying the relationships between structure and bitter taste in citrus phenolic glycosides. To their surprise, the product resulting from the hydrogenation of the bitter flavanone neohesperidin yielded an intensely sweet substance, neohesperidine DC (1–5). Since then, a large number of variants of the original sweetener have been synthesized. Some of these represent only small variations on the theme; others are more radical departures. All were made to gain a better understanding of taste-structure relationships to improve taste quality or raise solubility. From a practical standpoint, it appears that even the best of the new derivatives was not significantly better than the original compound. For this reason we will focus on neohesperidine DC, which is the only sweet dihydrochalcone currently in use.

At present, neohesperidine DC is receiving renewed interest because of its recent approval both as a sweetener and as a flavor ingredient. Today, there is an increasing trend to explore the use of sweetener blends in foods to (a) produce a sweetness profile closer to that of sucrose, (b) mask aftertaste, (c) improve stability, (d) meet cost restraints, and (e) reduce the daily intake of any particular sweetener. Recent developments have shown that if neohesperidine DC is used at low levels and in combination with other intense or bulk sweeteners, there can be significant technological advantages in terms of sweetness synergy, positive impact on taste quality, and flavor-enhancing and modifying properties.

II. ORIGIN AND PREPARATION

Neohesperidine DC is a glycosidic flavonoid. Flavonoids are a family of natural substances ubiquitous in the plant kingdom because they occur in all higher plants. Depending on their chemical structure, they are classified in different groups, among them are flavanones, flavones, chalcones, dihydrochalcones, and anthocyanins. Dihydrochalcones are open-ring derivatives of flavanones and are defined by the presence of two C_6 rings joined by a C_3 bridge. Several flavanone glycosides are unique to *Citrus* (6). Thus, the peels of oranges and grapefruit contain, as main flavanone glycosides, hesperidin and naringin, which have different applications as a pharmaceutical raw material and as a bittering agent for the food industry, respectively.

Starting material for the commercial production of neohesperidine DC is either neohesperidin, which can be extracted from bitter orange (*Citrus aurantium*), or naringin, which is obtained from grapefruit (*Citrus paradisii*). The synthesis from extracted neohesperidin involves hydrogenation in the presence of a catalyst under alkaline conditions. The synthesis from extracted naringin is based on the conversion to phloroacetophenone-4′-β-neohesperidoside, which can be condensed with isovanillin (3-hydroxy-4-methoxybenzaldehyde) to yield neohesperidin. The reactions involved have been described in detail (7–11) and are summarized in Fig. 1.

Attempts to produce sweet dihydrochalcones by microbiological conversion of flavanone glycosides have also been reported in the literature (12). A novel biotechnological approach to the synthesis of neohesperidine DC is being explored by Jonathan Gressel and coworkers at the Weizmann Institute in Israel (13–18). They have characterized and purified to homogeneity a 52-kDa rhamnosyl transferase from young pummelo leaves (*Citrus maxima*) that is able to catalyze the transfer of rhamnose from UDP-rhamnose to the C-2 hydroxyl group of glucose that is attached by means of C-7-O to naringenin or hesperetin. In other words, the flavanone 7-β-D-glucosides can be converted in one step to the corresponding flavanone 7-β-neohesperidosides, naringin, and neohesperidin, respectively. Gressel et al. have pointed out that if the gene for the enzyme could be found, it might, at least in principle, be possible to reintroduce it into cultured *Citrus* cells, which could then be used to produce neohesperidin (and, hence, neohesperidine DC) from hesperetin 7-β-D-glucoside. The latter compound would be obtained by partially hydrolyzing hesperidin (a readily available byproduct of the orange processing industry) to remove the rhamnose attached at position 6 of the glucose.

Although neohesperidine DC has not yet been found in nature, several structurally related dihydrochalcones have been identified so far in about 20 different families of plants (19). Some of them have been consumed for a long time as natural sweetening agents. Sweet dihydrochalcones have been detected for

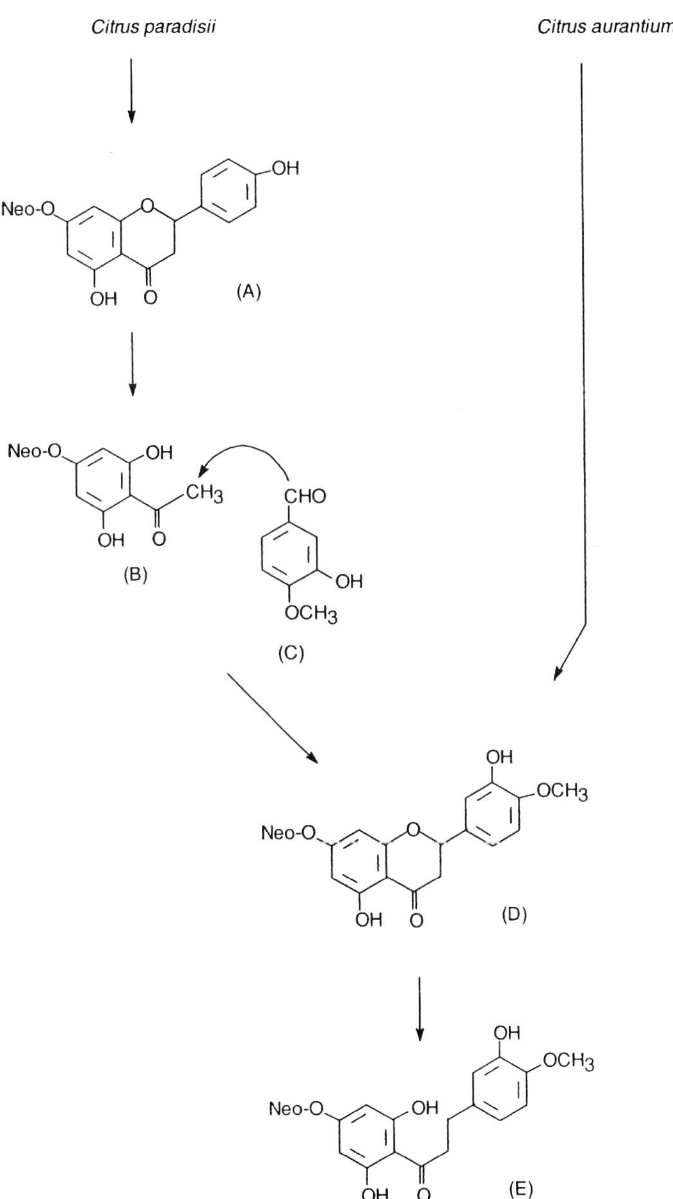

Figure 1 The process of producing neohesperidine DC from flavanones. (A) Naringin; (B) phloroacetophenone-4′-β-neohesperidoside; (C) isovanillin; (D) neohesperidin; (E) neohesperidine DC. (Neo-Neohesperidoside.)

example in the leaves of *Lithocarpus litseifolius*, a variety of sweet tea from China (20). Other sweet dihydrochalcones were detected in the fruit of *Iryanthera laevis*, which is used in the preparation of a Colombian sweet food (21). As a recent example of occurrence in processed foods, dihydrochalcone glycosides have been detected and quantified in different apple juices and jams at levels up to 15 ppm in an apple compote (22).

III. PHYSICAL PROPERTIES AND STABILITY

Neohesperidine DC is an off-white crystalline powder. The crystals are monoclinic, space group $P2_1$. X-ray diffraction demonstrated that the molecule is more or less J-shaped with the β-neohesperidosyl residue forming the curved part of the J and the hesperetin dihydrochalcone aglycone forming the linear segment (23, 24).

The solubility of neohesperidine DC in distilled water at room temperature is low (0.4–0.5 g/l), being freely soluble both in hot (80°C) water and aqueous alkali. As with other citrus flavonoids (hesperidin, naringin), it exhibits higher solubility in an alcohol-water mixture than in water or ethanol alone (25). Several methods for solubility enhancement have been described in the patent literature, such as preparation of the sodium and calcium salts (26) and combinations with water-soluble polyols such as sorbitol (27, 28). Solubility in water may also be enhanced when administered in propylene glycol or glycerol solutions (29). In any case, because it is used at low concentrations, the solubility in cold water is not a limitation for application in food and beverages (30).

Stability studies on neohesperidine DC have shown that the product can be stored for more than 3 years at room temperature without any sign of decomposition. It is slightly hygroscopic taking up water in a saturated environment up to a maximum of 15% (unpublished observations).

In liquid media and under certain conditions of high temperature and low pH, the glycosidic bonds are hydrolyzed, forming the aglycone hesperetin dihydrochalcone, glucose, and rhamnose (Fig. 2).

The stability of neohesperidine DC in aqueous model systems at various pHs and temperatures has been studied (31–33). Aqueous solutions at room temperature are quite resistant to hydrolysis to the free sugars and the aglycone as long as the pH does not fall below 2. (Even if hydrolysis were to occur, there would not be complete loss of sweetness because the aglycone, hesperetin dihydrochalcone, is itself sweet, although not very soluble). The degradation of neohesperidine DC was studied at pH values from 1 to 7 and temperatures ranging from 30° to 60°C for up to 140 days (33). The hydrolysis of neohesperidine DC could be represented as a first-order reaction across the range of temperatures and pHs tested. Optimum pH was 4, although half-life values indicate that no

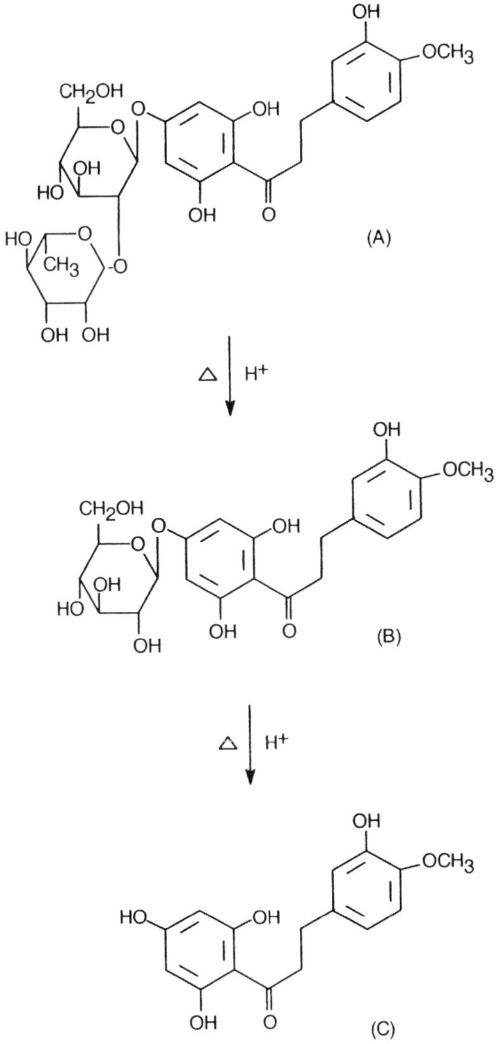

Figure 2 Neohesperidine DC degradation under strong acid conditions and high temperature. (A) Neohesperidine DC; (B) hesperetin dihydrochalcone 4′-β-glucoside; (C) hesperetin dihydrochalcone.

Table 1 Half-Lives of Neohesperidin Dihydrochalcone in Aqueous Buffered Solutions at pH 1 to 7 and Different Temperatures

	Half-life (days)		
pH	20°C	40°C	60°C
1	1360	55	3
2	2344	248	33
3	8268	1100	172
4	9168	1925	475
5	3074	976	357
6	2962	624	103
7	368	50	10

Source: Adapted from Ref. 33. Data at 20°C are extrapolated values using the Arrhenius plot.

stability problem would be expected in the pH range (2–6) (Table 1). The Arrhenius plot could be used to predict stability under temperature conditions other than those checked empirically. By use of this calculation model, it can be estimated that after 12 months at 25°C, the percentage of neohesperidine DC remaining unchanged would be 94.8% at pH 3.

It is widely accepted that buffered aqueous solutions are simplified systems and the data obtained indicate only general trends and do not consider potential interactions between neohesperidine DC and typical food constituents. Thus, the stability of neohesperidine DC has also been evaluated under processing and storage conditions in a number of complex foods.

The stability of neohesperidine DC has been studied during pasteurization of fruit-based soft drinks under different temperature and acidity conditions. No significant hydrolysis was detected in orange, lemon, apple, and pineapple even under severe conditions not used at industrial scale (1 hr at 90°C). Only at the lowest pH value tested (2.0) was there a significant loss of neohesperidine DC (8%) after 1 hr at 90°C (34).

With regard to long-term stability, no loss of neohesperidine DC was found after a 1-year storage period in a lemonade system, either in the dark or when exposed to light. Similarly, when samples were stored at 40°C for up to 3 months, no significant change was found in neohesperidine DC concentration (35).

No neohesperidine DC decomposition was noted under the temperature conditions prevailing during the fruit jam manufacturing process (102–106°C for 35–40 min), as judged by the identical values of nominal concentration included in the formulation and the neohesperidine DC concentration in jam, after processing, at storage time zero. A statistically nonsignificant degradation of 11% of the initial neohesperidine DC was observed after 18 months storage at room temperature (36).

Neohesperidine DC has been shown to remain stable after pasteurization and fermentation of milk and cold storage of yogurt. These stability properties, together with the lack of sweetener-related effects in the course of acidification, suggest that neohesperidine DC could be added, where appropriate, at an early stage of the yogurt manufacturing process (37).

The degradation of neohesperidine DC at high temperature and pH 6–7 has been studied to estimate losses during thermal processing of nonfermented milk-based products. An increase in the pH from 6 to 7 produced an increase in rate constants by a factor of 5. Loss of neohesperidine DC was less than 0.5% for the pasteurization and UHT sterilization conditions tested, and about 9–10% for in-container sterilization at 120°C for 10 min at pH 7 (38).

IV. SENSORY PROPERTIES AND APPLICATIONS

Several studies of the sweetness intensity of neohesperidine DC have been published. At or near threshold, neohesperidine DC is about 1800 times sweeter than sucrose on a weight basis (39). As concentration increases, the sweetness of neohesperidine DC decreases relative to that of sucrose, so that at the 5% sucrose level it is about 250 times sweeter. However, in other studies, higher levels of sweetness of 1000 and 600 times that of sucrose were reported at sucrose concentrations of 5 and 8.5%, respectively (25). From a practical standpoint, the maximum sweetness contribution to most food products from neohesperidine DC will not exceed that delivered by 3% sucrose. At this sucrose equivalent concentration, neohesperidine DC is described as 1500 times sweeter than sucrose (40).

Temporal characteristics of neohesperidine DC as the sole sweetener contrast with those of sugar and other sweeteners such as aspartame because of its long onset and persistence times, also eliciting a lingering licorice-like cooling aftertaste (41, 42). An explanation for the slow taste onset of neohesperidine DC is that some modifications of the molecule must occur within the oral cavity before the active glucophore is produced. Other explanations for the lingering taste would involve a strong and slowly reversible binding to the sweet receptor site with the neohesperidine DC molecule adopting a ''bent'' active conformation in elicitation of sweet taste (43–45). X-ray studies contradict these earlier suggestions (24) because it is proposed that the partially extended form of the dihydrochalcone is the sweet conformer.

Many attempts have been described to overcome these limitations by both chemical derivatization and combination with other substances (27, 28, 46–56). Thus, for example, recent studies (57) reported that complexation with β-cyclodextrin significantly reduces both the aftertaste and the sweet taste of neohesperidine DC. However, from a practical point of view, the most adequate way to overcome the negative slow onset of sweetness and lingering aftertaste is to blend it at low concentrations with other low-calorie sweeteners. With appropriate choice of blend concentrations, this approach proves to be a fully successful method of taking advantage of its synergy and flavor-modifying properties.

Substantial enhancement of sweetness in blends of neohesperidine DC with the relevant intense and bulk sweeteners have been reported with acesulfame K (58), sucralose (59), and according to more recent studies (40, 60) with the full range of low-calorie and bulk sweeteners. It, therefore, plays an important role as a minor component of sweetener blends, in which its contribution to the total sweetness would be not more than 10%, for any low-calorie foods in which intense sweeteners are normally used.

It is relatively common for sweeteners to modify and enhance flavor while also eliciting sweetness. In many cases, their flavor-modifying characteristics are perceived at concentrations above the threshold and could therefore be considered a consequence of sweetness, rather than a structure-induced effect. On the contrary, neohesperidine DC has been consistently shown to modify flavor and enhance mouthfeel, not only at suprathreshold, but also at subthreshold concentrations, being thus independent of sweetness induction (61–64). Functionality of neohesperidine DC as a flavor is closely parallel to that of other flavoring substances widely used in flavoring preparations: maltol and ethyl maltol. In addition, remarkable synergistic effects have been described between neohesperidine DC and maltol and/or ethyl maltol (65).

These flavoring effects of incorporating neohesperidine DC at levels below the sweetness threshold have been studied in detail in a range of sweet and savory products by using quantitative descriptive analysis sensory profiles. In all products tested, most of the statistically significant changes were considered beneficial for overall product quality. In general terms, the flavor-modifying effects of neohesperidine DC may be described as increases in mouthfeel perception and a general smoothing or blending of the individual elements of the product flavor profile (61).

It has been proposed that these flavor effects may be a consequence of the neohesperidine DC ability to reduce the perception of bitterness. The bitterness-reducing effects of neohesperidine DC were originally reported on limonin and naringin (39), the compounds responsible of the bitter taste of grapefruit juice. Thus, for example, at 1% sucrose equivalence neohesperidine DC increased the threshold of limonin 1.4-fold, being much more effective than sucrose in masking the bitterness of limonin. Today, bitterness amelioration has been described in other bitter-tasting substances, such as paracetamol, dextromethorphan, and other pharmaceuticals (66), or the formulation of special foods such as energy-boosting drinks (67).

The ability of neohesperidine DC to mask the unpleasant taste of many pharmaceuticals and to improve the organoleptic properties of other sweeteners led us to think about the potential use of this substance also in medicated feedstuffs. The positive effect of neohesperidine DC in feedstuffs for farm animals has been demonstrated, in particular when used as a flavor modifier in blends with saccharin because of the ability of neohesperidine DC not only to reduce saccharin aftertaste but also to mask the bitter taste of some components of feed. Surprisingly, neohesperidine DC has also been shown to act as an attractant for certain fish such as sea bass and rainbow trout, which suggests its use as an ingredient in new feeds for cultured fish (68).

V. TOXICOLOGY AND METABOLISM

The safety assessment of neohesperidine DC can be based on data from a number of toxicological studies conducted at the Western Regional Research Laboratory, Albany, California. These studies include subacute feeding trials in rats, a three-generation reproduction and teratogenicity study in rats, a 2-year chronic carcinogenicity/toxicity in rats, and a 2-year feeding study in dogs (69). The data, summarized by Horowitz and Gentili (11), showed no evidence of any increased incidence of tumors that could be associated with the ingestion of neohesperidine DC, as well as no adverse effect of toxicological significance even in the high-dose group.

Neohesperidine DC has been checked for mutagenicity in the Ames test; it is nonmutagenic, regardless of which of the various *Salmonella* tester strains is used (70–73). In mice, the compound causes no increase in the normal frequency of micronucleated polychromatic erythrocytes in bone marrow (74).

In 1985–86 a new, detailed study on the ‘‘subchronic (13-wk) oral toxicity of neohesperidine DC in rats’’ was carried out at the TNO-CIVO Toxicology and Nutrition Institute in the Netherlands (75). Neohesperidine DC was fed to groups of 20 male and 20 female Wistar rats at dietary levels of 0, 0.2, 1.0, and 5.0% for 91 days. Only at the 5% level were there any treatment-related effects (i.e., marginal changes in body weight and food consumption, cecal enlargement, and slight changes in some of the clinical chemistry variables). These phenomena were judged to be of little, if any, toxicological significance. Neither the low-dose nor intermediate-dose groups showed any compound-related effects, and none of the groups showed any ophthalmoscopical, hematological, or histopathological changes. It was concluded that the intermediate dose, which translates to about 750 mg of neohesperidine DC per kg body weight per day, was the no-effect level. On the basis of these and the earlier results, the Scientific Committee for Food of the European Community recognized neohesperidine DC as ''toxicologically acceptable'' in 1987 and assigned it an ADI of 0–5 mg/kg body weight (76). This ADI is adequate for a wide range of uses.

The metabolism of neohesperidine DC and that of flavonoid glycosides ingested in substantial amounts with ordinary foods (77) share many features and result partly in the formation of the same end-products (11). As with other flavonoid glycosides, it appears that metabolism is carried out largely by the action of intestinal microflora. After formation of the aglycone, hesperetin dihydrochalcone is split by bacterial glycosidases into phloroglucinol and dihydroisoferulic acid (representing ring A and B of the parent molecule). The latter compound is subsequently converted to a spectrum of metabolites that, like phloroglucinol, also result from the metabolism of certain naturally occurring flavonoids (78).

Excretion studies using [^{14}C] neohesperidine DC showed that, when oral doses of up to 100 mg/kg body weight were administered to rats, more than 90% of the radioactivity was excreted in the first 24 hr, primarily in the urine. After 24 hr, only traces of radioactivity could be detected in various tissues.

The caloric value of neohesperidine DC has been estimated to be not more than 2 cal/g, based on the assumption that the sugar residues are hydrolytically split and metabolized and that the aglycone is not extensively metabolized. Because of its high potency, neohesperidine DC would probably afford not more than 1/1000 as many calories as an equivalent amount of sucrose.

Neohesperidine DC has been proposed as a noncariogenic, nonfermentable sweetening agent, based on the finding that it is relatively inert to the action of cariogenic bacteria (79).

VI. REGULATORY STATUS

With the adoption and publication of the EU Sweeteners Directives (Directives 94/35/EC and 96/83/EC) (80, 81) and after their implementation in national food regulations, the use of neohesperidine DC as a sweetener is authorized in a wide range of foodstuffs (Table 2). Other countries such as Switzerland, the Czech Republic, and Turkey have adopted EU legislation.

As a flavor-modifying substance neohesperidine DC has also been included in the Directive 95/2/EC on Food Additives other than Colours and Sweeteners (82) for use in an additional number of foodstuffs, most of which, such as margarine, meat products, and vegetable protein products, are clearly not sweet.

During recent years the functionality of neohesperidine DC as a flavor and flavor modifier has steadily gained recognition in regulatory circles. Thus, neohesperidine DC has been recognized as GRAS by the Expert Panel of FEMA for use as a flavor ingredient in 16 food categories in 1993 (83) at levels of use

Table 2 Energy Reduced or with No Added Sugar Foodstuffs and Permitted Use Levels for Neohesperidin Dihydrochalcone in 94/35/EC and 96/83/EC Directives

Foodstuffs	Maximum level (ppm)
Water-based flavored drinks	30
Milk and milk-derivative–based drinks	50
Fruit-juice–based drinks	30
Water-based flavored desserts	50
Milk or milk-derivative–based preparations	50
Fruit-, vegetable-, egg-, cereal- and fat-based desserts	50
Breakfast cereals with a fiber content of more than 15% and containing at least 20% bran	50
"Snacks": certain flavors or ready to eat, prepacked, dry savory starch products and coated nuts	50
Confectionery with no added sugars	100
Cocoa- or dried-fruit–based confectionery	100
Starch-based confectionery	150
Cornets and waters for ice-cream	50
Cocoa-, milk-, dried-fruit or fat-based sandwich spread	50
Breath-freshening micro-sweet	400
Chewing gum	400
Drinks consisting of a mixture of a nonalcoholic drink and beer, cider, perry, spirit, or wine	30
Spirit drinks containing less than 15% alcohol by volume	30
Cider and perry	20
Alcohol-free beer or with an alcohol content not exceeding 1.2% vol	10
"Table beer" (original wort content less than 6%) except "Obergäriges Einfachbier"	10
Beers with a minimum acidity of 30 milliequivalents expressed as NaOH	10
Brown beer of the "oud bruin" type	10
Energy-reduced beer	10
Edible ices	50
Canned or bottled fruit	50
Jams, jellies, and marmalades	50
Sweet-sour preserves of fruit and vegetables	50
Fruit and vegetable preparations	50
Feinkostsalat	50
Sweet-sour preserves and semipreserves of fish and marinades of fish, crustaceans, and mollusks	30
Soups	50
Sauces and mustard	50
Fine bakery products for special nutritional uses	150
Complete formulas for weight control intended to replace total daily food intake or an individual meal	100
Complete formulas and nutritional supplements for use under medical supervision	100
Liquid food supplements/dietary integrators	50
Solid food supplements/dietary integrators	100
Food supplements/diet integrators based on vitamins and/or mineral elements, syrup-type or chewable	400

Table 3 GRAS Use Levels for Neohesperidin Dihydrochalcone as a Flavor Ingredient

Food category	Usual use (ppm)	Maximum use (ppm)
Frozen dairy	2	3
Soft candy	2	3
Gelatin and puddings	2	3
Soups	1	2
Nonalcoholic beverages	2	3
Fats and oils	4	4
Milk products	2	3
Fruit juice	2	3
Fruit ices	1	2
Processed vegetables	2	3
Condiments and relishes	2	3
Jams and jellies	2	3
Sweet sauce	2	3
Imitation dairy	3	4
Hard candy	2	4
Chewing gum	4	5

below the sweetness threshold as specified in Table 3. Other countries, such as Australia and New Zealand, have authorized the use of neohesperidine DC as an artificial flavoring without any use limitation.

VII. ANALYSIS IN FOODS

High-performance liquid chromatography (HPLC) is the most effective analytical method for precise and accurate analysis of neohesperidine DC, both as a raw material (84) and in foodstuffs (29, 85). Other reported methods such as thin-layer chromatography or ultraviolet spectrophotometry with and without chemical derivatization lack adequate selectivity and are only useful for quality control of the pure material. Neohesperidine DC has a different regulatory status, and hence different conditions of use, depending on the area. It is therefore important that it can be detected in products at the proposed level of use and that it be possible to analytically differentiate between flavoring and sweetening use levels.

HPLC methods to quantitate neohesperidine DC in foodstuffs have been reported (86–88). Although these previous methods are useful starting points, they were limited to one single product or studied foods and beverages with unrealistically high sweetener concentrations. An analytical method to detect and quantitate neohesperidine DC in foodstuffs has been developed and validated, yielding adequate results in terms of precision, accuracy, selectivity, and ruggedness to quantitate neohesperidine DC both at flavoring and sweetening use levels in soft drinks (89). The method has successfully been assayed also in complex foods such as dairy products, confectionery, and fat-based foods that require selective extraction of the sweetener with appropriate solvents (dimethyl sulfoxide, methanol, and their blends with water). Acceptable recoveries (>90%) were found in all tested samples both at flavoring and sweetening use levels. This method provides sufficient separation between the neohesperidine DC peak and the corresponding hydrolysis acid products.

Neohesperidine DC is used as a minor component of sweetener blends and therefore at very low concentrations. However, HPLC techniques allow detection and quantitation of neohesperidine DC at levels below those that are normally used for sweetening and flavoring purposes. Thus, detection and quantitation limits for neohesperidine DC, determined by the method based on the standard deviation of the response and the slope, are 0.2 and 0.7 mg/l, respectively. These values are below the minimum concentration, which shows a technological function in the final food.

Usually, extraction in dimethyl sulfoxide or alcohols is sufficient for selective extraction of neohesperidine DC; however, adsorption of neohesperidine DC onto Amberlite XAD and subsequent fractionation on Sephadex were judged to be essential steps for successful quantitation of neohesperidine DC in blackcurrant jam because of the fact that anthocyanin-related compounds may interfere with neohesperidine DC in crude extracts (36).

VIII. AVAILABILITY AND PATENT SITUATION

Neohesperidine DC is produced in industrial scale by Zoster, S. A. and marketed worldwide by Exquim, S. A. under the tradename of Citrosa®. Both companies belong to the pharmaceutical group Ferrer Internacional, S. A.

The original patents, which covered the manufacture of neohesperidine DC, have expired, although neohesperidine DC has a patent portfolio that covers its use and applications (63, 64, 68).

DEDICATION

This chapter is dedicated to the memory of Prof. Dr. Francisco Sabater Garcia.

REFERENCES

1. RM Horowitz, B Gentili. Dihydrochalcone derivatives and their use as sweetening agents. US Patent No. 3,087,821 (April 30, 1963).
2. RM Horowitz. Relations between the taste and structure of some phenolic glycosides. In: JB Harborne, ed. Biochemistry of Phenolic Compounds. New York: Academic Press, 1964, pp 545–571.
3. RM Horowitz, B Gentili. Taste and structure in phenolic glycosides, 1. Agric Food Chem 17:696–700, 1969.
4. RM Horowitz, B Gentili. Phenolic glycosides of grapefruit: A relation between bitterness and structure. Arch Biochem Biophys 92:191–192, 1961.
5. RM Horowitz, B Gentili. Flavonoids of citrus VI. The structure of neohesperidose. Tetrahedron 19:773–782, 1963.
6. RM Horowitz, B Gentili. Flavonoid constituents of citrus. In: S Nagy, PE Shaw, MK Veldhuis, eds. Citrus Science and Technology. Vol. 1. Westport, CT: The AVI Publishing Company, 1977, 397–426.
7. RM Horowitz, B Gentili. Conversion of naringin to neohesperidin and neohesperidin dihydrochalcone. US Patent No. 3,375,242 (March 26, 1968).
8. L Krbechek, G Inglett, M Holik, B Dowling, R Wagner, R Riter. Dihydrochalcones: Synthesis of potential sweetening agents. J Agric Food Chem 16:108–112, 1968.
9. GH Robertson, JP Clark, R Lundin. Dihydrochalcone sweeteners: Preparation of neohesperidin dihydrochalcone. Ind Eng Chem Prod Res Devel 13:125–129, 1974.
10. K Veda, H Odawara. Neohesperidin from naringin. Jpn Kokai 75:154–261, December 12, 1975 [Chem Abstr 84:180574, 1976].
11. RM Horowitz, B Gentili. Dihydrochalcone sweeteners from citrus flavanones. In: LO Nabors, RC Gelardi, eds. Alternative Sweeteners, 2nd ed. New York: Marcel Dekker, 1991, pp 97–115.
12. AHB Linke, S Natarajan, DE Eveleigh. A new screening agent to detect the formation of dihydrochalcone sweeteners by microorganisms. The Annual Meeting of the American Society of Microbiology, 1973, abstract No. 184, p 171.
13. E Lewinsohn, H Gavish, E Berman, R Fluhr, Y Mazur, J Gressel. Towards biotechnologically synthesized neohesperidin for the production of its dihydrochalcone sweetener. International Conference on Sweeteners. Los Angeles, CA: American Chemical Society, September 22–25, 1988, abstract No. 39.
14. E Lewinsohn, L Britsch, Y Mazur, J Gressel. Flavanone glycoside biosynthesis in Citrus. Plant Physiol 91:1323–1328, 1989.
15. E Lewinsohn, E Berman, Y Mazur, J Gressel. 7-glycosylation and (1-6) rhamnosylation of exogenous flavanones by undifferentiated Citrus cell cultures. Plant Science 61:23–28, 1989.
16. E Lewinsohn, L Britsch, Y Mazur, J Gressel. Flavanone glycoside biosynthesis in Citrus: Chalcone synthase, UDP-glucose-flavanone-7-O-glucosyl transferase and rhamnosyl transferase activities in cell-free extracts. Plant Physiol 91:1323–1328, 1989.
17. M Bar-Peled, E Lewinsohn, R Fluhr, J Gressel. UDP-rhamnose flavanone-7-O-glucoside-2′-O-rhamnosyl transferase, purification and characterization of an enzyme

catalyzing the production of bitter compounds in Citrus. J Biol Chem 266:20953–20959, 1991.

18. M Bar-Peled, R Fluhr, J Gressel. Juvenile-specific localization and accumulation of a rhamnosyl transferase and its bitter flavonoids in foliage, flowers and young fruits. Plant Physiol 103:1377–1384, 1993.
19. BA Bohm. The minor flavonoids. In: TH Harborne, ed. The Flavonoids. Advances in Research since 1980. New York: Chapman & Hall, 1988, pp 329–348.
20. N Rui-Lin, T Tanaka, J Zhow, O Tanaka. Phlorizin and trilobatin, sweet dihydrochalcone glucosides from leaves of *Lithocarpus litseifolius* (Hance) Rehd. (*Fagaceae*). Agric Biol Chem 46:1933–1934, 1982.
21. NL Garzón, SLE Cuca, VJC Martínez, M Yoshida, OR Gottlieb. The chemistry of Colombian *Myristicaceae*. Part 5. Flavonolignoid from the fruit of *Iryanthera laevis*. Phytochemistry 26:2835–2837, 1987.
22. FA Tomás-Barberán, C García Viguera, JL Nieto, F Ferreres, F Tomás-Lorente. Dihydrochalcone from apple juices and jams. Food Chemistry 46:33–36, 1993.
23. RY Wong, RM Horowitz. The X-ray crystal and molecular structure of neohesperidin dihydrochalcone sweetener. J Chem Soc Perkin Trans 1:843–848, 1986.
24. W Shin, SJ Kim, M Shin, SH Kim. Structure-taste correlations in sweet dihydrochalcone, sweet dihydroisocoumarin and bitter flavone compounds. J Med Chem 38: 4325–4335, 1995.
25. A Bär, F Borrego, O Benavente, J Castillo, JA del Río. Neohesperidin dihydrochalcone: Properties and applications. Lebens Wiss u Technol 23:371–376, 1990.
26. EB Westall, AW Messing. Salts of dihydrochalcone derivatives and their use as sweeteners. US Patent Nos. 3,984,394 (1976) and 4,031,260 (1977).
27. BK Dwivedi, PS Sampathkumar. Preparing neohesperidin dihydrochalcone sweetener composition. US Patent No. 4,254,155 (1981).
28. BK Dwivedi, PS Sampathkumar. Artificial sweetener composition and process for preparing and using same. US Patent No. 4,304,794 (1981).
29. F Borrego, I Canales, MG Lindley. Neohesperidin dihydrochalcone: State of knowledge review. Z Lebensm Unters Forsch 200:32–37, 1995.
30. MG Lindley, I Canales, F Borrego. A technical re-appraisal of the intense sweetener neohesperidin dihydrochalcone. In: FIE Conference Proceedings. Maarsen, The Netherlands: Expoconsult Publishers, 1991, pp 48–51.
31. GA Crosby, TE Furia. New sweeteners. In: TE Furia, ed. CRC Handbook of Food Additives. Vol. II. 2nd ed. Boca-Raton, FL: CRC Press, 1980, p 204.
32. GE Inglett, L Krbechek, B Dowling, R Wagner. Dihydrochalcone sweeteners–sensory and stability evaluation. J Food Sci 34:101–103, 1969.
33. I Canales, F Borrego, MG Lindley. Neohesperidin dihydrochalcone stability in aqueous buffer solutions. J Food Sci 57:589–591+643, 1993.
34. H Montijano, MD Coll, F Borrego. Assessment of neohesperidine DC stability during pasteurization of juice-based drinks. Int J Food Sci Technol 31:397–401, 1996.
35. H Montijano, F Borrego, FA Tomás-Barberán, MG Lindley. Stability of neohesperidine DC in a lemonade system. Food Chem 58:13–15, 1997.
36. FA Tomás-Barberán, F Borrego, F Ferreres, MG Lindley. Stability of the intense sweetener neohesperidin dihydrochalcone in blackcurrant jams. Food Chem 52:263–265, 1995.

37. H Montijano, FA Tomás-Barberán, F Borrego. Stability of the intense sweetener neohesperidine DC during yogurt manufacture and storage. Z Lebensm Unters Forsch 201:541–543, 1995.
38. H Montijano, FA Tomás-Barberán, F Borrego. Accelerated kinetic study of neohesperidine DC hydrolysis under conditions relevant for high temperature processed dairy products. Z Lebensm Unters Forsch 204:180–182, 1995.
39. DG Guadagni, VP Maier, JH Turnbaugh. Some factors affecting sensory thresholds and relative bitterness of limonin and naringin. J Sci Food Agric 25:1199–1205, 1974.
40. SS Schiffman, BJ Booth, BT Carr, ML Loose, EA Sattely-Miller, BG Graham. Investigation of synergism in binary mixtures of sweeteners. Brain Res Bull 50:72–81, 1996.
41. H Montijano, F Borrego. Neohesperidine DC. In: JM Dalzell, ed. LFRA Ingredients Handbook (Sweeteners). Surrey, UK: Leatherhead Food Research Association, Leatherhead, 1996, pp 181–200.
42. MO Portman, D Kilcat. Psychophysical characterization of new sweeteners of commercial importance for the EC food industry. Food Chem 56:291–302, 1996.
43. GE DuBois, GA Crosby, RA Stephenson, RE Wingard Jr. Dihydrochalcone sweeteners: Synthesis and sensory evaluation of sulfonate derivatives. J Agric Food Chem 25:763–771, 1977.
44. GA Crosby, GE DuBois, RE Wingard. The design of synthetic sweeteners. In: EJ Ariéns, ed. Drug Design. Vol 8. New York: Academic Press, 1979, pp 215–310.
45. JE Hodge, CE Inglett. Structural aspects of glycosidic sweeteners containing (1-2)-linked disaccharides. In: GE Inglett, ed. Symposium: Sweeteners. Westport, CT: AVI Publishing Co., 1974, pp 216–234.
46. GE DuBois, GA Crosby, JF Lee, RA Stephenson, PC Wang. Dihydrochalcone sweeteners. Synthesis and sensory evaluation of a homoserine dihydrochalcone conjugate with low aftertaste sucrose-like organoleptic properties. J Agric Food Chem 29:1269–1276, 1981.
47. GE DuBois, GA Crosby, P Saffron. Nonnutritive sweeteners: Taste-structure relationships for some new simple dihydrochalcones. Science 195:397–399, 1977.
48. GE DuBois, GA Crosby, RA Stephenson. Dihydrochalcone sweeteners. A study of the atypical temporal phenomena. J Med Chem 24:408–428, 1981.
49. GE DuBois, RA Stephenson. Sulfamo dihydrochalcone sweeteners. US Patent No. 4,283,434 (Aug. 11, 1981).
50. GE DuBois, RA Stephenson, GA Crosby. Alpha amino acid dihydrochalcones. US Patent No. 4,226,804 (Oct. 7, 1980).
51. GA Crosby, GE DuBois. Sweetener derivatives. US Patent No. 4,055,678 (Oct. 25, 1977).
52. GE DuBois, GA Crosby. Dihydrochalcone oligomers. US Patent No. 4,064,167 (Dec. 20, 1977).
53. GA Crosby, GE DuBois. Sweetener derivatives. US Patent No. 3,976,687 (Aug. 24, 1976).
54. RM Horowitz, B Gentili. Dihydrochalcone galactosides and their use as sweetening agents. US Patent No. 3,890,296 (June 17, 1975).

55. RM Horowitz, B Gentili. Dihydrochalcone xylosides and their use as sweetening agents. US Patent No. 3,826,856 (July 30, 1974).
56. GP Rizzi. Dihydrochalcone sweetening agents. US Patent No. 3,739,064 (June 12, 1973).
57. F Caccia, R Dispenza, G Fonza, C Fuganti, L Malpezzi, A Mele. Structure of neohesperidin dihydrochalcone/β-cyclodextrin inclusion complex: NMR, MS, and X-ray spectroscopy investigation. J Agric Food Chem 46:1500–1505, 1998.
58. GW von Rymon Lipinsky, E Lück. Sweetener mixture. US Patent No. 4,158,068 (1979).
59. MR Jenner. Sweetening agents. Eur Patent Application No. 0 507 598 A1 (1992).
60. HJ Chung. Measurement of synergistic effects of binary sweetener mixtures. J Food Sci Nutr 2:291–295, 1997.
61. MG Lindley, PK Beyts, I Canales, F Borrego. Flavor modifying characteristics of the intense sweetener neohesperidin dihydrochalcone. J Food Sci 58:592–594+666, 1993.
62. MG Lindley. Neohesperidin dihydrochalcone: Recent findings and technical advances. In: TH Grenby, ed. Advances in Sweeteners. Glasgow, UK: Chapman & Hall, 1996, pp 240–252.
63. R Foguet, A Cisteró, F Borrego. Body and mouthfeel potentiated foods and beverages containing neohesperidin dihydrochalcone. US Patent No. 5,300,309 (1994).
64. R Foguet, A Cisteró, F Borrego. Use of neohesperidin dihydrochalcone for potentiating body and mouthfeel of foods and beverages. Eur Patent Specification No. 0 500 977 B1 (1995).
65. L Engels, G Stagnitti. Flavor modifying composition. WO Patent No. 96/17527 (1996).
66. F Borrego, H Montijano. Anwendungsmöglichkeiten des Süssstoffs Neohesperidin Dihydrochalkon in Arneimitteln. Pharm Ind 57:880–882, 1995.
67. MA Bruna, RJ Miralles, JA Poch, FL Redondo. Compositions for alkalinizing and energy-boosting drinks. WO Patent No. 96/03059 (Feb. 8, 1996).
68. R Foguet, F Borrego, J Campos, JA Madrid, S Zamora. New feed for cultured fish. Eur Patent Application No. 0 655 203 A1.
69. MR Gumbmann, DH Gould, DJ Robbins, AN Booth. Toxicity studies of neohesperidin dihydrochalcone. In: JH Shaw, GG Roussos, eds. Proceedings, Sweeteners and Dental Caries. Washington, DC: Information Retrieval, 1978, pp 301–310.
70. RP Batzinger, SYL Ou, E Bueding. Saccharin and other sweeteners: Mutagenic properties. Science 198:944–946, 1977.
71. JP Brown, PS Dietrich, RJ Brown. Frameshift mutagenicity of certain naturally occurring phenolic compounds in the *Salmonella*/microsome test: Activation of anthraquinone and flavonol glycosides by gut bacterial enzymes. Biochem Soc Trans 5: 1489–1492, 1977.
72. JT MacGregor, L Jurd. Mutagenicity of plant flavonoids: Structural requirements for mutagenic activity in *Salmonella typhimurium*. Mutation Res 54:297–309, 1978.
73. JB Brown, PS Dietrich. Mutagenicity of plant flavonols in the *Salmonella*/mammalian microsome test: Activation of flavonol glycosides by mixed glycosidases from rat cecal bacteria and other sources. Mutation Res 66:223–240, 1979.

74. JT MacGregor. Mutagenicity studies of flavonoids *in vivo* and *in vitro*. Toxicol Appl Pharmacol 48:A47, 1979.
75. BAR Lina, HC Dreef-van der Meulen, DC Leegwater. Subchronic (13 wk) oral toxicity of neohesperidin dihydrochalcone in rats. Food Chem Toxicol 28:507–513, 1990.
76. Report of the Scientific Committee for Food on Sweeteners. Commission of the European Communities (CS/EDUL/-Final), December 11, 1987.
77. J Kuhnau. The flavonoids. A class of semiessential food components: Their role in human nutrition. World Rev Nutr Diet 24:117–191, 1976.
78. F DeEds. Flavonoids metabolism. Compr Biochem 125:417–423, 1971.
79. CW Berry, CA Henry. Baylor College of Dentistry, Meeting of the American Association for Dental Research, as reported in Food Chemical News, March 21, 1983, p 25.
80. European Parliament and Council Directive 94/35/EC of 30 June 1994 on sweeteners for use in foodstuffs. Off J Eur Comm (10.9.94) No. L237/3.
81. Directive 96/83/EC of the European Parliament and of the Council of 19 December 1996. Off J Eur Comm (19.2.97) No. L 48/16.
82. European Parliament and Council Directive 95/2/EC of 20 February 1995 on food additives other than colours and sweeteners. Off J Eur Comm (18.3.5) No. L61/1.
83. RL Smith, P Newberne, TB Adams, RA Ford, JB Hallagan. GRAS flavoring substances 17. The 17th publication by the Flavors and Extract Manufacturers' Association's Expert Panel on recent progress in the consideration of flavoring ingredients generally recognized as safe under the Food Additives Amendment. Food Technol 50:72–81, 1996.
84. MR Castellar, JL Iborra, I Canales. Analysis of commercial neohesperidin dihydrochalcone by high performance liquid chromatography. J Liq Chrom Rel Technol 20:2063–2073, 1997.
85. M Hausch. Simultaneous determination of neohesperidin dihydrochalcone and other sweetening agents by HPLC. Lebensmittelchemie 50:31–32, 1996.
86. R Schwarzenbach. Liquid chromatography of neohesperidin dihydrochalcone. J Chromatogr 129:31–39, 1996.
87. JF Fisher. A high-pressure liquid chromatographic method for the quantitation of neohesperidin dihydrochalcone. J Agric Food Chem 25:682–683, 1976.
88. JF Fisher. Review of quantitative analyses for limonin, naringin, naringenin rutinoside, hesperidin and neohesperidin dihydrochalcone in citrus juice by high performance liquid chromatography. Proc Int Soc Citric 3:813–816, 1976.
89. H Montijano, F Borrego, I Canales, FA Tomás-Barberán. Validated high-performance liquid chromatographic method for quantitation of neohesperidin dihydrochalcone in foods. J Chromatogr A 758:163–166, 1997.

7
Tagatose

Hans Bertelsen and Søren Juhl Hansen
Arla Foods Ingredients amba, Videbaek, Denmark

Rene Soegaard Laursen
Arla Foods Ingredients amba, Brabrand, Denmark

James Saunders
Biospherics Incorporated, Beltsville, Maryland

Kristian Eriknauer
Arla Foods Ingredients amba, Viby, Denmark

I. INTRODUCTION

D-tagatose is a low-calorie bulk sweetener with the following properties:

- It has 92% of the sweetness of sucrose.
- It has a reduced caloric value.
- It is noncariogenic.
- It is a prebiotic.
- It is a flavor enhancer.

D-Tagatose or tagatose is a ketohexose in which its fourth carbon is chiral and is a mirror image of the respective carbon atom of the common D-sugar, fructose. The CAS number for D-tagatose is 87-81-0. The empirical formula for D-tagatose is $C_6H_{12}O_6$. The molecular weight of D-tagatose is 180.16. The structural formula for D-tagatose is depicted in Fig. 1, along with that of D-fructose. Tagatose is a naturally occurring low-calorie bulk sweetener. Tagatose occurs in *Sterculia setigera* gum, a partially acetylated acidic polysaccharide (1). D-Tagatose is also found in heated cow's milk, produced from lactose (2) and occurs in various other dairy products.

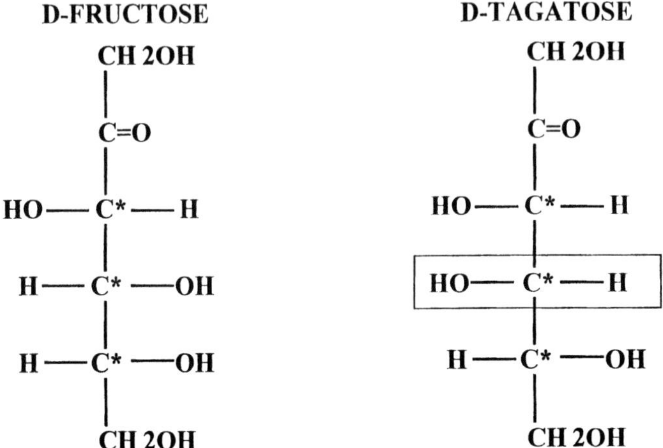

Figure 1 The fourth chiral carbon is a mirror image of fructose. *Designates chiral carbon.

On the basis of a similarity in sweetness and physical bulk to sucrose, D-tagatose is intended to be used as a reduced-calorie bulk sweetener. D-Tagatose is intended to be used as a reduced-calorie sugar replacement in ready-to-eat cereals, diet soft drinks, frozen yogurt/nonfat ice-cream, soft confectionery, hard confectionery, frosting, and chewing gum.

Biospherics Inc. of Beltsville, Maryland, patented the process in 1988 and in 1991 the use of D-tagatose in foodstuffs. In 1996, MD Foods Ingredients amba (now Arla Foods Ingredients amba) of Denmark bought the exclusive license to all patents and know-how pertaining to tagatose in foods and beverages and will be responsible for the production and commercialization of the sweetener.

II. PROCESS

The production of D-tagatose occurs in a stepwise manner, starting from the raw material lactose, which is then altered with the use of enzymes and various fractionating, isomerization, and purification techniques. Food-grade lactose is hydrolyzed to galactose and glucose by passing the solution through an immobilized lactase column.

The sugar mixture from the enzyme hydrolysis is fractionated by chromatography. The chromatographic separation of glucose and galactose is essential and is similar to the normal industrial separation of glucose and fructose, using the same type of Food and Drug Administration (FDA)–approved calcium-based cationic resins. The galactose fraction from chromatography is converted to D-tagatose under alkaline conditions by adding a suspension of technical-grade $Ca(OH)_2$ and, optionally, a technical-grade catalyst, $CaCl_2$. The reaction is stopped by adding technical-grade sulfuric acid (H_2SO_4). A limited number of side reactions are observed as part of the isomerization process of D-galactose. The reactions are well known and are typical of those that occur between all common hexose monosaccharides (like D-fructose and D-glucose) and any source of hydroxide ion. Tagatose is stable under the conditions of the isomerization process.

On removal of the gypsum formed, the resulting filtrate is further purified by means of demineralization and chromatography. The purified D-tagatose solution is then concentrated and crystallized to give a white crystalline product (>99% pure) (Fig. 2).

III. GASTROINTESTINAL FATE OF D-TAGATOSE

The rather small difference in chemical structure of D-tagatose compared with fructose has large implications on the overall metabolism of the sugars. The fructose carrier-mediated transport in the small intestine has no affinity for D-tagatose, and only approximately 20% of ingested D-tagatose is absorbed in the small intestine. The absorbed part is metabolized in the liver by the same pathway as fructose. The major part of ingested D-tagatose is fermented in the colon by indigenous microflora, resulting in the production of short-chain fatty acids (SCFAs).

A. Absorption in the Small Intestine

Among the different dietary monosaccharides, glucose and galactose are absorbed by an active energy-consuming transport mechanism. Fructose and xylose are absorbed by carrier-mediated, so-called facilitated diffusion. All other sugars are absorbed by passive absorption. Although D-tagatose and D-fructose are similar in terms of molecular structure, they are not taken up in the small intestine by the same carrier mechanism. In vitro experiments showed that D-tagatose does not inhibit the absorption of fructose or glucose (at concentrations up to 100 times greater than those of fructose) (3, 4). Binding of D-tagatose to the glucose and fructose carriers followed by transport across the mucosa can, therefore, be excluded.

The estimated absorption of radio labeled D-tagatose in rats is approximately 20% (5). Similarly, less than 26% of ingested D-tagatose was absorbed

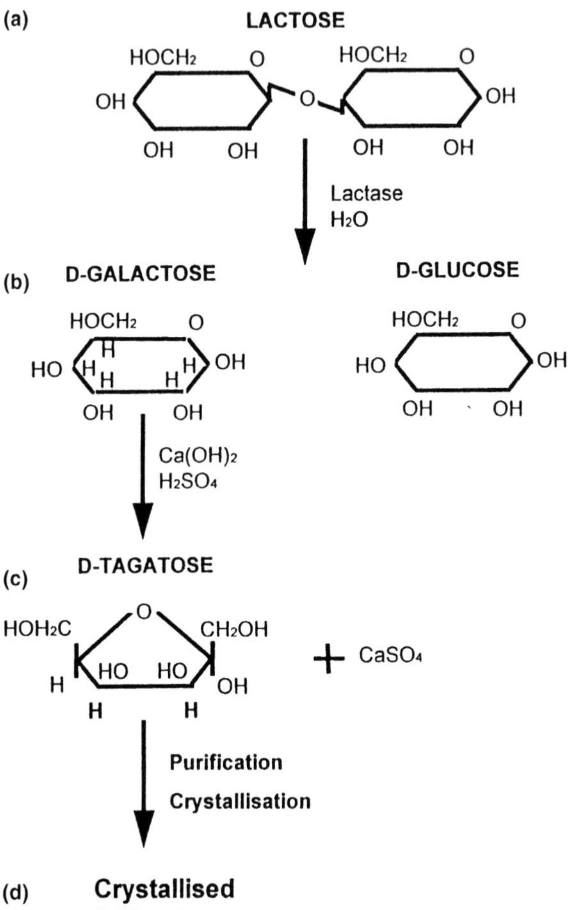

Figure 2 The two-step tagatose production process. (a) D-tagatose is derived from 99% lactose. (b) Step 1: An enzymatic hydrolysis of the lactose separates it into D-galactose and D-glucose. (c) Step 2: D-galactose is converted into D-tagatose using a chemical isomerization process. (d) The remaining process purifies and crystalizes D-tagatose.

in pigs according to measurements of the disappearance of D-tagatose from the digesta (6). Absorption rates of this magnitude are typical for monosaccharides and polyols of similar size that are absorbed by passive diffusion [e.g., mannitol (7) or L-fructose and L-gulose (8)]. Because the absorption of D-tagatose is likely to be even less than that of L-rhamnose, (because L-rhamnose has a slightly lower molecular weight and is slightly more lipophilic), the fractional absorption of D-tagatose in humans probably does not exceed 20%.

B. Excretion of D-Tagatose in Urine

A metabolic study with D-[U-^{14}C]-tagatose in rats demonstrated that approximately 4% of ingested D-tagatose or 20% of absorbed D-tagatose was excreted with the urine (5). The same study showed that 43% of an intravenous dose was excreted in the urine of unadapted rats over 48 hr, with most (>90%) being cleared by the kidneys within 6 hr after dosing.

In a pig study using 20% D-tagatose in the diet, 5% of the ingested D-tagatose was excreted in the urine, and this was independent of adaptation of D-tagatose (9). A human study with 30 g of D-tagatose showed 0.7–5.3% of ingested D-tagatose excreted in the urine (10).

C. Metabolism

The steps involved in the metabolism of fructose and D-tagatose are identical (11). However, an important difference occurs in the rate of metabolism at the reaction step, where fructose-1-P and D-tagatose-1-P are split by aldolase to glyceraldehyde and dihydroxyacetone phosphate. Although aldolase acts on both fructose and D-tagatose, evidence suggests that the rate at which aldolase cleaves D-tagatose-1-P is approximately 10% that for fructose-1-P (12) (Fig. 3).

Despite the low absorption of D-tagatose, one would expect a small transient increase in tagatose-1-P and concomitant decreases in Pi and adenosine triphosphate (ATP). This is because of the low affinity of aldolase b for tagatose-1-P. An increase of tagatose-1-P stimulates hexokinase activity, inhibits glycogen phosphorylase activity, and stimulates glycogen synthetase activity. The consequence of these effects on liver enzymes is an increase in the net deposition of glycogen in the liver. This has, in fact, been observed in the rat (13).

D. Fermentation

Unabsorbed D-tagatose is fermented by intestinal micro-organisms to SCFAs. No D-tagatose was found in feces of pigs ingesting a 10% D-tagatose diet (6). Similarly, no D-tagatose was found in human feces after a 30-g intake of D-tagatose

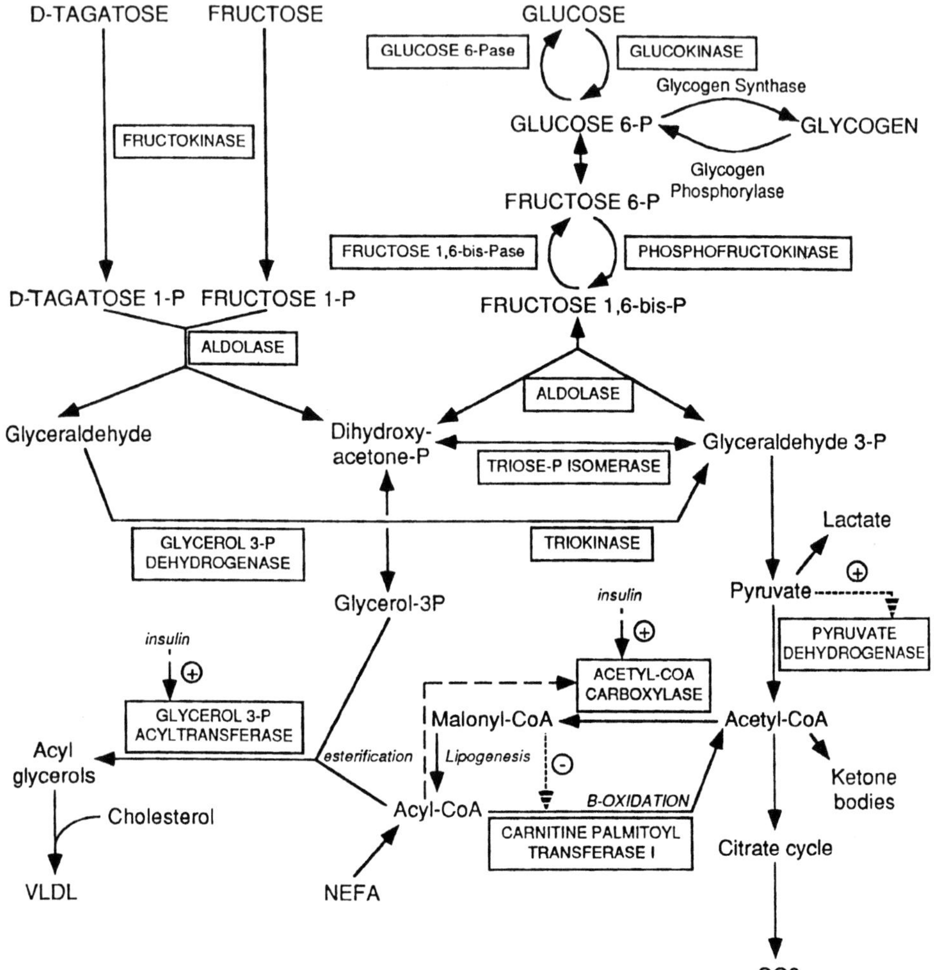

Figure 3 Fructose and D-tagatose use in the liver.

(10). In adapted rats fed a diet with 10% D-tagatose, about 2% of the ingested dose of ^{14}C-D-tagatose was recovered in feces (5). In unadapted rats, the recovery of D-tagatose in feces was much higher, approximately 25% of the ingested dose. Adaptation of the microflora in pigs, which results in increased in vitro fermentation of D-tagatose, is indicated by the finding of increased numbers of D-tagatose–degrading bacteria and a disappearance of watery stools after a few days of D-tagatose ingestion (5, 6).

IV. USE OF D-TAGATOSE BASED ON BIOLOGICAL PROPERTIES

D-Tagatose is a sweet-tasting monosaccharide with interesting nutritional and physiological properties. It is only partly absorbed in the small intestine, and the major part is fermented in the colon, where it is converted into biomass, SCFAs, CO_2, and H_2. D-Tagatose can be considered as a prebiotic on the basis of the promotion of beneficial bacteria and an increase in the generation of SCFAs, specifically an increased level of butyrate. Furthermore, the reduced absorption and special fermentation mean that the caloric value of D-tagatose is a maximum of 1.5 kcal/g. D-Tagatose consumption does not induce an increase of blood glucose or insulin levels and even blunts the glucose level when D-tagatose is taken before glucose or sucrose. This makes D-tagatose a desirable sugar substitute for people with diabetes. D-Tagatose is so slowly converted to organic acids by tooth plaque bacteria that it does not cause dental caries. It has satisfied the Swiss regulation as safe for teeth. The aforementioned properties suggest several dietary applications as products for people with diabetes, prebiotic foods, noncariogenic confections and low-calorie foods.

A. Caloric Value of D-Tagatose

Ingested D-tagatose is incompletely absorbed from the small intestine of animals and man. The fractional absorption is about 20% in rats and 25% in pigs. In humans, the fractional absorption is estimated at not more than 20% on the basis of data of a structurally related carbohydrate (L-rhamnose). The absorbed fraction of D-tagatose is readily metabolized through the glycolytic pathway yielding 3.75 kcal/g. The unabsorbed fraction of D-tagatose reaches the large intestine, where it is completely fermented by the intestinal microflora. The formed SCFAs are absorbed almost completely and are metabolized. The metabolic fate of D-tagatose resembles, therefore, that of other incompletely digested carbohydrates (e.g., polyols).

The energy value of D-tagatose was evaluated in two studies in rats and one study in pigs. A net metabolizable energy value of −0.12 kcal/g was obtained for D-tagatose in one of the rat studies (14). This may be explained by an inhibition of absorption of sucrose that was present in the basal diet in a very high concentration and/or by the relatively low amount of fermentable fiber in the basal diet. The second rat study suggested a metabolizable energy value of about 1.2 kcal/g (15). Most relevance was attached to the pig study because the digestive tracts of pigs and humans show many similarities. The pig study resulted in a net metabolizable energy value of 1.4 kcal/g for D-tagatose (16).

Estimation of the net metabolizable energy value of D-tagatose by the factorial method gave a range of 1.1–1.4 kcal/g. In this method, the energy contributed by each metabolic step is evaluated separately, taking into account data from all pertinent experiments (in vitro, in vivo, in humans, in experimental animals). The factorial approach also takes into account losses of energy that are caused indirectly by the fermentation of D-tagatose (e.g., increased fecal excretion of biomass and nonbacterial mass). Based on these studies, the FDA has approved the use of a factor of 1.5 kcal/g for calculating the caloric value of tagatose.

B. D-Tagatose and Diabetes

A clinical study (17) showed that oral intake of 75 g of tagatose gave no increase in plasma glucose or serum insulin in either normal persons or people with type 2 diabetes.

The preceding was confirmed in an eight-person (normal subjects) study, in which oral intake of 30 g of tagatose similarly did not lead to changes in plasma glucose and serum insulin (18).

Oral tagatose ($t = -30$ min) blunts the rise in plasma glucose and serum insulin seen after oral glucose or sucrose ($t = 0$) in normal and diabetic persons (17).

In the study with eight persons consuming 30 g of D-tagatose dissolved in water at $t = 0$ and being served a sucrose-rich lunch after 4 hr, the 30-g tagatose blunted the rise in plasma glucose and serum insulin after lunch compared with the same persons on either water or 30 g of fructose (18)

C. Tooth-Friendly Properties of D-Tagatose

D-Tagatose has been demonstrated to be noncariogenic in two studies using human volunteers. The evaluation consisted of two phases: (a) an evaluation of the cariogenicity of D-tagatose using telemetric techniques, and (b) an evaluation of the cariogenic potential involving subjects' adaptation to D-tagatose.

During the first phase of testing, a 10% aqueous solution of D-tagatose was tested for cariogenicity in human volunteers using intra oral plaque-pH-telemetry (19). Telemetrically recorded plaque-pH values after the ingestion of a substance at or above the pH limit of 5.7 can be regarded as a criterion of a low cariogenic potential of the tested food.

For the studies, six persons in generally good health served as test subjects. The subjects were asked not to alter their eating habits, and with the single exception of water rinses, they were instructed to refrain from all oral hygiene measures. Daily recordings of changes in pH were measured after subjects rinsed for 2 min with D-tagatose. Changes in pH were also measured after rinsing with 0.3 mol/l (10%) sucrose, which served as the positive control.

No critical decreases (i.e., <5.7) in the pH of interdental plaque caused by bacterial fermentation of D-tagatose were noted during either the 2-min D-tagatose rinsing periods or during the 30-min periods after rinsing with D-tagatose. The pH decreases that occurred subsequently to the 0.3 mol/l (10%) sucrose rinses provided adequate evidence of plaque metabolism and that the pH-telemetric equipment had functioned properly (Fig. 4).

In phase II, researchers tested whether D-tagatose would be fermented by plaque bacteria either before adaptation or after bacteria had a chance to adapt to D-tagatose during a period of frequent exposure. Any fermentation would cause more acidification in the plaque layers on exposure to D-tagatose. The same six subjects who participated in the acute D-tagatose telemetric evaluation also served as test subjects in this study. The subjects were asked not to alter their eating habits, and with the single exception of water rinses, they were instructed to refrain from all oral hygiene measures. During the individual periods of plaque accumulation and adaptation, the subjects rinsed five times a day. Each of the five rinsing applications consisted of two sequential 2-min rinses with 15 ml of 10% aqueous solutions of D-tagatose. Two-min sucrose rinses at 0.3 mol/l (10%) served as the positive control. Periods of adaptation to D-tagatose rinsing (from 3–7 days in duration) were arranged to determine whether adaptation influenced fermentation in the mouth.

The results showed no critical decrease in the pH (i.e., below pH of 5.7) of interdental plaque either during the rinsing periods or the 30-min periods after rinsing with D-tagatose. This contrasts with plaque pH after rinsing with sucrose, which always fell below the critical pH value of 5.7 because of the glycolytic

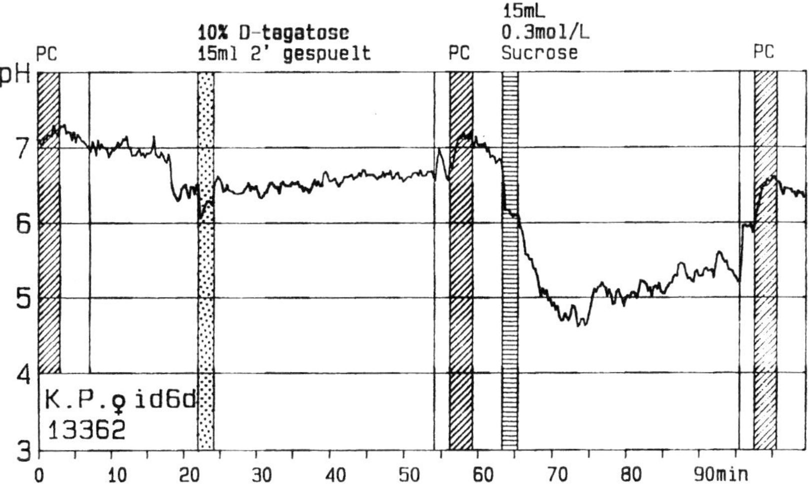

Figure 4 Measurement of oral pH in unadapted dental trial.

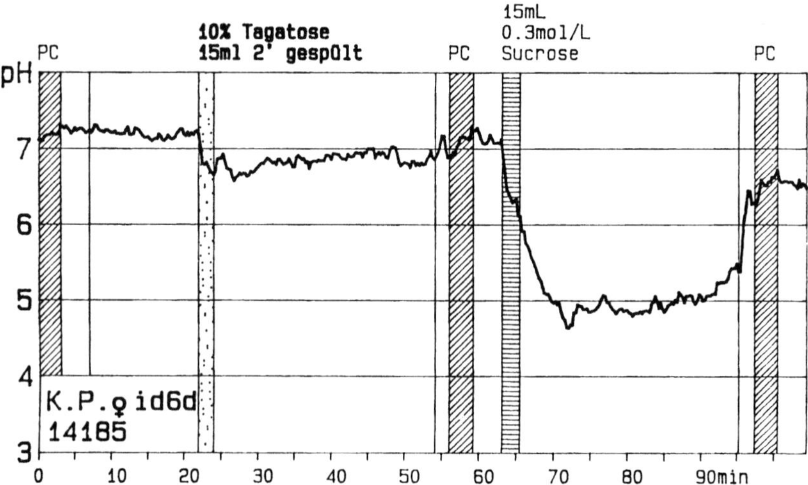

Figure 5 Measurement of oral pH in adapted dental trial.

production of bacterial acids. A comparison of the pH values of interdental plaque occurring in subjects who had been adapted to D-tagatose from 3–7 days shows that these values are similar to the pH values of interdental plaque in the same subjects when they were unadapted to D-tagatose. It can therefore be concluded that plaque layers having grown under the constant exposure to D-tagatose were not more acidified by a 10% D-tagatose rinse than unadapted plaque layers in the same volunteers (Fig. 5). D-Tagatose has thus been demonstrated to be noncariogenic in both studies.

D. Prebiotic Properties of D-Tagatose

The mucosal surfaces of the intestinal tract are one of the main sites of cell replication in the human body. In the colon, the epithelial cells are exposed not only to the circulation and to the endogenous secretions of other mucosal cells but also to the contents of the colonic lumen, which is rich in food residues, and to the metabolic products of the microflora (20). Epidemiological and animal studies suggest that dietary fat and protein may promote carcinogenesis in the colon, whereas increased fiber and complex carbohydrates in the diet may protect against colon cancer. Colonic luminal butyrate concentrations are postulated to be the key protective component of high-fiber diets against colon cancer (21).

Butyrate is one of the SCFAs that are the C2–C5 organic acids. These compounds are formed in the gastrointestinal tract of mammals as a result of anaerobic bacterial fermentation of undigested dietary components and are readily absorbed by the colonic epithelium. Dietary fiber is the principal substrate for the fermentation of SCFA in humans; however, intake of fiber is often low in a typical Western diet. Other undigested components, like starch and proteins, contribute to the production of SCFAs, as well as the low-molecular-weight oligosaccharides, sugars, and polyols that escape digestion and absorption in the small intestine. In the mammalian hind gut, acetate, propionate and butyrate account for at least 83% of SCFAs produced and are present in a nearly constant molar ratio 60:25:15 (21).

In pig studies, D-tagatose altered the composition and population of colonic microflora as evidenced by changes in the proportion of SCFAs produced. In vitro fermentation for 0–4 hr of colonic samples with 1% added D-tagatose from pigs adapted to D-tagatose for 17 days showed 46 mol% of butyrate, compared with a normal mol% of 17, resulting from in vitro fermentation for 0–4 hr of colonic samples from pigs fed the sucrose control diet (22).

Concentrations of butyrate in the cecum and colon of pigs were increased in a dose-response manner from ingestion of D-tagatose. Similar 12-hr in vitro incubation of intestinal samples from slaughtered pigs also showed a dose-response production of butyrate from ingested D-tagatose (23) (Fig. 6).

Increased in vivo concentrations of butyrate were seen in portal vein blood from both adapted and unadapted pigs. The appearance of butyrate in the portal

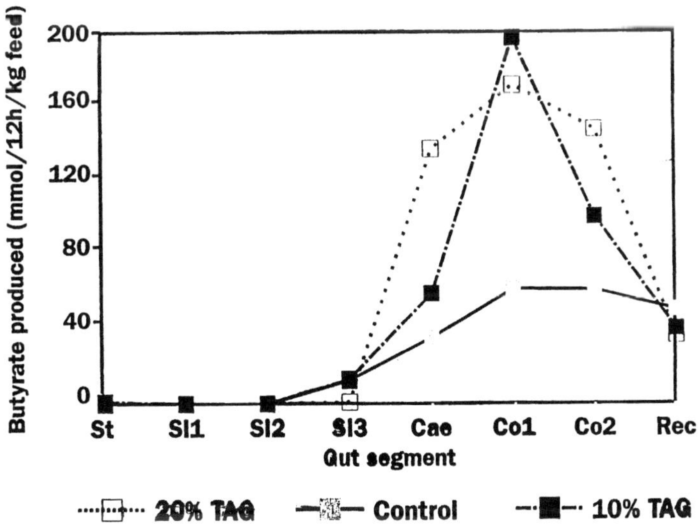

Figure 6 12-hour in vitro production of butyrate in pig gut segment slaughtered 6 hr after feeding.

vein of unadapted pigs showed an adaptation of butyrate production within the 12 hr of the experimental period (23).

None of the aforementioned studies in pigs, which sample colon contents and blood samples from the portal vein, can be performed in humans for ethical reasons. However, pigs and humans have similar gastrointestinal tracts and qualitatively the same types of indigenous intestinal bacteria. Therefore, the pig serves as a good model for humans in digestion and fermentation. To perform human studies, one has to rely on in vitro studies with D-tagatose added to human fecal slurries.

In human subjects fed a 10-g dose of D-tagatose three times daily throughout a 13-day test, the rate of in vitro fermentation (added level of D-tagatose at 1%) was greater with fecal samples of adapted volunteers than unadapted volunteers and, similarly, the mol% of butyrate was high (35% versus 22%) in a 4-hr incubation. After 48 hr, the unadapted fecal incubation also showed increased mol% of butyrate. Tagatose ingestion was characterized by changes in microbial population density and species. Pathogenic bacteria (such as coliform bacteria) were reduced, and specific beneficial bacteria (such as lactobacilli and lactic acid bacteria) were increased. The enrichment of lactobacilli is consistent with the screening of pure culture intestinal bacteria, which indicates a high frequency of D-tagatose fermentation in the tested *Lactobacillus* strains (24) (Fig. 7).

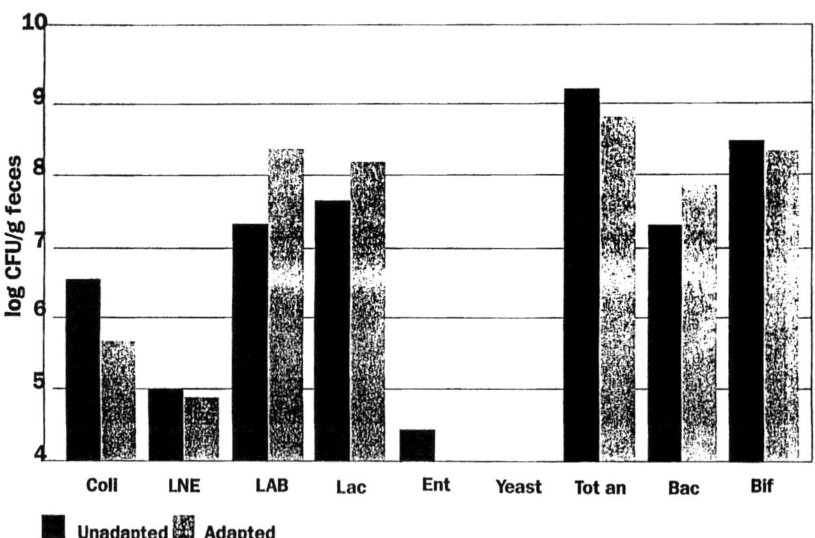

Figure 7 Influence of D-tagatose on bacterial composition in human feces.

The intestines of both pigs and humans harbor bacteria that produce butyrate as the fermentation end-product. D-Tagatose is a rather rare sugar for intestinal bacteria and is only fermented by a limited genera of bacteria, enterococci, lactobacilli, and, obviously, some butyrate-producing intestinal bacteria. Other fibers or undigested carbohydrates are either not a substrate for butyrate-producing bacteria or are competing for the delivered substrate favoring growth of other bacteria. A reasonable explanation for the butyrate-inducing effect of D-tagatose is that butyrate-producing bacteria are favored over most other bacteria when tagatose is supplied. The adaptation seen with ingestion of D-tagatose in both pigs and human volunteers is probably due to selection of butyrate-producing bacteria in the colon. The in vivo absorption study in pigs showed that the adaptation or selection of bacteria takes place within 12 hr. Similarly, the in vitro study with human feces indicates selection of bacteria within 48 hr of incubation.

Many studies have documented an important role of butyrate in the colon because it is the major and preferred fuel for the colonic epithelium (25) and plays an important role in the control of proliferation and differentiation of colonic epithelial cells (20). In contrast to the preceding effects on the normal epithelium, butyrate arrests growth of neoplastic colonocytes (26) and also inhibits the preneoplastic hyperproliferation induced by tumor promoters in vitro (21).

The lactobacilli and lactic acid bacteria are important inhabitants of the intestinal tract of man and animals with functional benefits like maintenance of the normal microflora, pathogen interference, exclusion and antagonism, immunostimulation and immunomodulation, anticarcinogenic and antimutagenic activities, deconjugation of bile acids, and lactase presentation in vivo (27). A number of studies on probiotic bacteria have shown that it is difficult for these bacteria to colonize the human colon; that is, after stopping the supply of probiotic bacteria in the diet, they disappear. This makes it more obvious to selectively feed the *Lactobacillus* already present in the colon.

On the basis of the preceding documented butyrate stimulation and selection of bacteria in the colon, D-tagatose is a promising prebiotic food bulk sweetener.

E. Human Tolerance

Extensive human clinical testing has been conducted on D-tagatose to ensure safety and tolerance (18, 28–34). Testing included a 14-day trial with 30 g given in a single daily dose, an 8-wk trial with a 75-g daily dosage given as three 25-g doses, and a 12-month trial with a 45-g daily dosage given as three 15-g doses. Because D-tagatose is malabsorbed, gastrointestinal effects similar to those of other undigested sugars would be expected. Unabsorbed molecules in the large intestine retain colonic fluid and increase water content in the feces, contributing to laxation or diarrhea, whereas fermentation by the microflora produces gas, leading to an increase in flatulence. As seen in clinical trials with tagatose consumption, a 30-g bolus dose was well tolerated by many subjects, and when gastrointestinal symptoms did occur, they were consistent with the type of effects produced by an undigested compound.

Tolerance to a 30-g divided dose of D-tagatose was investigated in two studies (32, 34); results demonstrated that most subjects tolerate this level of ingestion well. An analysis of all the data available for D-tagatose indicates that a bolus dose of 20 g or a divided daily dose of 30 g is well tolerated, with some sensitive individuals experiencing gastrointestinal symptoms including mild flatulence, laxation, diarrhea, and bloating. The gastrointestinal effects that were observed in all the clinical studies are consistent with those expected when consuming a malabsorbed compound, and there is recognition of interindividual variability in this response. It is expected that individuals will self-regulate their consumption of this type of food product on the basis of their own gastrointestinal response to the consumption.

V. PHYSICAL PROPERTIES

A. Sweetness

To determine the level of sweetness of D-tagatose relative to sucrose, a paired comparison test was conducted on D-tagatose and sucrose with six trained judges. This sweetness equivalency taste test was based on the difference threshold method (35). The sweetness level for D-tagatose was determined by means of a linear regression plot of the fraction of times that D-tagatose solutions were chosen as sweeter than 10% sucrose versus the percent D-tagatose concentration. The point at which judges could not distinguish which solution was sweeter (at 50% probability) resulted in equivalent sweetness of the D-tagatose solution to the 10% sucrose solution. This point was 10.8% D-tagatose. On the basis of this sweetness equivalence taste test, D-tagatose was determined to be 92% ($10/10.8 \times 100$) as sweet as sucrose in a 10% aqueous solution (Table 1).

B. Flavor Enhancer

To determine whether D-tagatose has specific flavor-enhancing properties, several tests were performed. These included subjective qualitative evaluation in various diet beverages and quantitative descriptive analysis profiling in cola beverages sweetened with aspartame:acesulfame-K blends and with sucralose. The purpose was to quantify the relative sweetness and mouth-feel–enhancing effects of D-tagatose.

Table 1 Relative Sweetness of Tagatose and Other Sugars and Polyols

	Relative sweetness
Sucrose	1
Fructose	0.8–1.7
Tagatose	0.92
Xylitol	0.8–1.0
Glucose	0.5–0.8
Erythritol	0.50
Sorbitol	0.4–0.7
Lactitol	0.4

At practical use levels, a combination of tagatose and aspartame provides impressive sweetness synergy. Tagatose is also synergistic with aspartame:acesulfame-K combinations. Consistent changes in flavor attributes have been observed across all sweetener systems. Tagatose speeds sweetness onset times and reduces bitterness in blends. The mouth-feel characteristics are improved (e.g., mouth drying is significantly reduced), the sweet aftertaste is significantly reduced, and the bitter aftertaste is reduced. The sensory contributions of tagatose were found to be universally beneficial.

From the subjective evaluation of aspartame:acesulfame-K sweetened lemon/lime soft drinks containing 0.20% tagatose we found a fresher and cleaner flavor profile with more depth of flavor. A cleaner aftertaste with no bitterness and a fuller, more syrupy mouth feel results.

A subjective evaluation of aspartame:acesulfame-K sweetened cola soft drinks containing 0.20% tagatose found lemon flavor notes to be enhanced. A more balanced flavor and sweetness resulted. Less lingering sweetness and enhanced mouth-feel were produced.

C. Crystal Form

Tagatose is a white crystalline powder with an appearance very similar to sucrose. The main difference is the crystal form. Tagatose has a tetragonal bipyramid form illustrated in Fig. 8.

Crystallization from aqueous solution results in anhydrous crystals in an α-pyranose form, with a melting point of 134–137°C. In solution, tagatose mutarotates and establishes an equilibrium of 71.3% α-pyranose, 18.1% β-pyranose, 2.6% α-furanose, 7.7% β-furanose, and 0.3% keto-form (36).

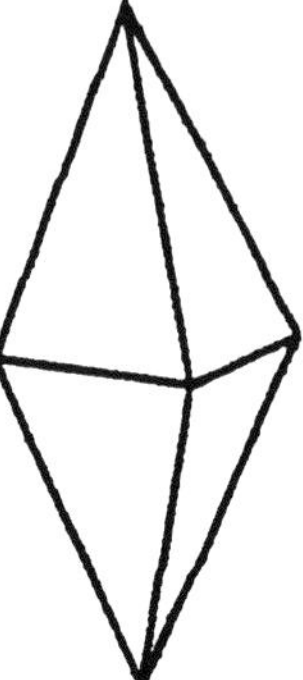

Figure 8 Tetragonal bipyramid crystal form of D-tagatose.

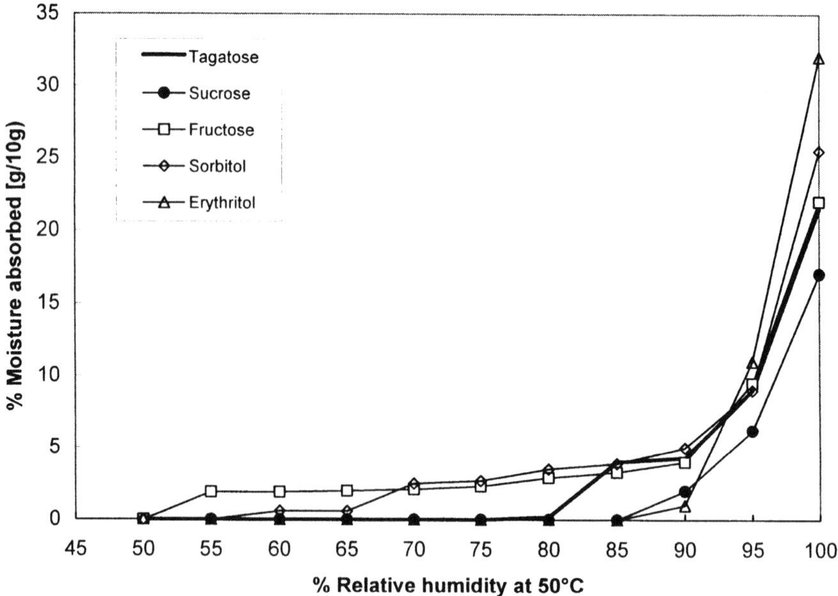

Figure 9 Moisture absorbency at different relative humidities.

D. Hygroscopicity

Tagatose is a nonhygroscopic product similar to sucrose (Fig. 9). This means that tagatose will not absorb water from its surrounding atmosphere under normal conditions and does not require special storage.

E. Water Activity

Water activity influences product microbial stability and freshness. Tagatose exerts a greater osmotic pressure, and hence, lower water activity than does sucrose at equivalent concentrations. The effect on water activity is similar to fructose (same molecular weight).

F. Solubility

Tagatose is soluble in water and similar to sucrose (Fig. 10). This makes it suitable for use in applications where it is substituted for sucrose. The same amounts produce nearly the same sweetness. Tagatose is suitable for applications like

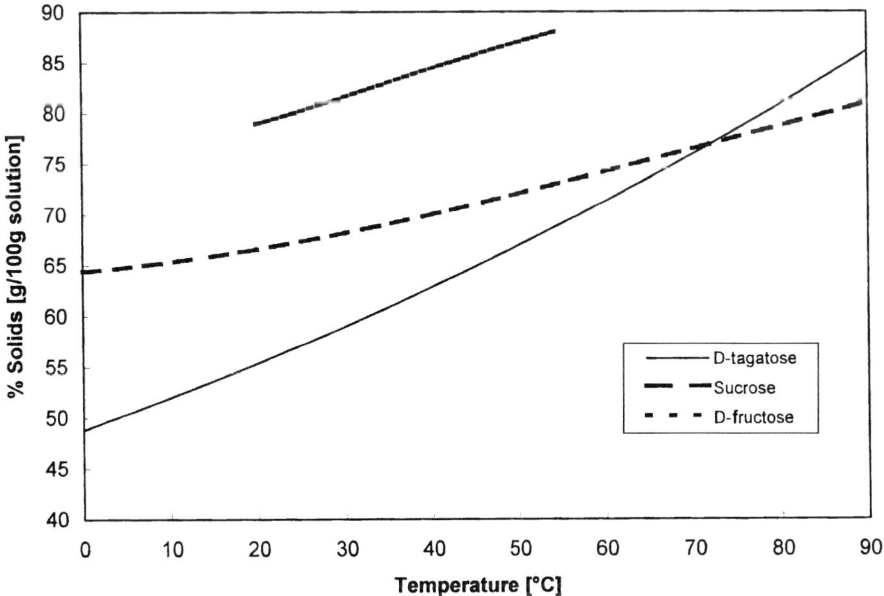

Figure 10 Solubility of tagatose, sucrose, and fructose in water at different temperatures.

hard-boiled candy, ice-cream, chocolate, soft drinks, and cereals. Compared with polyols, tagatose is more soluble than erythritol (36.7% w/w at 20°C) and less soluble than sorbitol (70.2% w/w at 20°C).

G. Viscosity

Tagatose solutions are lower in viscosity than sucrose solutions at the same concentrations but slightly higher than fructose and sorbitol (180 cP at 70% w/w and 20°C) (Fig. 11).

H. Heat of Solution

Tagatose has a cooling effect stronger than sucrose and slightly stronger than fructose (Table 2).

I. Chemical Properties

As a ketohexose, tagatose is a reducing saccharide and very active chemically. Tagatose takes part in the Maillard reaction, which leads to a distinct browning effect. It also decomposes (caramelizing) more readily than sucrose at high temperatures. At very low and high pH, tagatose is less stable and converts to various

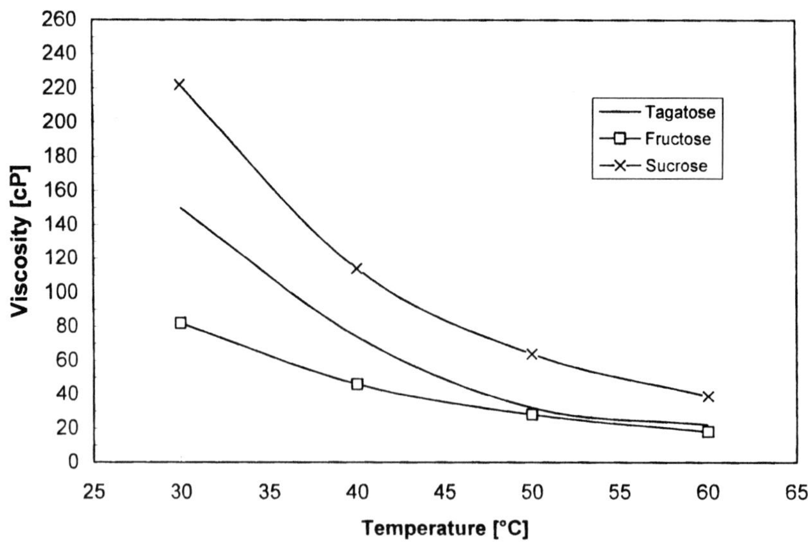

Figure 11 Viscosity at 70% w/w concentration and different temperatures.

Table 2 Heat of Solution of Tagatose, Sucrose, and Fructose at Different Temperatures

	Heat of solution [kJ/kg]	
	20°C	37°C
Sucrose	−18.2	−23.9
Fructose	−37.7	−62.3
Tagatose	−42.3	−84.1
Erythritol	−79.4	—
Sorbitol	−111	—

compounds. However, tagatose can be used satisfactorily in many different applications at high temperature when process time is kept short. It is relatively straightforward to produce hard-boiled candy under reduced pressure with tagatose.

VI. APPLICATIONS

A. Confectionery in General

Given its low caloric value, its bulking ability, and the same sweetness as sucrose, tagatose is well suited for confectionery products. In terms of flavor profile, confectionery products with tagatose will be close to products produced with sucrose.

B. Chocolate

Chocolate with reduced calories can favorably be produced using tagatose as a substitute for sucrose. The sensorial profiles are similar and calorie reductions of 20–25% are obtainable. Chocolate with tagatose can be produced with standard processing equipment used in the industry today.

C. Hard-Boiled Candies and Wine Gum

Noncariogenic hard-boiled candies with good flavor profile and stability can be produced with tagatose. When exposed to high temperatures, tagatose seems to have a behavior similar to that of fructose. Tagatose has a low glass transition temperature and promotes the Maillard browning effect. Cooked under vacuum and in combination with other sweeteners (50%–50%) tagatose is suitable for hard-boiled candies and wine gum.

D. Fondant

Because tagatose has a good ability to crystallize, it is well suited for producing fondants for low-calorie pralines.

E. Chewing Gum

Tagatose is suited for use in noncariogenic chewing gum, both for kernels and dragee.

F. Fudge

Tagatose's ability to crystallize makes it suitable for making fudge. Because tagatose caramelizes at a low temperature, the characteristic caramel flavor is easily obtained.

G. Caramel

Caramel produced with tagatose has a smooth and soft consistency. Browning and caramel flavor occur because of the low temperature of caramelizing. This makes tagatose suitable for caramel in chocolates.

H. Ice Cream

Tagatose can replace sucrose in a 1 : 1 ratio in ice-cream with good results.

I. Soft Drinks

In soft drinks it has been found that tagatose shows significant synergistic effects at even low doses with combinations of intense sweeteners. When blended with intense sweeteners, tagatose improves flavor and mouth-feel. It also seems that tagatose is able to stabilize the sweetness and flavor profile in soft drinks sweetened with blends of aspartame and acesulfame potassium. As aspartame degrades, sweetness synergy with acesulfame potassium is lost and the balance is shifted toward acesulfame potassium, which has a more bitter taste. Adding as little as 0.20% tagatose to the sweetener blend provides an element of stable sweetness, and the tagatose is able to mask the bitter taste of acesulfame potassium, thereby prolonging the shelf-life of the soft drink.

J. Breakfast Products

The prebiotic effect of tagatose can be well exploited in applications such as cereals with a dosage level of 15% and granola bars and fruit preparations for yogurts at 3%.

In conclusion, tagatose has three major areas of application:

1. Low-calorie bulk sweetener: Based on a significantly lower caloric value than sucrose, yet having a similar level of sweetness and similar physical bulking structure, tagatose can be used as a low-calorie bulk sweetener, replacing sugar and other sweeteners in different applications at different dose levels.
2. Prebiotic sweetener: Replacing sucrose in various applications and adding a prebiotic benefit.
3. Flavor enhancer: Tagatose added to sweetening systems based on potent sweeteners or high-intensity sweeteners such as aspartame alone or combinations of aspartame and acesulfame potassium improves the flavor profile and the mouth-feel.

VII. TOXICOLOGY AND REGULATORY ASPECTS

Arla Foods Ingredients amba is currently seeking its self-affirmed GRAS status for D-tagatose in the United States. All relevant studies to document the safety of tagatose have been performed at FDA-approved research institutes. The key articles documenting the safety of D-tagatose appear in the April 1999 issue of *Regulatory Toxicology and Pharmacology*.

The FDA has approved the use of a factor of 1.5 kcal/g for calculating the caloric value of tagatose.

VIII. CONCLUSION

With its broad range of properties, tagatose has a unique application profile and is set to be one of the major low-calorie bulk sweeteners of the future.

REFERENCES

1. EL Hirst, L Hough, JKN Jones. Composition of the gum of *Sterculia setigera*: Occurrence of D-tagatose in nature. Nature 163(4135):177, 1949.
2. E Troyano, I Martinez-Castro, A Olano. Kinetics of galactose and tagatose formation during heat-treatment of milk. Food Chem 45:41–43, 1992.
3. G Crouzoulon. Les proprietes cinetiques du flux d'entree du fructose a travers la bordure en brosse du jejunum du rat. Arch Int Physiol Biochim 86:725–740, 1978.
4. K Sigrist-Nelson, U Hopfer. A distinct D-fructose transport system in isolated brush border membrane. Biochim Biophys Acta 367:247–254, 1974.

5. JP Saunders, LR Zehner, GV Levin. Disposition of D-[U-^{14}C]-tagatose in the rat. Reg Toxicol Pharmacol 29:S46–S56, 1999.
6. HN Lærke, BB Jensen. D-Tagatose has low small intestinal digestibility but high large intestinal fermentability in pigs. J Nutr 129:1002–1009, 1999.
7. A Bar. Factorial calculation model for the estimation of the physiological caloric value of polyols. In: N Hosoya, ed. Proceedings of the International Symposium on Caloric Evaluation of Carbohydrates. Tokyo: The Japan Assoc. Dietetic & Enriched Foods, 1990, pp 209–252.
8. GV Levin, LR Zehner, JP Saunders, JR Beadle. Sugar substitutes: Their energy values, bulk characteristics, and potential health benefits. Am J Clin Nutr 62(suppl.): 1161S–1168S, 1995.
9. BB Jensen, A Laue. Absorption of D-tagatose and fermentation products of tagatose from the gastrointestinal tract of pigs. Internal report, 1998.
10. B Buemann, S Toubro, A Astrup. D-Tagatose, a stereoisomer of D-fructose increases hydrogen production in humans without affecting 24-hour energy expenditure, or respiratory exchange ratio. J Nutr 128:1481–1486, 1998.
11. R Rognstad. Pathway of gluconeogenesis from tagatose in rat hapatocytes. Arch Biochem Biophys 218(2):488–491, 1982.
12. HA Lardy. The influence of inorganic ions on phosphorylation reactions. In: WD McElroy, B. Glass, eds. Phosphorous Metabolism. Baltimore: Johns Hopkins, 1951, pp 477–499.
13. A Bar, BAR Lina, DMG de Groot, B de Bie, MJ Appel. Effect of D-tagatose on liver weight and glycogen content of rats. Reg Toxicol Pharmacol 29:S11–S28, 1999.
14. G Livesey, JC Brown. D-Tagatose is a bulk sweetener with zero energy determined in rats. J Nutr 126:1601–1609, 1996.
15. JP Saunders, GV Lewin, LR Zehner. Dietary D-tagatose lowers fat deposition and body weight gains of rats. Internal report, 1994.
16. H Jørgensen, HN Lærke. The influence of D-tagatose on digestibility and energy metabolism in pigs. Internal report, 1998.
17. T Donner, JF Wilber, D Ostrowski. D-Tagatose: A novel therapeutic adjunct for non-insulin dependent diabetes. Diabetes 45 (suppl. 2):125a, 1996.
18. B Buemann, S Toubro, JJ Holst, JF Rehfeld, A Astrup. D-Tagatose, a stereoisomer of D-fructose increases blood uric acid concentration. Metabolism, 49(8):969–976, 2000.
19. University of Zurich. Telemetric evaluation of D-tagatose provided by MD Foods Ingredients Amba, with regard to the product's qualification as being "safe for teeth." Zurich: Dental Institute, Clinic of Preventive Dentistry, Periodontology and Cariology Bioelectronic Unit, 1996.
20. IT Johnson. Butyrate and markers of neoplastic change in the colon. Eur J Cancer Prev 4:365–371, 1995.
21. OC Velázquez, HW Lederer, JL Rombeau. Butyrate and the colonocyte: Implications for neoplasia. Dig Dis Sci 41(4):727–739, 1996.
22. HN Johansen, BB Jensen. Recovery of energy as SCFA after microbial fermentation of D-tagatose (abstr). Int J Obes 21(suppl. 2):S50, 1997.
23. H Bertelsen, BB Jensen, B Buemann. D-Tagatose—A novel low calorie bulk sweetener with prebiotic properties. World Rev Nutr Dietetics 85:00, 1999.

24. BB Jensen, B Buemann, H Bertelsen. D-Tagatose changes the composition of the microbiota and enhances butyrate production by human fecal samples. International Symposia on Pro-Biotics and Pre-Biotics, Kiel, Germany, 11–12 June, 1998.
25. WEW Roediger. Role of anaerobic bacteria in the metabolic welfare of the colonic mucosa in man. Gut 21:793–798, 1980.
26. A Hague, J Butt, C Paraskeva. The role of butyrate in human colonic epithelial cells: An energy source or inducer of differentiation and apoptosis. Proc Nutr Soc 55: 937–943, 1996.
27. TR Klaenhammer. Functional activities of *Lactobacillus* probiotics: Genetic mandate. Int Dairy J 8:497–505, 1998.
28. B Buemann, S Toubro, A Raben, J Blundell, A Astrup. The acute effect of D-tagatose on food intake in humans. Br J Nutr 84:227–231, 2000.
29. B Buemann, S Toubro, A Astrup. Human gastrointestinal tolerance to D-tagatose. Reg Toxicol Pharmacol 29:S71–S77, 1999.
30. B Buemann, S Toubro, A Raben, A Astrup. Human tolerance to a single, high dose of D-tagatose. Reg Toxicol Pharmacol 29:S66–S70, 1999.
31. A Lee, DM Storey. Comparative gastrointestinal tolerance of sucrose, lactitol, or D-tagatose in chocolate. Reg Toxicol Pharmacol 29:S78–S82, 1999.
32. T Donner, JF Wilber, D Ostrowski. D-Tagatose, a novel hexose: Acute effects on carbohydrate tolerance in subjects with and without type 2 diabetics. Diabetes, Obesity Metab (accepted for publication), 1999.
33. T Donner, JF Wilber, D Ostrowski. D-Tagatose effects upon CHO and lipid metabolism in normal and diabetic subjects. Internal protocol and preliminary results, 1998.
34. BB Jensen, B Buemann. The influence of D-tagatose on bacterial composition and fermentation capacity of faecal samples from human volunteers. J Nutr (submitted for publication), 1999.
35. HT Lawless, H Heymann. Physiological and psychological foundations of sensory function. In: Sensory Evaluation of Food. New York: Chapman & Hall, 1998, pp 28–31.
36. GJ Wolff, E Breitmeier. C-NMR-Spektroskopische Bestimmung der Keto-Form in Währigen Lösungen der D-Fructose, L-Sorbose und D-Tagatose. Chem Z 103:232, 1979.

8
Neotame

W. Wayne Stargel, Dale A. Mayhew, C. Phil Comer, Sue E. Andress, and Harriett H. Butchko
The NutraSweet Company, Mount Prospect, Illinois

I. INTRODUCTION

Neotame (*N*-[*N*-(3,3-dimethylbutyl)-L-α-aspartyl]-L-phenylalanine 1-methyl ester) is a new high-intensity sweetener and flavor enhancer that is currently undergoing international regulatory consideration. Neotame is 7000 to 13,000 times sweeter than sucrose. It is derived from and is structurally similar to aspartame but is 30 to 60 times sweeter than aspartame, depending on the sweetness required in various food or beverage matrices. This zero-calorie sweetener has a clean sweet taste with no undesirable taste characteristics; neotame is functional and stable in a wide array of beverages and foods and should require no special labeling for phenylketonuria.

Neotame resulted from a long-term research program designed to discover high-intensity sweeteners with optimized performance characteristics. French scientists Claude Nofre and Jean-Marie Tinti invented neotame from a simple *N*-alkylation of aspartame (1, 2). The NutraSweet Company holds the rights to a wide range of patents related to neotame.

II. PHYSICAL CHARACTERISTICS AND CHEMISTRY

A. Structure and Synthesis

Neotame is manufactured from aspartame and 3,3-dimethylbutyraldehyde via reductive alkylation followed by purification, drying, and milling. The *N*-alkyl component of neotame, 3,3-dimethylbutanoic acid, occurs naturally in goat

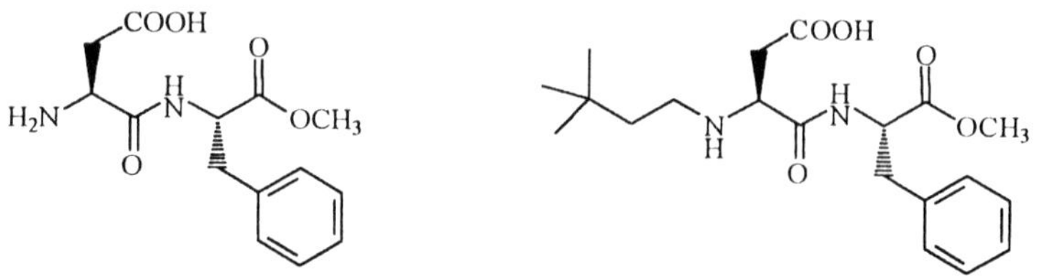

Aspartame

L-α-aspartyl-L-phenylalanine 1-methyl ester

Neotame

N-[N-(3,3-dimethylbutyl)-L-α-aspartyl]-L-phenylalanine 1-methyl ester

Figure 1 Comparison of chemical structures of neotame and aspartame.

cheese (3). The chemical structures of aspartame and neotame are compared in Fig. 1.

B. Dry Stability

Dry bulk chemical stability of neotame has been demonstrated for up to 4 years in studies that are continuing for a total of 5 years. Dry neotame is extremely stable, with the major degradation product being de-esterified neotame; de-esterified neotame is formed at extremely low levels by the simple hydrolysis of the methyl ester group from neotame. De-esterified neotame is also the major aqueous degradant product and the major in vivo metabolite of neotame. More strenuous conditions involving higher temperatures and humidity result in the formation of increased amounts of de-esterified neotame without significant amounts of other degradation products; thus, neotame maintains an excellent material balance even under adverse conditions. Fluorescent lighting and polyethylene packaging have no effect on the stability of neotame. Stability studies have confirmed that products containing dextrose, maltodextrin, and neotame are stable when stored for extended periods of time at relevant storage conditions of ambient temperature and humidity.

C. Stability in Food Applications

Neotame is stable under conditions of intended use as a sweetener across a wide range of food and beverage applications. Neotame's stability has been assessed through a systematic approach and demonstrated to be similar to that of aspartame with the exception of neotame's greater stability in baked and dairy goods. A diketopiperazine derivative is not formed.

Neotame maintains stability and functionality over a range of pHs, temperatures, and storage times representing relevant conditions of use. Traditionally, functionality and stability have been demonstrated for high-intensity sweeteners by developing data for a large number of categories. This approach generated extensive redundant data. For example, aspartame stability work was done for more than 30 different food categories, even though many of the categories have similar processing and storage conditions. A systematic, chemistry-based approach to stability testing was therefore established for neotame as determined by the key chemical and physical parameters that have an impact on the stability and functionality of a food additive (4).

1. The Matrix System for Stability Determinations

Functionality and chemical stability studies establish that neotame is functional as a sweetener and flavor enhancer in a variety of food applications when used in accordance with good manufacturing practices. The functionality of neotame was demonstrated with a three-dimensional food matrix representing the intended conditions of use in foods. On the basis of the experience with aspartame and knowledge of the chemistry of neotame, the three key properties of this food matrix are temperature, pH, and moisture; these determine the stability of neotame and thus its functionality under intended conditions of use. Studies on the functionality of aspartame have validated this matrix approach as a predictor of its stability and functionality in foods within the defined matrix. Similarly, if neotame is functional in foods at the edges of the matrix, it will be functional for all food applications that are within the limits of the matrix. The food applications tested in the matrix under commercially relevant processing and storage conditions are carbonated soft drinks, powdered soft drinks, baked goods (cake), yogurt, and hot-packed beverages. These foods represent the ranges of temperature, pH, and moisture relevant to neotame applications. As predicted by its chemistry, neotame's functionality and stability are similar to that of aspartame with the exception of greater stability in baked and dairy goods. The representative positions of these categories within the matrix are shown in Fig. 2.

Carbonated soft drinks (CSD), powdered soft drinks (PSD), cake, yogurt, and hot-packed still beverages make up most commercial applications for high-intensity sweeteners. Taken together, these five representative products along with tabletop products account for greater than 90% of all uses for a high-intensity sweetener. Thus, these formulations were investigated for stability over the expected shelf-life of the products. Additional applications selected to assess func-

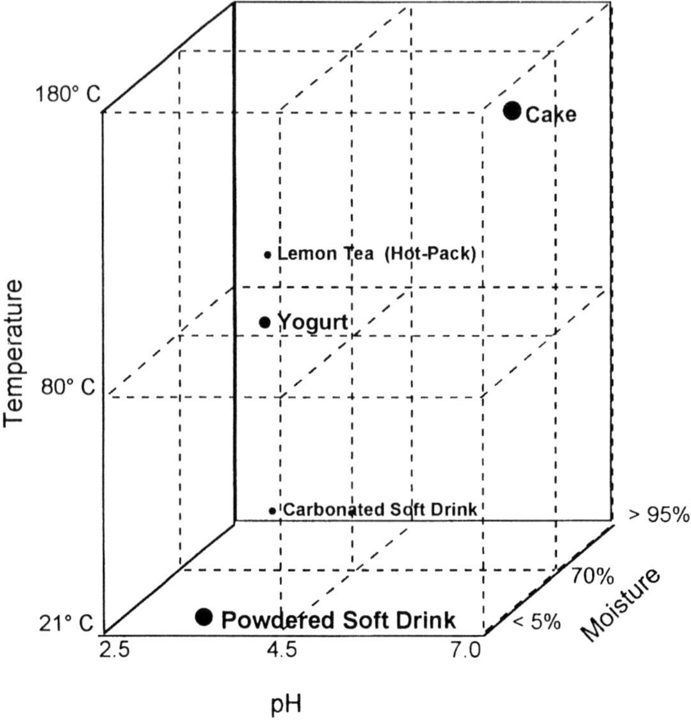

Figure 2 Matrix model for neotame applications.

tionality and stability of neotame included powdered tabletop preparations and chewing gum.

The greatest stability of neotame is in low moisture products such as tabletop preparations and PSD mixes; in these formulations, neotame shows little or no degradation after significant storage times. Neotame is also stable after normal commercial processing and storage times for products represented in the matrix such as hot-packed still beverages, cake, yogurt, and CSD.

Results for the product matrix model demonstrate that neotame is functional and stable as a sweetener in a wide range of product applications. The products included in the matrix were chosen to represent the wide range of chemical and physical conditions (e.g., pH, temperature, moisture) provided during the manufacture, packaging, and storage of commercially produced food and beverage products. In all products assessed in the matrix model, the stability of neotame was demonstrated to be comparable or improved relative to aspartame.

D. Degradation Products

1. Neotame is Not Subject to Diketopiperazine (DKP) Formation

Degradation studies demonstrate excellent mass balance for neotame. No DKP is formed from the intramolecular cyclization of the dipeptide moiety of neotame because of the presence of the *N*-akyl substitution on the aspartyl amino group. This results in excellent stability of neotame in baking applications. The possible formation of Maillard reaction products or in vitro nitrosation of neotame was also assessed to be negligible. Stability studies have confirmed that products containing dextrose, maltodextrin, and neotame are stable when stored for extended periods of time at relevant storage conditions of ambient temperature and humidity. Neotame is similarly inert to a number of food components such as flavoring agents and reducing sugars, including fructose.

2. De-esterified Neotame is the Major Degradant

The major route of degradation is the hydrolysis of the methyl ester moiety of neotame to form de-esterified neotame. De-esterified neotame is the only degradant formed to any extent and is also the major metabolite of neotame found in humans and animals. Under relevant conditions of use (pH 3.2 and 20°C), approximately 89% of neotame remained in mock beverage formulations after 8 weeks of storage. On the basis of product survey data, 90% of diet carbonated beverages are purchased and consumed within 8 weeks of production.

Carbonated soft drinks represent the largest use category for high-intensity sweeteners; for example, approximately 80% of all aspartame produced is used in carbonated soft drinks. The relative pattern for neotame use is not expected to deviate significantly from that of aspartame. Stability studies done with neotame in carbonated soft drinks at anticipated use levels did not result in detectable levels of degradants other than de-esterified neotame.

III. NEOTAME TASTE PROFILE

Neotame has a clean, sweet taste similar to sugar with no significant bitter, metallic, or other off-tastes. Moreover, this taste profile is maintained over the range of concentrations required in applications. Taste testing has shown that the sweetness of neotame increases as concentrations increase. Taste attributes other than sweetness remain at very low levels even as neotame concentrations increase.

A. Sweetness Intensity

1. Sweetness Related to Sucrose

"Sucrose equivalence" or "% SE" is the standardized sweetness intensity scale established in sweetener research with sucrose as the reference. An x% SE is equivalent in sweetness to an x% sucrose water solution. Neotame is functional as a sweetener in food applications at typical sweetness levels ranging from 3% SE (e.g., tabletop) to 10% SE (e.g., carbonated soft drinks). The concentration-response (C-R) curve for neotame in water was established by use of a trained sensory panel to evaluate the sweetness intensity of five solutions of neotame at increasing concentrations. The results are presented in Fig. 3.

The sweetness potency of neotame is many orders of magnitude greater than other high-intensity sweeteners. In water, at concentrations equivalent to

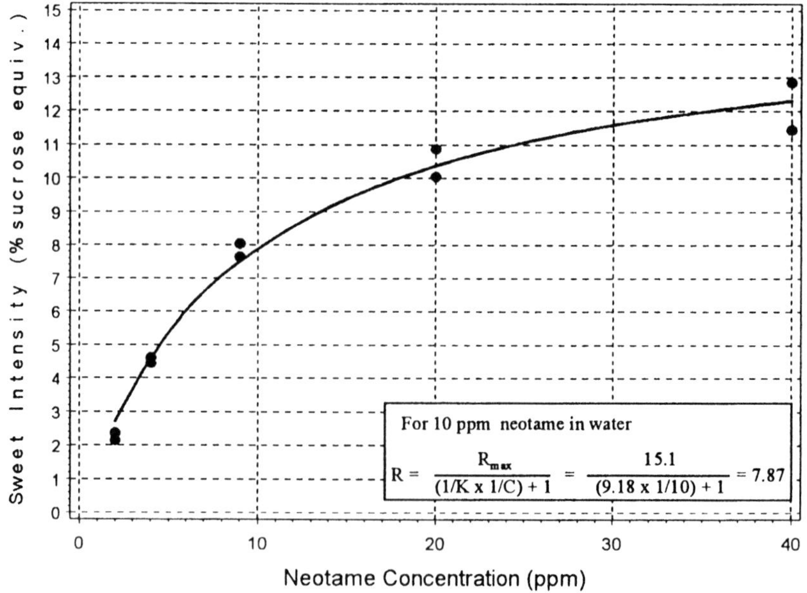

Figure 3 Neotame concentration-response curve in water. R = observed response; R_{max} = maximum observed response; C = sweetener concentration; 1/K = concentration that yields half-maximal response.

about 3% to 10% sucrose, neotame is 30 times sweeter than saccharin, about 60 to 100 times sweeter than acesulfame-K, and 300 to 400 times sweeter than cyclamate (5). Because of its remarkable sweetness potency, neotame will be used in food at considerably lower concentrations than other high-intensity sweeteners.

Neotame reaches a maximum sweetness intensity of 15.1% SE in water and 13.4% SE in a cola drink. In contrast, sweeteners such as acesulfame-K, cyclamate, and saccharin attain their maximum sweetness intensity (plateau) at, respectively, 11.6% SE, 11.3% SE, and 9% SE in water. Hence, neotame can be used as a stand-alone sweetener in a broader range of applications than can acesulfame-K, cyclamate, and saccharin, which are potentially limited in their use as sole sweeteners at higher concentrations (5).

2. Sweetness Intensity of Neotame

Neotame is 30 to 60 times sweeter than aspartame, depending on the sweetener application and desired intensity of sweetness. For example, sensory evaluation of neotame formulated in cola-flavored carbonated soft drinks indicates a sweetness intensity ratio 31 times that of aspartame in similar products.

3. Use Levels Associated with Flavoring Ingredients

As is the case with other high-intensity sweeteners, the high-intensity sweetness for neotame varies, depending on the specific food application and conditions of preparation. However, given its remarkable sweetness, neotame will be used in lower concentrations than other high-intensity sweeteners. In fact, consumer exposure to neotame is much lower than that for flavoring ingredients such as vanillin, cinnamon, and menthol, which are commonly used in foods and beverages.

B. Sensory Profile of Neotame Versus Sucrose

Neotame exhibits a clean taste profile at a number of use levels relevant to product applications. Neotame functions effectively as a sweetener and flavor enhancer in foods and beverages. A trained descriptive panel evaluated a number of different sweeteners, including neotame and sucrose, at comparable sweetness levels in water. The taste profile of neotame is similar to that of sucrose, with the predominant sensory characteristic of neotame being a very clean sweet taste. Other taste attributes such as metallic flavors, bitterness, sour, and salty are all below threshold levels. The flavor profile of neotame in water at a concentration of 10 mg/L compared with an 8% sucrose solution is shown in Fig. 4.

The sweetness temporal profile is a key to the functionality of a sweetener and is complementary to its taste profile. It demonstrates the changes in the perception of sweetness over time. Every sweetener exhibits a characteristic onset time of response and extinction time. Most high-intensity sweeteners display a

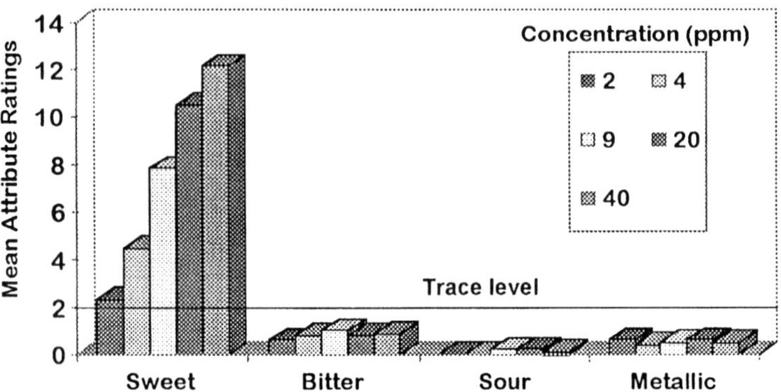

Figure 4 Taste profile of neotame at various concentrations in water.

significantly longer linger than sucrose. As shown in Fig. 5, the sweetness temporal profile of neotame in water is close to that of aspartame, with a slightly slower onset but without significantly longer linger.

Like other intense sweeteners, the temporal profile for neotame is application specific. In confectionery applications such as sugar-free chewing gum, neotame's sweet linger positively extends both sweetness and flavor. In other applications such as powdered soft drinks, the temporal parameters for neotame, specifically its onset and linger, are similar to those of sucrose. The onset and linger of sweetness can be modified by combining neotame with small amounts of commonly used flavorings or additives such as emulsifiers or acids. The ability to modify onset and extinction of sweetness by formulation is a distinct advantage and unique characteristic of neotame.

Neotame can be used as a sweetener either alone or blended with other sweeteners in foods and beverages. Sensory evaluations in carbonated soft drinks, powdered soft drinks, cakes, yogurt, hot-packed lemon tea, chewing gum, and as a tabletop sweetener in hot coffee and iced tea demonstrate the functionality of neotame in different food and beverage matrices.

C. Flavor-Enhancing Property

Flavor enhancers are substances that can be added to foods to supplement, enhance, or modify the original taste and/or aroma of a food without imparting a characteristic taste or aroma of its own (6). Sensory testing demonstrated flavor enhancement for neotame at very low subsweetening concentrations.

Flavor-enhancing properties combine with intense sweetness to provide unique functionality to neotame as a food additive. The flavor synergy of neotame in fruit-based drinks reduces the amount of juice required to maintain the mouth-

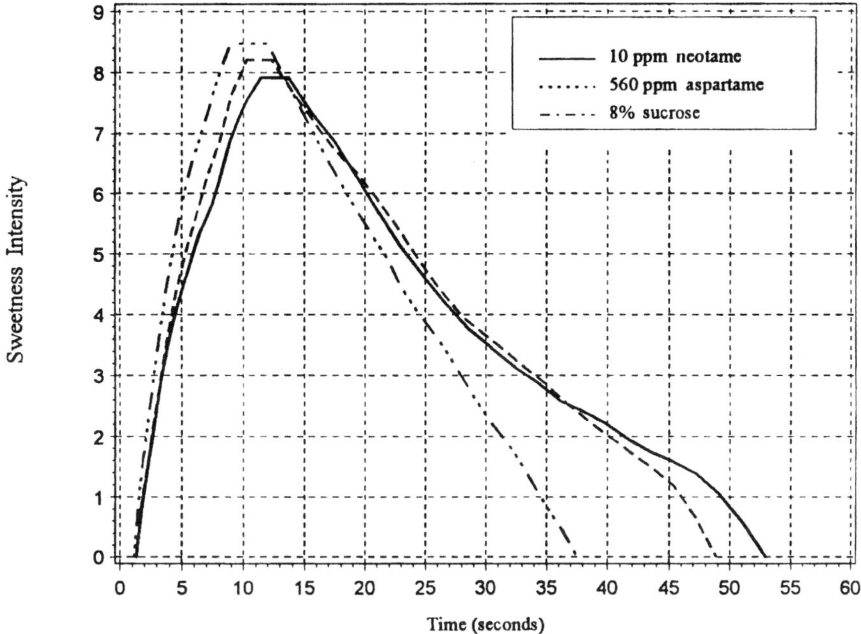

Figure 5 Comparative temporal profile of neotame vs. sugar and aspartame at isosweet concentrations in water.

feel of a higher juice level. Neotame maintains tartness, which allows a reduction in acid use. Neotame allows a significant reduction in mint flavoring required for chewing gum with no decrease in flavor intensity. Furthermore, neotame masks bitter tastes of other food ingredients and sweeteners. This unique combination of properties allows for both improved taste of products and a reduction in requirements for other added flavors.

IV. UTILITY IN FOOD APPLICATIONS AND BEVERAGE SYSTEMS

A. Solubility

The solubility of neotame as determined in water, ethyl acetate, and ethanol at a range of temperatures illustrates how neotame behaves in various food matrices. Neotame is soluble in ethanol at all temperatures tested. The solubility of neotame increases in both water and ethyl acetate with increasing temperature. This solubility may create opportunities for food and beverage manufacturers to use neo-

Table 1 Solubility of Neotame (grams/100 grams solvent)

Temperature	Water	Ethyl acetate	Ethanol
15°C	1.06	4.36	>100
25°C	1.26	7.70	>100
40°C	1.80	23.8	>100
50°C	2.52	87.2	>100
60°C	4.75	>100	>100

tame in a wide range of liquid delivery systems. Table 1 shows the solubility of neotame in water, ethyl acetate, and ethanol.

The potency, solubility, and stability of neotame offer new opportunities to design efficient delivery systems to handle neotame and to assure that it is evenly distributed in food applications. Neotame can be delivered in a wide range of product forms commonly used in food production processes including solid and liquid delivery systems.

B. Dissolution

Neotame dissolves rapidly in aqueous solutions because of the low use levels required for sweetening. For example, the dissolution rate of neotame was determined in water; the percentages of neotame dissolved from 50-mg samples stirred into 900 ml of deionized water at 37°C were determined by ultraviolet absorbance at 1 through 5 minutes. The results indicate that at 2 minutes, 93.0% of the total neotame had dissolved. Within 5 minutes, 98.6% of the total amount of neotame had dissolved.

C. Admixture Potential

Blending sweeteners to achieve synergy is a common industry practice. Synergy can be defined as occurring when the sweetness of a mixture of sweeteners is greater than the sum of their respective, individual sweetness intensities. Blends or combinations of two or more sweeteners are being used more frequently to achieve the desired level of sweetness in food and beverage products. Neotame is compatible in admixture with other sweeteners, including sugar. Blending neotame with sugar allows significant calorie reduction in products compared with sugar alone without compromising taste.

The taste profile of sweetened foods and beverages can be tailored to specific taste requirements by blending. Sweetener blends offer taste advantages such as lower off-flavors and more sucrose-like temporal profiles (quicker onset and shorter sweet linger) than individual sweeteners. In the same way, blends of neotame with other sweeteners will provide opportunities for food manufacturers to formulate better tasting products. The flavor enhancement quality of neotame, which masks bitter flavors even at subsweetening levels, makes neotame in admixture blends a desirable choice with other sweeteners that possess potentially undesirable or more complex taste profiles.

V. SAFETY

A. Projected Consumption/Exposure

The mean consumption of neotame for all users is estimated to be 0.02 mg/kg body weight/day. Consumption for 90th percentile users of neotame is estimated to be 0.05 mg/kg body weight/day. These anticipated consumption levels of neotame provide extremely wide margins of safety when evaluated in light of results from animal safety studies. Anticipated levels of neotame consumption are based on known 14-day estimates of aspartame consumption (7–9), a conservative sweetening intensity ratio for neotame of 31 times that of aspartame, and the potential for neotame to replace 50% of current market share for aspartame.

B. Safety Evaluation of Neotame

Neotame has been subjected to extensive investigations in toxicological studies to establish the safety of its use in foods (10, 11). Numerous studies, including subchronic, chronic, and special studies, have been done in rats, mice, dogs, and rabbits. In addition, long-term feeding studies were done in rodents to evaluate carcinogenic potential; teratogenicity studies were done in rats and rabbits to determine the potential for effects on the fetus; a two-generation study was done in rats to determine the potential for reproductive effects. Studies including bacterial and mammalian systems evaluated mutagenic potential of both neotame and its major metabolite. Chronic and carcinogenicity studies in the rat included in utero exposure to neotame, ensuring that animals were dosed throughout the duration of the studies from conception to study termination. Safety studies were done according to the general principles in current U.S. FDA guidelines (12) and other international guidelines.

Neotame was generally administered in the diet, because this route of exposure is the most relevant to human consumption. Diets were formulated and concentrations adjusted regularly to provide dosages in mg/kg/day rather than dosing at fixed percentages of neotame in the diet. Safety studies were performed with those strains for which the most complete databases were available; for example, study strains included the Sprague-Dawley–derived CD® rat, the CD-1 mouse, the beagle dog, and the New Zealand white rabbit. Microscopic examinations were done on tissues from all animals at all doses in all key toxicity studies. In addition, standard and supplemental parameters were evaluated for evidence of immunotoxicity or neurotoxicity. Findings in these studies are summarized in the following.

C. General Toxicology and Carcinogenicity

The safety of neotame was established in dietary studies in rats, mice, and dogs. These studies used a wide range of dose levels for neotame. Test species evaluations included clinical observations, body weight, body weight gain, food and water consumption, hematology, clinical chemistry, urinalysis parameters, ophthalmologic examinations, electrocardiograms (in dogs), and assessments of gross necropsy findings and microscopic evaluations.

Neotame was well-tolerated in all subchronic studies and was without adverse effects even at high doses (3000 mg/kg/day in the rat, 8000 mg/kg/day in the mouse, and 1200 mg/kg/day in the dog). There was no test article-related mortality, and there were no changes in appearance or behavior at any time during these subchronic studies at dietary doses 24,000 to 160,000 times the anticipated 90th percentile level of human consumption.

Food refusal or spillage was observed when neotame was present at higher concentrations in the diet. Diets containing extremely high levels of neotame (providing 1200 to 8000 mg/kg/day) were immediately disliked and partially rejected. The immediate decrease in food consumption generally followed by accommodation to near control values suggests poor palatability of the diet. Furthermore, the effect of neotame on food consumption was a function of the concentration of neotame in food and not the dosage of neotame consumed. For example, dogs consumed a higher dosage of neotame when offered a diet containing 3.5% neotame than when offered a diet containing 5% neotame. Special dietary preference studies in rats confirmed that sufficiently high concentrations of neotame reduced palatability of diet. Poor palatability of neotame-containing diet was observed, particularly in rats and mice, as small but consistent reductions in food consumption as animals approached the body weight plateau of adulthood.

One-year dietary studies were done in rats and dogs; 2-year carcinogenicity studies were done in rats and mice. In rats, both the 1-year and carcinogenicity studies included parental exposure before mating and in utero exposures for offspring; thus, animals used in safety studies were exposed to neotame from conception, during prenatal and postnatal development, and throughout the duration of the studies. The highest maximum tolerated doses (MTD) in the 1-year studies were 1000 mg/kg/day in rats and 800 mg/kg/day in dogs. There was no target organ toxicity in chronic studies in either the rat or dog. In addition, there was no evidence of target organ toxicity or carcinogenicity when neotame was administered for 2 years to rats with in utero exposure at dosages up to 1000 mg/kg/day, or to mice at dosages up to 4000 mg/kg/day. All nonneoplastic findings were consistent with those expected of aging rodents.

The no-observed adverse effect levels (NOAELs) were the highest doses tested in definitive toxicology studies in rats, mice, and dogs. On the basis of the results of the chronic toxicity and carcinogenicity studies, the NOAEL is at least 1000 mg/kg body weight/day in rats, 4000 mg/kg body weight/day in mice, and 800 mg/kg body weight/day in dogs. Given that consumer exposure to neotame even at the 90th percentile is so small, margins of safety based on each of these species are tens of thousands times greater than projected human exposure.

D. Reproduction and Teratology

Sprague-Dawley rats were fed doses of neotame up to 1000 mg/kg for 10 weeks in males and 4 weeks in females before pairing. Dosing continued throughout pairing, gestation, and lactation to weaning of the F_1 offspring at day 21 of age. At approximately 4 weeks of age, animals were selected for the F_1 generation and were exposed for 10 more weeks before being bred to produce the F_2 generation. The F_2 generation was raised to day 21. Consumption of high doses of dietary neotame by animals through two successive generations did not result in any effects on reproductive performance, postnatal development, or development of the embryo or fetus.

In a teratology study, female Sprague-Dawley rats were fed a neotame-containing diet to provide doses up to 1000 mg/kg/day for 4 weeks before pairing and throughout gestation until day 20 after mating. There were no effects on clinical signs, food consumption, body weights, or weight gains or any evidence of treatment-related effects on the dams or on their litters. Examination of the fetuses for external, visceral, and skeletal abnormalities revealed no fetotoxic or teratogenic effects.

No teratogenicity occurred in rabbits dosed with neotame up to 500 mg/kg/day by oral gavage between days 6 and 19 after mating. There were no test article–related clinical observations or postmortem effects in dams or in litters. There were no teratogenic effects of neotame on fetal examination. The rat and rabbit reproductive and teratology studies demonstrated that neotame even at high doses had no effect on reproduction or development of the embryo or fetus.

E. Genetic Toxicology

Neotame was not mutagenic in the Ames assay in five strains of *Salmonella typhimurium* and a strain of *Escherichia coli* when tested at concentrations of up to 10,000 μg/plate with or without metabolic activation with liver S9 fraction from Aroclor-induced rats. Neotame produced no evidence of mutagenic activity in the mouse lymphoma cell gene mutation assay with and without metabolic activation at concentrations ranging up to 800 μg/ml. No chromosomal aberrations were detected in Chinese hamster ovary cells after exposure to concentrations of neotame up to 500 μg/mL without metabolic activation and up to 1000 μg/mL with metabolic activation.

Neotame was administered by oral gavage to mice at doses up to 2000 mg/kg body weight in a bone marrow micronucleus assay. Treatment with neotame did not induce any changes in the ratio of polychromatic erythrocytes to total erythrocytes or in the frequency of micronucleated polychromatic erythrocytes in this assay system. No deaths, clinical signs of toxicity, or changes in body weight occurred.

In addition to the mutagenicity evaluations with neotame, the major in vivo metabolite that is also the major in vitro degradant, de-esterified neotame, was not mutagenic in the Ames/*Samonella* assay and the xanthine-guanine phosphoribosyl transferase mutation assay in Chinese hamster ovary cells. On the basis of the results of in vivo and in vitro genotoxicity testing, neither neotame nor de-esterified neotame, the major metabolite and degradant, has the potential for mutagenicity or clastogenicity.

F. Metabolism

Data from both humans and animals show that neotame has an excellent metabolic and pharmacokinetic profile. All metabolites identified in humans were also present in species used in safety studies. Neotame is rapidly, but incompletely, absorbed in all species. Radiolabeled doses of neotame are completely eliminated. The major route of metabolism is to de-esterified neotame. Methanol is produced in equimolar quantities during the formation of de-esterified neotame whether through metabolism or degradation; the maximum amount of methanol exposure at projected 90th percentile consumption of neotame is negligible compared with levels of methanol considered to be safe (13). Neotame does not induce liver microsomal enzymes. On the basis of studies with albumin, neither neotame nor de-esterified neotame is likely to interfere with binding to plasma proteins.

Neotame and de-esterified neotame have short plasma half-lives with rapid and complete elimination. Peak plasma concentrations of neotame and de-esterified neotame after an oral dose occur approximately at 0.5 hour and within 1 hour, respectively. Neither neotame nor de-esterified neotame accumulate after repeated dosing in humans or toxicology species. Absorbed neotame is rapidly excreted in the urine and feces. In radiolabeled studies, most of the total radioactivity was excreted as de-esterified neotame in the feces in all species and likely reflects largely unabsorbed neotame in the gastrointestinal tract. A comparison of metabolism and pharmacokinetic data in animals and humans demonstrates that the species used in the toxicology studies are relevant for predicting human safety.

G. Overview of Clinical Studies

The results from clinical studies demonstrate that neotame is safe and well tolerated in humans even in amounts well above projected chronic consumption levels. These clinical studies included single doses ranging from 0.1 to 0.5 mg/kg body weight, repeated doses of 0.25 mg/kg hourly for 8 hours totaling 2 mg/kg/day, and daily doses including 1.5 mg/kg/day for up to 13 weeks in healthy populations, including male and female subjects and doses of 1.5 mg/kg/day in a population of male and female individuals with non-insulin-dependent diabetes mellitus (type 2). Specifically, there were no clinically significant or neotame-related changes in any vital signs, electrocardiogram results, or ophthalmologic examinations. There were no neotame-related observations of clinical significance in any biochemical, hematological, physiological, or subjective findings. There were no statistically significant differences between neotame and placebo treatments in reported adverse experiences. Furthermore, neotame had no significant effect on plasma glucose or insulin concentrations or glycemic control in the population with non-insulin-dependent diabetes mellitus.

The 0.25 mg/kg hourly dose in the repeated dose study is equivalent to consuming about 1 liter of beverage sweetened with 100% neotame every hour for an 8-hour period. The high dose of 1.5 mg/kg/day in the 13-week study is equivalent to consuming about 6 liters of beverage sweetened with neotame daily for 13 weeks. Thus, the results of the clinical studies with neotame with doses up to 40 times the projected 90th percentile consumption clearly confirm the safety of neotame for use by the general population.

VI. PATENTS

Neotame was invented by the French scientists Claude Nofre and Jean-Marie Tinti and patented in 1992 (1, 2). The NutraSweet Company holds the rights to a wide range of patents related to neotame.

VII. REGULATORY STATUS

Neotame is currently undergoing regulatory review for use as a general sweetener and flavor enhancer. A petition limited to tabletop use was submitted to the U.S. Food and Drug Administration (FDA) in December, 1997 (10), and a petition for general use of neotame was submitted to FDA in December, 1998 (11). Dossiers for the registration of neotame are being submitted to various regulatory agencies worldwide.

VIII. CONCLUSION

Neotame is a new, noncaloric, high-intensity sweetener and flavor enhancer. Neotame is structurally related to and chemically derived from aspartame but is 30 to 60 times sweeter. Neotame is 7000 to 13,000 times sweeter than sucrose. Neotame should not require special labeling for phenylketonuria and does not degrade to a diketopiperazine. Neotame can favorably modify and enhance flavor and taste at or below sweetening levels. The small amounts of neotame required for sweetening will reduce the cost of delivering and handling sweeteners.

On a sweetness equivalency basis versus existing sweetener alternatives, neotame offers the potential to deliver improved cost structure because of its high sweetness potency and low levels of use.

Consumer exposure to neotame in foods is estimated to be 0.05 mg/kg/day at the 90th percentile level of consumption. Clinical studies have demonstrated neotame to be safe and well tolerated in amounts up to 40 times the projected 90th percentile of use. In addition, neotame does not alter glycemic control in subjects with non-insulin-dependent diabetes mellitus. Margins of safety for neotame in the various species used for safety testing are tens of thousands fold greater than estimated human exposure levels. Thus, the very large safety margins in animals and the demonstrated safety in humans at multiples many times 90th percentile consumption levels establish the safety of neotame for its intended use as a sweetener and flavor enhancer. The technical and functional qualities of neotame make this new high-intensity sweetener and flavor enhancer desirable in a wide variety of food and beverage preparations.

REFERENCES

1. C Nofre, JM Tinti. In quest of hyperpotent sweeteners. In: M Mathlouthi, JA Kanters, GG Birch, eds. Sweet-Taste Chemoreception. London: Elsevier, pp 205–236, 1991.
2. D Glaser, J Tinti, C Nofre. Evolution of the sweetness receptor in primates. I. Why does alitame taste sweet in all prosimians and simians, and aspartame only in Old World simians? Chem Senses 20(5):573–584, 1995.
3. A Pierre, JL Le Quere, MH Famelart, A Riaublanc, F Rousseau. Composition, yield, texture and aroma compounds of goat cheeses as related to the A and O variants of alpha s1 casein in milk. Lait 78:291–301, 1998.

4. MW Pariza, SV Ponakala, PA Gerlat, S Andress. Predicting the functionality of direct food additives. Food Technol 52(11):56–60, 1998.
5. GE DuBois, DE Walters, SS Schiffman, ZS Warwick, BJ Booth, SD Pecore, K Gibes, BT Carr, LM Brands. Concentration-response relationships of sweeteners. In: DE Walters, FT Orthoefer, GE DuBois, eds. Sweeteners Discovery, Molecular Design, and Chemoreception. American Chemical Society, Washington DC, 1991.
6. FDA. Flavor enhancers. 21 CFR 170.3(o)(11). 1999.
7. HH Butchko, FN Kotsonis. Acceptable daily intake vs actual intake: the aspartame example. J Am Coll Nutr 10(3):258–266, 1991.
8. HH Butchko, C Tschanz, FN Kotsonis. Postmarketing surveillance of food additives. Reg Toxicol Pharmacol 20:105–118, 1994.
9. HH Butchko, FN Kotsonis. Acceptable daily intake and estimation of consumption. In: C Tschanz, HH Butchko, WW Stargel, FN Kotsonis, eds. The Clinical Evaluation of a Food Additive: Assessment of Aspartame. Boca Raton, FL: CRC Press, 1996, pp 43–53.
10. FDA. Monsanto Co.: Filing a food additive petition. Federal Register 63(27):6762, 1998.
11. FDA. Monsanto Co.: Filing a food additive petition. Federal Register 64(25):6100, 1999.
12. FDA. Toxicological principles for the safety assessment of direct food additives and color additives used in food: "Redbook II." Washington, DC: Food and Drug Administration, 1993.
13. FDA. Food additives permitted for direct addition to food for human consumption; dimethyl dicarbonate. Federal Register 61:26786–26788, 1996.

9
Saccharin

Ronald L. Pearson
PMC Specialties Group, Inc., Cincinnati, Ohio

The name saccharin is aptly derived from the Latin *saccharum* for sugar. It is commercially available in three forms: acid saccharin, sodium saccharin, and calcium saccharin. These forms have been variously determined to be 200–800 times sweeter than sucrose, depending on the saccharin concentration (1). Coupled with its low cost, this intense sweetness makes saccharin, in all its forms, the most cost-effective of all currently commercialized sweeteners. The American manufacturing process uses as its starting material purified, commercially manufactured methyl anthranilate, an ester that is found in a number of fruit juices, notably grape. As a pioneer alternative sweetener, saccharin has had a difficult history, requiring considerable persistence and effort to overcome its detractors. It is the only sugar substitute that has been in use for more than a century and worldwide is the largest volume alternative sweetener produced. Saccharin has withstood the test of time and has earned a rightful place in our food and beverages.

I. DISCOVERY

Saccharin was discovered by chemists Ira Remsen and Constantine Fahlberg in May 1878. Fahlberg had received his doctorate at Leipzig in 1873 and was visiting Johns Hopkins University in Baltimore. There, under Professor Remsen's tutelage, he worked on the oxidation of *o*-toluene-sulfonamide as part of Remsen's ongoing research project (2). To the surprise of both chemists, an elemental analysis revealed that the oxidation product was the condensed heterocycle *o*-sulfobenzimide rather than the expected *o*-sulfamoylbenzoic acid. Several weeks later, Fahlberg discovered the intense sweetness of the new compound after spilling a solution on his hands and later eating bread at dinner. This serendipity was duly noted in the investigators' original report (3): ''Sie schmeckt angenehm süss, sogar süsser als der Rohrzucker.'' (It tastes pleasantly sweet, even sweeter than cane sugar.)

Remsen had a personal disdain for commercial ventures. However, Fahlberg, an experienced sugar chemist, aggressively pursued the commercial potential of the new compound. He named it ''Fahlberg's saccharin'' and obtained a U.S. patent (4) without informing Remsen. This professional rudeness was a source of extreme irritation to Remsen, who summarized his feelings in 1913: ''I did not want (Fahlberg's) money, but I did feel that I ought to have received a little credit for the discovery'' (2).

II. EARLY PRODUCTION

Fahlberg displayed the sweetening power of saccharin in a London exposition in 1885. This saccharin was manufactured in a pilot plant Fahlberg operated in New York. Two German patents were issued in 1886 to Fahlberg; these were assigned to Fahlberg, List and Company. The plant, under the name of Saccharinfabrik A.G., was moved to Westerhusen, Germany. In 1900, Fahlberg reported an annual production of 190,000 kg. In 1902, partly at the insistence of beet sugar producers, saccharin production in Germany was brought under strict control, and saccharin was attainable only through pharmacies (5).

John F. Queeny became acquainted with saccharin sometime in the late 1890s. At that time, he was the purchasing agent for Meyer Brothers Drug Company in St. Louis, Missouri, which was importing the sweetener from Germany. Queeny tried to persuade his employer to manufacture the sweetener rather than import it, but Meyer Brothers did not foresee saccharin as a profitable venture (6). In 1901, Queeny took his personal savings of $1,500 and, together with $3,500 obtained from Liquid Carbonic, he began to manufacture saccharin himself. He chose his wife's maiden name in naming his new venture the Monsanto Chemical Company.

Early production of saccharin was a touch-and-go proposition with rickety old equipment. Furthermore, saccharin was targeted by several antagonists, the most aggressive of which was the German cartel known as the Dye Trust. The Dye Trust waged a price war and cut the price from $4.50 a pound to $1.00 a pound (6).

Opposition to saccharin also arose from domestic sources. A prominent food editor of the time, Alfred W. McCann, wrote, ''Saccharin is as false and scarlet as the glow of health transferred from the rouge pot to the cheek of a baud'' (6). Saccharin was also criticized for having no food value or calories, the very characteristic that distinguished it from sugar. Saccharin remained approved for use in the United States partly through the influence of President Theodore Roosevelt, who was enraged when told by a federal health official that it might be banned. The president was quoted in 1906 as saying that "anyone who says saccharin is injurious to health is an idiot!" (7). Support from the president continued, with Mr. Queeny receiving the following letter, dated July 7, 1911:

> I always completely disagreed about saccharin both as to the label and as to its being deleterious . . . I have used it myself for many years as a substitute for sugar in tea and coffee without feeling the slightest bad effects. I am continuing to use it now: Faithfully yours, T. Roosevelt (7).

Saccharin use increased during World War I and immediately thereafter as a result of sugar rationing. This was particularly evident in Europe (8). By 1917, saccharin was a common tabletop sweetener in America and Europe. The sweetener was introduced to the Far East in 1923. The consumption of saccharin continued between the World Wars, with an increase in the number of products in which it was used. The shortage of sugar during World War II again produced a significant increase in saccharin use, firmly establishing a need for this alternative sweetener.

In the early 1950s, the Maumee Chemical Company began to manufacture saccharin by a novel process developed by O. F. Senn and G. F. Schlaudecker (9, 10), who helped found Maumee Chemical Company. Maumee's introduction of calcium saccharin as a alternate soluble form soon followed. In 1966, the Sherwin Williams Company acquired Maumee and sold this business to PMC Specialties Group, Inc. in 1985.

III. CURRENT PRODUCTION

A number of companies around the world manufacture saccharin. Many manufacturers use the basic synthetic route described by Remsen and Fahlberg (Fig. 1). Toluene is treated with chlorosulfonic acid to produce ortho- and para-toluenesulfonyl chloride. Subsequent treatment with ammonia forms the corresponding toluenesulfonamides. The ortho-toluenesulfonamide is separated from its para isomer (this separation is alternatively performed on the sulfonyl chlorides). The *ortho*-toluenesulfonamide is then oxidized to ortho-sulfamoylbenzoic acid, which on heating is cyclized to saccharin (5).

PMC Specialties Group, the sole U.S. producer, currently uses improved Maumee chemistry, producing saccharin from purified, manufactured methyl anthranilate, a substance naturally occurring in grapes (11).

The Maumee process originally diazotized anthranilic acid by adding an equivalent of sodium nitrite and sulfuric acid (Fig. 2). The resulting diazo solution

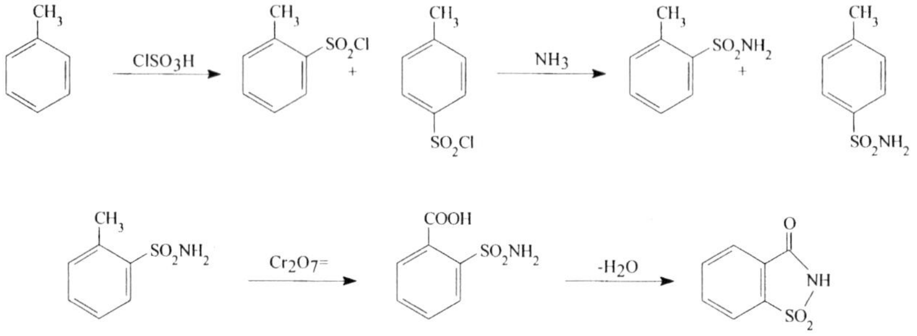

CH_3 $ClSO_3H$ CH_3 SO_2Cl + CH_3 SO_2Cl NH_3 CH_3 SO_2NH_2 + CH_3 SO_2NH_2

CH_3 SO_2NH_2 $Cr_2O_7^{=}$ COOH SO_2NH_2 $-H_2O$ O NH SO_2

Figure 1 Remsen-Fahlberg process.

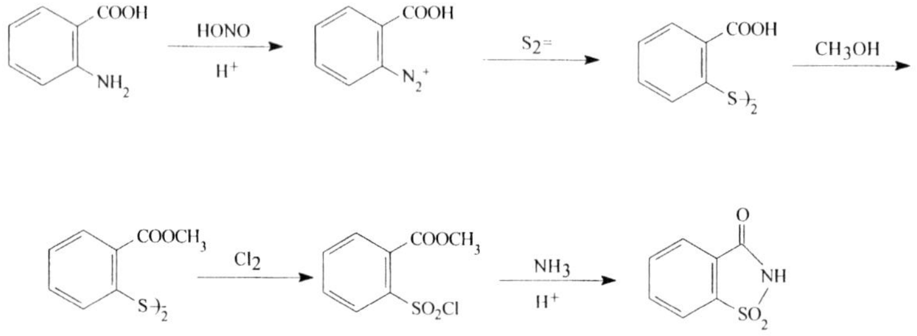

COOH NH_2 HONO H^+ COOH N_2^+ $S_2^{=}$ COOH $S)_2$ CH_3OH

$COOCH_3$ $S)_2$ Cl_2 $COOCH_3$ SO_2Cl NH_3 H^+ O NH SO_2

Figure 2 Original Maumee process.

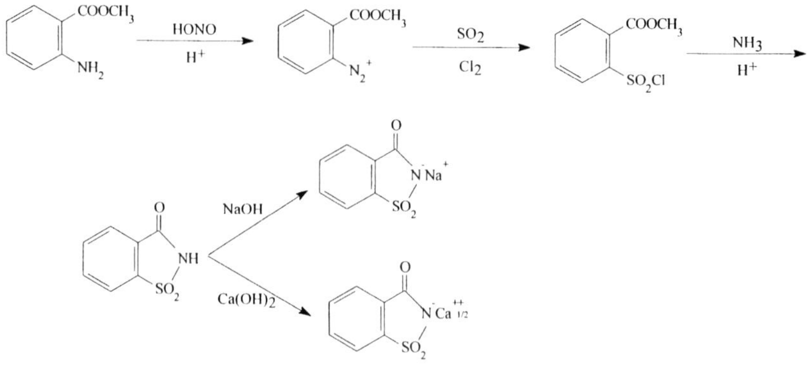

$COOCH_3$ NH_2 HONO H^+ $COOCH_3$ N_2^+ SO_2 Cl_2 $COOCH_3$ SO_2Cl NH_3 H^+

O NH SO_2 NaOH O N^-Na^+ SO_2 $Ca(OH)_2$ O $N^-Ca^{++}_{1/2}$ SO_2

Figure 3 Continuous Maumee process.

was added to sodium disulfide; the product was collected, washed, and dried. The disulfide was then esterified by treatment with methanol in sulfuric acid and oxidized with gaseous chlorine to *ortho*-carbomethoxybenzenesulfonyl chloride. Amidation with excess ammonia yielded ammonium saccharin, and neutralization with sulfuric acid produced insoluble acid saccharin (12).

Today, the Maumee process has evolved into an efficient continuous process (Fig. 3). Methyl anthranilate is diazotized to form 2-carbomethoxybenzenediazonium chloride. Sulfonation followed by oxidation yields 2-carbomethoxybenzenesulfonyl chloride. Again, amidation of this sulfonylchloride, followed by acidification, forms insoluble acid saccharin. Subsequent addition of sodium hydroxide or calcium hydroxide produces the soluble sodium and calcium forms, respectively (13).

IV. PROPERTIES

Acid saccharin exists as a white odorless, crystalline powder. It is a moderately strong acid, with only slight solubility in water. However, its intense sweetness allows even slight aqueous solubility to be quite sufficient for all sweetener applications. There are three commercially available saccharin forms: (a) acid, (b) sodium, and (c) calcium. Sodium saccharin is the most commonly used form because of its high solubility, high stability, and superior economics. All are

Table 1 Properties of Saccharin Forms

	Acid saccharin	Sodium saccharin	Calcium saccharin
Molecular formula	$C_7H_5NO_3S$	$C_7H_4NO_3SNa \cdot 2H_2O$	$[C_7H_4NO_3S]_2Ca \cdot 2H_2O$
Molecular weight	183.18	241.20	440.48
C.A.S. Registry No.	81-07-2	128-44-9	6485-34-3
Melting point (°C)	228–229	>300	>300
Appearance	White solid	White solid	White solid
pKa[a]	1.30	—	—
Solubility (g/100 g water) at:[b]			
20°C	0.2	100	37
35°C	0.4	143	82
50°C	0.7	187	127
75°C	1.3	254	202
90°C	—	297	247

[a] From Ref. 15.
[b] From Refs. 16, 44.

manufactured to meet Food Chemicals Codex and U.S. Pharmacopeia/National Formulary specifications (14a,b). Their properties are listed in Table 1.

Several additional salts of saccharin have been reported. These include silver, ammonium, cupric, lithium, magnesium, zinc, and potassium. Although all of these are intensely sweet, none is available commercially. Substitutions on the nitrogen of saccharin eliminates its sweetness, whereas carbon substitution gives unpredictable sweetness results (17a,b), although no substituent imparts a greater sweetness than that of saccharin itself (18).

X-ray crystallography has shown that the acid form of saccharin exists as dimers formed by hydrogen bonding between the imide hydrogen and the keto oxygen (19a,b).

V. SWEETENING POWER AND ADMIXTURE POTENTIAL

Saccharin is approximately 300 times as sweet as sugar dissolved in water at 7 wt% concentration. Coupled with sodium saccharin's low price (approximately \$3.00/lb), this sweetening factor means that a penny's worth of saccharin has the sweetening power of 1 pound of sugar. The relative economics of current commercial sweeteners are compared in Table 2, showing that all the saccharin forms are by far the least expensive per sweetness equivalent.

In 1921, Paul observed that saccharin's sweetening power relative to sucrose increases with decreasing concentration (20). Recent data were used in Fig. 4, which again demonstrates the dramatic increase in sweetening potency, relative to sucrose, with decreasing sweetener concentration (1). He found a similar concentration effect with dulcin, another alternative sweetener used at the time, but that is no longer approved. Paul made the remarkable observation that when these two sweeteners are blended, each sweetener retains its higher sweetening power in the presence of the other. For example, to achieve a sweetness equal to a 9% sucrose solution, individually 450 mg/l of saccharin or 1250 mg/l of dulcin is required. But this 9% sucrose solution is also equaled with a blend containing 190 mg/l of saccharin plus 120 mg/l of dulcin, only 310 mg/l in total. Individually, the 190 mg/l of saccharin is equivalent to a 6% sucrose solution, whereas the 120 mg/l of dulcin is equivalent to a 3% sucrose solution. Thus by blending the two sweeteners, Paul was able to "add" their individual sweetening equivalencies and achieve the same sweetness with less total sweetener. This effect is more completely demonstrated by Paul's data presented in Table 3.

Today, this phenomenon of increasing sweetening power with decreasing concentration has been established for many sweeteners (21, 22), and additive effects have been reported for blends of saccharin with several other sweeteners. In addition, the blend is frequently sweeter than would be predicted by the additive effect. This enhanced effect is termed "synergistic." Table 4 lists various

Table 2 Economics of Commercial Sweeteners

Sweetener	Sweetness/lb (1) compared with sucrose at a concentration of 3 wt%	7 wt%	Cost ($/lb)	Relative cost $/sweetness sodium saccharin = 1.0 3%	7%
Sodium saccharin $2H_2O$	500	300	2.75[a]	1.0	1.0
Acid saccharin	—	370	5.00[a]	—	1.5
Calcium saccharin $2H_2O$	500	300	5.25[a]	1.9	1.9
Sodium cyclamate	29	29	2.27[b]	14.2	8.5
Aspartame	231	159	30.00[b]	23.6	20.6
Acesulfame K	188	97	37.00[b]	35.8	41.6
HFCS[c] (55%)	1.0	1.0	0.25[b]	45.5	27.3
Sucralose	664	558	—[d]	—	—
Alitame	4170	2690	—[d]	—	—
Sucrose	1.0	1.0	0.36[e]	65.5	39.3

[a] PMC Specialties Group, Fall/1998.
[b] Typical price, Fall/1998.
[c] High fructose corn syrup.
[d] Currently not being marketed, Fall/1998.
[e] Chemical Marketing Reporter, 12/7/98.

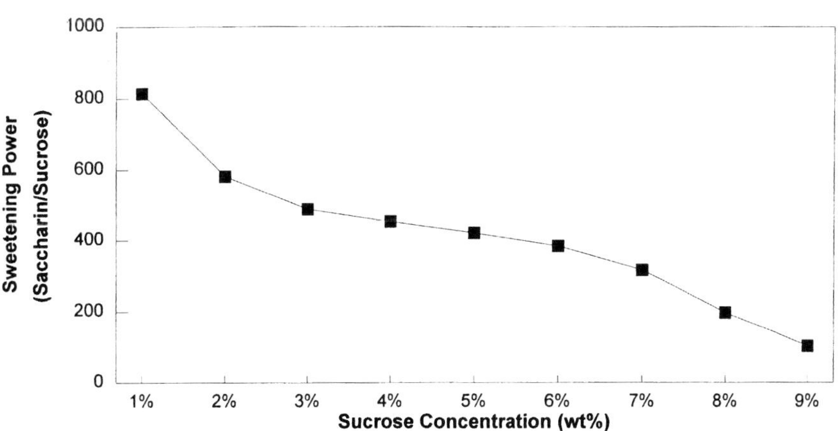

Figure 4 Sweeting power of saccharin vs. concentration. (From Ref. 1.)

Table 3 Additive Effect of Saccharin-Dulcin Blends

Sucrose concentration		Saccharin		Dulcin		Blend		
(g/l)	(weight %)	Conc. (mg/l)	Sweetening power (saccharin/sucrose)	Conc. (mg/l)	Sweetening power (dulcin/sucrose)	Saccharin (mg/l)	Dulcin (mg/l)	Total (mg/l)
10	1	20	500	30	333	—	—	—
20	2	30	667	55	364	—	—	—
30	3	55	545	120	250	—	—	—
40	4	100	400	290	138	55	30	85
50	5	150	333	450	111	55	55	110
60	6	190	316	665	90	100	55	155
70	7	280	250	855	82	150	55	205
80	8	370	216	1050	76	190	55	245
90	9	450	200	1250	72	190	120	310
100	10	535	187	1430	70	280	120	400

Solutions were of equivalent sweetness.
Source: Ref. 20.

Table 4 Saccharin Admixture Potential

Binary mixture saccharin +	Sweetness enhancement	Reference
Aspartame	Synergistic	21
Acesulfame K	Additive	21
Cyclamate	Synergistic	21
Sucralose	Synergistic	23
Alitame	Synergistic	21
Sucrose	Synergistic	21
Fructose	Synergistic	21
Dulcin	Additive	20

binary blends with saccharin and their reported effect on sweetness intensity. Whether synergistic or merely additive, sweetener blends provide an obvious economic advantage.

Typically, saccharin is formulated with either other sweeteners or masking agents to avoid a bitter aftertaste perceived by certain individuals only at higher concentrations. Many high-intensity alternative sweeteners follow a similar trend of becoming more bitter as their concentration increases (24). Fortunately, the bitter taste threshold concentration is higher than the sweet taste threshold concentration (24). In addition, the bitter threshold concentration for saccharin (25) depends on the formulation constituents as does its sweet taste threshold concentration. Only 25% of the population is able to detect this aftertaste (26), which is apparently intrinsic to the saccharin molecule itself rather than to the presence of a potent impurity (27). This use of multiple sweeteners allows the food formulator to take advantage of (a) improved economics offered by saccharin, (b) any synergy or additive effect among the sweeteners, and (c) avoidance of undesirable taste consequences caused by any single sweetener constituent.

The earliest blends were composed of saccharin and dulcin (28). Later, before its U.S. ban in 1969, sodium cyclamate was used extensively with saccharin (29) in the rapidly growing diet soft drink market. This combination is synergistic and mutually beneficial because cyclamate alone also has objectionable taste characteristics. The taste profile of a 10:1 cyclamate:saccharin blend has been found to be similar to that of sucrose (30). This cyclamate:saccharin blend purportedly possessed a clean overall taste similar to sucrose and was very cost-effective.

More recently, blends of saccharin and aspartame have been used in low-calorie soft drinks (31). Binary saccharin blends with other dipeptides (32, 33), acesulfame K (34), sucralose (23), and ternary blends with cyclamate and aspartame have also been reported (35–37). Even sucrose or fructose and saccharin

Table 5 Flavor Profile and Cost Comparison of Saccharin/Aspartame Blends at 4% Sucrose Equivalency in Water

	100% APM	70% CS / 30% APM	70% NS / 30% APM	30% CS / 70% APM	30% NS / 70% APM
Concentration (mg/liter)	180	77 / 33	77 / 33	33 / 77	33 / 77
Cost for one million liters ($)	11,880	889	466	381	200
		2,178	2,178	5,082	5,082
		3,067	2,644	5,463	5,282

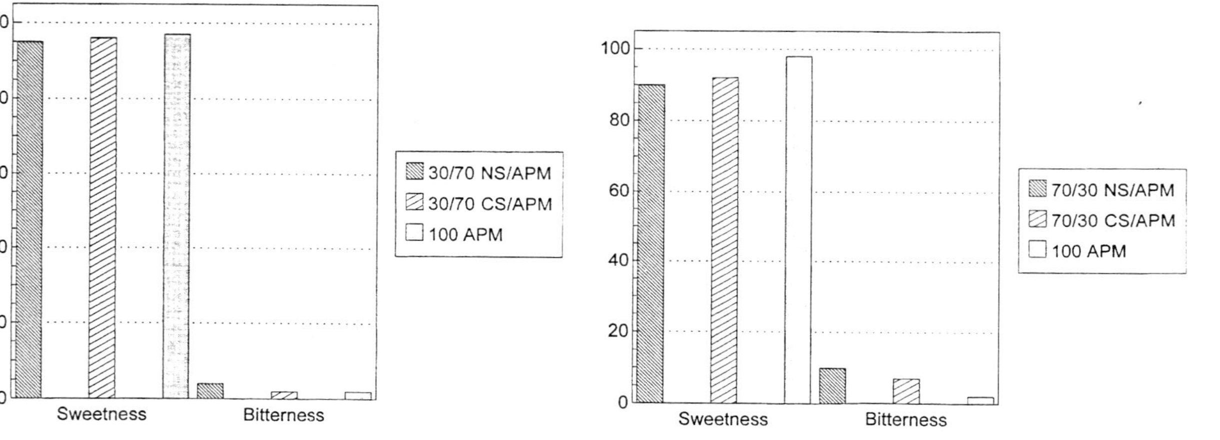

Table 6 Flavor Profile and Cost Comparison of Saccharin/Aspartame Blends at 10% Sucrose Equivalency in Water

	100% APM	100% CS	70% CS / 30% APM	70% NS / 30% APM	30% CS / 70% APM	30% NS / 70% APM
Concentration (mg/liter)	600	300	196 / 84	196 / 84	84 / 196	84 / 196
Cost for one million liters ($)	39,600	3,465	2,264 5,544	1,186 5,544	970 12,936	508 12,936
			9,808	6,730	13,906	13,444

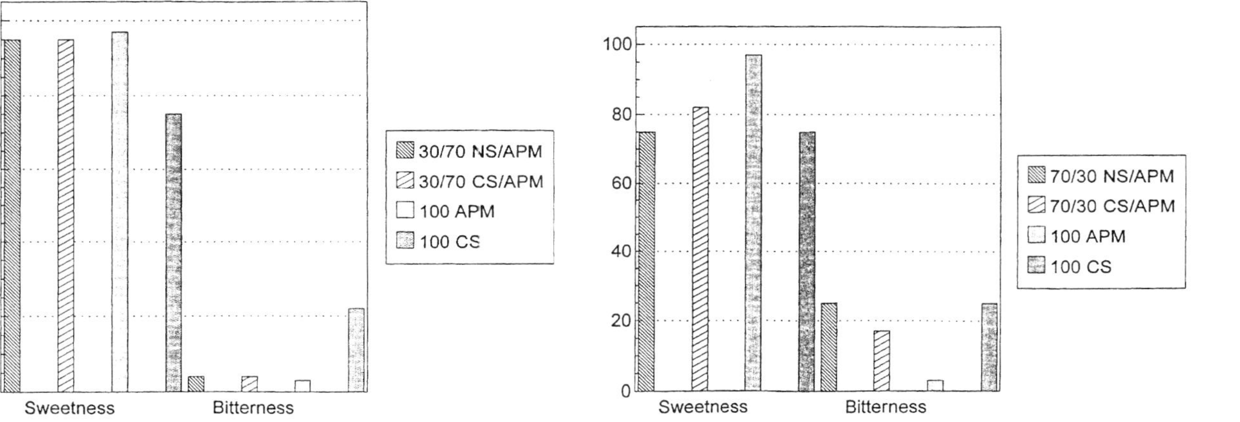

blends can provide reduced-calorie products with good taste characteristics (38–40) at lower cost.

The blending of either sodium saccharin (NS) or calcium saccharin (CS) with aspartame (APM) is an example of the economic advantage that can be achieved without attendant loss of taste aesthetics. Tables 5 and 6 display the lack of increased bitterness as saccharin is raised to 30% of the saccharin/aspartame blend, even when matching a 10% sucrose solution. The economic incentive is dramatic because the 30% saccharin/70% aspartame blends have a cost ranging between one third to one half that of aspartame alone.

Increasing the saccharin content to 70% did result in increased bitterness; therefore, there is a limit to how much saccharin can be formulated into a blend. Substitution of the calcium ion for the sodium ion gave equal sweetness with marginally less bitterness.

A second approach to the use of saccharin is the use of ''masking agents.'' A popular tabletop sweetener uses cream of tartar and dextrose as masking agents (41). Many of these are described in reviews by Bakal (35) and Daniels (42), with new versions continually being added to the list (43).

Most food and beverage literature assessing the improved taste of saccharin blends has focused on sodium saccharin. Recently, the calcium salt of saccharin has been found to possess a shorter, cleaner aftertaste with less bitterness (44). In blends with other sweeteners, calcium saccharin may offer improved taste aesthetics.

No single alternative sweetener available today closely matches the full taste attributes of natural sugar. Moreover, these alternative sweeteners all have one or more additional disadvantages such as poor solution stability, high cost, or objectionable taste characteristics. By blending saccharin salts with other sweeteners, the food chemist can economically provide low-calorie foods and beverages with optimized taste characteristics while minimizing any single sweetener's weakness (45). This multiple sweetener concept is more completely described in Chapter 24.

VI. STABILITY AND SHELF-LIFE

In its bulk form, saccharin and its salts show no detectable decomposition over periods as long as several years (44). Another major advantage of saccharin and its salts is high stability in aqueous solutions over a wide pH range. With saccharin, the food and beverage formulator does not have to compromise taste aesthetics by altering the pH to minimize sweetener hydrolysis. Thus, taste is maintained over a longer shelf-life.

In 1952, DeGarmo and coworkers studied the stability of saccharin in aqueous solutions (46). They found that saccharin solutions buffered at pHs ranging

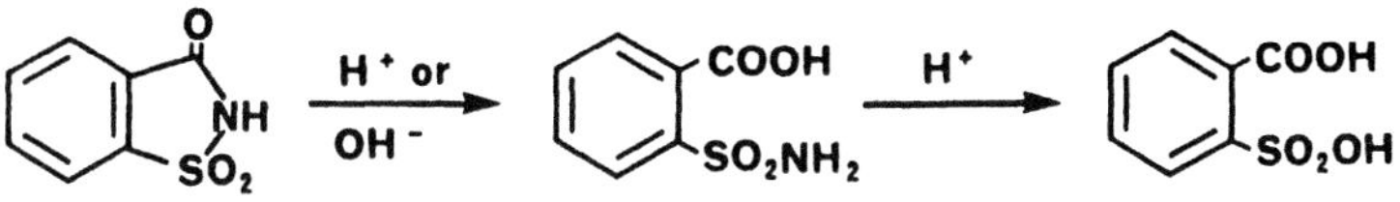

$$\text{Saccharin (C}_6\text{H}_4\text{–CO–NH–SO}_2\text{)} \xrightarrow[\text{OH}^-]{\text{H}^+ \text{ or}} \text{C}_6\text{H}_4(\text{COOH})(\text{SO}_2\text{NH}_2) \xrightarrow{\text{H}^+} \text{C}_6\text{H}_4(\text{COOH})(\text{SO}_2\text{OH})$$

Figure 5 Hydrolysis of saccharin.

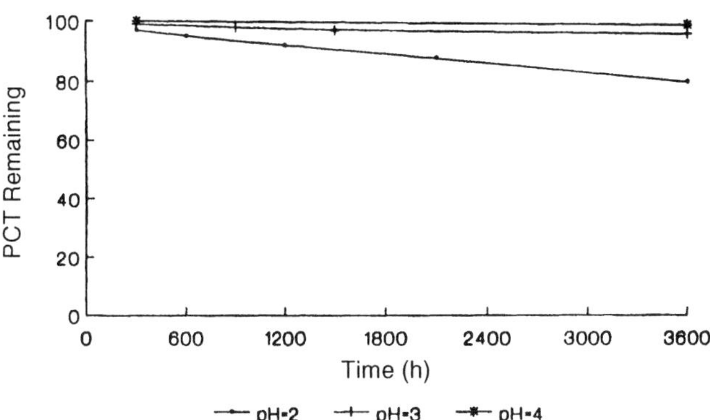

Figure 6 Saccharin hydrolysis at 20°C (68°F).

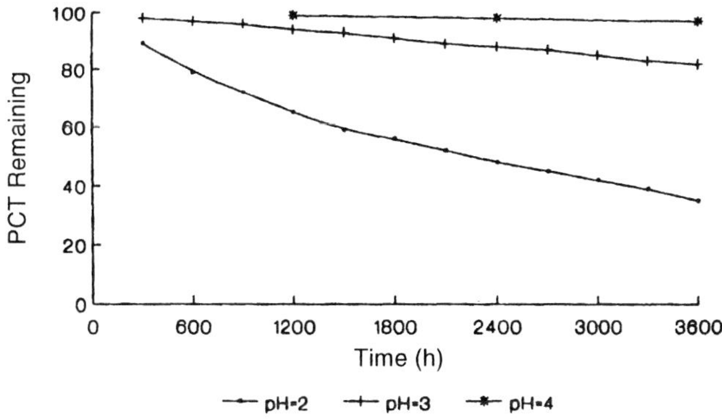

Figure 7 Saccharin hydrolysis at 40°C (104°F).

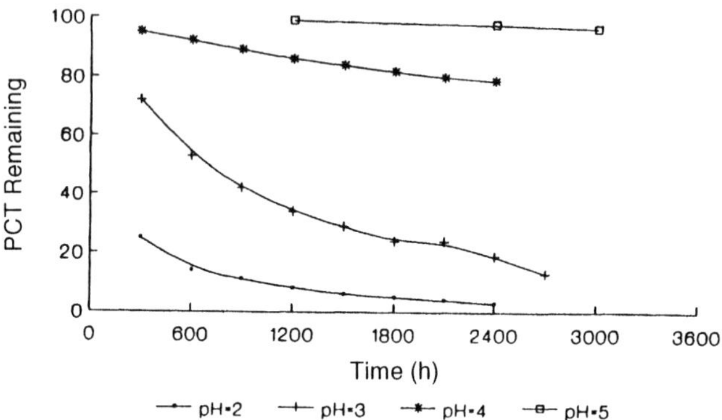

Figure 8 Saccharin hydrolysis at 80°C (176°F).

from 3.3–8.0 were essentially unchanged after heating for 1 hr at 150°C. More recent work that uses high-performance liquid chromatographic techniques has confirmed their findings (44). Only under severe laboratory conditions of high temperature, high and low pH, over an extended period does saccharin hydrolyze to a measurable extent. The only hydrolysis products are 2-sulfobenzoic acid and 2-sulfamoylbenzoic acid (Fig. 5) (46). Hydrolysis curves for sodium saccharin at various temperatures and pHs are shown in Figures 6, 7, and 8 (44).

VII. FUNCTIONALITY

Over the last century saccharin and its salts have found their way into a variety of beverages, foods, cosmetics, and pharmaceuticals. Its primary function is to provide sweetness safely and economically without the attendant calories. Saccharin is used as a noncaloric sweetener in the following foods and beverages:

Soft drinks, fruit juice drinks, other beverages, and beverage bases or mixes
Tabletop sweeteners in tablet, powder, or liquid form
Processed fruits
Chewing gum and confections
Gelatin desserts, juices, jams, and toppings
Sauces and dressings

In addition, over the past century saccharin has found use in a variety of nonfood applications, including (a) nickel electroplating brightener (47), (b) agricultural chemical intermediate (48, 49), (c) animal food sweetener (50), (d) pharmaceutical chemical intermediate (51), (e) chemical intermediate for a mammalian repellant (52, 53), (f) biocide intermediate (54), (g) anaerobic adhesive accelerator (55, 56), (h) personal care products (57), and (i) pharmaceutical taste modifier (58). The chemistry of the molecule saccharin has been the subject of several reviews (59–61).

VIII. METABOLISM

Saccharin is not metabolized by humans. No metabolism has been detected in studies that used modern analytical techniques. Treatment of rats by a variety of means has attempted to optimize the detection of minor metabolic pathways, but all have failed to reveal biotransformation. Most data on the biotransformation of saccharin demonstrates that saccharin is excreted unchanged, predominately in the urine, in both humans and laboratory animals (62). Low levels of metabolites reported in three early studies represent artifacts resulting from inadequate techniques and/or impurities in the saccharin used (63).

IX. TOXICOLOGICAL EVALUATION

Saccharin has been the subject of extensive scientific research and debate. It is one of the most studied ingredients in the food supply. Although the totality of available research indicates saccharin is safe for human consumption, there has been controversy over its safety. The controversy rests primarily on findings of bladder tumors in some male rats fed high doses of sodium saccharin (64–67).

Recent saccharin research has concentrated on understanding the mechanism of saccharin-related tumor formation in the male rat. The saccharin-related tumors observed in male rats are the result of a sequence of events that begins with the administration of high doses of sodium saccharin. This leads to the formation of a urinary milieu conducive to the formation of a calcium phosphate–containing precipitate that is cytotoxic to the urothelium. This, in turn, leads to a regenerative hyperplasia, which persists over the animal's lifetime if the administration of sodium saccharin continues, resulting in the formation of bladder tumors. The no effect level for the hyperplastic response to sodium saccharin is 1% of diet, and this is identical to the no-effect level for the hyperplastic response to sodium saccharin, tumor promotion, and the tumors observed in the two-generation bioassays (68). Administration of high doses of other sodium salts (e.g., sodium ascorbate) produces effects similar to those of high doses of sodium saccharin (69, 70).

Extensive research in human populations has shown no increased risk of bladder cancer in humans. More than 30 human studies have been completed, supporting saccharin's safety at human levels of consumption (71).

In 1997, a special International Agency for Research on Cancer (IARC) panel determined the bladder tumors in male rats resulting from the ingestion of high doses of sodium saccharin are not relevant to man. In 1998, IARC downgraded saccharin from a Group 2B substance, possibly carcinogenic to humans, to Group 3. Substances may be placed in Group 3 when there is strong evidence that the mechanism of carcinogenicity in experimental animals does not operate in humans. This was the first time IARC had considered mechanistic data (72).

In 1996, the Calorie Control Council submitted a petition to the U.S. National Toxicology Program (NTP) requesting that, under the program's new criteria allowing for the consideration of mechanistic data, saccharin be delisted from the NTP's *Ninth Report on Carcinogens*. In May 2000, the NTP released the 9th edition and announced that saccharin had been delisted (73).

X. REGULATORY

Saccharin is a widely used noncaloric sweetener available for use in more than 100 countries. In the United States, it is approved for use under an interim food additive regulation, permitting use for special dietary and certain technological purposes (74). In 1977, the U.S. Food and Drug Administration proposed a ban on saccharin as a result of studies reporting bladder tumors in some male rats fed high doses of sodium saccharin. The U.S. Congress passed a moratorium preventing the proposed ban. The moratorium has been extended seven times. The current moratorium is in effect until May 1, 2002. In 1991, the FDA formally withdrew its 1977 proposed ban on saccharin. On December 21, 2000, U.S. President Bill Clinton signed legislation removing the saccharin warning label that had been required on saccharin-sweetened foods and beverages since 1977 (75).

In 1993, the Joint Food and Agriculture/World Health Organization Expert Committee on Food Additives reviewed saccharin and doubled its acceptable daily intake. The Committee noted that the animal data which earlier raised questions about saccharin are not considered relevant to man (76). Saccharin also has been evaluated and determined safe by the European Union's Scientific Committee for Food (77).

XI. CONCLUSION

Saccharin has been used to sweeten foods and beverages for more than a century and remains an important sweetener. It is one of the most researched food ingredients available, having been studied in rats, mice, hamsters, monkeys, and humans and determined to be safe for human consumption.

REFERENCES

1. GE DuBois, et al. Sweeteners: Discovery, Molecular Design, and Chemoreception. American Chemical Society, Washington, D.C., 1991, p 274.
2. GB Kauffman, PM Priebe. Ambix 25:Pt III 191–207, 1978.
3. C Fahlberg, I Remsen. Berichte 12:469–473, 1879.
4. C Fahlberg, U.S. patent 319,082 (1885).
5. DS Tarbell, AT Tarbell. J Chem Educ 55:161–162, 1978.
6. J Forrestal. Faith, Hope and $5,000. New York: Simon & Schuster, 1977, pp 17–22.
7. EM Whelan. The Conference Board Magazine 16:54, 1977.
8. T Smith. Saccharin. American Council on Science and Health, 1979, p 14.
9. OF Senn. U.S. Patent 2,667,503 (1954).
10. GF Schlaudecker. U.S. Patent 2,705,242 (1955).
11. P Bedoukian. Perfumery and Flavoring Synthetics. New York: Elsevier, 1967, pp 41–47.
12. Chemical Engineering 61:128, 1954.
13. Chemical and Engineering News 41:76–78, 1963.
14a. Food Chemical Codex, 3rd ed. Washington, DC: National Academy Press, 1981.
14b. U.S. Pharmacopeia XXIII, National Formulary 18. Rockville, MD: United States Pharmacopeial Convention, Inc., 1994.
15. IH Pitman, H-S Dawn, T Higuchi, AA Hussain. J Chem Soc B:1230, 1969.
16. A Salant. Handbook of Food Additives. Cleveland: The Chemical Rubber Co., 1968, pp 503–512.
17a. OJ Magidson, SW Gorbatschow. Chem Ber 56B:1810–1817, 1923.
17b. GH Hamor. Science 134:1416, 1961.
18. K Othmer. Encyclopedia of Chemical Technology, 4th ed. Vol. 23, New York: Wiley-Interscience, 1997, pp 564–566.
19a. JCJ Bart. J Chem Soc B 376–382, 1968.
19b. Y Okaya. Acta Cryst B 25:2257, 1969.
20. T Paul. Chem Zeit 45:38–39, 1921.
21. SS Schiffman, et. al. Brain Res Bull 38(2):105–120, 1995.
22. MA Amerine, RM Pangborn, EB Roessler. Food Science and Technology. New York: Academic Press, 1965, pp 86–103.
23. PK Beyts, Z Latymor. U.S. Patent 4,495,170 (1985).
24. SS Schiffman et al. Brain Res Bull 36(5):505–513, 1995.
25. H Schmandke, K Hoppe, B Gassmann. Nahrung 29:417–420, 1985.
26. FJ Helgren, MJ Lynch, FJ Kirchmeyer. J Am Pharm Assoc 44:353, 1955.
27. CP Rader, SG Tihanyi, FB Zienty. J Food Sci 32:357–360 1967.
28. T Paul. Chem Zeit 45:705–706, 1921.
29. FJ Helgren. U.S. Patent 2,803,551 (1957).
30. KM Beck. Food Prod Dev 9:47–54, 1975.
31. D Scott. U.S. Patent 3,780,189, (1973).
32. TM Brennan, ME Hendrick. U.S. Patent 4,411,925 (1983).
33. TM Brennan, ME Hendrick. U.S. Patent 4,399,163 (1983).

34. G Von Ryman Lipinski, E Luck. U.S. Patent 4,158,068 (1979).
35. AI Bakal. Chem Ind 18:700, 1983.
36. H Lim, CS Setser, SS Kim. J Food Sci 54:625, 1989.
37. TP Finucane. Canadian Patent 1,043,158 (Chem Abstr 90:102244h).
38. L Hyvonen, P Koivistoinen, A Ratilainen. J Food Sci 43:1580, 1978.
39. L Hyvonen, R Kurkela, P Koivistoinen, A Ratilainen. J Food Sci 43:1577, 1978.
40. L Hyvonen, R Kurkela, P Koivistoinen, A Ratilainen. J Food Sci 43:251, 1978.
41. ME Eisenstadt. U.S. Patent 3,625,711 (1971).
42. R Daniels. Sugar Substitutes and Enhancers. Park Ridge, NJ: Noyes Data Corporation, 1973, pp 80–103.
43. JA Riemer. U.S. Patent 5,336,513 (1994).
44. PMC Specialties Group. Unpublished data.
45. A Porter. Chem Ind 18:696–699, 1983.
46. O De Garmo, GW Ashworth, CM Eaker, RH Much. J Am Pharm Assoc 41:17, 1952.
47. GA DiBari. Metal Finishing, 57th Guidebook-Directory. Metals and Plastics Publications, Inc., Hackensack, NJ. 1989, pp 244–265.
48. S Seki, et al. U.S. Patent 3,629,428 (1971).
49. G Levitt U.S. Patent 4,238,621 (1980).
50. MR Kare, GK Beauchamp. Dukes' Physiology of Domestic Animals, 9th ed., Ithaca, NY: Cornell University Press, 1977, pp 713–730.
51. JG Lombardino. U.S. Patent 4,289,879 (1981).
52. Chem Marketing Rep: 17 (November 14, 1988).
53. GT Hollander, M Blum. U.S. Patent 4,661,504 (1987).
54. RL Wakeman, JF Coates. U.S. Patent 3,280,137 (1966).
55. VK Krieble. U.S. Patent 3,046,262 (1962).
56. VK Krieble. U.S. Patent 3,218,305 (1965).
57. LG Peterson, LA Sanker, JG Upson. U.S. Patent 5,370,864 (1994).
58. The Merck Index, 12th Ed., White House Station, NJ: Merck & Co., Inc., 1996, p 8460.
59. LL Bambas. The Chemistry of Heterocyclic Compounds, Vol. 4, New York: Interscience 1952, pp 318–353.
60. H Hettler. Advances in Heterocyclic Chemistry, Vol. 15, New York: Academic Press, 1973, pp 233–276.
61. JL Lowe. Chem Eng Monogr 15:285–305, 1982.
62. TW Sweatman, AG Renwick. The tissue distribution and pharmacokinetics of saccharin in the rat. Toxicol Appl Pharmacol 55:18–31, 1980.
63. AG Renwick. The metabolism of intense sweeteners. Xenobiotica 16(10/11):1057–1071, 1986.
64. MO Tisdel, et al. Long-term feeding of saccharin in rats. In: Inglett, ed. Symposium Sweeteners, Westport, Ct: Avi Publishing Co., Inc., 1974, pp 145–158.
65. JM Taylor, MA Weinberger, L Friedman. Chronic toxicity and carcinogenicity of the urinary bladder of sodium saccharin in the in utero-exposed rat. Toxicol Appl Pharmacol 54:57–75, 1980.
66. DL Arnold, et al. Long-term toxicity study of orthotoluene sulfonamide and sodium saccharin in the rat. Toxicol Appl Pharmacol 52:113–152, 1980.

67. GP Schoenig, et al. Evaluation of the dose response and in utero exposure of saccharin in the rat. Food Chem Toxic 23(4/5):475–490, 1985.
68. SM Cohen. Saccharin safety update. In: Low-Calorie Sweeteners: Proceedings of a Seminar in New Delhi, India, Sponsored by the International Sweeteners Association and the Confederation of Indian Food Trade & Industry. February 5–6, 1997.
69. LB Ellwein, SM Cohen. The health risks of saccharin revisited. Crit Rev Toxicol 20:311–326, 1990.
70. SM Cohen, TA Anderson, LM de Oliveira, LL Arnold. Tumorigenicity of sodium ascorbate in male rats. Cancer Res 58:2557–2561, 1998.
71. M Elcock, RW Morgan. Update on artificial sweeteners and bladder cancer. Reg Toxicol Pharmacol 17:35–43, 1993.
72. JM Rice, et al. Rodent tumors of urinary bladder, renal cortex, and thyroid gland. In: IARC Monographs Evaluations of Carcinogenic Risk to Humans Toxicological Sciences 49:166–171, 1999.
73. National Toxicology Program (NTP). 9th Report on Carcinogens. U.S. Government Printing Office, 2000.
74. 21 CFR §180.37. Saccharin, ammonium saccharin, calcium saccharin, and sodium saccharin.
75. H.R. 4577. Making appropriations for the Departments of Labor, Health and Human Services, and Education, and related agencies for the fiscal year ending September 30, 2001, and for other purposes. Signed by the President (effective date) December 21, 2000.
76. Joint FAO/WHO Expert Committee on Food Additives (JECFA). Evaluation of Certain Food Additives and Contaminants. Forty-first report. WHO Technical Report Series. World Health Organization, Geneva, Switzerland.
77. European Parliament and Council Directive 94/35/EC of 30 June 1994 on sweeteners for use in foodstuffs. Official Journal of the European Communities. No. L 237/3, September 10, 1994.

10

Stevioside

A. Douglas Kinghorn, Christine D. Wu, and Djaja Djendoel Soejarto
University of Illinois at Chicago, Chicago, Illinois

I. INTRODUCTION

Stevia rebaudiana, a South American plant, is the source of the highly sweet *ent*-kaurenoid diterpene glycosides, stevioside, rebaudioside A, and several other sweet-tasting analogs. Although stevioside, the major sweet constituent of this species, was first isolated in impure form in the first years of the 20th century, this compound was not used extensively as a sweetener until its development for this purpose by a group of Japanese manufacturers in the early 1970s. Currently, the largest use of stevioside remains in Japan, although this compound is used increasingly in South Korea. For this compilation, we have had access not only to the previous versions of this chapter (1, 2) but also to other recent reviews on stevioside, rebaudioside A, and "sugar-transferred" stevioside (3–5). Particular emphasis in this chapter will be made on the latest additions to our knowledge of stevioside.

II. PRODUCTION

S. rebaudiana (Bertoni) Bertoni is a member of a New World genus of 150–300 species that belong to the tribe Eupatorieae of the family Asteraceae (sunflower family). The plant may reach a height of 80 cm when fully grown and is native to Paraguay in the Department of Amambay, in particular the slopes and valleys of the Cordillera of Amambay on the border of Paraguay with Brazil (2). Although high concentration levels (10% or more) of stevioside and other sweet diterpene glycosides occur in dried *S. rebaudiana* leaves, this type of sweet compound appears to be rare in the genus *Stevia* as a whole. When 110 *Stevia* leaf herbarium specimens were analyzed organoleptically and chemically for the presence of sweet compounds, stevioside was found in only one species in addition to *S. rebaudiana*, namely, *S. phlebophylla* A. Gray, a Mexican species that may now be extinct (2).

The commercialization of *S. rebaudiana* leaves for sweetening and flavoring purposes has been quite rapid since first being introduced to Japan. More than 100 food and beverage items containing *S. rebaudiana* sweeteners are now commercially available in Japan, and in a recent year, about 200 metric tons of purified stevioside and other sweetener products were prepared from about 2000 metric tons of dried plant leaves (6) (Table 1). As has been the case for some time (2), most *S. rebaudiana* leaves for the Japanese market are cultivated in the People's Republic of China, especially in Fujian, Zhejiang, and Guangdong provinces (2). Cultivation of *S. rebaudiana* for the Japanese market also occurs in Taiwan, Thailand, and Vietnam. Stevioside now occupies 40% of the sweetener market in Korea, with most of this being used in the sweetening of the alcoholic beverage, *soju*, and produced from leaves of *S. rebaudiana* grown in the People's Republic of China (7). Production also occurs in southern Brazil for the market in that country (3). In recent years, publications have appeared describing the possibility of establishing the cultivation of *S. rebaudiana* in additional countries such as Canada (5), the Czech Republic (8), India (9), and Russia (10). Information on the propagation of *S. rebaudiana* from cuttings and seed and cultivar development has been documented recently (5).

Methods for the production of extracts of *S. rebaudiana* containing stevioside in varying degrees of purity have been summarized (1, 2). Much of this type of information has appeared in the patent literature, especially from Japan. Most

Table 1 Use of *Stevia rebaudiana* Extracts in the Japanese Food Industry in 1995[a]

Food item	Percent of total
Japanese-style pickles	28.1
Beverages and yogurt	17.1
Dried seafoods	12.6
Ice cream and sherbert	12.6
Tabletop sweeteners and miscellaneous	7.4
Soy sauce and soy paste	6.3
Mashed and steamed fish and meat	4.9
Seasonings	4.8
Confectionery and bread	3.5
Seafoods boiled down with soy sauce	2.7

[a] As estimated by Maruzen Pharmaceuticals Co., Ltd. (11).

methods for the purification of stevioside entail initial extraction into an aqueous solvent, followed by refinement involving one or more of selective extraction into a polar organic solvent, decolorization, precipitation, coagulation, adsorption, ion exchange, and crystallization (1, 2).

III. PHYSICAL AND ORGANOLEPTIC PROPERTIES, SOLUBILITY, AND STABILITY

Although chemical work to determine the structural nature of stevioside began at the turn of the past century, more than 60 years elapsed before the structure of this compound was finally elucidated. A group at the National Institutes of Health in Bethesda, Maryland, was able to show that one sugar unit of stevioside occurs as a D-glucopyranosyl functionality attached β to a carboxyl group, whereas the second is a sophorose [2-*O*-(β-D-glucopyranosyl)-D-glucose] unit attached β to an alcoholic group on the aglycone (1, 2). The structure and stereochemistry of steviol (*ent*-13-hydroxykaur-16-en-19-oic acid), the aglycone obtained on the enzymatic hydrolysis of stevioside, were finally determined in 1963 (2). Isosteviol is produced from stevioside by acidic hydrolysis and is a ring-D rearranged isomer of steviol (2, 3). During the 1970s, additional sweet steviol glycoside analogs of stevioside were identified as *S. rebaudiana* leaf constituents, comprising rebaudiosides A and B, steviolbioside, rebaudioside C (= dulcoside B), rebaudiosides D and E, and dulcoside A (1, 2). Some evidence exists that rebaudioside B and steviolbioside are not native *S. rebaudiana* constituents but are formed by partial hydrolysis during the extraction process (3, 4). The structures of the sweet *ent*-kaurene glycosides of *S. rebaudiana* leaves are shown in Fig. 1. The sweet *ent*-kaurene glycosides occur at very high concentration levels in dried *S. rebaudiana* leaves, with approximate values for the four most abundant compounds being stevioside (5–10% w/w), rebaudioside A (2–4% w/w), rebaudioside C (1–2% w/w), and dulcoside A (0.4–0.7% w/w) (2).

In pure form, stevioside is a white crystalline material with a melting point of 196–198°C, an optical rotation of −39.3 degrees in water, an elemental composition of $C_{38}H_{60}O_{18}$, and a molecular weight of 808.88 (2). Stevioside is only sparingly soluble in water but is highly soluble in ethanol. Rebaudioside A is considerably more water soluble than stevioside because it contains an additional glucose unit in its molecule (2).

Stevioside is a stable molecule at 100°C when maintained in solution in the pH range 3–9, although it decomposes quite readily at alkaline pH levels of greater than 10 under these conditions (1). Both stevioside and its analog, rebaudioside A, have been found to be stable when formulated in acidulated beverages that were stored at 60°C for 5 days (2). Detailed stability profiles have been determined for stevioside when treated with dilute mineral acids and en-

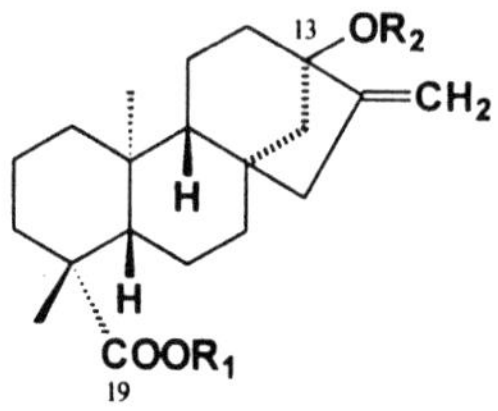

	R_1	R_2
Stevioside	β-glc	β-glc^2-β-glc
Rebaudioside A	β-glc	β-glc^2-β-glc 3 β-glc
Rebaudioside B	H	β-glc^2-β-glc 3 β-glc
Rebaudioside C	β-glc	β-glc^2-α-rha 3 β-glc
Rebaudioside D	β-glc^2-β-glc	β-glc^2-β-glc 3 β-glc
Rebaudioside E	β-glc^2-β-glc	β-glc^2-β-glc
Dulcoside A	β-glc	β-glc^2-α-rha
Steviolbioside	H	β-glc^2-β-glc

Glc = D-glucopyranosyl; rha = L-rhamnopyranosyl

Figure 1 Structures of the eight sweet-tasting glycoside constituents of *Stevia rebaudiana* leaves.

zymes (2). A large number of analytical methods are available for the determination of purity and stability of stevioside and rebaudioside A (1, 2).

Stevioside has been rated as possessing about 300 times the relative sweetness intensity of 0.4% w/v sucrose, although its sweetness intensity varies with concentration and the compound exhibits some bitterness and an undesirable aftertaste (1). The sweetness intensities (i.e., sweetening power relative to sucrose, which is taken as = 1) of the other seven *S. rebaudiana* sweet principles have been determined as follows: rebaudioside A, 250–450; rebaudioside B, 300–350; rebaudioside C, 50–120; rebaudioside D, 250–400; rebaudioside E, 150–300; dulcoside A, 50–120; and steviolbioside, 100–125 (2). Rebaudioside A, the second most abundant *ent*-kaurene glycoside occurring in the leaves of *S. rebaudiana* is better suited than stevioside for use in foods and beverages because it is not only more water soluble but it also exhibits a pleasanter taste (2, 4).

Many attempts have been made to improve the sweetness hedonic parameters of stevioside by formulation with a variety of flavor-masking and sweetness-enhancing agents (1, 2). Efforts have been made to produce strains of *S. rebaudiana* that have a higher ratio of rebaudioside A to stevioside compared with wild Paraguayan populations to harness the preferential properties of rebaudioside A compared with stevioside (2). In another type of strategy, enzymatic transglycosylation of stevioside has led to analogs with improved taste profiles over the parent substance (2, 4). In this procedure, modification of the sugar moiety at C-19 of stevioside is conducted with enzymic transglycosylation, and cyclomaltodextrin-glycanotransferase (CGTase) is used to catalyze *trans*-α-1,4-glycosylation (4). A transglycosylated (''sugar-transferred'') product of stevioside is sold commercially in Japan, produced by shortening the α-glucosyl chain of the mixture of compounds obtained by CGTase treatment using β-amylase (4).

IV. AVAILABILITY AND COST

It is apparent that the natural area of distribution of *S. rebaudiana* has been considerably reduced in recent years in its native habitat in the Sierra of Amambay in northeastern Paraguay, where the plant is found in natural grasslands on mountain slopes and valleys occurring at altitudes of 500–700 m above sea level. The reason for this is that many of the subtropical forests surrounding the natural populations of *S. rebaudiana* have been exploited for their timber, with the newly cleared land then used for other agricultural purposes. In addition, thousands of *S. rebaudiana* plants have been transplanted to other areas to start large-scale plantations. Because *S. rebaudiana* is sensitive to changes in its environment, the net effect is that the germplasm of this species may be threatened at present (2).

As noted earlier in this chapter, by far the largest cultivation areas where *S. rebaudiana* is currently produced are in the People's Republic of China. The price of *S. rebaudiana* products has become much cheaper in recent years. For example, sweet products made from *S. rebaudiana* leaves containing high levels of stevioside were initially priced at 100,000 Japanese yen (equivalent to $833 U.S. at 120 yen per $) but have been reduced to less than 10,000 Japanese yen (equivalent to $8.33 U.S.) (11).

V. USE AND ADMIXTURE POTENTIAL

Several authors have indicated that *S. rebaudiana* leaves have been used to sweeten bitter beverages such as maté (*Ilex paraguayensis* St.-Hil.) in Paraguay for centuries (reviewed in 2). However, it has been suggested that this plant was not of particularly great significance to the indigenous Guarani Indians of Paraguay, because they placed more value on the use of honey as a sweetener (2). In recent times, the use of *S. rebaudiana* leaves to sweeten maté or foods was seen to be practiced only sporadically in Paraguay (2). Lewis (12) has documented certain previously unpublished memoranda concerning the early use of *S. rebaudiana* leaves as a sweetener and concluded that the sweet properties of *S. rebaudiana* leaves have been known to local populations in Paraguay since at least before 1887. This knowledge only slowly disseminated to the extended population, both because of the rarity of the plant within its distribution range and a comparative lack of interest by the colonists (2, 12). For several years, teas made from *S. rebaudiana* leaves have been prescribed by physicians in Paraguay for the treatment of diabetes (2). *S. rebaudiana* products are now approved as additives by the Food National Codes of both Paraguay and Argentina (13). In neighboring Brazil, after the initial introduction of two *S. rebaudiana* products (2), extracts of the leaves of this plant (purified to contain a minimum of 60% stevioside) and pure stevioside (free from steviol and isosteviol) have been approved for use in foods and beverages, dietetic foods and beverages, chewing gum, medicines, and oral hygiene products (14). In addition, stevioside was regulated for use in Brazil in soft drinks, although strict labeling requirements were put forth (15).

S. rebaudiana extracts were introduced commercially for sweetening purposes in Japan in 1976 (2). Since then, Japanese commercialization of *S. rebaudiana* has been quite rapid, and this has been attributed to certain factors. First, sucrose consumption has been curtailed in Japan because of health concerns related to dental caries, obesity, and diabetes. Second, certain artificial sweeteners were banned or severely restricted in Japan in the 1960s. Third, there has been a perception among the Japanese government regulatory agencies and the public alike that synthetic compounds are inherently more harmful than naturally occurring substances (2). Extracts of *S. rebaudiana* containing stevioside have particular advantages that have contributed to their development as widely used noncaloric sucrose substitutes in Japan. For example, stevioside has been found to suppress the pungency of sodium chloride, a universal preservative and flavoring agent that is commonly added to Japanese-style vegetables, dried seafoods, soy sauce, and miso (bean paste) products. Stevioside is relatively stable under normal elevated temperatures involved in food processing and does not turn brown on heating or ferment during use. The compound does not precipitate at the acid pH levels characteristic of many soft drinks (1, 2). At present, three categories of products are produced from *S. rebaudiana* on the Japanese market, namely, "stevia extracts," "rebaudioside A-enriched stevia extracts," and "sugar-transferred stevia extracts," (16). "Stevia extracts" contain at least 80% of total steviol glycosides, inclusive of stevioside, rebaudioside A, rebaudioside C, and dulcoside A, and are specified to contain not more than 20 ppm of heavy metals (determined as Pb) and not more than 2 ppm of arsenic (determined as As_2O_3). "Rebaudioside A-enriched stevia extracts" contain larger amounts of rebaudioside A than "stevia extracts," whereas the "sugar-transferred stevia extracts" are treated with CGTase and β-amylase, as mentioned in an earlier section, and contain more than 85% of total steviol glycosides (16). Table 1 provides estimates of the major categories of Japanese foods in which *S. rebaudiana* products were used in 1995.

Stevioside was initially approved as a food additive in South Korea in 1984, and it must be of at least 98% purity for use after being dried for 2 hr at 100°C (17). Stevioside is permitted for use in distilled liquors, unrefined rice wines, confectionery, soy sauce, and pickles, although not so far in bread, baby foods, dairy products, and as a tabletop sweetener (7, 17). Its use in liquor (particularly *soju*) in Korea has been permitted since 1991 for domestic use only (7, 17). Recently, there has been a considerable degree of scrutiny by governmental authorities in Korea concerning the safety of stevioside, particularly in terms of whether this compound is hydrolyzed to steviol in the approximately 45 types of *soju* on the market (18).

Products manufactured from *S. rebaudiana* leaves also have some uses in the People's Republic of China. Teas are prepared from the plant using either hot or cold water and are recommended for increasing the appetite, as a digestant, for losing weight, for keeping young, and as a sweet-tasting low-calorie tea (2).

There is an active market for *S. rebaudiana* products in the United States, which was estimated as being worth about $10 million in 1998. These products are sold under the category of "dietary supplements," either in the form of the powdered leaf, as liquid extracts, or in sachets. Stevioside has not received approval as a sucrose substitute by the U.S. Food and Drug Administration, and *S. rebaudiana* products have not been accorded "Generally Recognized as Safe" status either. Although several U.S. herbal manufacturers began to use *S. rebaudiana* in their products in the 1980s, an import ban on the plant was effected in 1991 but was rescinded in 1994 (19). Herbal products containing *S. rebaudiana* are manufactured in Italy and are exported for retail sale to other European countries, although, to date, stevioside has not been approved as a food additive by any country in the European Union (20).

S. rebaudiana extracts containing high proportions of stevioside are available in Japan in combination with glycyrrhizin, with which it is synergistic, resulting in the improvement of taste quality for both natural sweeteners. In addition, stevioside has been found to be synergistic with aspartame, cyclamate, and acesulfame K, but not with saccharin (1).

VI. METABOLISM

Only limited data are available on the in vitro and in vivo metabolism of stevioside and other *S. rebaudiana* sweet constituents. An initial investigation in which stevioside and rebaudioside A were degraded to steviol by rat intestinal flora in vitro was reviewed previously (1). Steviol has also been found as a major metabolite of stevioside when a tritiated form of the compound was fed to Wistar rats at an oral dose of 125 mg/kg. The biological half-life of stevioside was estimated to be 24 hr, and 125 hr after compound administration, the highest percentages of radioactivity were found in the feces, followed by expired air and urine. It was concluded that although a portion of orally administered stevioside was excreted unchanged in the feces of the rat, most of it was degraded by the intestinal bacterial flora to steviol, steviolbioside, and glucose, which were then absorbed in the cecum. Absorbed glucose was metabolized and excreted in the expired air as carbon dioxide and water, whereas steviol was conjugated in the liver and excreted into the bile. It was also inferred from the results of biliary and fecal excretion that enterohepatic circulation of steviol occurred (2). More recently, the distribution of [^{131}I]-stevioside has been reported in rats after IV administration, and within 2 hr, 52% of the radioactivity was present in the bile, which consisted of [^{131}I]-iodosteviol (47% of total radioactivity), unchanged [^{131}I]-stevioside (37%), and an unidentified metabolite (15%) (21).

Despite the widespread use of *S. rebaudiana* extracts containing stevioside as sweeteners in foods in Japan, the metabolism of this substance in humans does not seem to have been investigated to date.

VII. SAFETY STUDIES

A. Toxicity Studies in Rodents

Previous acute, subacute, and chronic toxicity studies carried out in Japan in rats on *S. rebaudiana* extracts with various stevioside levels have not resulted in the demonstration of any significant toxic effects (1, 2). In one of several toxicity studies carried out more recently, stevioside was fed to F344 rats at various levels (0.31, 0.62, 1.25, 2.5, and 5%) for 13 weeks. A concentration of 5% in the diet was considered to be the maximum tolerable dose of stevioside for a proposed 2 year carcinogenicity study in rats. Although lactic dehydrogenase and single-cell necrosis of the liver were increased in male animals treated with stevioside, these were not considered specific because of the lack of clear dose responses (22). In a 2 year chronic oral toxicity study performed in male and female Wistar rats fed a diet containing 85% pure stevioside (0, 0.2, 0.6, and 1.2%), it was concluded that the maximum no-effect level of stevioside was equivalent to 1.2% of the diet. The rats did not show any treatment-related changes in growth, general appearance, and clinical biochemical values relative to controls. It was projected from this study that an acceptable intake of stevioside in humans would be 7.938 mg/kg/day (23). In a carcinogenicity study, stevioside of high purity (95.6%) was added to the diet of male and female F344 rats at concentrations of 0, 2.5, and 5% for 2 years. Histopathological examination showed that there was no significantly altered development of neoplastic or non-neoplastic lesions attributable to any organ or tissue, except for a decreased incidence of mammary adenomas in females and a reduced incidence of chronic nephropathy in males. Although there was a significant decrease in the final survival rate of males treated at the 5% dietary incorporation level, it was concluded that stevioside is not carcinogenic in rats under the experimental conditions used (24).

The acute toxicity of stevioside and of steviol, the aglycone obtained from this sweetener on enzymatic hydrolysis, has been investigated in the mouse and hamster, as well as the rat (both males and females). Stevioside was not lethal to any of the test animals at doses up to 15 g/kg body weight (25). The LD_{50} values for steviol in hamsters (which were more susceptible to this compound than either mice or rats) was 5.20 and 6.10 g/kg body weight for males and females, respectively. Death was attributed to acute renal failure, and severe degeneration of the proximal tubular cells was observed histopathologically. The LD_{50} value for steviol in mice and rats was >15 g/kg body weight in both cases (25).

B. Genetic Toxicity Studies

Partially purified extracts and pure sweet constituents (stevioside, rebaudiosides A-C, dulcoside A, steviolbioside) of *S. rebaudiana* have been tested extensively for their mutagenic activity. None of these substances has been reported to be mutagenic when evaluated with *Salmonella typhimurium* strains TA98, TA100, TA1538, and TM67 or *Escherichia coli* strain WP2, either in the presence or absence of metabolic-activating systems (1, 2).

However, it was shown in the mid-1980s that steviol (Fig. 2), the aglycone of all of the *S. rebaudiana* sweet constituents including stevioside, is mutagenic in a forward mutation assay using *S. typhimurium* TM677 in the presence of a metabolic-activating system derived from a 9000 × *g* supernatant fraction from the livers of Aroclor-1254–pretreated rats. Unmetabolized steviol was inactive in this system (2). These observations were confirmed independently by two other groups (26, 27). Moreover, metabolically activated steviol gave positive re-

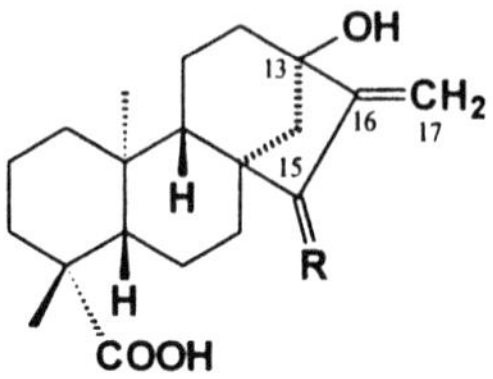

	R
Steviol	H_2
15α-Hydroxysteviol	α-OH, β-H
15-Oxosteviol	O

Figure 2 Structures of steviol, 15α-hydroxysteviol, and 15-oxosteviol.

sponses in chromosomal aberration and gene mutation tests but was not active in the reverse mutation assay (Ames test) using *S. typhimurium* strains TA97, TA98, TA100, and TA102 (26).

Several steviol analogs have been assessed for mutagenicity under the same conditions as used for steviol, and it is apparent that the C-13 tertiary hydroxyl group and the C-16, C-17 exomethylene group must be present to elicit a mutagenic effect (2). 15α-Hydroxysteviol (Fig. 2) was generated as a major in vitro metabolite under the same experimental conditions used in the treatment of steviol but was not mutagenic either in the presence or absence of the activating system (2). Moreover, a chemical oxidation product of 15α-hydroxysteviol, namely, 15-oxosteviol (Fig. 2) was proposed as a weak direct-acting mutagen. 15-Oxosteviol was also bactericidal but was not detected as a metabolite of steviol (2). However, the activity of this compound as a direct-acting mutagen was later disputed (27). In a more comprehensive study on the potential genotoxicity of steviol, this compound when metabolically activated produced confirmed activity in the forward mutation assay using *S. typhimurium* TM677, as well as chromosomal aberrations and gene mutations in a Chinese hamster lung fibroblast cell line (28). Steviol was also weakly positive in the umu test using *S. typhimurium* TA1535/pSK1002 either with or without metabolic activation. However, stevioside was not mutagenic in any of the assays used, whereas steviol, even in the presence of the S9 metabolic activating system, was inactive in a number of additional mutagenicity assays (28). It has been shown recently that in *S. typhimurium* strain TM677, steviol induces mutations of the guanine phosphoribosyltransferase (*gpt*) gene (29).

Accordingly, although recent studies have confirmed the mutagenic activity of metabolically activated steviol in a forward mutation assay on an unmutated gene, an actual active metabolite of steviol responsible for this activity has not yet been structurally elucidated. However, it has been pointed out by investigators at the National Institute of Health Sciences in Tokyo that the mutagenicity observed in in vitro experiments on activated steviol does not appear to be significant biologically, given the lack of carcinogenicity found in F344 rats after a 2-year feeding study of stevioside. It was specifically demonstrated during this study that steviol was generated by enzymatic hydrolysis and the levels of this aglycone of stevioside were quantitated by high-performance liquid chromatography (24).

C. Effects on Reproduction

It was first documented in 1968 that the Paraguayan Matto Grosso Indians use teas produced from the leaves and stems of *S. rebaudiana* for their contraceptive and antifertility effects and that an aqueous extract of *S. rebaudiana* leaves led to reduced fertility in female rats (1). These effects were not duplicated under similar conditions in experimental animals examined in other laboratories (1). However, a further report has appeared more recently from Brazil that suggests a deleterious effect on the fertility of mature female mice as a result of the intragastric administration of teas prepared from *S. rebaudiana* leaves. Fertility was reduced by 20% and 40%, respectively, by treatment with 1% and 5% infusions, during the 12-day period before mating, using small groups of mice (up to seven in a group). The 1% infusion when given during the mating period reduced the number of uterine implants but had no effect if given before mating began. Because stevioside was not used as part of the protocol, the authors of the study were unclear as to whether the *S. rebaudiana* sweet glycosides or some other leaf constituents were responsible for the observed effects (30). In contrast, two reports have appeared in which no adverse endocrine effects were seen in male rats fed *S. rebaudiana* leaf extracts (31, 32).

Stevioside and steviol have also been evaluated for their effects on reproduction. Stevioside administered to male and female rats at concentrations of up to 3% of the diet produced no abnormal responses in mating performance or fertility of any of the groups, and no external, internal, or skeletal anormalities were seen in the fetuses (2). More recently, stevioside was administered to Wistar rats by gavage at daily doses up to 1 g/kg body weight, from days 6 through 15 of pregnancy. After death at day 20, it was determined that stevioside caused no fetal abnormalities and no toxic signs in the pregnant rats and fetuses (33). However, when steviol was studied in pregnant hamsters, it was determined that doses of 0.75 and 1.0 g/kg body weight were toxic to both dams and fetuses. The dams experienced a decrease of body weight and increased mortality relative to controls, and the number of live fetuses per litter and the mean fetal weight decreased. However, no dose-related teratogenesis was detected. The levels of steviol administered were proportional to about 80 times that of an acceptable intake of stevioside for humans (7.938 mg/kg body weight per day) (34).

No follow-up has been published on a compound of apparently unknown structure that was referred to as ''dihydroisosteviol'' and reported in 1960 to exhibit a statistically significant antiandrogenic effect in a chick-comb bioassay at a dose of 3.0 mg/comb (2). ''Dihydroisosteviol'' was not found to be effective in inhibiting the action of testosterone at total dose levels of 5 and 20 mg per animal when injected subcutaneously in 28-day-old Charles River male castrated rats. The origin of ''dihydroisosteviol'' was not stipulated in this study, and no compound with such a structure seems to have been determined subsequently as a metabolic or degradative product of any of the *S. rebaudiana* sweet constituents (2).

D. Effects on Carbohydrate Metabolism

Mention has been made earlier in this chapter of the use of *S. rebaudiana* extracts in Paraguay as a remedy for diabetes. Hypoglycemic effects in humans and rats using *S. rebaudiana* extracts have been claimed in several scientific publications from South America, thereby providing some credence for the medicinal use of this plant as a hypoglycemic agent. However, a number of other investigators have found no significant effects on blood glucose levels in rats (1, 2).

Stevioside has been found to affect monosaccharide transport in an isolated, perfused liver preparation (2). Glucose absorption was inhibited in hamsters fed 2.5 g/kg/day stevioside for 12 weeks, which was attributed to both a decrease in intestinal Na^{+}/K^{+}-ATPase activity and a decreased absorptive area in the intestines (35). Stevioside (0.2 mM), when given orally both with fructose (0.2 mM) and alone to rats fasted for 24 hr, led to increased glycogen deposition in the liver (36). Steviol (1 mM) inhibited glucose intake and altered the morphology of intestinal absorptive cells (37).

E. Other Biological Activities

A hypotensive effect in human subjects was noted in a Brazilian study wherein *S. rebaudiana* tea was administered daily for 30 days (2). It was mentioned also in the previous version of this chapter that stevioside resulted in alterations of mean blood pressure and renal function when tested on normal Wistar rats (2). In the last decade, there has been increasing evidence that *S. rebaudiana* extracts and stevioside have demonstrable vasoactive properties. When aqueous *S. rebaudiana* extracts (2 ml of an extract corresponding to 66.7 g *S. rebaudiana*/100 ml) were administered to normal Wistar rats for 40 and 60 days, such chronic administration led to hypotension, diuresis, and natiuresis. For the 60-day treated group, an increase in renal plasma flow was observed (38). In Wistar rats, stevioside given as an infusion up to 16 mg/kg body weight/hr was found to be secreted by way of the renal tubular epithelium to induce a fall in the renal tubular reabsorption of glucose (39). The effect of stevioside (>90% purity; 16 mg/kg body weight/hr infusion) on renal function in both normal and hypertensive rats was investigated using clearance techniques, and, in an analogous fashion to *S. rebaudiana* extracts, this compound was found to induce hypotension, diuresis, and natiuresis (40). On the basis of the results obtained, stevioside has actually been suggested for the clinical treatment of hypertension in humans (40).

Urinary enzyme levels, changes in blood urea nitrogen (BUN) and plasma creatinine levels, and ultrastructural changes in the kidney have been examined after a single high dose of 1.5 g/kg body weight of stevioside in rats administered by subcutaneous injection. The levels of BUN and plasma creatinine both increased, as did urinary glucose. Histopathological examination showed nephrotoxicity in the proximal convoluted tubules of the kidney rather than the glomeruli or other tubules, and it was concluded that the mechanism of toxicity did not involve lipid peroxidation (41).

A series of studies on the effects of stevioside and several of its glycosidic and nonglycosidic derivatives on cellular and subcellular metabolism have been performed and reviewed by Brazilian investigators. The activities of these compounds have been probed in systems such as rat liver mitochondria, rat renal cortical tubules, human and rabbit erythrocytes, and rabbit reticulocytes. In general, *S. rebaudiana* glycosidic constituents such as stevoside and steviolbioside are either inactive in these systems or else found to be less active than their hydrolytic products steviol and isosteviol (2).

VIII. CARIOGENICITY POTENTIAL

Pure stevioside and rebaudioside A were tested for cariogenicity in an albino rat model. Sixty Sprague-Dawley rats were colonized with *Streptococcus sobrinus* and were divided into groups fed basal diet 2000 supplemented with either 0.5% stevioside, 0.5% rebaudioside A, 30% sucrose, or no compound addition. All four groups were killed after 5 weeks of feeding, and viable *S. sobrinus* counts were enumerated, and caries was evaluated according to Keyes' technique. It was concluded that stevioside and rebaudioside were not cariogenic under the conditions of the study (42). In a recent in vitro study, the eight sweet constituents of *S. rebaudiana* (stevioside, rebaudiosides A-E, dulcoside A, steviolbioside) and two hydrolytic products of stevioside (steviol and isosteviol) were tested against a panel of cariogenic and periodontopathic oral bacteria. Their antibacterial activity and their ability to inhibit sucrose-induced adherence, glucan binding, and glucosyltransferase (GTF) activity were evaluated. None of these compounds suppressed the growth or acid production of the cariogenic organism, *Streptococ-*

cus mutans, or affected sucrose-induced adherence or GTF activity. However, rebaudiosides B, C, and E, steviol, and isosteviol inhibited the glucan-induced aggregation of mutans streptococci to some extent and could provide oral health benefits by interference with cell surface functions of cariogenic bacteria (43).

IX. HUMAN EXPOSURE

S. rebaudiana extracts have been used for more than a century in Paraguay to sweeten beverages. Also, they have been used increasingly for about a quarter of a century in Japan, with the human consumption of stevioside in 1996 being on the order of 200 metric tons (6). In South Korea, about 115 metric tonnes of stevioside were consumed in 1995 (18). No evidence of adverse reactions caused by the ingestion of these materials has appeared in the scientific literature, and, on this basis, *S. rebaudiana* extracts and stevioside do not seem to present a potential toxicity risk for humans at the low consumption levels used in sweetening.

X. REGULATORY STATUS

S. rebaudiana extracts containing stevioside are approved as food additives in Japan, South Korea, Brazil, Argentina, and Paraguay and are used in herbal preparations or dietary supplements in other countries, inclusive of the People's Republic of China (where most of this plant is produced for commerce) and in the United States and western Europe.

Stevioside has been reviewed by both the Joint Food and Agriculture Organization/World Health Organization Expert Committee on Food Additives (JECFA) and the Scientific Committee for Food of the European Union (EU) and determined not acceptable as a sweetener on the basis of presently available data, which are considered insufficient (44, 45).

REFERENCES

1. AI Bakal, L O'Brien Nabors. Stevioside. In: L O'Brien Nabors, RC Gelardi, eds. Alternative Sweeteners. New York: Marcel Dekker, 1986, pp 295–307.
2. AD Kinghorn, DD Soejarto. Stevioside. In: L O'Brien Nabors, RC Gelardi, eds. Alternative Sweeteners, 2nd ed., Revised and Expanded. New York: Marcel Dekker, 1991, pp 157–171.
3. JR Hanson, BH De Oliveira. Stevioside and related sweet diterpenoid glycosides. Nat Prod Rep 10:301–309, 1993.

4. O Tanaka. Improvement of taste of natural sweeteners. Pure Appl Chem 69:675–683, 1997.
5. JE Brandle, AN Starratt, M Gizjen. *Stevia rebaudiana*: Its agricultural, biological, and chemical properties. Can J Plant Sci 78:527–536, 1998.
6. Anonymous. Mini mini marketing—stevioside. Tech J Food Chem Chemicals 9: 140, 1996.
7. JH Seon. Stevioside as natural sweetener. Report of the Pacific R & D Center, Seoul, Korea, 1995, pp 1–8.
8. A Nepovim, H Drahosova, P Valicek, T Vanek. The effect of cultivation conditions on the content of *Stevia rebaudiana* plants cultivated in the Czech Republic. Czech Rep Pharm Pharmacol Lett 8:19–21, 1998.
9. MV Chalapathi, S Thimmegowda, S Sridhara, VRR Parama, TG Prasad. Natural non-calorie sweetener stevia (*Stevia rebaudiana* Bertoni)—a future crop of India. Crop Res 14:347–350, 1997.
10. O Dzyuba, O Vseross. *Stevia rebaudiana* (Bertoni) Hemsley—a new source of natural sweetener for Russia. Rastit Resur 34:86–95, 1998.
11. K Mizutani. Maruzen Pharmaceuticals Co., Ltd., Fukuyama, Hiroshima, Japan. Private communication.
12. WH Lewis. Early uses of *Stevia rebaudiana* (Asteraceae) leaves as a sweetener in Paraguay. Econ Bot 46:336–339, 1992.
13. Anonymous. Yerba dulce. Prensa Aromática, 2(No. 11):9, 1997.
14. Anonymous. Diáro Oficial, Brazil, Resolution No. 14, January 26, 1988.
15. Anonymous. Diáro Oficial, Brazil, Resolution No. 67, April 8, 1988.
16. Anonymous. Stevia extract. In: Japan Food Additive Association, ed. Voluntary Specifications of Non-Chemically Synthesized Food Additives, 2nd ed. Tokyo: Japan Food Additive Association, 1993, pp 119–124.
17. Anonymous. Stevioside. In: Ministry of Health and Welfare, ed. Standard for Food Additives. Seoul, Korea: Ministry of Health and Welfare, 1996, pp 268–270.
18. J Kim. Seoul National University, Seoul, Korea. Private communication.
19. J O'Neill. Stevia: a sweetener that's not a sweetener. New York Times, March 2, 1999.
20. GM Ricchiuto. Specchiasol Industria Erboristica, Bussolengo, Verona, Italy. Private communication.
21. VN Cardoso, MF Barbosa, E Muramoto, CH Mesquita, MA Almeida. Pharmacokinetic studies of [I^{131}]-stevioside and its metabolites. Nucl Med Biol 23:97–100, 1996.
22. Y Aze, K Toyoda, K Imaida, S Hayashi, T Imazawa, Y Hayashi, M Takahashi. Subchronic oral toxicity study of stevioside in F344 rats. Eisei Shikensho Hokoku 109:48–54, 1991.
23. L Xili, B. Chengjiny, X Eryi, S Reiming, W Yeungming, S Haodong, H Zhiyian. Chronic oral toxicity and carcinogenicity of stevioside in rats. Food Chem Toxicol 30:957–965, 1992.
24. K Toyoda, H Matsui, T Shoda, C Uneyama, K Takada, M Takahashi. Assessment of the carcinogenicity of stevioside in F344 rats. Food Chem Toxicol 35:597–603, 1997.
25. C Toskulkao, L Chaturat, P Temcharoen, T Glinsukon. Acute toxicity of stevioside,

a natural sweetener, and its metabolite, steviol, in several animal models. Drug Chem Toxicol 20:31–44, 1997.
26. M Matsui, K Matsui, T Nohmi, H Mizusawa, M Ishidate. Mutagenicity of steviol: an analytical approach using the southern blotting system. Eisei Shikensho Hokoku 107:83–87, 1989.
27. E Procinska, BA Bridges, JR Hanson. Interpretation of results with the 8-azaguanine resistance system in *Salmonella typhimurium*: no evidence for direct acting mutagenesis by 15-oxosteviol, a possible metabolite of steviol. Mutagenesis 6:165–167, 1991.
28. M Matsui, K Matsui, Y Kawasaki, Y Oda, T Noguchi, Y Kitagawa, M Sawada, M Hayashi, T Nohmi, K Yoshihira. Evaluation of the genotoxicity of stevioside and steviol using six *in vitro* and one *in vivo* mutagenicity assays. Mutagenesis 11:573–579, 1996.
29. M Matsui, T Sofuni, T Nohmi. Regionally targeted mutagenesis by metabolically activated steviol: DNA sequence analysis of steviol-induced mutants of guanine phosphoribosyltransferase (*gpt*) gene of *Salmonella typhimurium* TM677. Mutagenesis 11:565–572, 1996.
30. BA Portello Nunes, NA Pereira. Efieto da Caá-hee [*Stevia rebaudiana* (Bert.) Bertoni] sobre a fertilidad de animais experimentais. Rev Bras Farm 89:46–50, 1988.
31. D Sincholle, P Marcorelles. Étude de l'activité anti-androgénique d'un extrait de *Stevia rebaudiana*. Plantes Méd Phytothèr 23:282–287, 1989.
32. RM Oliveira-Filho, OA Uehara, CASA Minetti, LBS Valle. Chronic administration of aqueous extract of *Stevia rebaudiana* (Bert.) Bertoni in rats: endocrine effects. Gen Pharmacol 20:187–191, 1989.
33. M Usami, K Sakemi, K Kawashima, M Tsuda, Y Ohno. Tetratogenicity study of stevioside in rats. Eisei Shikensho Hokoku 113:31–35, 1995.
34. C Wasuntarawat, P Temcharoen, C Toskulkao, P Munkornkarn, M Suttajit, T Glinsukon. Developmental toxicity of steviol, a metabolite of stevioside, in the hamster. Drug Chem Toxicol 21:207–212, 1998.
35. C Toskulkao, M Sutheerawattananon. Effects of stevioside, a natural sweetener, on intestinal glucose absorption in hamsters. Nutr Res 14:1711–1720, 1994.
36. MO Hubler, A Bracht, AM Kelmer-Bracht. Influence of stevioside on hepatic glycogen levels in fasted rats. Res Commun Chem Pathol Pharmacol 84:111–118, 1994.
37. C Toskulkao, M Sutheerawattananon, P Piyachaturawat. Effects of steviol, a metabolite of stevioside, on glucose absorption in everted hamster intestine *in vitro*. Toxicol Lett 80:153–159, 1995.
38. MS Melis. Chronic administration of aqueous extract of Stevia rebaudiana in rats: renal effects. J Ethnopharmacol 47:129–134, 1995.
39. MS Melis. Renal excretion of stevioside in rats. J Nat Prod 55:688–690, 1992.
40. MS Melis. Stevioside effect on renal function in normal and hypertensive rats. J Ethnopharmacol 36:213–217, 1992.
41. C Toskulkao, W Deechakawan, V Leardkamolkarn, T Gliusukon. The low calorie natural sweetener stevioside: nephrotoxicity and its relationship to urinary enzyme excretion in the rat. Phytother Res 8:281–286, 1994.
42. S Das, AK Das, RA Murphy, IC Punwani, MP Nasution, AD Kinghorn. Evaluation

of the cariogenic potential of the intense natural sweeteners stevioside and rebaudioside A. Caries Res 26:363–366, 1992.
43. CD Wu, SA Johnson, R Srikantha, AD Kinghorn. Intense natural sweeteners and their effects on cariogenic bacteria. Annual Meeting of the American Association for Dental Research, Minneapolis, MN, March 4–7, 1998.
44. Joint FAO/WHO Expert Committee on Food Additives. Toxicological Evaluation of Certain Food Additives. WHO Food Additives Series, Geneva, 42:119–143, 1999.
45. European Commission, Brussels Scientific Committee on Food. Opinion on Stevioside as a Sweetener. Adopted June 17, 1999.

11
Sucralose

Leslie A. Goldsmith and Carolyn M. Merkel
McNeil Specialty Products Company, New Brunswick, New Jersey

I. INTRODUCTION

The high-quality sweetness of sucralose (SPLENDA® Brand Sweetener) was discovered as a consequence of a research program conducted at Queen Elizabeth College at the University of London during the 1970s (1). Hough and his colleagues, with the support of Tate & Lyle PLC, showed that the selective chlorination of sugar could result in intensely sweet compounds (2). This discovery led to a series of studies eventually exposing sucralose (1,6-dichloro-1,6-dideoxy-β-D-fructofuranosyl-4-chloro-4-deoxy-α-D-galactopyranoside) (Fig. 1) as the most promising candidate as an ideal sweetener.

The selective chlorination of the sucrose molecule produced remarkable changes to the sweetness intensity and stability of sucrose, without compromising taste quality. Sucralose has a pleasant sweet taste similar to sucrose and has no unpleasant aftertaste. Sucralose is a white, crystalline, nonhygroscopic, free-flowing powder. The sweetener is highly soluble in water, ethanol, and methanol and has negligible effect on the pH of solutions. The viscosity of sucralose solutions is similar to that of sugar. Sucralose exerts negligible lowering of surface tension.

The chlorination of sucrose in the 1 and 6 positions of the fructose moiety and the inversion and chlorination of the 4 position on the glucose moiety causes the remarkable stability of sucralose. The resulting glycosidic linkage of sucralose is significantly more resistant to acid and enzymatic hydrolysis than that of the parent compound. The resistance of the glycosidic bond is responsible for the inability of mammalian species to digest the molecule and metabolize it as an energy source. Therefore, sucralose is noncaloric. This resistant bond is also

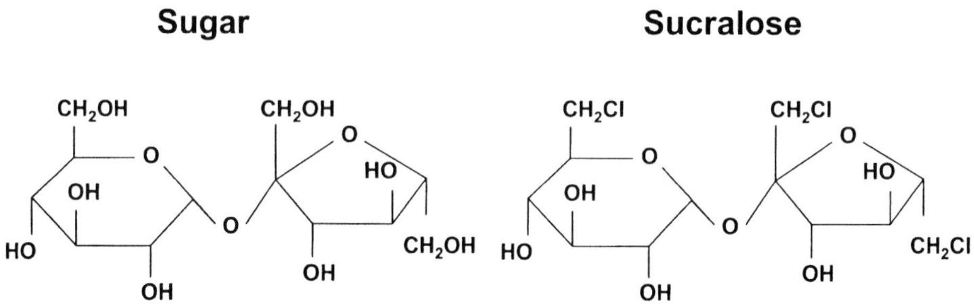

Figure 1 Chemical structure of sugar and sucralose.

the reason that microorganisms responsible for plaque formation cannot use the sweetener, and thus sucralose is noncariogenic.

The safety of sucralose has been established through one of the most thorough testing programs ever performed. Scientists and regulators around the world have evaluated more than 100 studies conducted over 20 years. Those studies were designed to determine the biological fate of sucralose and to evaluate its potential to have any adverse effects. Sucralose is safe for all individuals. Sucralose has been approved in more than 40 countries, and people around the world consume the sweetener daily.

II. SENSORY CHARACTERISTICS OF SUCRALOSE

A. Sweetness Intensity

The sweetness factor curves for sucralose and three other high-potency sweeteners, aspartame, saccharin, and acesulfame-K, are shown in Fig. 2. Sucrose is the sweetness reference used in the construction of the curves. The sweetness factor curves express the number of times a compound is sweeter than an isosweet concentration of sucrose. The curves demonstrate a general decrease in sweetness concentration for each high-potency sweetener relative to increasing sucrose concentration. Sucralose displayed the greatest sweetness factor or highest potency among the sweeteners examined. Sucralose is approximately 750 times sweeter than sucrose at a concentration equisweet to a 2% sucrose solution. At 9% sweetness equivalence, sucralose is approximately 500 times sweeter than sucrose. As a general guideline, sucralose is regarded as being 600 times sweeter than sucrose. Because of the intense sweetness of sucralose, an extremely small amount is necessary to achieve the desired sucrose isosweetness level. As an example, approximately 20 mg of sucralose per liter is needed to equal the sweetness of a 2% sucrose solution. At 5% sucrose equivalency, a level at which a normal cup

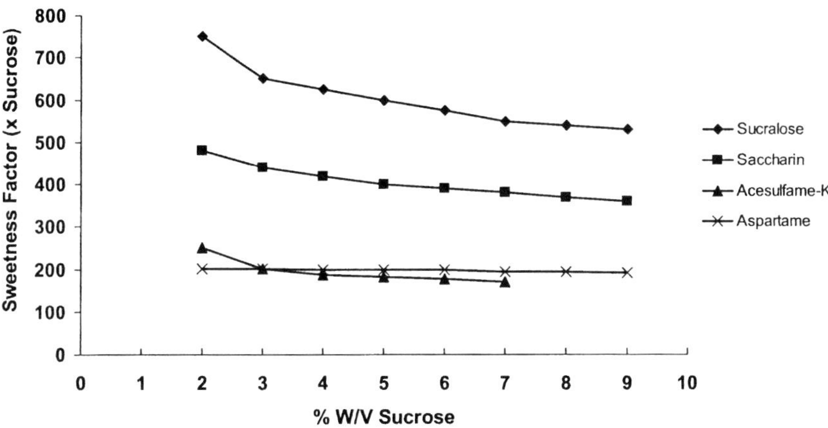

Figure 2 Sweeteness intensity.

of coffee is typically sweetened, approximately 83 mg of sucralose per liter would be required. It is important to note that the actual use levels of sucralose are influenced by characteristics such as the pH, temperature, and viscosity of the food or beverage system in which sucralose is incorporated.

B. Temporal Qualities

An understanding of the temporal properties of sucralose as they relate to other sweeteners is of interest to food and beverage formulators. The temporal profiles of the sweeteners relate to the onset of sweetness, time to maximum intensity, and rate of sweetness decay. At an aqueous isosweet concentration, the time to maximum sweetness intensity is similar for sucrose and the other high-potency sweeteners tested (Table 1).

Table 1 Temporal Sweetness Characteristics of Sucralose

	Maximum intensity (0–12 scale)	Time to maximum sweetness (sec)	Sweetness duration time (sec)
Sucrose	7.6	4.1	66.1
Sucralose	7.8	5.0	75.4
Aspartame	7.5	6.2	76.7
Saccharin	7.5	3.1	77.2
Acesulfame-K	7.5	4.9	77.4

Deviations from sucrose become apparent when studying the rate of decline in sweetness with individual high-potency sweeteners. Under the test conditions used, sucralose and aspartame had a similar but slower rate of sweetness decay than sucrose. It should be remembered that different temporal properties might emerge as a result of the concentration of the sweetener used as well as other formulation issues.

C. Taste Quality

The myriad of products around the world that are sweetened with sucralose demonstrate the taste quality of sucralose. In addition, sucralose is available globally in consumer tabletop forms (e.g., granular, tablet, packet), as a sugar replacement. Scientifically, the flavor attributes of isosweet concentrations of sucralose and sucrose in aqueous solution were identified and quantified using a sip and spit procedure. The sweeteners were profiled at a moderate sweetness level of 5%. The panelists evaluated a 10-ml sample by holding the sample in their mouth for 5 seconds and then expectorating. During this time, intensity estimates were made of several taste attributes using an unstructured line scale. Sweet and non-sweet aftertastes were evaluated 60 seconds after expectoration. The experimental series was repeated twice, and the scores averaged across replicates.

Several of the more important flavor attributes are presented in Fig. 3. This comparison demonstrates that sucralose is a high-quality sweetener with a flavor profile similar to that of its sugar origin. At a 90% confidence interval, no signifi-

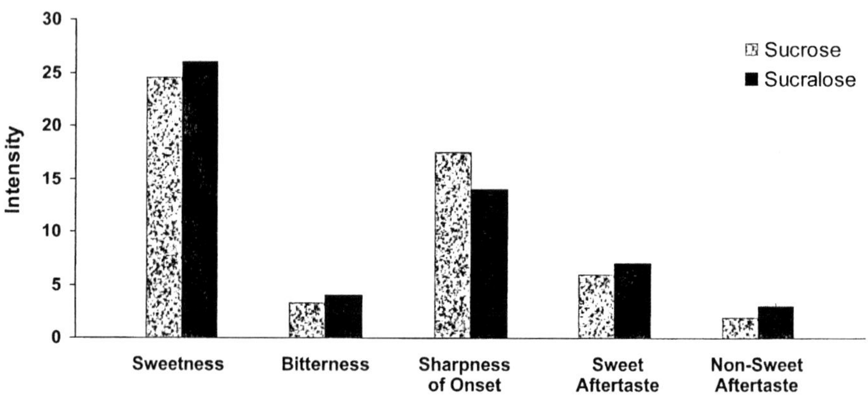

Figure 3 Flavor attributes of sugar and sucralose.

cant difference was found between the intensity of the flavor attributes evaluated in either sweetener. Bitterness was very low in both sucralose and sucrose. The rate or sharpness of sweetness onset and persistence of sweet and nonsweet aftertaste were similar under the conditions of this test. The food or beverage system in which they are studied or the concentration of the sweetener evaluated has an impact on these flavor attributes (3).

D. Sucralose Synthesis

Sucralose is derived from sucrose in a process that substitutes chlorine atoms for hydroxyl group on three sites. In addition, the hydroxyl group of the glucose moiety is isomerized to form sucralose as shown in Fig. 1.

III. SAFETY AND REGULATORY STATUS OF SUCRALOSE

A. Safety

More than 100 scientific studies have been conducted over the past 20 years to evaluate the safety of sucralose for human consumption. Recently, the safety database for sucralose was published in a peer-reviewed supplement of *Food and Chemical Toxicology*, ''Sucralose Safety Assessment'' (4). No evidence exist that the consumption of sucralose or its hydrolysis products would cause any untoward effects. Sucralose is nontoxic and does not hydrolyze or dechlorinate after ingestion. A small amount of hydrolysis of sucralose can be found in products, depending on pH, time, and temperature. The animal studies clearly demonstrate the overall safety of sucralose even under lifelong, high-dose test conditions that would exaggerate any health effects. Studies included evaluation of animals that were exposed to sucralose from conception throughout normal life span and with amounts that far exceed the probable maximum human consumption. In addition, the hydrolysis products of sucralose were subjected to almost the same level of testing as sucralose, including a separate cancer study. Grice and Goldsmith presented a complete overview of the safety data in *Food and Chemical Toxicology* (5).

Metabolism studies indicate that the dog, rat, mouse, and man metabolize sucralose similarly. Therefore, the results of the safety studies conducted on sucralose in animals can be extrapolated to man with confidence.

Sucralose was determined to be nontoxic after acute exposure in rodents up to 16,000 mg/kg body weight. This represents about 16,000 times the projected average intake of sucralose, or the consumption of approximately 16,000 cans of carbonated soft drinks per day.

Sucralose and its hydrolysis products demonstrated no carcinogenic activity, as was shown in three lifetime studies in animals. Two of these studies, a mouse study and a rat study, used concentrations of up to 3% of the diet as sucralose. No adverse effects were seen in rats or mice at doses equivalent to 1500 mg/kg body weight/day, or approximately 1500 times the projected average human consumption.

A two-generation reproduction study in rats, teratology studies in rats and rabbits, special studies in rats related to sperm function, and genetic toxicity studies showed that sucralose and the hydrolysis products had no effect on fertility or reproduction and no teratological effects. Specific studies in mice and monkeys at doses hundreds of times the maximum amount of anticipated human consumption have shown that sucralose and the hydrolysis products had no effect on the nervous system.

Special studies have been performed in rats to evaluate palatability of diets containing sucralose and the effect of sucralose on body weight gain and organ weights in that species. These studies demonstrated that rats dislike the taste of diets that contain high concentrations of sucralose and therefore consume less food than concurrent control animals. In addition, the use of the consumed diet is impacted by the sucralose present, which is poorly absorbed from the gastrointestinal tract and has high osmotic activity. Consequently, over time, the rats supplied diets containing high concentrations of sucralose weighed less than control animals on standard diet. The weights of a number of organs, most notably the thymus and cecum, were affected in the animals consuming the diets with high amounts of sucralose.

Gavage administration and special diet administration studies were used to determine the cause for the effects on body weight gain and organ weight. The gavage studies avoided the confusing variable of altered taste by delivering sucralose directly into the rat's stomach through a plastic tube. Special diet studies accounted for the aforementioned attributes of sucralose incorporation into the diet by correcting for the exact amount of food the animals consumed and the animal's subsequent growth. These studies showed that sucralose had no adverse effect on animal growth or any particular organ weight or function. Particular emphasis was paid to the immune system, because the thymus gland is an important component of that system. The evaluation of the immune system included a special study that demonstrated that sucralose had no adverse effect on immune function. Lord and Newberne (6) have investigated the relationship between sucralose consumption and cecal enlargement in great detail. They showed that the cecal enlargement caused by sucralose is similar to that caused by other poorly absorbed osmotically active carbohydrates and is not of toxicological concern.

The safety of sucralose in man has been evaluated in a number of clinical studies, including studies with individuals who have diabetes. In one study, daily doses of 500 mg, which is equivalent in sweetness to two-thirds of a pound of sugar a day, were fed to 77 volunteers. There were no physical, biochemical, or hematological changes in the volunteers in this study or any of the other clinical studies conducted. Sucralose was found to have no effect on blood glucose, blood fructose, and/or insulin secretion in healthy animals, and it was therefore predicted that no effect in these parameters would be seen in healthy human volunteers or individuals with diabetes. A clinical trial that used 1000 mg/day for 3 months in healthy volunteers demonstrated no adverse effect on glucose homeostasis. Subsequent studies with individuals having type 1 or type 2 diabetes have shown no abnormalities in glucose control. The most important of these studies was a 3 month, five-center study, in which 136 patients were carefully evaluated and monitored following a protocol developed with regulatory authorities. After dosing with 667 mg/day for 3 months to individuals with type 2 diabetes, no adverse effect on glucose homeostasis was detected, particularly in parameters that monitored the long-term blood glucose levels.

B. Regulatory Status

Before the submission of the sucralose database to regulatory agencies, an independent panel of 16 internationally recognized scientific experts was asked to review the sucralose toxicological database and projected human intake data. They were to provide a detailed review on the adequacy of the toxicological data and evaluate the safety of sucralose for human consumption. The panel included experts in toxicology, oncogenicity, clinical toxicology, genetic toxicology, metabolism, biochemistry, reproductive toxicology, physiology, nutrition, hematology, pediatrics, risk assessment, neurotoxicology, and immunology. The expert panel on sucralose concluded that on the basis of the evaluation of the data presented to them and regulatory agencies and on the discussions and deliberations of the entire panel ''sucralose is safe for its intended purpose of use'' (7).

Regulatory agencies around the world have concluded that sucralose is safe. At present, more than 40 countries and international bodies have endorsed the safety of sucralose, including. South Africa, Japan, China, Saudi Arabia, Russia, Australia, New Zealand, Brazil, Argentina, Mexico, Canada, and the United States. In June 1990, the Joint FAO/WHO Expert Committee on Food Additives (JECFA), assigned a permanent Acceptable Daily Intake (ADI) for sucralose of 15 mg/kg body weight/day. Because the Expected Daily Intake (EDI) for sucralose at the 90th percentile as determined by the U.S. Food and Drug Administration (FDA) is 1.6 mg/kg/day (8), the assignment of an ADI value greater than 2.0 mg/kg/day allows virtually unrestricted use of sucralose as a sweetener. The approval granted sucralose by the U.S. FDA in April 1998 was the broadest initial approval ever granted a new food additive. Subsequently, in August 1999, the FDA approved sucralose as a general-purpose sweetener.

C. Cariogenicity

Sprague-Dawley rats infected with *Streptococcus mutans* were provided with sucrose or sucralose along with a diet that contained no additional sucrose. After

35 days, the level of infection of *S. mutans* was determined and the sulcal caries scored. Rats fed sucralose had a similar number of total viable flora; however, the level of *S. mutans* of the total viable flora was decreased by up to 20-fold compared with the sucrose control. Sucralose-consuming rats had less than 50% of the sulcal caries evident in the sucrose control.

Rats were desalivated and inoculated with *S. mutans* and *Actinomyces viscosus* and fed diets that contained either 56% sucrose or sucralose at 93 mg/100 g. The number of coronal lesions did not differ significantly among groups, but the severity of the lesions was significantly lower after 35 days of feeding the sucralose diet compared with the sucrose control.

Substitution of sucralose for sucrose resulted in substantially fewer lingual and proximal lesions. Although similar levels of root surface exposures were achieved in the test and control diet, only rats fed sucralose remained free of root surface caries. These data show that sucralose is noncariogenic (9).

In a second study, it was demonstrated that when sucralose was the sole source of carbon, the growth of 10 strains of oral bacteria and plaque could not be supported. When sucralose was incorporated in a liquid medium containing glucose or sucrose, all organisms tested displayed similar acid production compared with a control without sucralose. Sucralose was shown, using ^{14}C-labeled sucralose, to inhibit the formulation of the glucan and fructan polymers found in plaque. These data show that oral bacteria do not metabolize sucralose and that sucralose has a noncompetitive and reversible inhibitory action on the enzymes necessary to synthesize glucan and fructan polymers (10). The results of these two studies demonstrate that sucralose is noncariogenic and will not contribute to the development of caries.

IV. CHEMICAL AND STABILITY CHARACTERISTICS OF SUCRALOSE

A. Physicochemical Characteristics of Sucralose

Physicochemical characteristics of sucralose that have particular relevance to the potential use of this sweetener in food and beverage applications have been evaluated. Physiochemical properties, such as solubility, are important to understand, because they have an impact on how a food ingredient can be used in the food manufacturing environment.

B. Solubility

Solubility was measured in a thermostatically controlled Wheaton jacketed glass vessel. The temperature range studied was 20°–60°C. As Fig. 4 shows, sucralose

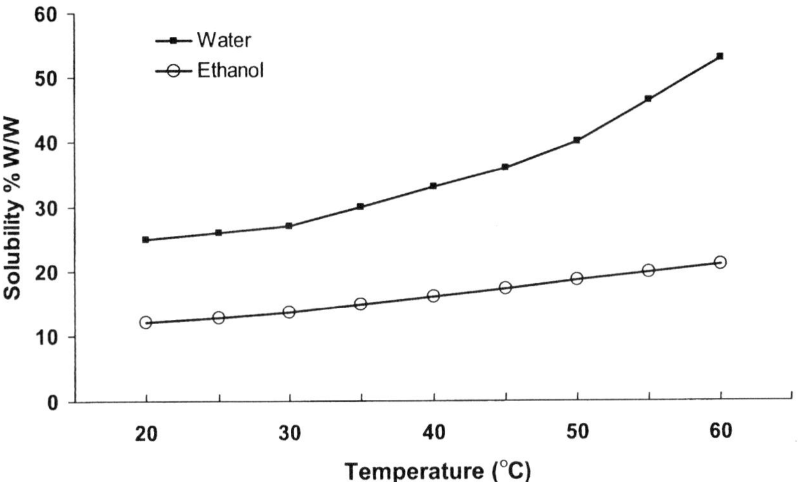

Figure 4 Sucralose solubility in water and ethanol.

is freely soluble in water and ethanol. As the temperature increases, so does the solubility of sucralose in each of the solvents.

The solubility of sucralose in water ranges from 28.3 g/100 ml at 20°C to 66 g/100 ml at 60°C. The solubility of sucralose in ethanol was found to range from 9.5 g/100 ml at 20°C to 18.9 g/100 ml at 60°C. These data demonstrate that sucralose will be easy to use in food operations and that special techniques or equipment to solubilize the sweetener are unnecessary (11).

C. Dynamic Viscosities

The dynamic viscosities of aqueous sucralose solutions were measured from 20°C to 60°C and from 10–50% w/w concentrations within the solubility range. A Rheomat 30 viscometer was used and was fitted with a concentric cylinder head suitable for solutions of low viscosity. The dynamic viscosities of distilled water and 10% w/w aqueous sucralose solutions at different shear rates are shown in Fig. 5. Variations in viscosity over the four shear rates of 512, 691, 939, and 1280 sec^{-1} were insignificant, demonstrating that the sucralose solutions are Newtonian in behavior. Dynamic viscosities of aqueous sucralose solutions measured at 939 sec^{-1} are shown in Fig. 6. Comparison of these data with available data for the viscosity of sucrose solutions shows that sucralose and sucrose have very similar viscosities.

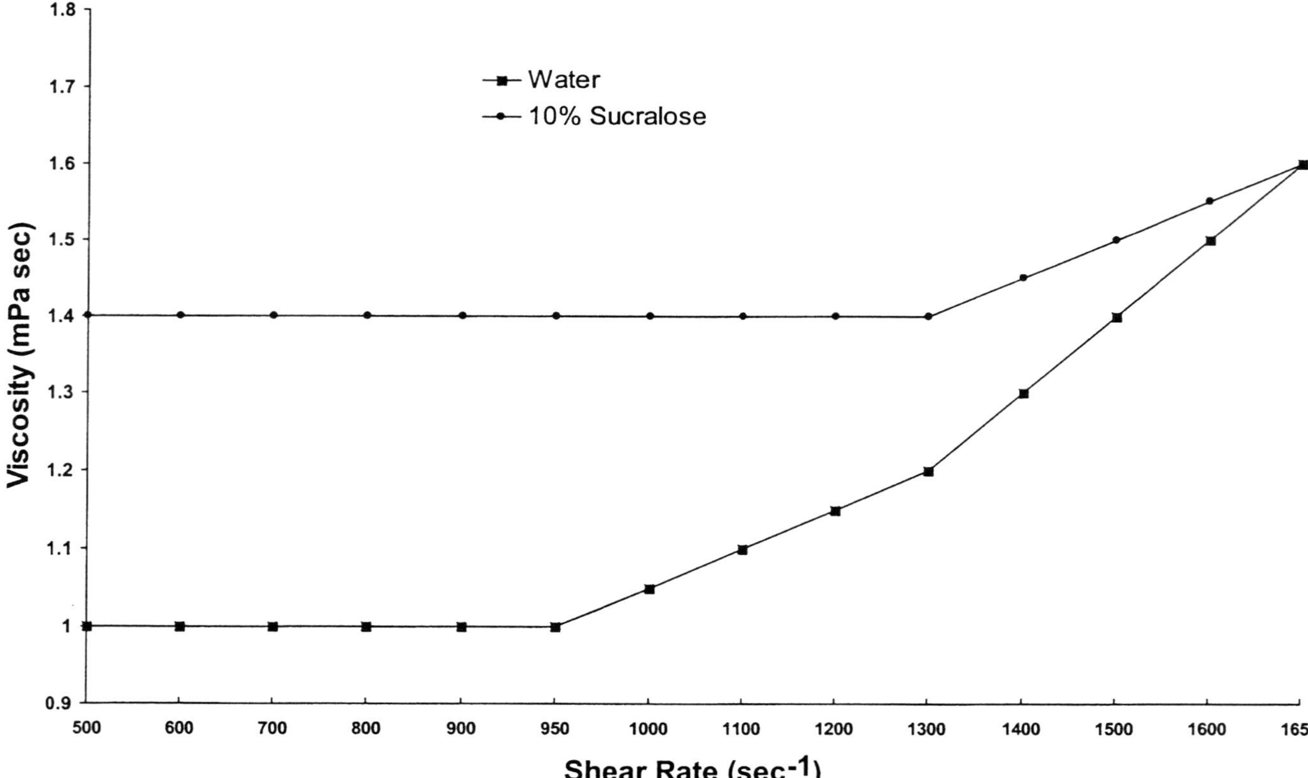

Figure 5 Viscosities of water and 10% sucralose at different shear rates.

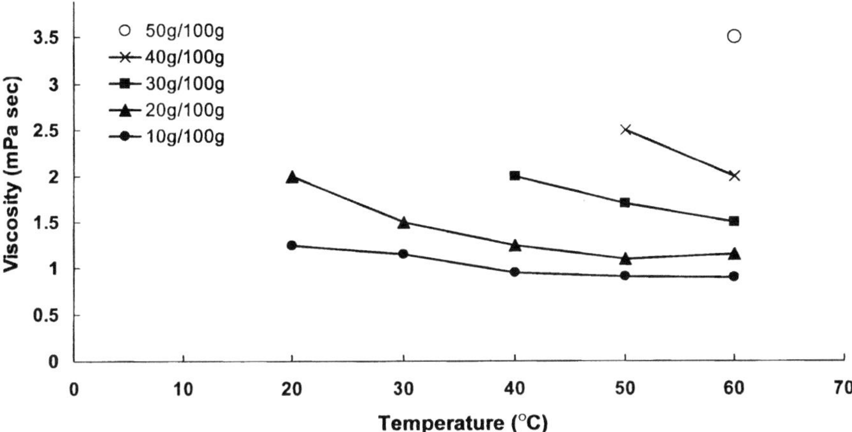

Figure 6 Viscosities of aqueous sucralose solutions measured at 939 sec^{-1}.

The low viscosities and Newtonian behavior of sucralose solutions show that viscosity will not create mixing or dispersion problems in normal food unit operations or processes.

D. Surface Tension

The surface tension of aqueous sucralose solutions held at 20°C was measured with a Kruss model K8600 tensiometer by use of the ring method. Dilute sucralose solutions, 0.1% and 1%, exhibited negligible lowering of surface tensions. The reduction was 0.8 milliNewtons per meter (mN/m) to 3.1 mN/m, respectively. This negligible reduction in surface tension shows that sucralose is not a surfactant, and it will not cause foaming in liquid products.

E. Refractive Index

Although valid and precise high-performance liquid chromatography (HPLC) and gas chromatography (GC) analytical methods for the determination of sucralose are available, more rapid and simpler techniques may be useful to food manufacturers for quality assurance and quality control purposes.

The use of refractive index as a simple means of determining the concentration of sucralose in simple aqueous solutions was evaluated with the use of a Bellingham and Stanley model 60/95 Abbe refractometer. Aqueous solutions of sucralose were prepared at 5, 10, 15, 20 and 25% and tested at 20°C.

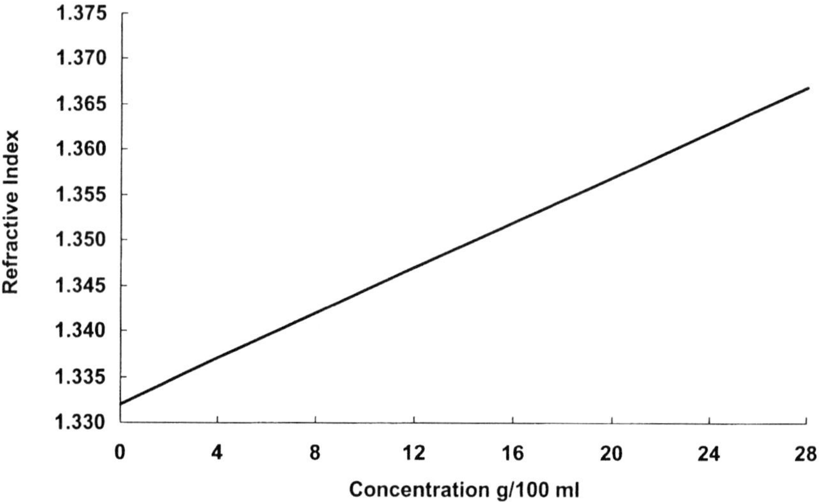

Figure 7 Refractive index of aqueous sucralose solutions.

As Fig. 7 shows, the refractive index of the sucralose solutions is linear with respect to concentration over the range of the concentrations studied. Refractive index can be used to determine sucralose concentration in simple solutions. These physicochemical characteristics demonstrate that sucralose will function well within typical unit operations and processes that are used in the food industry.

F. Chemical Interactions

In the process of chlorination of the sucrose molecule to synthesize sucralose, the primary reaction sites on the molecule are occupied, and this reduces the chemical reactivity of the compound.

The reactivity of sucralose was evaluated in a model system study in which 1% solutions of sucralose were prepared with 0.1% of one of the following:

Bases (niacinamide, monosodium glutamate)
Oxidizing and reducing agents (hydrogen peroxide, sodium metabisulfite)
Aldehydes and ketones (acetaldehyde, ethyl acetoacetate)
Metal salt (ferric chloride)

These classes of compounds and the specific compounds were selected as typical food, beverage, and pharmaceutical system chemicals, and they are representative of many compounds found in formulations. The solutions were held at 40°C for 7 days, at pH 3, 4, 5, or 7, and analyzed for sucralose content by an HPLC

Table 2 Sucralose Interaction Study: Sucralose Retention (%) After 7 Days at 40°C

Sample	pH 3.0	pH 4.0	pH 5.0	pH 7.0
Control	100	99.8	98.9	99.5
Sucralose and hydrogen peroxide	100		99.7	
Sucralose and sodium metabisulfite	99.9		99.9	
Sucralose and acetaldehyde	100		100	
Sucralose and ethyl acetoacetate	98.8		100	
Sucralose and ferric chloride	95.9		98.0	
Sucralose and niacinamide		100		100
Sucralose and monosodium glutamate		99.8		100

technique. Table 2 shows the sucralose retention in each of the solutions. As can be seen from the data, there are no chemical interactions of concern. These data demonstrate that sucralose is chemically inert, and therefore chemical interactions of sucralose with ingredients in formulations are not likely to occur.

G. Stability of the Ingredient

Stability of the dry pure sucralose, as well as the aqueous solutions of sucralose, have been evaluated to determine the shelf life and proper storage and distribution conditions for the ingredient. Because of the physicochemical characteristics and the inherent stability of the molecule in aqueous solution, two forms of the ingredient, dry and a liquid concentrate, are available.

H. Stability of Dry Neat Sucralose

The stability of dry neat sucralose has been measured at several temperatures to assess the shelf life and the storage conditions required to properly store and distribute this form of the ingredient. The first indication of the breakdown of neat sucralose is the development of a mild discoloration (tan/pink), which is accompanied by a very small release of HCl. The development of color occurs before the detection of a loss of sucralose by HPLC analysis. Time and temperature conditions that have impact on the appearance of slight color change have been shown to be affected by packaging materials and container head space. The mild discoloration is a quality issue, and proper storage of dry sucralose is necessary to ensure that sucralose will be of the proper quality for its incorporation in products.

As an example, dry neat sucralose was sealed in glass vials, stored at 75°F (23.8°C), 86°F (30°C), 104°F (40°C), and 122°F (50°C), and monitored for the development of color. The time to first discoloration was 18 months at 75°F; 3 months at 86°F; 3 weeks at 104°F; and less than 1 week at 122°F. As these data show, dry sucralose decomposition kinetics is non-Arrhenius. However, dry neat sucralose has sufficient shelf life if the product is properly packaged, stored, and handled at appropriate temperatures. Abuse of the ingredient is a quality issue, which requires effective materials management techniques to ensure the full shelf life of the ingredient.

I. Stability of Sucralose Liquid Concentrate

The solubility and inherent aqueous stability of sucralose allows for a sucralose liquid concentrate for industrial use. A liquid product form allows operational efficiencies in manufacturing.

A liquid concentrate product is commercially available as a 25% (w/w) sucralose solution with a dual preservative system of 0.1% sodium benzoate and 0.1% potassium sorbate, buffered (0.02 M citrate) to maintain a pH of 4.4. No chemical or physical changes have been found in a sucralose liquid concentrate product stored for greater than 3 years at 68°F (20°C) and 104°F (40°C). No loss of sucralose, sodium benzoate, or potassium sorbate is analytically detectable, the solution maintains its clarity, no color development occurs, and there are no changes in pH. Triple challenge microbiological studies on fresh and aged liquid sucralose concentrate have clearly demonstrated that the liquid ingredient will not be a vector for microbiological contamination when good manufacturing practices are observed.

This product form provides the end user with an ingredient system that is extremely stable, chemically and microbiologically, and compatible with most food unit operations.

J. Stability of Sucralose in Model Systems

The use of sucralose in food products was first evaluated by the use of model food systems. Subsequent sucralose use in commercial products has provided practical data verifying the reliability of the model system. The model food systems are simplified to reduce the complexity of food analysis, food interactions, and to provide a milieu in which the inherent stability of a food ingredient can be understood.

K. Model Dry Food Systems

Two model dry systems were used to evaluate the stability of sucralose in dry food systems: sucralose dry blended with maltodextrin and sucralose co-spray dried with maltodextrin.

Sucralose was dry blended with maltodextrin in the following ratios: 75:25, 50:50, and 25:75 (sucralosemaltodextrin). As the amount of sucralose was decreased, the time to discoloration increased by factors of 1.5, 2, and 8, respectively, compared with neat sucralose. The dispersion of sucralose in dry mix applications, typically at use levels of less than 1%, also markedly increases the stability of sucralose. The dry stability improvement of sucralose with dilution has been shown in products in the marketplace to provide sufficient stability for dry applications. For example, commercially available powdered gelatin mix in retail packages was stored for 6 months at 35°C, followed by 2 years at room temperature. No loss of sucralose was detected by HPLC.

To determine the appropriateness of using sucralose in spray-drying operations and in spray-dried products, sucralose was spray-dried with maltodextrin in 50:50 and 25:75 sucralosemaltodextrin ratios. As the amount of sucralose decreased, the time to first discoloration was increased by factors of 10 and 30 times, respectively. These data demonstrate that the use of sucralose in spray-drying and in spray-dried applications is appropriate.

These studies demonstrate that sucralose can be used in dry food applications with no expectation of discoloration when the food products are handled in normal food distribution systems.

L. Model Aqueous Food Systems

Simple aqueous solutions systems studies and radiolabeled studies in beverages and baked goods were performed to demonstrate the inherent stability of sucralose in liquid and aqueous food systems.

Before the introduction of commercial products, the aqueous stability of sucralose was determined over various pHs, temperatures, and times in model systems studies.

Fig. 8 shows the stability data for sucralose solutions stored at 100°C. After 1 hour there was 2% loss of sucralose at pH 3, 5, and 7. After 2 hours, there was 2% loss at pH 5, a 3% loss at pH 7, and a 4% loss of sucralose at pH 3. These data were collected using an HPLC technique that has a coefficient of variation of ±2%.

Because a loss of 10–15% of a sweetener is necessary before a loss of sweetness can be detected, consumers will not detect these losses of sucralose as a loss of sweetness (12). These data show that sucralose will withstand heat-processing conditions that are typical of most food processes and that sucralose will be able to withstand the rigors of rework or reprocessing.

Fig. 9 shows the stability of sucralose solutions stored at 30°C for up to 1 year at pH 3.0, 4.0, 6.0, and 7.5. Again an HPLC technique that has a coefficient of variation ±2% was used for analysis. No measurable loss of sucralose was found after 1 year of storage at pH 4.0, 6.0, and 7.5. At pH 3.0 there was less than 4% loss of sucralose after 1 year of storage (13). This level of stability provides food manufacturers and consumers with food products that have a consistent level of sweetness throughout their shelf life.

The stability data in model aqueous systems was used to predict the efficacy

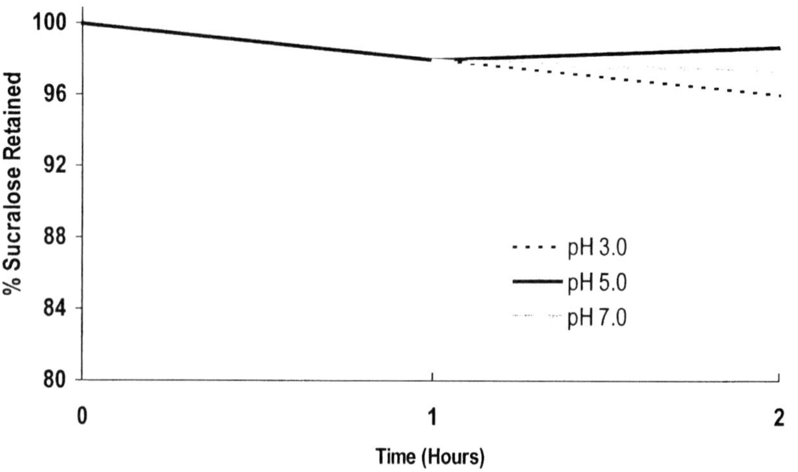

Figure 8 Aqueous stability of sucralose, effect of pH at 100°C.

of using sucralose in aqueous food systems. Since commercialization, sucralose has been used to formulate high-quality food products that maintain their sweetness throughout the food-processing and distribution system. This stability provides the consumer the benefit of high-quality products that are consistent in their sweetness profile and level, regardless of their processing or shelf life.

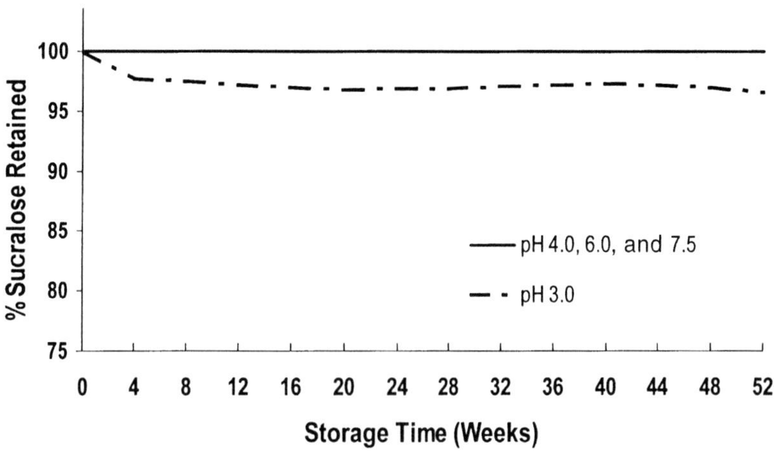

Figure 9 Aqueous stability of sucralose, effect of pH at 30°C.

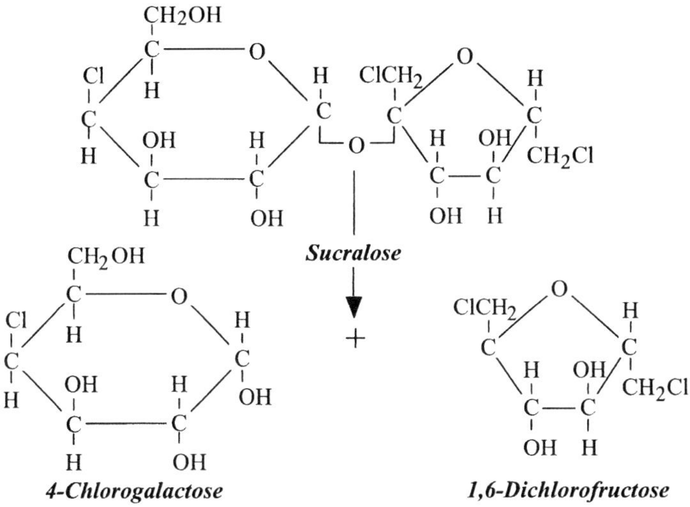

Sucralose

4-Chlorogalactose + *1,6-Dichlorofructose*

Figure 10 Breakdown of sucralose to component moieties.

These studies and all the stability studies on sucralose demonstrate that sucralose, in aqueous systems, will slowly break down to its component moieties, 4-chlorogalactose (4-CG) and 1,6-dichlorofructose (1,6-DCF), as shown in Fig. 10. This breakdown is depends on pH, temperature, and time.

M. HP Formation

To fully delineate the mechanism of breakdown in food systems, a ^{36}Cl-labeled sucralose aqueous stability study was conducted. Radiolabeled sucralose was prepared with a specific activity of 1.027 microcuries (μc) per milligram and a radiochemical purity of 99%. This material was then formulated into glycine/HCl buffers at pH 2.5, 3.0, and 3.5, or in a cola carbonated soft drink formulated at pH 3.0 and 3.5. Sucralose was added at 200 parts per million concentration. Buffers and colas were stored at 30°C or 40°C for 1 year and sampled at 0, 8, 16, 26, and 52 weeks.

Samples of the buffers and colas were analyzed with thin-layer chromatography (TLC) and liquid scintillation counting. The amounts of sucralose, 4-chlorogalactose, and 1,6-dichlorofructose were calculated after correcting for the amount of 4-chlorogalactose and 1,6-dichlorofructose present in the radiolabeled sample analyzed on day 0.

The pattern of radioactivity of samples from the buffers and the colas was consistent with the pattern of radioactivity that was found in the ^{36}Cl-labeled sucralose sample analyzed at day 0. No new compounds were separated or identified. No evidence of the presence of inorganic chloride was found in the buffer or cola products.

Fig. 11 shows the stability data of the pH 3.0 cola, which reacted similarly to the buffered solution (not shown). The sample had been stored at 40°C for 1 year, and the degradation of the sample is consistent with the stability data discussed earlier. The increase in 4-chlorogalactose and 1,6-dichlorofructose is consistent with the knowledge that sucralose breaks down to only these two products.

These data from the ^{36}Cl-labeled sucralose study were then used to determine whether all the sucralose could be accounted for as the hydrolysis products and sucralose. Mass balance, for the purpose of this study, is defined as the percent of sucralose accounted for by residual sucralose and the hydrolysis products and is shown in Table 3. These data show that all of the sucralose was accounted for as sucralose, 4-chlorogalactose, or 1,6-dichlorofructose, and there is no indication of any other sucralose breakdown products at these pHs. No other compounds were formed, and the 4-chlorogalactose and 1,6-dichlorofructose did not interact or break down further.

A second radiolabeled stability study was performed in baked goods. The objective of this study was to demonstrate the stability of sucralose in a variety of common baked goods prepared under typical baking conditions. For this study, ^{14}C-labeled sucralose was used to minimize the difficulties associated with the recovery and detection of low levels of sucralose in the presence of carbohydrates

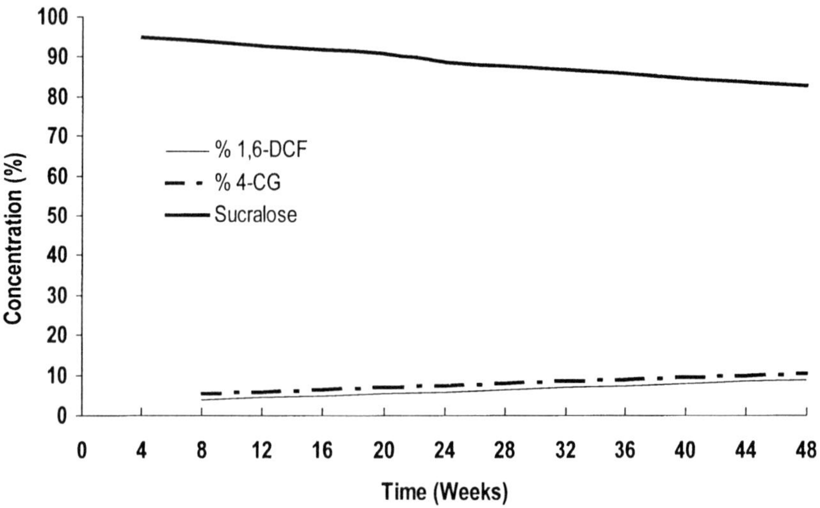

Figure 11 pH 3.0 cola 40°C stability.

Table 3 Percent Mass Balance in ^{36}Cl-Labeled Sucralose Stability Study at 40°C Storage

Solution	8 wk	16 wk	26 wk	52 wk
Aqueous buffer				
pH 2.5	99.8	98.9	98.8	100.0
pH 3.0	99.9	99.9	99.6	100.1
pH 3.5	100.2	100.0	100.0	100.1
Cola				
pH 3.0	99.9	102.0	99.8	100.8
pH 3.5	100.0	99.7	100.1	100.1

found in baked goods. Cakes, cookies, and graham crackers were selected for this study, because they represent a cross section of common bakery ingredients and typical baking conditions encountered in the baking industry.

The baked goods were prepared under simulated baking time and temperature conditions. The sucralose was recovered by aqueous extraction, and the extracts were cleaned up by a methanol precipitation before TLC analysis.

The total radioactivity expected for 100% recovery of the ^{14}C-labeled material from each baked product was calculated and compared with the actual level of radioactivity recovered in the aqueous extract. This comparison demonstrated that essentially 100% of the sucralose radioactivity that was added to each formulation before baking was recovered by the extraction of the baked goods.

With the demonstration of complete recovery of the radiolabled material, the extracts were subjected to TLC using two solvent systems independently as eluents. The solvent systems were selected because of their ability to separate sucralose from its potential breakdown products. The amount of radioactivity was quantitatively determined. The quantity and distribution of radioactivity contained in each baked product extract was compared with the quantity and distribution of a sucralose standard solution.

The most significant findings of this study were that no products other than sucralose were found under the conditions of this experiment, and the TLC distribution of radioactivity from the baked goods extracts corresponded almost exactly with that of the sucralose standard. These findings demonstrate that sucralose is a sweetener that is suitable for its intended use as a high-potency sweetener in baked goods (14).

These model system stability studies and the subsequent data from actual products clearly demonstrate that sucralose is a remarkably stable ingredient that has a wide range of applicability. Furthermore, it is unlikely that stability will in any way restrict the use of sucralose in formulations.

N. Analytical Methodology

Significant efforts have been made to develop analytical methods for the recovery and analysis of sucralose from food products. HPLC has been shown to be the most effective analytical method for precision and accuracy in the analysis of sucralose.

As with much of food analysis, the extraction method is the key to acceptable recoveries. The solubility of sucralose in water and in alcohol facilitates the selective extraction of sucralose from food products. When required, additional sample clean up and concentration of sucralose is achieved using solid phase extraction (SPE) techniques. A nonpolar C-18 SPE cartridge has been most effective in most of the applications evaluated. An Alumina SPE cartridge has been most effective for dairy products.

Reversed-phase HPLC using a C-18 column with a mobile phase of 70% water and 30% methanol is used in most analyses. Occasionally, acetonitrile is used in place of methanol. Injection volumes typically in the range of 100–150 ul are used, but they vary with the concentration of sucralose in the sample being evaluated.

Although several options exist regarding detection, refractive index is typically the most appropriate. Refractive index provides a detection system that has fewer problems with interference than ultraviolet detection and can be used effectively when temperature is controlled (13).

V. USE OF SUCRALOSE IN FOOD APPLICATIONS

A. Food Categories Included in U.S. Food Additive Petition

Because of the excellent physicochemical characteristics and the remarkable stability of sucralose, a petition, broad in scope to allow the use of sucralose, was developed and submitted to the U.S. FDA in February 1987. This petition covered many traditional categories of use for high-potency sweeteners and several nontraditional use categories. Table 4 lists the Code of Federal Regulation (CFR) categories for which sucralose use was petitioned. These categories are broad in nature and are not product specific. The data submitted in the sucralose food additive petition clearly demonstrated that sucralose has both the safety profile and the stability requirements for the effective use of the sweetener in all of the petitioned categories. Therefore, the FDA approved sucralose for use in these categories in April 1998. Recognizing the unique combination of safety, estimated use level, and stability,the FDA approved the use of sucralose as a general-purpose sweetener in August 1999. These data and regulatory actions established

Table 4 Initial Application Categories in Sucralose Food Additive Petition (U.S.)

- Baked goods and baking mixes
- Beverages and beverage bases
- Chewing gum
- Coffee and tea
- Confections and frostings
- Dairy products analogs
- Fats and oils (salad dressings)
- Frozen dairy desserts and mixes
- Fruits and water ices
- Gelatins and puddings
- Jams and jellies
- Milk products
- Processed fruits and fruit juices
- Sugar substitutes
- Sweet sauces, toppings, and syrups

Table 5 Suggested Sucralose Usage Levels in Selected Applications

Product category	Recommended sucralose usage level (ppm)	Product category	Recommended sucralose usage level (ppm)
		Beverages	
Frozen desserts	100–150	Ready to drink	150–250
Ice cream type	75–150	Powdered (as consumed)	200–250
Ice milk type	100–200	Carbonated	175–225
Sherbert type	150–250	Fountain syrup	1700–1900
Water ice			
Baked goods		Syrups	
Cake	500–600	Pancake	1000–1500
Cookies	300–350	Chocolate	1000–1300
Apple pie filling	175–225	Fruit products	
Brownies	200–300	Jams and jellies	400–850
Graham cracker	200–300	Canned fruit	250–300
Cheesecake	200–250	Powdered desserts	
Dairy products		Instant pudding	250–350
Plain yogurt	100- 150	Chewing gum	500–500
Fruit yogurt	175–225	Pourable salad dressing	100–150
Flavored milk	60–100		

the relative limitless uses of sucralose governed by its physicochemical characteristics and high-quality taste.

B. Typical Sucralose Usage Levels

Extensive development in support of commercialization of sucralose demonstrates the usefulness of sucralose in sweetening high-quality food and beverage products. The incorporation of sucralose into hundreds of products in countries around the world substantiates the data developed in those studies. Table 5 lists suggested use levels for sucralose in a range of applications. The actual use level in specific formulations varies depending on sweetness level desired, other ingredients used in the formulation, and flavor systems used.

VI. PATENT STATUS

An extensive patent portfolio is related to sucralose that covers its use, methods of synthesis, and other intellectual property. Patents related to sucralose were first issued in the United Kingdom in 1979, Canada in 1981, and the United States in 1982.

VII. CONCLUSION

Sucralose is a high-quality, high-potency, noncaloric sweetener that is derived by the selective chlorination of sucrose. The resulting product is extraordinarily stable with excellent physicochemical characteristics that permit the use in a wide variety of applications. The safety database has shown sucralose to be safe for its intended uses for human consumption. These characteristics provide the food, beverage, and pharmaceutical industry with a unique opportunity to improve existing products and develop totally new products that will meet the ever-growing consumer demand for good-tasting, high-quality products.

REFERENCES

1. L Hough, RA Khan. Trends Biochem Sci 3:61–63, 1978.
2. MR Jenner. Progress in sweeteners. In: T Grenby, ed. London: Elsevier, 1989, pp 121–141.
3. SG Wiet, PK Beyts. Psychosensory characteristics of the sweetener sucralose. Proceedings of the 49th Annual Meeting of the Institute of Food Technologists, New Orleans, LA, 1988.

4. Sucralose Safety Assessment. Food Chem Toxicol 38(2):S1–S129, 2000.
5. HC Grice, LA Goldsmith. Sucralose—An overview of the toxicity data. Food Chem Toxicol 38(2):S1–S6, 2000.
6. GH Lord, PM Newberne. Renal mineralization—A ubiquitous lesion in chronic rat studies. Food Chem Toxicol 18(6):449–455, 1990.
7. Anonymous. U.S. Food Additive Petition 7A3987, 1987, pp A000755–A000783.
8. Anonymous. U.S. Federal Register, National Archives and Records Administration, No. 64, April 3, 1998.
9. WH Bowen, DA Young, SK Pearson. The effects of sucralose on coronal and root-surface caries. J Dent Res August: 1487–1487, 1990.
10. DA Young, WH Bowen. The influence of sucralose on bacterial metabolism. Dent Res August: 1480–1484, 1990.
11. MR Jenner, A Smithson. Physico-chemical properties of the sweetener sucralose. J Food Sci 54:1646–1649, 1989.
12. Anonymous. U.S. Food Additive Petition 7A3987, 1987, pp 739–754.
13. V Mulligan, ME Quinlan, MR Jenner. General procedures for the analytical determination of the high intensity sweetener sucralose in good products. Proceeding of the 49th Annual Meeting of the Institute of Food Technologists, New Orleans, LA, 1988.
14. RL Barndt, G Jackson. Food Technol 44(1):62–66, 1990.

12
Less Common High-Potency Sweeteners

A. Douglas Kinghorn
University of Illinois at Chicago, Chicago, Illinois

César M. Compadre
University of Arkansas, Little Rock, Arkansas

I. INTRODUCTION

In this chapter, recent progress on the study and use of several less commonly encountered naturally occurring and synthetic sweet substances will be described. Several of the natural sweeteners have served as template molecules for extensive synthetic modification. Greater emphasis will be provided for those compounds that have commercial use as a sweetener in one or more countries, and priority will be given to those sweet compounds for which there is recent published information. Following the format of the earlier versions of this chapter (1, 2), substances that modify the sweet taste response will also be mentioned.

II. NATURALLY OCCURRING COMPOUNDS

A. Glycyrrhizin

Glycyrrhizin, which is also known as glycyrrhizic acid, is an oleanane-type triterpene glycoside whose use as a sweetener has been reviewed (1–5). This compound (Fig. 1), a diglucuronide of the aglycone, glycyrrhetinic acid, is extracted from the rhizomes and roots of licorice (*Glycyrrhiza glabra* L. Fabaceae) and other species in the genus *Glycyrrhiza*. In its natural form, the compound occurs in *G. glabra* in yields of 6–14% as a mixture of metallic salts. Well-established procedures are available for the extraction and purification of glycyrrhizin from the plant. Conversion of glycyrrhizin to ammoniated glycyrrhizin (by treatment of a hot-water extract of licorice root with sulfuric acid followed by neutralization with dilute ammonia) results in a more water-soluble compound that is reasonably stable at elevated temperatures (1–3). Glycyrrhizin has been rated as approximately 50–100 times the sweetness of sucrose, although it has a slow onset of sweet taste and a long aftertaste. Ammoniated glycyrrhizin has similar hedonic properties to glycyrrhizin. Its sweetness potency, which is about 50 times that of sucrose, is increased in the presence of sucrose (1, 2). There have been several attempts to modify the sensory parameters of glycyrrhizin by synthetically modifying its carbohydrate units (2, 5). Recently, it has been found that the monoglucuronide of glycyrrhizic acid (MGGR; Fig. 1) is 941 times sweeter than sucrose and is sweeter than several other glycyrrhizin monoglycosides (4–6). MGGR may be produced from glycyrrhizin by enzymatic hydrolysis using a microbial enzyme from *Cryptococcus magnus* (4) and is now used commercially in Japan to sweeten dairy products such as yogurt, chocolate milk, and soft drinks flavored with fruits (7). Other derivatives of glycyrrhizin that are sweeter than the parent compound and MGGR are esters of 3-*O*-β-glycyrrhetinic acid with L-aspartyl dipeptide derivatives (8) and with α-amino acids (9). Periandrins I–IV are additional naturally occurring oleanane-type triterpene glycosides that were isolated by Hashimoto and colleagues from the roots of *Periandra dulcis* Mart. (Fabaceae) (Brazilian licorice) in the early 1980s. These compounds are 90–100 times sweeter than sucrose but occur in the plant in low yields and are somewhat difficult to purify from bitter substances with which they co-occur (1, 2). A fifth compound in this series, periandrin V, which varies structurally from periandrin I as a result of the substitution of a terminal β-D-glucuronic acid unit by a β-D-xylopyranosyl moiety, exhibits about twice the sweetness potency

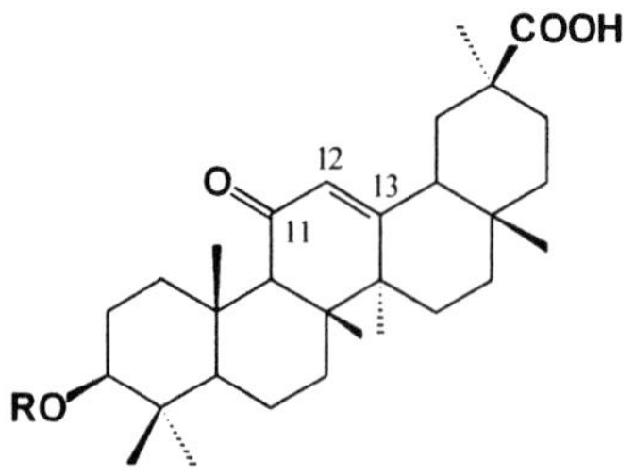

COOH
O
12
13
11
RO

	R
Glycyrrhizin	β-glcA^2-β-glcA
MGGR	β-glcA

Figure 1 Structures of glycyrrhizin and MGGR. (GlcA = D-glucuronopyranosyl).

of the other periandrins (10). While it was at one time thought that glycyrrhizin was the sweet principle of the leaves of *Abrus precatorius* L. (Fabaceae), five novel cycloartane-type triterpene glycosides, abrusosides A-E, proved to be responsible for this sweetness. Although these compounds are pleasantly sweet and their water-soluble ammonium salts were rated as up to 100 times the sweetness potency of sucrose, they produce a long-lasting sweet sensation and occur in the leaves in less than 1% w/w combined yield (11, 12).

Partially purified *Glycyrrhiza* extracts, which contain at least 90% w/w glycyrrhizin, are used widely in Japan for sweetening and flavoring foods, beverages, oriental medicines, cosmetics, and tobacco (1, 2). In 1994, it was reported that about 9000 tons of licorice are imported into Japan each year, mainly from Afghanistan, Pakistan, the People's Republic of China, and Russia (4). In the United States, ammoniated glycyrrhizin is included in the generally recognized as safe (GRAS) list of approved natural flavoring agents, not as an approved sweetener (13). There are many applications for this compound as a flavorant, flavor modifier, and foaming agent. Although it is useful for incorporation into confectionery and dessert items, ammoniated glycyrrhizin is only used in carbonated beverages that do not have too low a pH because this substance tends to precipitate at pH levels less than 4.5 (1).

There is a very extensive literature on biological activities of glycyrrhizin other than its sweetness as exemplified by its antiallergic, anti-inflammatory, antitussive, and expectorant actions. Unfortunately, the widespread use of glycyrrhizin and ammoniated glycyrrhizin by humans has been found to lead to pseudoaldosteronism, which is manifested by hypertension, edema, sodium retention, and mild potassium diuresis (1, 2, 14, 15). The 11-oxo-$\Delta^{12,13}$-functionality in ring C of glycyrrhetinic acid has been attributed as the part of the molecule responsible for this untoward activity (2). Glycyrrhetinic acid is known to inhibit the renal enzyme 11β-hydroxysteroid dehydrogenase, which catalyzes the inactivation of cortisol to cortisone (14). The Ministry of Health in Japan has issued a caution stipulating that glycyrrhizin consumption should be limited to 200 mg/day when used in medicines (2). Similarly, the Dutch Nutrition Information Board has advised against daily glcyrrhizin intake in excess of 200 mg, corresponding to about 150 g of licorice confectionery (3). Glycyrrhizin, at a level of 0.5–1%, has been shown to inhibit plaque formation mediated by *Streptococcus mutans*, a cariogenic bacterial species (2). As a consequence, it has been suggested that glycyrrhizin is suitable for wider use as a vehicle and sweetener for medications used in the oral cavity (2).

B. Mogrosides

Mogrosides IV and V are the principal sweet cucurbitane-type triterpene glycoside constituents of the dried fruits of the Chinese plant *lo han kuo*. This species, a member of the family Cucurbitaceae, was accorded the binomial *Momordica grosvenorii* Swingle in 1941, which was then changed to *Thladiantha grosvenorii* (Swingle) C. Jeffrey in 1979. However, more recently, the plant name has been changed again to *Siraitia grosvenorii* (Swingle) C. Jeffrey. Chemical studies on this plant did not begin until the 1970s (2).

The structures of mogrosides IV and V (Fig. 2) were established by Takemoto and colleagues after extensive chemical and spectroscopic studies (2). The most abundant sweet principle of *lo han kuo* dried fruits is mogroside V, which may occur at concentration levels of more than 1% w/w (2). Mogroside V is a polar compound, because it contains five glucose residues, which readily permit its extraction into either water or 50% aqueous ethanol. Aqueous solutions containing mogroside V are stable when boiled (1, 2). Mogrosides IV and V have been rated as being in the ranges 233–392 and 250–425 sweeter than sucrose by human taste panels, depending on the concentrations at which they were tested (2, 16). Several other minor sweet and nonsweet cucurbitane-type triterpenoid glycoside constituents have been reported from both the dried and fresh fruits of *S. grosvenorii* (17, 18). Of these, siamenoside I was rated as much as 563 times sweeter than sucrose at a concentration of 0.010% w/v (17).

Extracts made from *lo han kuo* fruits have long been used by local populations in Kwangsi province in southern regions of the People's Republic of China for the treatment of colds, sore throats, and minor stomach and intestinal complaints (2). Preparations made from *lo han kuo* are available in Chinese medicinal herb stores in several countries including the United States. Recently, a major corporation in the United States has filed a patent application for the use of ex-

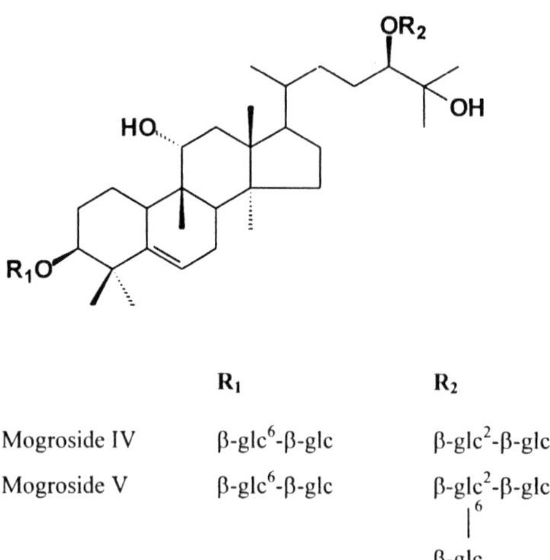

	R_1	R_2
Mogroside IV	β-glc^6-β-glc	β-glc^2-β-glc
Mogroside V	β-glc^6-β-glc	β-glc^2-β-glc (6 → β-glc)

Figure 2 Structures of mogroside IV and mogroside V. (Glc = D-glucopyranosyl.)

tracts from *S. grosvenorii* and other *Siraitia* species as a sweet juice (19). It was estimated that the domestic demand for *lo han kuo* fruits containing mogroside V for food, beverages, and medicinal products in Japan in 1987 was 2 metric tons, representing a sales volume of 40 million yen (2).

Safety studies on mogroside V have not been extensive to date. The compound has been found to be nonmutagenic when tested in a forward mutation azaguanine assay with *Salmonella typhimurium* strain TM677 (2). Mogroside V produced no mortalities when administered by oral intubation at doses up to 2 g/kg body weight in acute toxicity studies in mice, and an aqueous extract of *lo han kuo* fruits exhibited an LD_{50} in mice of >10 g/kg body weight (2). There appear to have been no adverse reactions among human populations who have ingested aqueous extracts of *lo han kuo*, which would be expected to contain substantial quantities of mogroside V. Therefore, *lo han kuo* extracts containing mogroside V are worthy of wider application for sweetening purposes in the future because of their apparent safety in addition to favorable sensory, stability, solubility, and economic aspects.

C. Phyllodulcin

Phyllodulcin is produced from its naturally occurring glycosidic form by enzymatic hydrolysis when the leaves of *Hydrangea macrophylla* Seringe var. *thunbergii* (Seibold) Makino (Saxifragaceae) and other species in this genus are crushed or fermented. Phyllodulcin (Fig. 3) is a dihydroisocoumarin and was isolated initially in 1916 and structurally characterized in the 1920s. In 1959, this sweet compound was found to have 3*R* stereochemistry (2). In recent work, phyllodulcin from the unprocessed leaves of its plant of origin has been found to be a 5:1 mixture of the *R* and *S* forms (20). Phyllodulcin has been detected in the leaves of *H. macrophylla* subsp. *serrata* var. *thunbergii* at levels as high as 2.36% w/w (2). Several patented methods are available for the purification of phyllodulcin. In one such procedure, after initial extraction from the plant with methanol or ethanol, hydrangenol (a nonsweet analog of phyllodulcin) and pigment impurities were removed after pH manipulations and extraction with chloro-

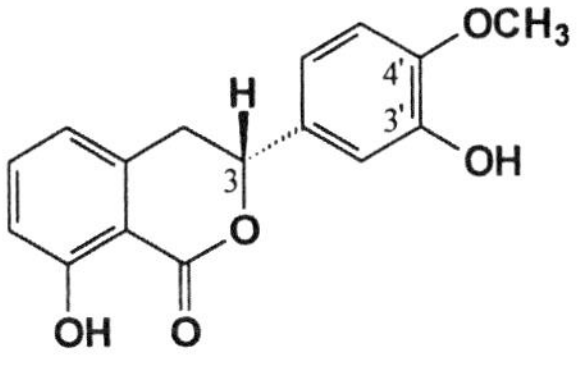

Figure 3 Structure of phyllodulcin.

form. Phyllodulcin was then selectively extracted in high purity at pH 10 with a nonpolar solvent (2). The relative sweetness of phyllodulcin has been variously reported as 400 and 600–800 times sweeter than sucrose, although the compound exhibits a delay in sweetness onset and a licorice-like aftertaste (2). There have been extensive attempts to modify the phyllodulcin structure to produce compounds with improved sensory characteristics. As a result of such investigations, it has been established that the 3-hydroxy-4-methoxyphenyl (isovanillyl) unit of phyllodulcin must be present for the exhibition of a sweet taste, but the phenolic hydroxyl group and the lactone function can be removed without losing sweetness (2, 16, 21). To date, the phyllodulcin derivatives that have been produced synthetically seem to have limitations in terms of their water solubility, stability, and/or sensory characteristics (2). Phyllodulcin remains an attractive target for total synthesis, with a number of additional procedures reported recently (22–24).

The fermented leaves of *H. macrophylla* var. *thunbergii* (Japanese name *amacha*) are used in Japan to produce a sweet tea that is consumed at *Hamatsuri*, a Buddhist religious festival (2). This preparation is listed in volume XIII of the *Japanese Pharmacopeia* and is used also in confectionery and other foods for its cooling and sweetening attributes (20). A 1987 estimate indicated that demand for extracts of *Hydrangea* species containing phyllodulcin was 1 metric ton, with a value of 15 million yen (2). Pure phyllodulcin has been found to be nonmutagenic in a forward mutation assay and also not acutely toxic to mice when administered by oral intubation at up to 2 g/kg body weight (2). The low solubility in water and the sensory shortcomings of phyllodulcin that have been mentioned would seem to limit the prospects of this natural product from becoming more widely used as a sweetener in the future.

D. Sweet Proteins

1. Thaumatin

Thaumatins I and II are the major sweet proteins obtained from the arils of the fruits of the West African plant *Thaumatococcus daniellii* (Bennett) Benth. (Marantaceae), with altogether five different thaumatin molecules now known (thaumatins I, II, III, and a and b) (2, 25). Thaumatin I, composed of 207 amino acid residues, of molecular weight 22,209 daltons, has a relative sweetness of between 1600 and 3000 when compared with sucrose on a weight basis (2, 25). Thaumatin protein (which is known by the trade-name of Talin® protein) was comprehensively reviewed by J.D. Higginbotham in the First Edition of *Alternative Sweeteners* in terms of botany, production, biochemistry, physical characteristics, sensory parameters, sweetness synergy with other substances, applications (including flavor potentiation and aroma enhancement effects), safety assessment, cariogenic evaluation, and regulatory status (2) and has been featured in other reviews on sweet proteins (25–27). Accordingly, this sweet protein is not considered in depth in this chapter. However, the recent literature on thaumatin I is extensive and includes x-ray crystallographic studies (28, 29), investigations on sites in the molecule responsible for the sweet taste of this protein (e.g., 30), and methods for the production of this sweet substance using recombinant strains of fungi and bacteria (e.g., 31–33).

Talin® protein was initially permitted as a food additive in Japan in 1979 (2). In 1987 there was an estimated Japanese demand of 200 kg of thaumatin, which was valued at a price of 350,000 yen per kilogram (2). Despite the fact that Talin® protein has been approved as a sweetener in Australia and several European countries (2, 34), it now appears that the major use of this product in the future will be as a flavor enhancer. In the United States, thaumatin has been accorded GRAS status as a flavor adjunct for a number of categories, including milk products, fruit juices and ices, poultry, egg products, fish products, processed vegetables, sweet sauces, gravies, instant coffee and tea, and chewing gum.

2. Monellin

The sweet protein isolated from the fruits of another African plant, *Dioscoreophyllum cumminsii* (Stapf) Diels (Menispermaceae), has been called "monellin," a substance containing two polypeptide chains of molecular weight 11,086 daltons (2, 25). "Monellin" is actually monellin 4, one of five sweet proteins, named monellins 1–5, that have been isolated from this plant (2). Isolation procedures on buffered aqueous extracts of *D. cumminsii* fruit pulp enable 3–5 g of protein to be purified per kilogram of fruit, with a sweetness intensity relative to 7% w/v sucrose of 1500–2000 times (2). Monellin is somewhat costly to produce, and its plant of origin is difficult to propagate. In addition, the compound has a slow onset of taste, along with a persistent aftertaste, and its sweet effect is both thermolabile and pH sensitive (2). No toxicological data appear to have been published for this compound (2).

Despite the fact that monellin is not used commercially as a sweetener, there has been considerable interest in this compound in the scientific literature, particularly in regard to its crystal structure (25, 35), chemical synthesis (25, 36), recombinant production (37, 38), and molecular mechanism of sweet taste (39).

3. Other Sweet Proteins

Mabinlins I and II are sweet proteins produced by the seeds of *Capparis masakai* Lévl (Capparidaceae), a plant that grows in Yunnan province in the People's Republic of China, and used in traditional Chinese medicine to treat sore throats and as an antifertility agent (2). Children are reported to chew the seed meal of *C. masakai* both as a result of its sweet taste and because it imparts a sweet taste to water drunk later (2). The molecular weights of mabinlins I and II were originally reported as 11,600 and 10,400 daltons, respectively (2). In early work, these sweet albumins were found to occur in a combined yield of 13% w/w of the dry weight of the defatted seeds, with mabinlin I being the major component (2). Altogether, five homologs have been isolated from impure mabinlin protein, using sequential extraction with 0.5 M NaCl solution, ammonium sulfate fractionation, CM-Sepharose ion-exchange chromatography, and gel filtration (25). Mabinlins I and II are reported to be less sweet than thaumatin and monellin, although similar in sweetness quality (2). The sweet taste of mabinlin II persists on incubation at 80°C for 48 hr at pH 6, although the sweetness of mabinlin I is lost after 30 min when stored under the same conditions (2). However, a 0.1% solution of mabinlin II synthesized chemically using a solid-phase method exhibited an astringent-sweet taste (40). Owing to its heat stability, most recent interest has focused on mabinlin II among the *C. masakai* seed proteins, and this substance has been produced by cloning and DNA sequencing (41). To date, no pharmacological studies appear to have been performed on any of the mabinlin proteins.

Pentadin is a sweet protein with an estimated molecular weight of about 12,000 daltons and a sweetness potency of about 500 times sweeter than sucrose on a weight basis, which was reported from the fruits of the African plant *Pentadiplandra brazzeana* Baillon (Pentadiplandraceae) about 10 years ago. More recently, a second sweet protein, brazzein, has been described from the same plant source as pentadin (42). Brazzein was isolated and the amino acid sequence characterized by Ming and Hellekant and consists of 54 amino acid residues with a molecular weight 6473 daltons. This substance was rated as 2000 times and 500 times sweeter, respectively, than 2% w/v and 10% w/v sucrose, with a more sucroselike temporal profile than other protein sweeteners (42). The sweetness is not destroyed by storage at 80°C for 4 hr (42). In subsequent work, four disulfide bonds in brazzein were determined by mass spectrometry and amino acid sequencing of the cystine-containing residues present (43). The compound has been synthesized by a solid-phase method, and the product was identical to natural brazzein (44). Complete ^{1}H and partial ^{13}C NMR information (45) and preliminary x-ray crystallographic data (46) have been obtained for brazzein. Accordingly, brazzein has considerable promise for future commercialization as a naturally occurring sweetening agent, because of its favorable taste profile and thermostability.

A final sweet protein that has been discovered recently is curculin from the fruits of *Curculigo latifolia* (Hypoxidaceae) (25, 47). This will be discussed later in the chapter in the section on sweetness-enhancing agents.

E. Miscellaneous Highly Sweet Plant Constituents

1. Hernandulcin

Hernandulcin (Fig. 4) is a bisabolane sesquiterpene, which was isolated initially as a minor constituent of a petroleum ether-soluble extract from the aerial parts

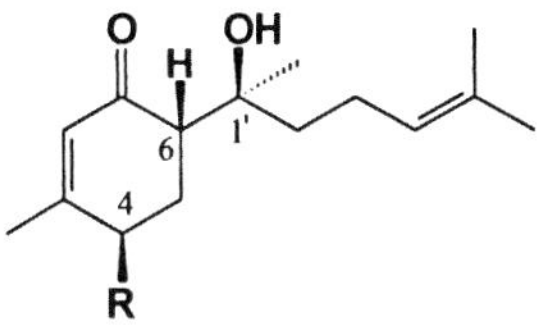

	R
Hernandulcin	H
4β-Hydroxyhernandulcin	OH

Figure 4 Structures of hernandulcin and 4β-hydroxyhernandulcin.

of the herb *Lippia dulcis* Trev. (Verbenaceae), collected in Mexico. This plant was known to be sweet by the Aztec people, according to the Spanish physician Francisco Hernández, who wrote a monograph entitled *Natural History of New Spain* between 1570 and 1576. The *L. dulcis* sweet constituent was named in honor of Hernández and was rated as 1000 times sweeter than sucrose on a molar basis when assessed by a taste panel (2). Racemic hernandulcin was synthesized by a directed aldol condensation from two commercially available ketones and the naturally occurring (6*S*,1′*S*)-diastereomer was produced in the laboratory from (*R*)-limonene (2). It has been concluded by analog development that the C-1′ hydroxyl and the C-1 carbonyl groups of hernandulcin represent the AH and B groups in the Shallenberger model of sweetness, and the C-4′, C-5′ double bond is a third functionality necessary for the exhibition of a sweet taste (2). In the last few years, several different synthetic routes have been proposed for hernandulcin [mainly in its (±)-racemic form], and the natural (+)-form was produced in both hairy root and shoot cultures of *Lippia dulcis* (reviewed in 12). A recollection of *Lippia dulcis* from Panama afforded (+)-hernandulcin in a high yield at the flowering stage (0.15% w/w) (48). Also obtained from this Panamanian sample was a second sweet substance, (+)-4β-hydroxyhernandulcin (Fig. 4), in which the 4β-OH group provides a possible linkage position for sugars and other polar moieties to synthesize more water-soluble hernandulcin analogs (13, 48). Racemic hernandulcin was not a bacterial mutagen and was not acutely toxic for mice in preliminary safety studies (2). The high sweetness potency of hernandulcin is marred by an unpleasant aftertaste and some bitterness, and the molecule is thermolabile (2, 12).

2. Rubusoside

Tanaka and coworkers have determined the *ent*-kaurene diterpene rubusoside (Fig. 5) to be responsible for the sweet taste of the leaves of *Rubus suavissimus*

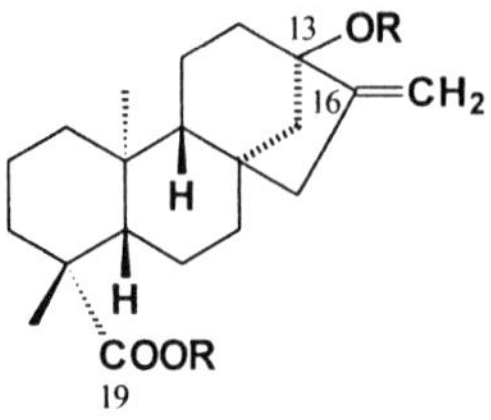

R = β-glc

Figure 5 Structure of rubusoside. (Glc = D-glucopyranosyl.)

S. Lee (Rosaceae), a plant indigenous to southern regions of the People's Republic of China (2). Rubusoside is extractable from the plant with hot methanol and occurs in high yield in the leaves of *R. suavissimus* (>5% w/w) (2). When evaluated at a concentration of 0.025%, rubusoside was rated as possessing 114 times the sweetness of sucrose, although its quality of taste sensation was marred by some bitterness. Several minor analogs of rubusoside have been isolated and characterized from *R. suavissimus* leaves, with some found to taste sweet and others bitter or neutral tasting (49). A considerable amount of work has been performed on the 1,4-α-transglucosylation of rubusoside, using a cyclodextrin-glucanotransferase-starch (CGTase) system produced from *Bacillus circulans* to produce improved sweeteners based on this parent compound (50).

A sweet tea called *tian-cha*, prepared from the leaves of *R. suavissimus* is consumed as a summer beverage in Guangxi province in the People's Republic of China. Also, during festivals, local populations mix aqueous extracts with rice to make cakes. Furthermore, the tea made from the leaves of *R. suavissimus* has been used in folk medicine to treat diabetes, hypertension, and obesity (2). More recently, *R. suavissimus* leaves have begun to be used as a "health-giving" food ingredient in Japan because in addition to rubusoside and the other minor sweet principles, antiallergic ellagitannins are also present (5). In an acute toxicity experiment on rubusoside, the LD_{50} was established as about 2.4 g kg body weight when administered orally to mice (2). Subsequently, in a subacute toxicity study, rubusoside was incorporated into the diet of mice for 60 days at a dose of one-tenth the LD_{50}, and no distinct toxicity or side effects were observed (2). However, the aglycone of rubusoside is steviol (*ent*-13-hydroxykaur-16-en-19-oic acid), which has been shown to be mutagenic in bacterial forward mutation assays when metabolically activated (2). The mutagenicity of steviol is discussed more fully in Chapter 10.

3. Baiyunoside

Baiyunoside (Fig. 6) is a labdane-type diterpene glycoside based on the aglycone, (+)-baiyunol, which was first isolated in 1983 by Tanaka and coworkers from

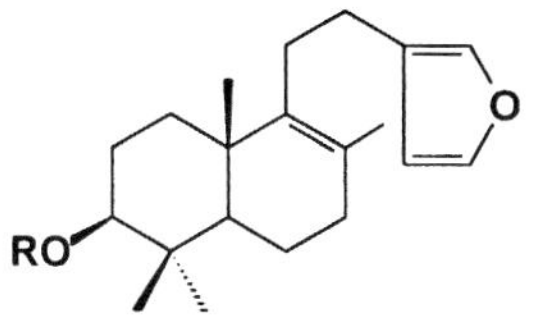

R = β-glc^{2}-β-xyl

Figure 6 Structure of baiyunoside. (Glc = D-glucopyranosyl; xyl = D-xylopyranosyl.)

a plant used in Chinese traditional medicine, namely, *Phlomis betonicoides* Diels (Labiatae) (2). This butanol-soluble compound was found to be about 500-fold sweeter than sucrose and to possess a lingering aftertaste lasting more than 1 hr (2). Synthetic routes are available for both (±) and (+)-baiyunol, and a general glycosylation procedure has been developed for baiyunol (2). More recently, Nishizawa's group has synthesized a large number of baiyunoside analogs, some of which were as sweet or sweeter than the parent compound (51). No safety studies appear to have been performed thus far on baiyunoside.

4. Steroidal Saponins

Osladin is a steroidal saponin constituent of the fern *Polypodium vulgare* L. (Polypodiaceae), which was isolated and structurally characterized by Herout and coworkers in 1971 (2). The stereochemistry of the aglycone of osladin was defined by Havel and Cerny in 1975, although the configuration of the rhamnopyranosyl unit at C-26 was not determined. However, when the compound assigned as osladin by Herout et al. was synthesized by Yamada and Nishizawa, it was not found to be sweet (52, 53). It turned out that the C-22*S*, C-25*R*, C-26*S* stereochemistry originally proposed for osladin required reassignment as C-22*R*, C-25*S*, C-26*R* (Fig. 7) (52, 53). Moreover, the original sweetness intensity value for osladin relative to sucrose was revised downward from 3000 to 500 (53).

A related steroidal saponin, polyposide A was isolated as the major sweet principle of the rhizomes of *Polypodium glycyrrhiza* D.C. Eaton (Polypodiaceae) (2). This plant is known by the common name of "licorice fern" and is a North American species native to the Pacific northwest. Polypodioside A was originally assigned as the $\Delta^{7,8}$ analog of osladin, with the configurations of the sugar substituents being determined using spectroscopic methods (2). The aglycone of polypodoside A was identified as the known compound, polypodogenin, a compound whose stereochemistry was proposed by Czech workers (2). However, a need was felt to re-examine polypodoside A (Fig. 7), given the uncertainty in the stereochemistry of osladin referred to previously, and synthetic interconversion

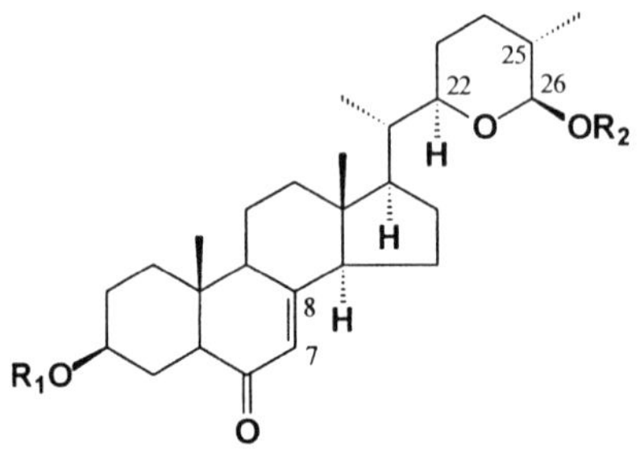

	R_1	R_2	Other
Osladin	β-glc²-α-rha	α-rha	7,8-dihydro
Polypodoside A	β-glc²-α-rha	α-rha	-

Figure 7 Structures of osladin and polypodoside A. (Glc = D-glucopyranosyl; rha = L-rhamnopyranosyl.)

work also led to a structural revision from C-22*S*, C-25*R*, C-26*S* to C-22*R*, C-25*S*, C-26*R* for this sweet *P. glycyrrhiza* constituent (54). Polypodoside A was found to be nonmutagenic and not acutely toxic for mice when dosed by oral intubation at up to 2 g/kg body weight (2). In subsequent sensory tests using a small human taste panel, polypodoside A was assessed as exhibiting 600 times the sweetness intensity of a 6% sucrose solution but also revealed a licorice-like off-taste and a lingering aftertaste (2). Thus, the potential of polypodoside A for commercialization is marred by its relative insolubility in water, its sensory characteristics, and difficulties in collecting *P. glycyrrhiza* rhizomes (2). Despite the fact that osladin has now been subjected to total synthesis, it is probable that this compound will have similar limitations to polypodoside A in terms of its potential commercial prospects, as previously noted (1, 2).

5. Pterocaryosides A and B

Pterocarya paliurus Batal. (Juglandaceae) is a plant whose leaves are used by local populations in Hubei province, People's Republic of China, to sweeten foods. Two sweet-tasting secodammarane saponins, designated pterocaryosides A and B (Fig. 8), were isolated from an extract of the leaves and stems of *P. paliurus* (55). These compounds differ structurally in only the nature of the attached sugar (a D-quinovose unit in pterocaryoside A compared with an L-arabinose unit in pterocaryoside B). Pterocaryosides A and B were shown not to be toxic in bacterial mutagenesis and mouse acute toxicity tests and were rated as about 50 and 100 times sweeter than 2% w/v sucrose, respectively (55). Ptero-

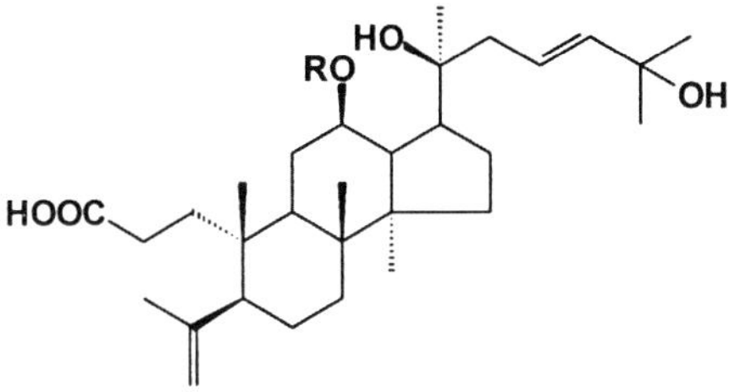

HO
RO
OH
HOOC

	R
Pterocaryoside A	β-qui
Pterocaryoside B	α-ara

Figure 8 Structures of pterocaryosides A and B. (Qui = D-quinovopyranosyl; ara = L-arabinopyranosyl.)

caryosides A and B, in being the first sweet-tasting secodammarane sweeteners to have been discovered, represent a new sweet-tasting chemotype and may serve as useful lead compounds in the future for synthetic optimization (55). Additional work on *Pterocarpa (Cyclocarya) paliurus* carried out in the People's Republic of China has led to the isolation of several sweet dammarane glycosides (56).

III. SYNTHETIC COMPOUNDS

In addition to the well-known synthetic sweeteners described in other chapters of this book, activity in this area has yielded some of the most intensely sweet compounds known to mankind and has improved our understanding of the structural requirements for the sweet-tasting response.

A. Oximes

Perillartine (Fig. 9), the α-*syn*-oxime of perillaldehyde, has been known to be highly sweet since 1920 and is reported to be up to 2000 times sweeter than sucrose (1, 2). In contrast, perillaldehyde itself (Fig. 9), the major constituent of the volatile oil of *Perilla frutescens* (L.) Britton (Labiatae), is only slightly sweet. Perillartine is used commercially in Japan as a replacement for maple syrup or licorice for the sweetening of tobacco, but more widespread use of this compound for sweetening has been restricted by a limited solubility in water, an appreciably bitter taste, as well as a menthol-licorice off-taste that accompanies sweetness (1, 2).

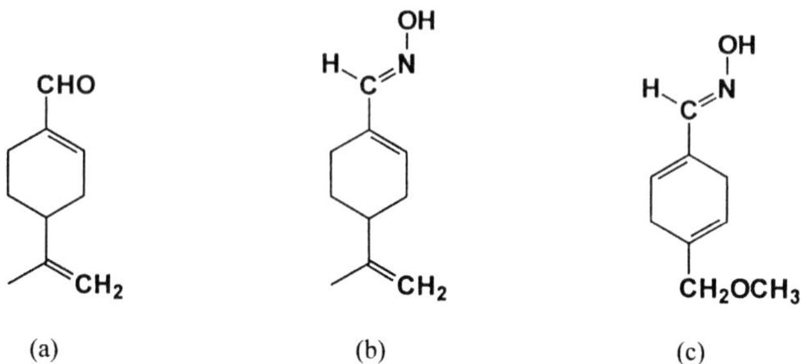

Figure 9 Structures of (a) perillaldehyde, (b) perillartine, and (c) SRI oxime V.

The intense sweetness and structural simplicity of perillartine have promoted the synthesis of numerous analogs (1, 2, 57). This work has not only led to a better understanding of the functional groups in compounds of the oxime class that confer sweetness and bitterness, but has also led to the development of several improved sweet compounds. One of the most promising of such derivatives is SRI oxime V (Fig. 9). This compound is 450 times sweeter than sucrose on a weight basis and exhibits much improved water solubility when compared with perillartine.

SRI oxime V has no undesirable aftertaste and is stable above pH 3 (1, 2). This substance was shown not to be a bacterial mutagen in the Ames assay and exhibited an LD_{50} of >1 g/kg body weight in the rat after a single oral dose (2). The compound is readily absorbed and metabolized, with excretion nearly quantitative within 48 hr after administration to the rat, dog, and rhesus monkey. The major metabolites of SRI oxime V were found to be products resulting from the oxidation of either the methoxymethyl or the aldoxime moieties, as well as those occurring after thioalkylation and glucuronidation (2). Subchronic toxicity tests on this compound conducted in rats with a diet containing 0.6% SRI oxime V for 8 wk revealed no apparent toxic effects. It has been suggested that SRI oxime V shows such promise as an artificial sweetener that a chronic toxicity evaluation is warranted (2).

B. Urea Derivatives

Dulcin (*p*-ethoxyphenylurea) has been known to be sweet for more than a century. The compound is about 200 times sweeter than sucrose and was briefly marketed as a sucrose substitute in the United States. Commercial use of this compound was discontinued after it was found to be toxic to rats at a low dose. Dulcin has also been found to be mutagenic (1, 2).

Another group of sweet ureas of more recent interest are the carboxylate-solubilized *p*-nitrophenyl derivatives, which were discovered by Peterson and

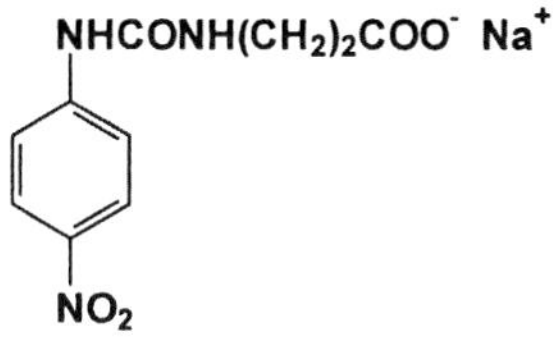

NHCONH$(CH_2)_2COO^-$ Na^+

NO_2

Figure 10 Structure of suosan.

Müller (1, 2). Suosan, the sodium salt of *N*-(*p*-nitrophenyl)-*N′*-(β-carboxyethyl)-urea (Fig. 10), is representative of this series and has been reported to be about 350 times sweeter than sucrose, although it has significant bitterness. Other compounds in this class are even sweeter than suosan (1, 2). Structure-sweetness relationships have been investigated for the sweet-tasting arylureas (1, 2).

Combination of the structures of cyanosuasan and aspartame has led to the development of superaspartame (Fig. 11) with a sweetness potency of 14,000 times that of sucrose (57). The observation that a replacement of the ureido moiety (NHCONH) with a thioureido moiety (NHCSNH) increases sweetness potency has led to the development of compounds such as the thio derivative of superaspartame, which is reported to be 50,000 times sweeter than sucrose. The sweetness potentiation induced by sulfur has been reviewed by Roy (58).

Structure-activity relationship investigations in the suosan series of sweeteners has been extended to include additional replacements for the carboxyl group. Tetrazole analogs have been prepared and were found to be sweet. However, both the urea and thiourea tetrazolyl analogs exhibited reduced potency compared with the carboxyl-containing compounds (59).

C. *N*-Alkylguanidines

Nofre and coworkers reported the synthesis of a series of (phenylguanidino)- and {[1-phenylamino)ethyl]amino}-acetic acid derivatives with varying sweetness

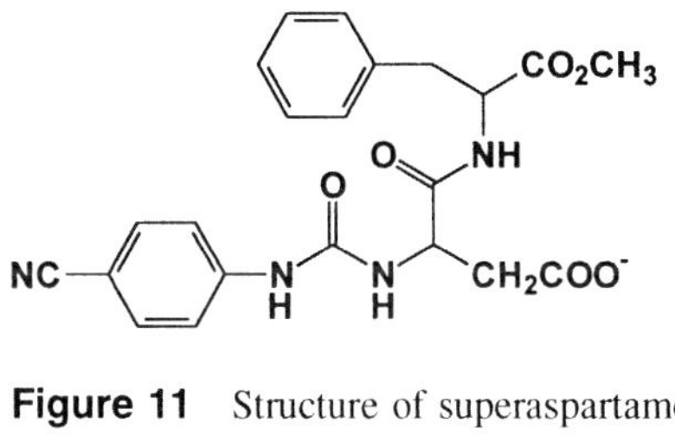

CO_2CH_3 NH O O NC N H N H CH_2COO^-

Figure 11 Structure of superaspartame.

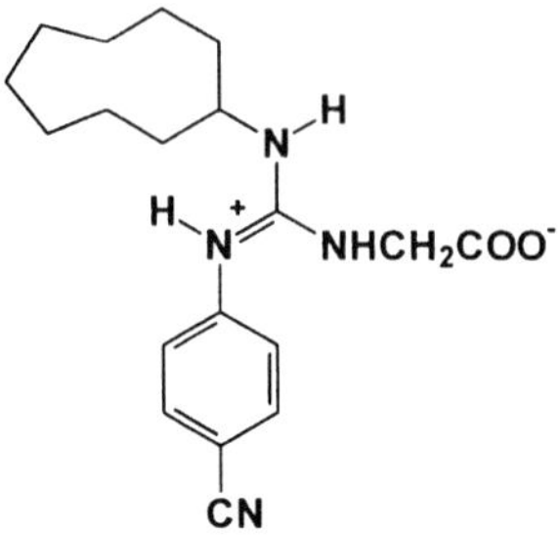

Figure 12 Structure of sucrononic acid in the zwitterionic form.

potencies (2). Sucrononic acid (Fig. 12), an *N*-cyclononylguanidine derivative, represents the most potent synthetic sweetener, with a reported sweetness potency of 200,000 times that of 2% sucrose solution (57). Recently, monoclonal antibodies against the *N*-alkylguanidine family of highly sweet-tasting compounds have been developed (60). These antibodies may be useful probes in the study of sweet taste chemistry and in the identification of novel sweet taste ligands from combinatorial chemical libraries.

D. Miscellaneous Compounds

1. Tryptophan Derivatives

The sweetness of derivatives of the amino acid tryptophan was discovered by Kornfield and his coworkers in 1968, when it was observed that racemic 6-trifluoromethyltryptophan has an intensely sweet taste (1, 2). Additional studies demonstrated that these compounds are sweet when in the D-form, with 6-chloro-D-tryptophan being some 1000 times sweeter than sucrose. The L-form of this compound is tasteless but has been found to produce antidepressant activity (2).

It was reported by Finley and Friedman that racemic *N*′-formyl and *N*′-acetyl derivatives of kynurenine, an intermediate in the metabolism of tryptophan, are approximately 35 times sweeter than sucrose and elicit an immediate sweet taste on contact with the tongue (2). The 6-chloro derivative of kynurenine [3-(4-chloroanthraninoyl)-DL-alanine] (Fig. 13) has been reported to be 80 times sweeter than sucrose and to possess no significant aftertaste or off-flavor (2).

2. Trihalogenated Benzamides

2,4,6-Tribromobenzamidines, which are substituted at the C-3 position by a carboxyalkyl or a carboxyalkoxy group, are intensely sweet. For example 3-(3-carbamoyl-2,4,6-tribromophenyl)propionic acid (Fig. 14) was rated 4000 times

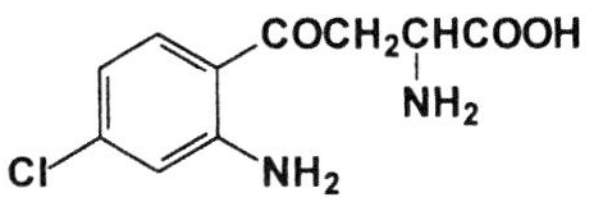

Figure 13 Structure of 3-(4-chloroanthraniloyl)-DL-alanine.

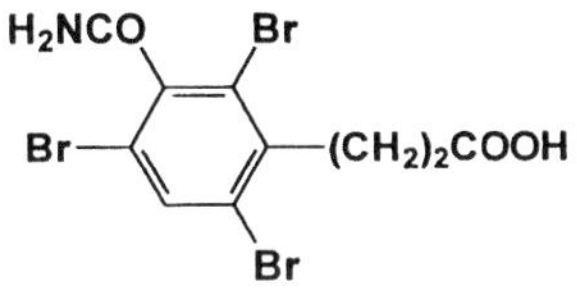

Figure 14 Structure of 3-(3-carbamoyl-2,4,6-tribromophenyl)propionic acid.

sweeter than sucrose. The compound has a slow onset and a slightly lingering aftertaste, as well as some bitterness (2). Within this compound class, the intensity of sweet taste depends on the chain length of the carboxyalkyl or carboxyalkoxy group (2). The acute toxicities of several tribromobenzamides have been determined in mice, and the results were comparable with analogous data obtained for saccharin and aspartame (2).

IV. SWEETNESS MODIFYING SUBSTANCES

A. Sweetness Inducers and Enhancers

1. Sweet Proteins

Miraculin is a tasteless basic glycoprotein constituent of the fruits of *Richardella dulcifica* (Schumach. & Thonning) Baehni [formerly *Synsepalum dulcificum* (Schumach. & Thonning) DC. (Sapotaceae) (miracle fruit), which has the propensity of making sour or acidic materials taste sweet (1, 2). Native miraculin occurs as a tetramer of a large polypeptide unit, constituted by 191 amino acid residues, with a carbohydrate content of 13.9% and an overall molecular weight of 24,600 daltons (25). The complete amino acid sequencing of miraculin monomer was accomplished by Kurihara and coworkers (2), and the purification, biochemistry, and biological properties of this glycoprotein have been subjected to review (25). Miracle fruit concentrate was at one point commercially available in the United States as an aid in dieting but was removed from the market because FDA approval as a food additive was never obtained (2).

Curculin, a sweet-tasting protein from *Circuligo latifolia* (Hypoxidaceae), mentioned earlier in this chapter, also possesses sweetness-enhancing effects. The sweet taste of curculin disappears a few minutes after being held in the mouth, but a sweet taste occurs on the subsequent application of water to the mouth (24). The primary structure of curculin monomer contains 114 amino acids, of molecular weight 12,491, with native curculin being a dimer of two monomeric peptides connected through two disulfide bridges (24, 47). Because curculin is stable at 50°C for a year, studies directed toward the commercialization of this sweet-tasting/sweetness-enhancing protein are in progress (24).

2. Other Sweetness Enhancers

A number of plant constituents have sweetness-enhancing properties, including cynarin, chlorogenic acid, and caffeic acid (2). Arabinogalactin (larch gum) is capable of enhancing the sweetness potency and taste qualities of saccharin, cyclamate, and protein sweeteners such as thaumatin and monellin (1, 2). Recently, a series of oleanane-type triterpene glycoside esters, strogins 1–5, were described from the leaves of *Staurogyne merguensis* Wall. (Acanthaceae). Three of these compounds, strogins 1, 2, and 4, elicit a sweet taste in water when held in the mouth, in similar manner to miraculin and curculin, whereas two of the compounds in this series, strogins 3 and 5, were inactive in this regard (61).

B. Sweetness Inhibitors

1. Phenylalkanoic Acids

Substituted phenoxyalkanoic acids are potent inhibitors of the sweet taste response without disrupting the taste cell membranes, and their effects are immediately reversible (62). The sodium salt of 2-(4-methoxyphenoxy)propionic acid (Fig. 15) is commercially available under the trade name of Cypha® to modulate excessive sweetness in formulated products (62). This compound has been found to be a natural constituent of roasted Colombian Arabica coffee beans and has been granted GRAS status for use in confectionery and frostings, soft candies, and snack products, for use up to levels of 150 ppm. A 100-ppm solution of this

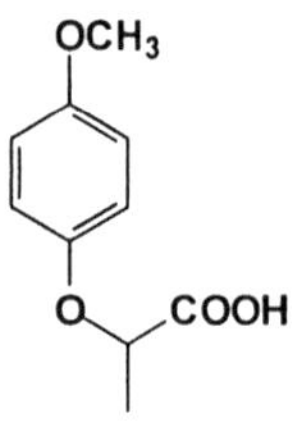

OCH_3 … O … COOH

Figure 15 Structure of 2-(4-methoxyphenoxyl)propionic acid.

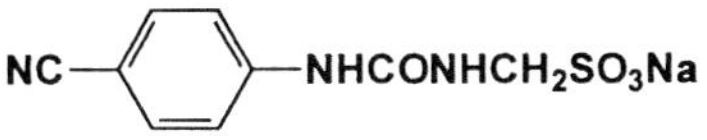

NC—C_6H_4—$NHCONHCH_2SO_3Na$

Figure 16 Structure of *N*-(4-cyanophenyl)-*N′*-[(sodiosulfo)methyl]urea.

compound is able to reduce the sweetness of a 12% sucrose solution to the perceived level of 4% sucrose. It also inhibits the sweetness of other nonsucrose bulk and intense sweeteners, but it does not have an impact on salty, bitter, or sour tastes (62, 63).

Interestingly, the structural modification of suosan (Fig. 10) led to the discovery that *N*-(4-cyanophenyl)-*N′*-[(sodiosulfo)methyl]urea (Fig. 16) inhibits the sweet taste of a variety of sweeteners. High-potency sweeteners with a slow onset and lingering sweet taste were the least inhibited. This compound can also antagonize the bitter taste responses to caffeine and quinine, although it has no effect on the sour (citric acid) or salty (NaCl or KCl) taste response (64).

2. Arylalkylketones

Several arylalkylketones and arylcycloakylketones have also been discovered to inhibit the sweet taste of sucrose and other bulk and intense sweeteners. One such compound, the commercially available 3-(4-methoxylbenzoyl)propionic acid (Fig. 17), is capable of reducing the sweetness of 40% w/v aqueous sucrose by over a sixfold margin when present at a 2% w/w concentration at pH 7 compared with when it is absent in the formulation. This compound and its analogs are recommended for use in soft puddings, infused vegetables, and other food products (2).

3. Triterpene Glycoside Sweetness Inhibitors

Considerable recent progress has been made in the characterization of plant-derived triterpene glycosides, particularly from three species, *Gymnema sylvestre* R. Br. (Asclepiadaceae), *Ziziphus jujuba* P. Miller (Rhamnaceae), and *Hovenia dulcis* Thunb. (Rhamnaceae) (65). The parent compounds from these

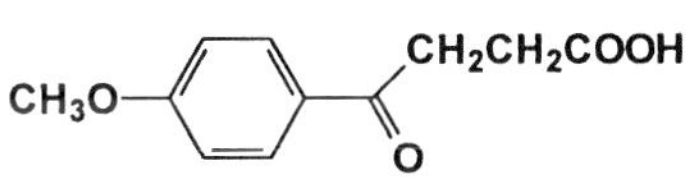

CH_3O—C_6H_4—C(=O)—CH_2CH_2COOH

Figure 17 Structure of 3-(4-methoxybenzoyl)propionic acid.

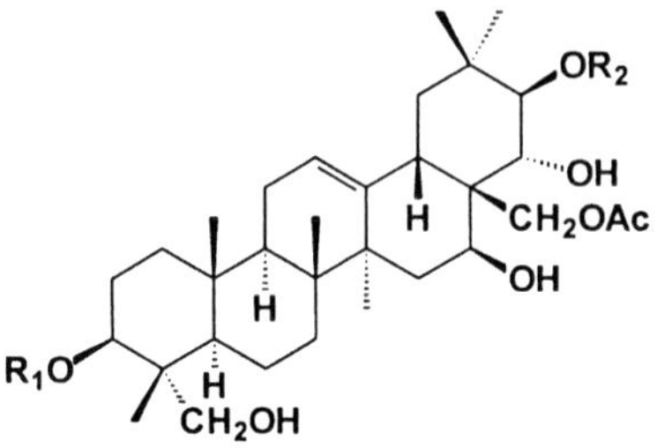

R_1	R_2
β-glcA	tga

Figure 18 Structure of gymnemic acid I. (GlcA = D-glucuronopyranosyl; tga = tiglic acid.)

three plants are, respectively, gymnemic acid I (Fig. 18), zizyphin (Fig. 19), and hoduloside I (Fig. 20). Nearly 20 analogs of gymnemic acid I (a group of oleanane-type triterpene glycosides) have been reported from *G. sylvestre* leaves, with most of these compounds being less potent as sweetness-inhibitory substances than gymnemic acid I itself (65). In addition, a sweetness-inhibitory peptide of 35 amino acids, gurmarin, has been isolated from the leaves of *G. sylvestre* (24, 65, 66), and 10 additional triterpene glycoside sweetness inhibitors,

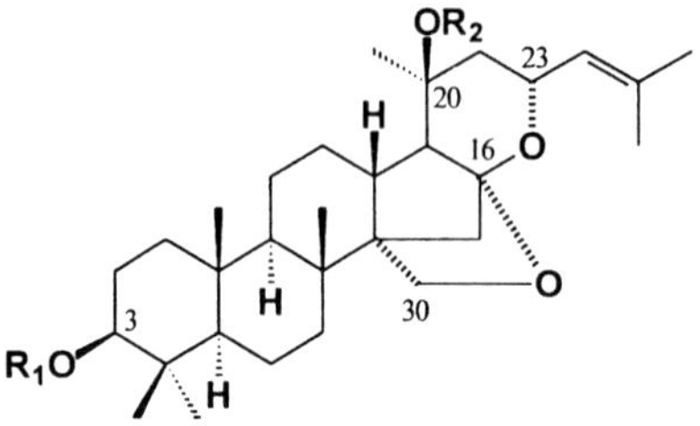

R_1	R_2
α-ara^4-α-rha	α-rha^2-Ac, with Ac at position 3

Figure 19 Structure of ziziphin. (Ara = L-arabinopyranosyl; rha = L-rhamnopyranosyl.)

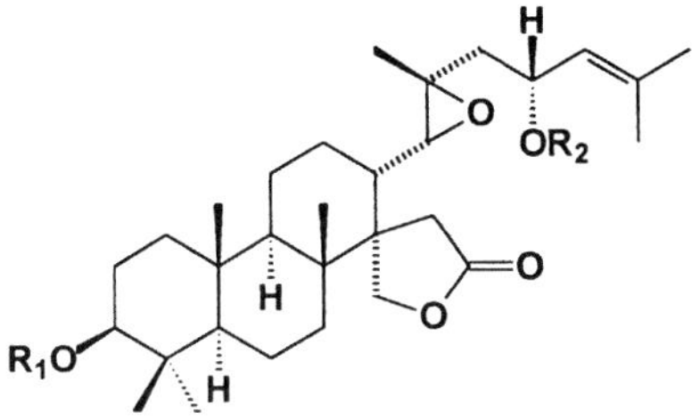

R_1	R_2
β-glc[2]-α-rha	β-glc

Figure 20 Structure of hoduloside I. (Glc = D-glucopyranosyl; rha = L-rhamnopyranosyl.)

alternosides I–X, have been isolated and characterized from the roots of *Gymnema alternifolium* (67). About 10 sweetness-inhibitory dammarane-type triterpenoid sweetness inhibitors have now been isolated and purified from the leaves of *Z. jujuba*, with all of these being equal to or less potent than ziziphin itself (65). Several dammarane-type sweetness inhibitors have also been reported in the leaves of *H. dulcis*, with the parent compound, hoduloside I, being one of the most potent (65). However, ziziphin and hoduloside I are somewhat less potent than gymnemic acid I as sweetness inhibitory agents (65). Recently, a series of novel antisweet oleanane-type triterpene glycosides, sitakisosides I–XX, has been reported from the stems of *Stephanotis lutchinensis* Koidz var. *japonica* (68–70).

REFERENCES

1. L O'Brien Nabors, GE Inglett. A review of various other alternative sweeteners. In: L O'Brien Nabors, RC Gelardi, eds. Alternative Sweeteners. New York: Marcel Dekker, 1986, pp 309–323.
2. AD Kinghorn, CM Compadre. Less common sweeteners. In: L O'Brien Wabors, RC Gelardi, eds. Alternative Sweeteners, 2nd ed., Revised and Expanded. New York: Marcel Dekker, 1991, pp 151–171.
3. GR Fenwick, J Lutomski, C Nieman. Liquorice, *Glycyrrhiza glabra* L.—composition, uses, and analysis. Food Chem 38:119–143, 1990.
4. K Mizutani. Biological properties, production, and use of chemical constituents of licorice. In: C-T Ho, T Osawa, M-T Huang, RT Rosen, eds. Food Phytochemicals for Cancer Prevention II, Symposium Series No. 547. Washington, DC: American Chemical Society, 1994, pp 322–328.

5. O Tanaka. Improvement of taste of natural sweeteners. Pure Appl Chem 69:675–683, 1997.
6. K Mizutani, T Kuramoto, Y Tamura, N Ohtake, S Doi, N Nakaura, O Tanaka. Sweetness of glycyrrhetic acid 3-*O*-D-monoglucuronide and the related glycosides. Biosci Biotech Biochem 58:554–555, 1994.
7. K Mizutani. Maruzen Pharmaceuticals Co., Ltd., Fukuyama, Hiroshima, Japan. Private communication.
8. T Machinami, T Suami. Glycyrrhetinic acid derivatives and low-calorie sweeteners containing them. Jpn Kokai Tokkyo Koho, JP10028548, Mar 2, 1998, 6 pp.
9. R Nakamura, Y Amino, T Takemoto. Glycyrrhetinic acid derivatives and sweeteners containing them. Jpn Kokai Tokkyo Koho, JP11000131, Jan 6, 1999, 3 pp.
10. R Suttisri, M-S Chung, AD Kinghorn, O Sticher, Y Hashimoto. Periandrin V, a further sweet triterpene glycoside from *Periandra dulcis*. Phytochemistry 34:405–408, 1993.
11. Y-H Choi, RA Hussain, JM Pezzuto, AD Kinghorn, JF Morton. Abrusosides A-D, four novel sweet-tasting triterpene glycosides from the leaves of *Abrus precatorius*. J Natl Prod 52:1118–1127, 1989.
12. AD Kinghorn, N Kaneda, N-I Baek, EJ Kennelly, DD Soejarto. Noncariogenic intense natural sweeteners. Med Res Rev 18:347–360, 1998.
13. RL Smith, P Newburne, TB Adams, RA Ford, JB Hallogan, and the FEMA Expert Panel. GRAS flavoring substances 17. Food Technol 50:72–81, 1996.
14. GJ de Klerck, MG Nieuwenhuis, JJ Beutler. Hypokalaemia and hypertension associated with the use of liquorice flavoured chewing gum. BMJ 314:731–732, 1997.
15. JJ Chamberlain, IZ Abolnik. Pulmonary edema following a licorice binge. West J Med 167:184–185, 1997.
16. AD Kinghorn, F Fullas, RA Hussain. Structure-activity relationships of highly sweet natural products. In: A-ur-Rahman, ed. Studies in Natural Product Chemistry, Vol. 15, Structure and Chemistry (Part C). Amsterdam: Elsevier, 1995, pp 1–41.
17. K Matsumoto, R Kasai, K Ohtani, O Tanaka. Minor cucurbitane-glycosides from fruits of *Siraitia grosvenori* (Cucurbitaceae). Chem Pharm Bull 38:2030–2032, 1990.
18. J Si, D Chen, Q Chang, L Shen. Isolation and determination of cucurbitane-glycosides from fresh fruits of *Siraitia grosvenorii*. Zhiwu Xuebao 38:489–494, 1996.
19. CM Fischer, HJ Harper, WJ Henry Jr, MK Mohlenkamp Jr, K Romer, RL Swaine Jr. Sweet beverages and sweetening compositions. PCT International Application WO9418855, Sep 1, 1994, 33 pp.
20. M Yoshikawa, T Murakami, T Ueda, H Shimoda, J Yamahara, H Matsuda. Development of bioactive functions in *Hydrangea Dulcis Folium*. VII. Absolute stereostructures of 3*S*-phyllodulcin, 3*R*- and 3*S*-phyllodulcin glycosides, and 3*R*- and 3*S*-thunberginol H glycosides from the leaves of *Hydrangea macrophylla* Seringe var. *thunbergii* Makino. Heterocycles 50:411–418, 1999.
21. A Bassoli, L Merlini, G Morini, A Vedani. A three-dimensional receptor model for isovanillic sweet derivatives. J Chem Soc Perkin Trans II:1449–1454, 1998.
22. T Naoki, T Nakano, K. Goto, T Seisho. Syntheses of dihydroisocoumarins, (±)-hydrangenol and (±)-phyllodulcin, utilizing an annelation reaction of enami-

nones with ethyl acetoacetate. Studies on the β-carbonyl compounds connected with the β-polyketides. XIII. Heterocycles 35:289–295, 1993.

23. SV Kessar, Y Gupta, S Singh. *o*-Toluate carbanions: facile synthesis of hydrangenol and phyllodulcin. Indian J Chem, Sect B 32:668–669, 1993.
24. A Ramacciotti, R Fiaschi, E Napolitano. Highly enantioselective syntheses of natural phyllodulcin. J Org Chem 61:5371–5374, 1996.
25. Y Kurihara. Characteristics of antisweet substances, sweet proteins, and sweetness-inducing proteins. Crit Rev Food Sci Nutr 32:231–252, 1992.
26. EC Zemanek, BP Wasserman. Issues and advances in the use of transgenic organisms for the production of thaumatin, the intensely sweet protein from *Thaumatococcus daniellii*. Crit Rev Food Sci Nutr 35:455–466, 1995.
27. M Szwacka, S Malepszy. Plant sweet taste-inducing and taste-modifying proteins. Biotechnologia 1:64–82, 1998.
28. CM Ogata, PF Gordon, AM De Vos, SH Kim. Crystal structure of a sweet-tasting protein, thaumatin I, at 1.65 Å resolution. J Mol Biol 228:893–908, 1992.
29. T-P Ko, J Day, A Greenwood, A McPherson. Structures of three crystal forms of the sweet protein thaumatin. Acta Crystallogr Sect D Biol Crystallogr 50D:813–825, 1994.
30. T Saumi, L Hough, T Machinami, N Watanabe, R Nakamura. Molecular mechanisms of sweet taste. 7. The sweet protein thaumatin I. Jpn Food Chem 60:277–285, 1997.
31. I Faus, C Patino, JL Del Rio, C Del Moreal, HS Barroso, J Blade, V Rubio. Expression of a synthetic gene encoding the sweet-tasting protein thaumatin in the filamentous fungus *Penicillium roquefortii*. Biotechnol Lett 19:1185–1191, 1997.
32. Q Liu, Z He, J Cao, B Li. Induction of thaumatin gene into tobacco through *Agrobacterium*-mediated system. Yichuan 20:24–27, 1998.
33. F-J Moralejo, R-E Cordoza, S Gutierrez, JF Martin. Thaumatin production in *Aspergillus awamori* by use of expression cassettes with strong fungal promoters and high gene dosage. Appl Environ Microbiol 65:1168–1174, 1999.
34. B Muermann. The new European food additive legislation. Part 2. Sweeteners in food. Ernaehr-Umsch 45:206–208, 1998.
35. L-W Hung, M Kohmura, Y Ariyoshi, S-H Kim. Structure of an enantiomeric protein, D-monellin, at 1.8 Å resolution. Acta Crystallogr Sect D Biol Crystallogr D54:494–500, 1998.
36. M Ota, Y Ariyoshi. Solid-phase synthesis of single-chain monellin, a sweet protein. Biosci Biotechnol Biochem 62:2043–2045, 1998.
37. IH Kim, KJ Lim. Large-scale purification of recombinant monellin from yeast. J Ferment Bioeng 82:180–182, 1996.
38. K Kondo, Y Miura, H Sone, K Kobarashi, H Iijima. High-level expression of a sweet protein, monellin, in the food yeast *Candida utilis*. Nat Biotechnol 15:453–457, 1997.
39. T Suami, L Hough, T Machinami, N Watanabe. Molecular mechanisms of sweet taste. Part 6: the sweet protein, monellin. Jpn Food Chem 56:275–281, 1996.
40. M Kohmura, Y Ariyoshi. Chemical synthesis and characterization of the sweet protein mabinlin II. Biopolymers 46:215–223, 1998.

41. S Nirasawa, Y Masuda, K Nakaya, T Kurihara. Cloning and sequencing of a cDNA encoding a heat-stable sweet protein, mabinlin II. Gene 181:225–227, 1996.
42. D Ming, G Hellekant. Brazzein, a new high-potency thermostable sweet protein from *Pentadiplandra brazzeana* B. FEBS Lett 355:106–108, 1994.
43. M Kohmura, M Ota, H Izawa, D Ming, G Hellekant, Y Ariyoshi. Assignment of the disulfide bonds in the sweet protein brazzein. Biopolymers 38:553–556, 1996.
44. H Izawa, M Ota, M Kohmura, Y Ariyoshi. Synthesis and characterization of the sweet protein brazzein. Biopolymers 39:95–101, 1996.
45. JE Caldwell, F Abildgaard, D Ming, G Hellekant, JL Markley. Complete ^{1}H and partial ^{13}C resonance assignments at 37 and 22 °C for brazzein, an intensely sweet protein. J Biomol NMR 11:231–232, 1998.
46. I Kohki, M Ota, Y Ariyoshi, H Sasaki, M Tanokura, M Ding, J Caldwell, F Abilgaard. Crystallization and preliminary X-ray analysis of brazzein, a new sweet protein. Acta Crystallogr Sect D Biol Crystallogr D52:577–578, 1996.
47. H Yamashita, S Theerasilp, T Aiuchi, K Nakaya, Y Nahamura, Y Kurihara. Purification and complete amino acid sequence of a new type of sweet protein with taste-modifying activity, curculin. J Biol Chem 265, 15770–15775, 1990.
48. N Kaneda, I-S Lee, MP Gupta, DD Soejarto, AD Kinghorn. (+)-Hydroxyhernandulcin, a new sweet sesquiterpene from the leaves and flowers of *Lippia dulcis* Trev. J Nat Prod 55:1136–1141, 1992.
49. K Ohtani, Y Aikawa, R Kasai, W-H Chou, K Yamasaki, O Tanaka. Minor diterpene glycosides from sweet leaves of *Rubus suavissimus*. Phytochemistry 31:1553–1559, 1992.
50. K Ohtani, Y Aikawa, H Ishikawa, R Kasai, S Kitahata, K Mizutani, S Doi, M Nakaura, O Tanaka. Further study on the 1,4-α-transglucosylation of rubusoside, a sweet steviol-bisglucoside from *Rubus suavissimus*. Agric Biol Chem 55:449–453, 1991.
51. H Yamada, M Nishizawa. Syntheses of sweet-tasting diterpene glycosides, baiyunoside and analogs. Tetrahedron 48:3021–3044, 1992.
52. H Yamada, M Nishizawa. Total synthesis of intensely sweet saponin, osladin. SYNLETT 54–56, 1993.
53. H Yamada, M Nishizawa. Synthesis and structure revision of intensely sweet saponin osladin. J Org Chem 60:386, 1995.
54. M Nishizawa, H Yamada, Y Yamaguchi, S Hatakayama, I-S Lee, EJ Kennelly, J Kim, AD Kinghorn. Structure revision of polypodoside A: major sweet principle of *Polypodium glycyrrhiza*. Chem Lett 1555–1558, 1994; Erratum ibid 1579, 1994.
55. EJ Kennelly, L Cai, L Long, L Shamon, K Zaw, B-N Zhou, JM Pezzuto, AD Kinghorn. Novel highly sweet secodammarane glycosides from *Pterocarya paliurus*. J Agric Food Chem 43:2602–2607, 1995.
56. DJ Yang, ZC Zhong, ZM Xie. Studies on the sweet principles from the leaves of *Cyclocarya paliurus*. Acta Pharm Sin 27:841–844, 1992.
57. JM Tinti, C Nofre. Design of sweeteners. A rational approach. In: DE Walters, FT Othoefer, GE Dubois, eds. Sweeteners: Discovery, Molecular Design, and Chemoreception, Symposium Series No. 450. Washington, DC: American Chemical Society, 1991, pp 88–99.

58. G Roy. A review of sweet taste potentiation brought about by divalent oxygen and sulfur incorporation. Crit Rev Food Sci Nutr 31:59–77, 1992.
59. WH Owens. Tetrazoles as carboxylic acid surrogates in the suosan sweetener series. J Pharm Sci 79:826–828, 1990.
60. JM Anchin, S Nagarajan, J Carter, MS Kellogg, GE DuBois, DS Linthicum. Recognition of superpotent sweetener ligands by a library of monoclonal antibodies. J Mol Recognit 10:235–242, 1997.
61. A Hiura, T Abakane, K Ohtani, R Kasai, K Yamasaki, Y Kurihara. Taste-modifying triterpene glycosides from *Staurogyne merguensis*. Phytochemistry 43:1023–1027, 1996.
62. MG Lindley. Phenoxyalkanoic acid sweeteners. In: DE Walters, FT Othoefer, GE Dubois, eds. Sweeteners: Discovery, Molecular Design, and Chemoreception, Symposium Series No. 450. Washington, DC: American Chemical Society, 1991, pp 251–260.
63. C Johnson, GG Birch, DB MacDougall. The effect of the sweetness inhibitor 2-(4-methoxyphenoxy)propanoic acid (sodium salt) (Na-PMP) on the taste of bitter-sweet stimuli. Chem Senses 19:349–358, 1994.
64. GW Muller, JC Culbertson, G Roy, J Zeigler, DE Walters, MS Kellogg, SS Schiffman, ZS Warwick. Carboxylic acid replacement structure-activity relationships in suosan-type sweeteners. A sweet taste antagonist. 1. J Med Chem 15:1747–1751, 1992.
65. R Suttisri, I-S Lee, AD Kinghorn. Plant-derived triterpenoid sweetness inhibitors. J Ethnopharmacol 47:9–26, 1995.
66. K Kamei, R Takano, A Miyasaka, T Imoto, S Hara. Amino acid sequence of sweet taste suppressing peptide (gurmarin) from the leaves of *Gymnema sylvestre*. J Biochem 111:109–112, 1992.
67. K Yoshikawa, H Ogata, S Arihara, H-C Chang, J-D Wang. Antisweet natural products. XIII. Structures of alternosides I–X from *Gymnema alternifolium*. Chem Pharm Bull 46:1102–1107, 1998.
68. K Yoshikawa, H Taninaka, Y Kan, S Arihara. Antisweet natural products. X. Structures of sitakisosides I–V from *Stephanotis lutchinensis* Koidz. var. *japonica*. Chem Pharm Bull 42:2023–2027, 1994.
69. K Yoshikawa, H Taninaka, Y Kan, S Arihara. Antisweet natural products. XI. Structures of sitakisosides VI–X from *Stephanotis lutchinensis* Koidz. var. *japonica*. Chem Pharm Bull 42:2455–2460, 1994.
70. K Yoshikawa, A Mizutani, Y Kan, S Arihara. Antisweet natural products. XIII. Structures of sitakisosides XI–XX from *Stephanotis lutchinensis* Koidz. var. *japonica*. Chem Pharm Bull 45:62–67, 1997.

13
Erythritol

Milda E. Embuscado and Sakharam K. Patil
Cerestar USA, Inc., Hammond, Indiana

I. INTRODUCTION

Erythritol is a unique member of the polyol family. Polyols or polyhydric alcohols are polyhydroxyl compounds derived through the hydrogenation of their parent reducing sugars (1). These compounds do not have an aldehyde group, therefore, they do not undergo Maillard reaction and are relatively stable to heat and changes in pH. Erythritol is a small linear sugar alcohol with four carbon atoms (Fig. 1). It belongs to the acyclic alcohols, a group of sugar alcohols. Its small molecular size gives erythritol its extraordinary physicochemical, nutritional, and physiological properties, such as lower caloric value and higher digestive tolerance compared with other polyols. These properties will be discussed in more detail later in this chapter.

Erythritol is a naturally occurring substance. It is found in a variety of foods such as grapes, pears, melons, and mushrooms and in fermented products such as soy sauce, sake, and wines (Table 1). The concentration of erythritol in foods and fermented products ranges from 22–1500 mg per liter or kg (2). Erythritol is a minor component of blood and amniotic fluids of cows and other mammals (3). It can also be found in the human body, for instance, in the lens tissue (4), in cerebrospinal fluid (5), in seminal plasma (6), and in the blood serum (7). Erythritol is the main polyol in the human urine with concentration ranging from 10–30 mg/l (2).

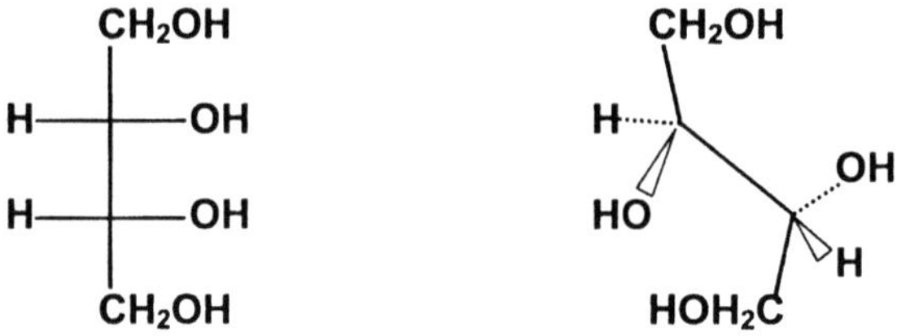

CH_2OH
H—C—OH
H—C—OH
CH_2OH

CH_2OH
H
HO
OH
H
HOH_2C

Figure 1 Molecular structure and chemical formula of erythritol.

II. PRODUCTION

Erythritol was first isolated from algae, lichens, and grasses more than 80 years ago (8). It was also prepared through fermentation by use of *Aspergillus niger* (9) and *Penicillium herquei* (10). In 1943, a German patent was granted to Reppe and Schnabel for the synthesis of erythritol from 2-butene-1,4-diol (8). Erythritol can also be synthesized from periodate-oxidized starch (11) or dialdehyde starch (12). Basically, there are two main methods of preparing erythritol: through chemical synthesis (e.g., reduction of meso-tartrate or oxidation of 4,6-0-ethylidene-D-glucose) and through fermentation. The chemical process is complex and costly because it involves several intricate steps to get the final product and the starting material is expensive. Fermentation is a simpler process requiring only a few steps and is less expensive because the initial substrate is low in cost and readily available.

Table 1 The Natural Occurrence of Polyols in Various Foods

Foods	Erythritol content
Wine	130–300 mg/l
Sherry wine	70 mg/l
Sake	1550 mg/l
Soy sauce	910 mg/l
Miso bean paste	1310 mg/kg
Melons	22–47 mg/kg
Pears	0–40 mg/kg
Grapes	0–12 mg/kg

Source: Reprinted with permission from Advances in Sweeteners, Chapter 8, J Goossens and M Gonze, TH Grenby (ed), p. 153, © 1996 Aspen Publishers, Inc.

The commercial production of erythritol uses a completely biotechnological process using enzymes and osmophilic yeasts or fungi (2, 13–18). All other polyols are prepared by the catalytic hydrogenation of a precursor; for example, xylose from xylan is used to manufacture xylitol. *Moniliella*, *Trigonopsis*, or *Torulopsis* are some of the microorganisms that can convert glucose to erythritol in relatively high yields. The basic process is outlined in Fig. 2. Wheat starch or cornstarch is the usual starting material. These are hydrolyzed primarily to glucose and other carbohydrates in lower concentration. An inoculum of the osmotolerant microorganism is added to the substrate, which ferments glucose to a mixture of erythritol and minor amounts of glycerol and ribitol. One of the yeasts used, *Moniliella*, can thrive at high sugar concentration and at the same time produce erythritol. This characteristic of *Moniliella* is advantageous because a concentrated substrate can be used as a starting material, and glucose can be added continuously to the fermentation tank without adversely affecting the

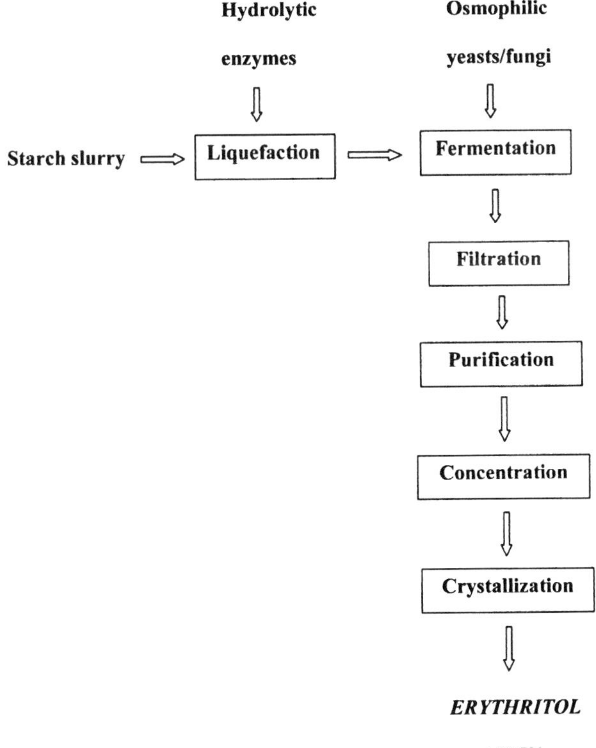

Figure 2 Commercial production of erythritol.

growth of the microorganism and its efficiency to produce erythritol. The resulting concentrated fermentation broth facilitates the isolation and purification of erythritol. The separation process is simpler and less energy-intensive. Conversion yields of 40–50% have been reported using the fermentation process. Cerestar, a company of Eridania Béghin-Say, is producing erythritol commercially with an entirely biotechnological process (17).

Several methods of improving erythritol productivity have been carried out. Strains of *Trichosporonoides* sp. have been isolated from honeycomb that can give high erythritol yields and low amounts of glycerol from glucose or sucrose substrates (19). A high erythritol-producing strain of *Trichosporon* sp. was also isolated from honeycomb (20). The strain produced 141 g of erythritol per liter of medium at 35°C after 3 days of incubation. The fermentation medium consisted of 30% glucose and 4% corn steep liquor. Induced mutation has also been used to improve the strain of *Aureobasidium* sp. so its conversion ratio would increase (18). This method resulted in yields between 43–52%. Using this microorganism, the fermentation broth did not foam, substrate as high as 83.3% can be used, and only small amounts of by-products were synthesized. Using this microorganism and under certain fermentation conditions by-products were eliminated (18). Another study dealt with the effect of increased osmotic pressure on the rate of erythritol production using *Trigonopsis variabilis* (21). The rate was increased from 0.09–1.9 g/g-day when the osmotic pressure was increased from 1.3–3.9 Kpa. It is most likely that more research studies will be undertaken in the future to improve erythritol production. This can be done either through the use of bioengineering to improve the present strains of erythritol-producing osmotolerant yeasts or fungi or through formulation of enzyme cocktails that would be more efficient in converting glucose or other sugars to erythritol.

III. PHYSICOCHEMICAL AND FUNCTIONAL PROPERTIES

A. Sweetness

Tables 2 and 3 summarize the properties of erythritol, other polyols, and sucrose. The sweetness of erythritol is around 65% that of sucrose. Its sweetness profile is similar to sucrose with slight acidity and bitterness but with no detectable aftertaste. Most polyols have sweetness levels less than sucrose. Xylitol is the only polyol that has the closest degree of sweetness to sucrose, but its cooling effect is more intense than erythritol. Maltitol is another polyol, which has a higher degree of sweetness than erythritol. The rest of the polyols have levels of sweetness less than or equal to erythritol. It must be remembered that sweetness in itself is not the only important taste criteria but also persistence of sweetness, presence or absence of aftertaste, and the sweetness profile (i.e., how close it is to sucrose). For bulk sweeteners like erythritol, other nonsweet flavor attributes

Table 2 Properties of Erythritol Compared with Other Polyols and Sucrose

Sugar	Sweetness (sucrose = 1)	Heat of solution (kJ/kg)	Cooling effect	Viscosity cp, 25°C	Hygroscopicity
Glycerol C3 MW = 92	0.60		N/A	954	High
Erythritol C4 MW = 122	0.53–0.70	−180	Cool	Very low Insoluble at 70%	Very low
Xylitol C5 MW = 152	0.87–1.00	−153	Very cool	Very low	High
Mannitol C6 MW = 182	0.50–0.52	−121	Cool	Low Insoluble at 70%	Low
Sorbitol C6 MW = 182	0.60–0.70	−111	Cool	Low 110 cp at 70% solution	Median
Maltitol C12 MW = 344	0.74–0.95	−79	None	High	Median
Isomalt (Palatinit) C12 MW = 344	0.35–0.60	−39	None	High	Low
Lactitol C12 MW = 344	0.35–0.40	−53	Slightly cool	Very low	Median
Sucrose C12 MW = 342	1.00	−18	None	Low High at 70% solution	Median

Source: Ref. 13 with modification.

Table 3 Properties of Erythritol Compared with Other Polyols and Sucrose

Sugar	Melting point (°C)	Tg (°C)[a]	Solubility g/100g H_2O (25°C)	Heat stability (°C)	Acid stability
Glycerol	17.8	−65	∞	Decomposes at 290	
Erythritol	126	−53.5 (eutectic)	37–43	>160	2–12
Xylitol	94	−46.5	63	>160	2–10
Mannitol	165	−40	18–22	>160	2–10
Sorbitol	97	−43.5	70–75	>160	2–10
Maltitol	150	−34.5	60–65	>160	2–10
Isomalt (Palatinit)	145–150	−35.5	25–28	>160	2–10
Lactitol	122		55–57	>160	>3
Sucrose	190	−32	67	Decomposes at 160–186	Hydrolyzes at acidic and alkaline pH

Source: Ref. 13 with modification.
[a] From Ref. 22.

and mouthfeel are also important quality factors. These quality factors will determine whether an ingredient will be an acceptable and a successful alternative sweetener.

The level of sweetness of erythritol can be increased through blending with an intense sweetener like aspartame or acesulfame K. Table 4 shows the ratios

Table 4 Percent Synergy for Mixtures of Bulk and Intense Sweeteners

	Bulk sweetener–intense sweetener sweetness contribution ratio[a]					
Blend	1–99	5–95	15–85	85–15	95–5	99–1
Erythritol–Aspartame (% synergy)[b]	−3	−7	10	30[c]	25[c]	24[c]
Erythritol–Acesulfame K (% synergy)[b]	12	8	19[c]	32[c]	31[c]	27[c]

[a] Expected sweetness (sucrose equivalent value, SEV) = 10.
[b] % synergy = 100 × [{SE mixture/SE (100% erythritol + 100% intense sweetener) ÷ 2} −1], where SE is the panel's sweet intensity rating. The equation is based on Ref. 23.
[c] Significant at $P < 0.05$.

used to obtain the desired sweetness contributions from erythritol and from an intense sweetener. It also shows the percent synergy at these ratios. For example, to attain 1:99 sweetness ratio, a weight of erythritol that would give an equivalent of 1% sweetness to the mixture was added to preweighed aspartame contributing 99% of the sweetness. These were dissolved in water, and the intensity of sweetness was determined by a trained panel. Results show that there is a synergistic effect when erythritol is combined with aspartame to obtain 85:15, 95:5, and 99:1 sweetness blends (Table 4). With acesulfame K, this synergy was observed starting with a 15:85 sweetness ratio and at higher levels of erythritol. This intensification of sweetness was not obtained when sucrose was used instead of erythritol (24). Blending erythritol with intense sweeteners like aspartame or acesulfame K has a lot of potential applications in beverage and food formulations.

B. Cooling Effect

All polyols exhibit negative heats of solution. Energy is needed to dissolve the polyol crystals, thus they absorb the surrounding energy, resulting in a lowering of the temperature or a cooling of the solution. This property is likewise observed when the dry powder of polyol is dissolved in the mouth. This creates a cooling sensation in the mouth. The degree of cooling primarily depends on the magnitude of the heat of solution. As shown in Table 2, the cooling effect ranges from none to very cool, which can be directly related to the magnitudes of the heat of solution. It appears that this cooling effect is only slightly detected or not perceived at all if the heat of solution values are equal to or greater than 79 kJ/kg. Heats of solution lower than this value give a cooling effect while values at about −153 kJ/kg exhibit a strong cooling sensation (e.g., xylitol). As shown in Table 2, this cooling effect is not observed for maltitol, isomalt, and sucrose and only slightly for lactitol. Erythritol imparts a moderate cooling effect in the mouth. This property is advantageous for food and pharmaceutical applications when the cooling effect is desired or is an intrinsic property of the product such as in food formulations containing peppermint or menthol. The cooling sensation is also beneficial in pharmaceutical preparations requiring soothing effects (e.g., lozenges, cough drops, throat medication, breath mints).

C. Solubility and Hygroscopicity

Erythritol is moderately soluble in water unlike sucrose and other polyols (xylitol, sorbitol, maltitol, and lactitol), which are quite soluble in water. Mannitol and isomalt are less soluble in water than erythritol. Erythritol's reduced affinity for water is reflected in its sorption isotherm (Fig. 3). Figure 3 also shows the isotherm of sucrose for comparative purposes. The sorption isotherms were determined using an SGA-100 water sorption analyzer (VTI Corporation, FL). The change in weight of a completely dried erythritol powder was monitored at 25°C

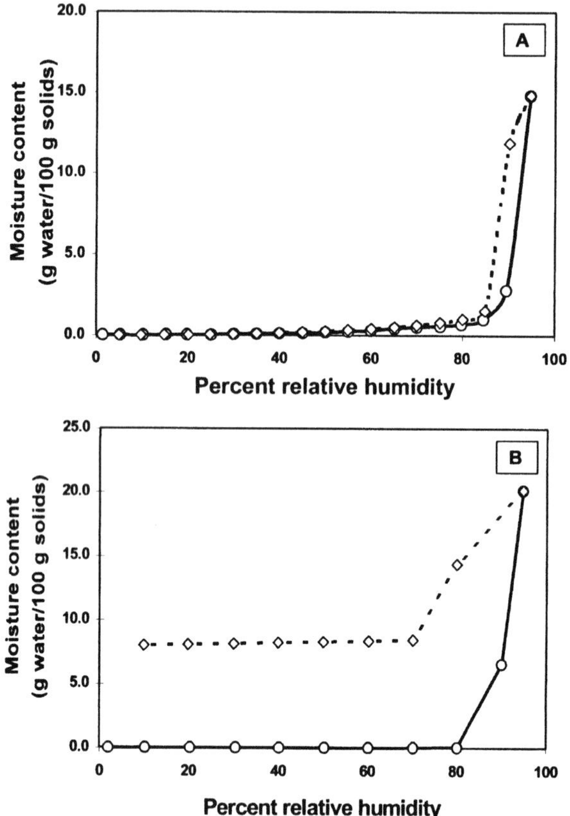

Figure 3 Sorption isotherms of (A) erythritol and (B) sucrose at 25°C (Courtesy of VTI Corporation, FL). ○——○, Adsorption; ◊- -◊, desorption.

when the relative humidity was increased from 5–95% and then back to 5% in 5% steps. Erythritol went into solution above 90% relative humidity, whereas sucrose did at a lower relative humidity (around 84%). To determine the exact deliquescence point, erythritol was exposed to relative humidities between 88–94% in a stepwise fashion. Results show that erythritol starts to deliquesce between 92–94% relative humidities. When erythritol was exposed to decreasing relative humidities, the sample lost most of its moisture, especially when the relative humidity reached 85%. The adsorption and desorption isotherm profile of erythritol is quite different from that of sucrose (Fig. 3). As the relative humidity was reduced to 75%, sucrose lost some of the moisture it absorbed at higher relative humidity. At about 75% relative humidity, the moisture content of sucrose stabilized and remained the same even when the relative humidity was reduced down to 5%. The adsorption and desorption properties of erythritol have important implications in product formulation and storage.

The solubility of erythritol limits its applications to certain types of food preparations. However, this property is desirable in applications where moisture pickup of the product on storage should be held at a minimum (e.g., fruit pieces, fruit bars, pastries). An increased rate of moisture absorption for these types of products usually results in loss in quality (softening, stickiness) or even microbiological deterioration. When erythritol is used as an ingredient or as a coating material, the moisture pickup of the product is retarded or is held to a minimum. Replacement of hygroscopic sugars with erythritol may also have a desirable impact on the choice of packaging material. In place of expensive multilayered laminates, a simple and less expensive packaging material may suffice to protect the product. In addition, a product coated with erythritol is more stable when directly exposed to relative humidities between 85–90% compared with products coated with sugars. It is also worth mentioning that the presence of erythritol to control the water activity (Aw) inhibited the growth of *Staphylococcus aureus* even at high Aw levels (0.92–0.94) (25). Other solutes, such as sodium chloride, potassium chloride, sucrose, glucose, sodium lactate, and sodium acetate, required a lower Aw to inhibit the growth of *S. aureus* (25).

D. Other Characteristics

The hydrogenation of mono-, di-, and oligosaccharides converts the aldehyde group of the corresponding sugars to a primary alcohol and its ketonic function into a secondary alcohol. This chemical change brings about improved chemical stability, modified physicochemical and functional properties, and reduced bioavailability or slow digestibility for polyols. Erythritol is stable at high temperatures and at a wide pH range (Table 3). Like other polyols, it does not undergo a Maillard or browning reaction because it does not have a reactive aldehyde group. This is beneficial in food applications, where maintaining the delicate flavor and the intrinsic qualities of the products is important. Food products with sugars and proteins or amino acids undergo browning discoloration when exposed to heat and oftentimes develop a bitter taste.

IV. METABOLISM

A. Caloric Value of Erythritol

The metabolic pathway of erythritol is different from glucose, sucrose, and other polyols. Erythritol is rapidly absorbed in the small intestine through passive diffusion because of its small molecular volume. Glucose and sucrose are also readily absorbed in the small intestine, but they are metabolized to produce energy and carbon dioxide. Erythritol, on the other hand, is not metabolized and is rapidly excreted unchanged in the urine. Figure 4 shows the rate of $^{13}CO_2$ excretion and the cumulative $^{13}CO_2$ excretion after ingestion of 25 g of ^{13}C-labeled glucose, lactitol, or erythritol. There was no detectable $^{13}CO_2$ excretion for erythritol, whereas for glucose the maximum level was reached after 2–3.5 hr of oral administration (26). The rate of $^{13}CO_2$ for lactitol was slower than for glucose, and it reached its peak 6 hr after ingestion.

The H_2 excretions of ^{13}C-labeled glucose, lactitol, or erythritol are shown in Fig. 5. The H_2 excretions for erythritol and glucose were low compared with lactitol. Lactitol produced a significant amount of H_2 after 2 hr of ingestion. These results support the findings that after an oral intake of erythritol, it is not metabolized systemically or completely by the gut microflora (26). More than 90% of ingested erythritol is excreted in the urine (27).

As shown in Figure 5, the small amount of unabsorbed erythritol may reach the large intestine and may be fermented by the colonic microorganisms to volatile fatty acids (VFA) and CH_4 or H_2. VFA contribute to additional energy on absorption. The total caloric value of erythritol was calculated using the factorial approach on the basis of its metabolic fate in the body and amounted to a maximum value of 0.2 calories/g (28, GRAS affirmation petition). Other polyols (xylitol, sorbitol, and mannitol) are likewise absorbed by diffusion but at a slower rate. Maltitol and isomalt are partially hydrolyzed and then slowly absorbed by the body. The slow rate of absorption produces adverse effects such as high osmotic load, which can cause abdominal pain, diarrhea, and flatulence, especially when high amounts of polyols are ingested. The unabsorbed polyols are fermented by the gut microflora on reaching the large intestine. Lactitol is not absorbed from the small intestine. It passes into the large intestine, where it is completely fermented. On the basis of their metabolic pathways, the energy values of polyols were determined (Table 5). Erythritol has the lowest caloric value, and maltitol and hydrogenated maltodextrins have the highest energy values among the polyols. The U.S. Food and Drug Administration issued a no-objection letter for the value of 0.2 calories/g for erythritol. For the remaining polyols, the FDA permits energy values of 1.6 for mannitol to 3.0 calories/g for hydrogenated starch hydrolysates, the highest caloric value for polyols.

B. Cariogenicity and Acidogenicity

Noncariogenicity and nonacidogenicity are important properties of polyols. Table 6 summarizes the dental health properties of erythritol and other polyols. The criteria used to assess the noncariogenic property of a substance are its nonfer-

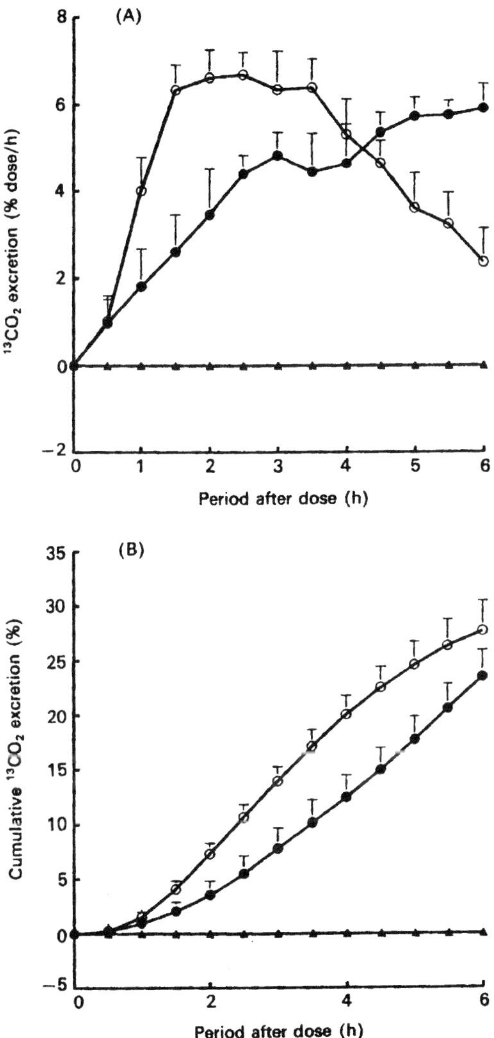

Figure 4 (A) Rate of $^{13}CO_2$ excretion (% dose/hr) after ingestion of 25 g ^{13}CO-labeled glucose (○), lactitol (●), or erythritol (▲). (B) Cumulative amount of excretion (% administered dose) after $^{13}CO_2$ ingestion of 25 g ^{13}CO-labeled glucose (○), lactitol (●), or erythritol (▲) in human volunteers. (From Ref. 26.)

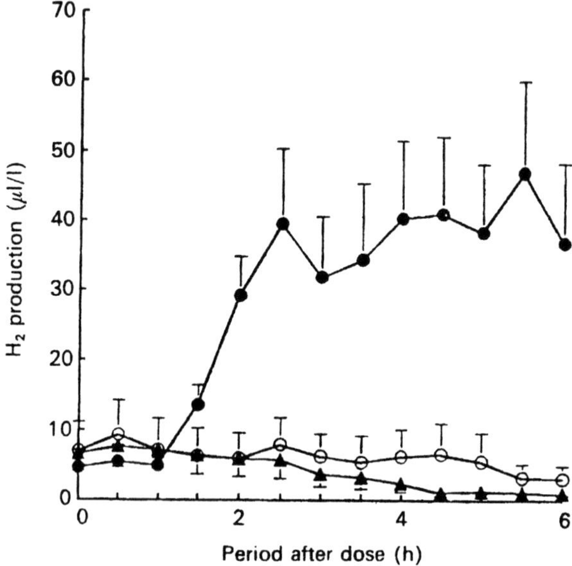

Figure 5 H_2 excretion after ingestion of 25 g ^{13}CO-labeled glucose (○), lactitol (●), or erythritol (▲) in healthy volunteers. (From Ref. 26.)

Table 5 Energy Value of Undigestible Sugars [kJ/g (kcal/g)]

	Condition of testing	
Undigestible sugar	Fasting	Postprandia
Erythritol	0–1.0 (0–0.2)	—
Xylitol	8.4–10.9 (2.0–2.6)	13.8–16.3 (3.3–3.9)
Sorbitol	8.4–10.9 (2.0–2.6)	13.8–16.3 (3.3–3.9)
Mannitol	6.3–7.9 (1.5–1.9)	—
Maltitol	11.7–13.4 (2.8–3.2)	14.6 (3.5)
Isomalt (Palatinit)	10.0–12.1 (2.4–2.9)	—
Lactitol	—	5.85–10.4 (1.4–2.5)
Hydrogenated maltodextrins (Maltidex)	11.7–13.4 (2.8–3.2)	14.6 (3.5)

Source: Ref. 13 and GRAS affirmation petition (GRASP 7 GO422).

Table 6 Assessment Criteria for the Dental Health Related Properties of Polyols and Results

Polyol	Oral acid fermentation (pH-telemetry)	Insoluble glucan formation	Human clinical caries studies available	Rat caries studies	Safe for teeth/ tooth-friendly
Erythritol	No[a,b]	No[b]	?	Low	Yes[b]
Xylitol	No	No	Yes	Low	Yes
Sorbitol	Low/no	No	Yes	Intermediate/ low	Yes
Mannitol	No	No		Low	Yes
Isomalt	No	No	No	Low	Yes
Lactitol	No	No	No	Low	Yes
Maltitol	No	No	No	Low	Yes
HSH[c]	Low/yes[c]	No	Yes	Low	Yes[c]

[a] From Ref. 30.

[b] From Refs. 1 and 33.

[c] Depends on product composition.

Source: Reprinted with permission from Advances in Sweeteners, Chapter 4, H Schiweck and SC Ziesenitz, TH Grenby (ed), p. 76, © 1996 Aspen Publishers, Inc.

mentability by oral microorganisms, nonacidogenic property, and the absence of glucan formation in in vitro rat caries studies (29). Fermentation of sugars by oral bacteria produces acid, which has an adverse effect on the tooth enamel below a critical pH. For a substance to be classified as nonacidogenic, the critical pH value in humans as measured by plaque pH telemetry should be equal to or greater than 5.7. Erythritol, like other polyols, is considered tooth-friendly (Table 6). It is not used as a substrate for lactic acid production or for plaque polysaccharide synthesis by oral streptococci (30). The acid production of *Streptococcus mutans*, a plaque-forming bacterium, in 3% solutions of erythritol or other polyols under controlled in vitro conditions is shown in Figure 6. Erythritol and xylitol did not support acid production by this microorganism for more than 36 hr of incubation. The noncariogenic property of erythritol makes it an important ingredient for tooth-friendly products (29).

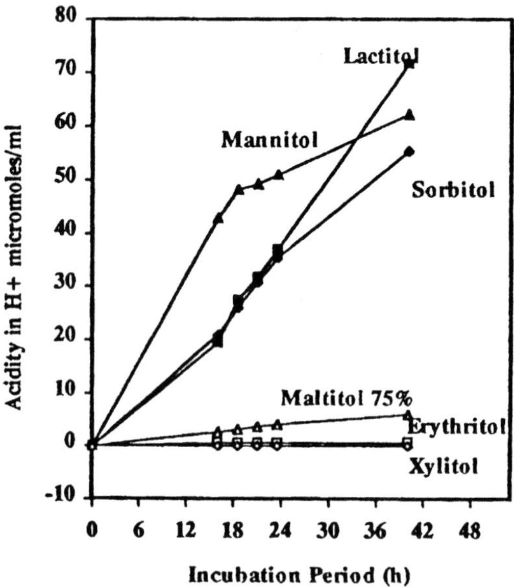

Figure 6 Development of the acidity by *Streptococcus mutans* strains growing on a 3% solution of erythritol or another polyol under controlled in vitro conditions. Cultures with an initial bacteria count of 3×10^9 were incubated in test tubes at 37°C in a biological buffering system. Acidity is expressed as total H^+ concentration in micromoles/ml accumulating over time period. (Reprinted with permission from Advances in Sweeteners, Chapter 8, J Goossens and M Gonze, TH Grenby, ed., p. 164, © 1996 Aspen Publishers, Inc.)

Table 7 Summary of Pivotal Toxicity Studies on Erythritol

Title of study	Erythritol dosage	Duration	Findings
1. Oral toxicity (34) Animal: Wistar rats (male and female)	0, 5, 10%	4 weeks	No significant toxicological effects
2. Subchronic oral toxicity (35) Animal: mice, rats	0, 5, 10, 20%	90 days	No treatment-related abnormalities based on histopathological examination
3. Embryotoxicity and teratogenicity (36) Animal: rats	0, 2.5, 5, 10%	Day 0 to 21 of gestation	No fetotoxic, embryotoxic, or teratogenic effects
4. Two-generation reproduction study (37) Animal: rats	0, 2.5, 5, 10%	Two generations	No adverse effect on fertility and reproductive performance of parents and on the development of their progeny
5. Teratology study (38) Animal: rabbits	1.0, 2.24, 5.0 g/kg intravenously	Daily from days 6 to 18 of gestation	No effect on reproductive performance or in fetal development
6. Chronic oral toxicity (39) Animal: dogs	0, 2, 5, 10%	53 weeks	Daily consumption of up to 3.5 g/kg body weight was well tolerated
7. Mutagenicity (40) Organism/cell line: *Salmonella typhimurium* *Escherichia coli* Chinese hamster fibroblast cell	15.8 to 50,000 μg/plate		Negative in reverse mutation assays and negative in chromosome aberration tests
8. Chronic toxicity and carcinogenicity (41) Animal: rats (male and female)	0, 2, 5, 10%	104–107 weeks	No evidence of tumor-inducing or tumor-promoting effect

V. SAFETY AND DIGESTIVE TOLERANCE

On the basis of numerous safety studies following FDA's Redbook guidelines, the safety of erythritol has been established. These include in vitro tests for mutagenicity and clastogenicity; subchronic toxicity tests in mice, rats, and dogs; chronic toxicity/carcinogenicity studies in rats and dogs; a multigeneration study in rats, and two teratogenicity studies in rats and rabbits (27, 31, 32). Table 7 contains a summary of recent toxicity studies, which overwhelmingly support the safety of erythritol as a food ingredient. On the basis of the entire safety data package on erythritol, it is concluded that erythritol is safe for its intended use in foods (27).

The digestive tolerance of erythritol compared with other polyols is given in Table 8. Erythritol is highly tolerated by humans and by animals because it is such a small molecule that it is immediately absorbed by the body and excreted within 24 hr (>80%). This rapid absorption at high concentration prevents the accumulation of unabsorbed products in the large intestine, which may cause side effects such as diarrhea, abdominal pain, and flatulence. Erythritol exhibits high digestive tolerance even at high dosage levels.

Table 8 Digestive Tolerance Doses of Undigestible Sugars

Undigestible sugar	Condition of testing: Single load	Throughout the day	Chronic exposure	Dose (g)
Erythritol	—[a]			30–55
Xylitol			—[b]	60–80
Sorbitol	+	−	−	10
	−	+	+	30–40
Maltitol	+	−	−	20
	−	+	+	60
Isomalt	+	−	−	20–30
(Palatinit)	−	+	+	50
Lactitol	−	+	+	50

[a] Single oral dosage did not alter plasma or urine osmolarity or electrolyte balance. Erythritol is well tolerated by the digestive tract (42).

[b] Seven-day high-dosage ingestion using a double-blind, two-way crossover study in 12 healthy male volunteers; sucrose used as a control. Erythritol was well tolerated and without any side effects (32).

Source: Ref. 12 with modification.

VI. FOOD AND PHARMACEUTICAL APPLICATIONS

Erythritol is a unique low-calorie bulk sweetener with unique physicochemical and functional properties. It is moderately sweet (about 70% as sweet as sucrose), has a mild cooling effect, and contributes more smoothness than sucrose, sorbitol, maltitol, xylitol, isomalt, and mannitol. It can be used in a number of formulations because of its unique properties (2, 14–18, 43). Erythritol has a low calorie value (maximum, 0.2 cal/g) and can significantly reduce the energy values of food products when used to replace sugars. It does not promote tooth decay and can be used in tooth-friendly products. It is highly tolerated without gastrointestinal side effects under intended conditions of use. It is also suitable for diabetic food preparations because it does not affect blood glucose or insulin levels even during a 2-week daily administration to patients with diabetes (44). Table 9 summarizes some potential applications of erythritol. It also includes the patents granted for various uses of erythritol. Some of the notable applications of erythritol are in chewing gum, candies, chocolate, lozenges, fondant, fudge, bakery products, beverages, and as a sugar replacement (such as a tabletop sweetener).

Table 9 Maximum Use Levels of Erythritol that are GRAS

Application	Specific product	Maximum use level (%)	Number of patents granted[a]	
			United States	Others
Sugar substitute	Tabletop sweetener	100	2	4
Confectionery	Soft candies	40	1	4
	Hard candies	50	1	2
Chewing gum		60	2	4
Low-calorie beverages	Soft drinks, fruit juice	1.5		1
Chocolate	Plain/milk chocolate	40		1
Fermented milk				2
Coating				2
Dehydrated food				
Bakery products	Fat cream in cookies, cakes, pastries	60		
Dietetic products	Cookies, wafers	7		

[a] For the period 1969–03/99.

VII. REGULATORY STATUS

Erythritol has been approved and marketed in Japan since 1990. It is used in candies, chocolates; soft drinks; chewing gum; yogurt; fillings and coatings in cookies, jellies, and jams; and as a sugar substitute in Japan. In early 1997, the Food and Drug Administration accepted the generally recognized as safe (GRAS) affirmation and dental health claim petitions for erythritol. The GRAS status of erythritol was based on the large body of published safety data including animal toxicological and clinical studies. The consumption of erythritol from its natural occurrence in foods is estimated to be 25 mg/person/day in the United States and is 106 mg/person/day in Japan. On the basis of the extensive food safety database and other related studies, an independent food safety expert panel concluded that erythritol is GRAS under the conditions of its intended use (27, 31). The maximum use levels of erythritol based on current good manufacturing practice are summarized in Table 9.

ACKNOWLEDGMENTS

We thank J.C. Troostembergh and P. de Cock (Cerestar Europe) and F. Turnak (Cerestar USA) for critically reviewing our manuscript and providing us with several references. We acknowledge the valuable studies on erythritol undertaken by our European colleagues. We also thank S. Jackson for acquiring important articles from various sources; C. Austin, J. Rorher, and E. Hakmiller (Cerestar USA) for their contribution; E. Embuscado for her computer search on erythritol and providing several key references. ME thanks I. Abbas (Cerestar USA) for guidance and for giving her the opportunity to study an exciting and fascinating group of alternative sweeteners, the polyols.

REFERENCES

1. PM Olinger, VS Velasco. Opportunities and advantages of sugar replacement. Cereal Foods World 41:110–113, 116–117, 1996.
2. J Goossens, M Gonze. Nutritional properties and applications of erythritol: a unique combination? In: TH Grenby, ed. Advances in Sweeteners. Gaithersburg, MD: Aspen Publishers, 1995, pp 150–186.
3. GP Roberts, A MacDiarmid, P Gleed. The presence of erythritol in the fetal fluid of fallow deer. Res Vet Science 20:154–256, 1976.
4. M Tomana, JT Prchal, L Cooper Garner, HW Skala, SA Barker. Gas chromatographic analysis of lens monosaccharides. J Lab Clin Med 103:137–142, 1984.

5. C Servo, J Pabo, E Pitkaenen. Gas chromatographic separation and mass spectrometric identification of polyols in human cerebrospinal fluid and plasma. Acta Neurol Scand 56:104–110, 1977.
6. P Storset, O Stooke, E Jellum. Monosaccharides and monosaccharide derivatives in human seminal plasma. J Chrom 145:351–357, 1978.
7. J Roboz, DC Kappatos, J Greaves, JF Holland. Determination of polyols in serum by selected ion monitoring. Clin Chem 30:1611–1615, 1984.
8. S Budavari, ed. The Merck Index. Whitehouse Station, NJ: Merck Research Laboratories, 1996, pp 624, 763–764, 912, 979, 1490–1491, 1517–1518, 1723–1724.
9. JL Yuill. Production of i-erythritol by *Aspergillus niger*. Nature 162:652, 1948.
10. JA Galarraga, KG Neill, H. Raistrick. Biochemistry of microorganisms: the colouring matters of *Penicillium herquei* Bainier and Sartory. Biochem J 61:456–464, 1955.
11. A Jeanes, CS Hudson. Preparation of meso-erythritol and d-erythronic lactone from periodate-oxidized starch. J Org Chem 20:1565–1568, 1955.
12. FH Otey, JW Sloan, CA Wilham, CL Mehltretter. Erythritol and ethylene glycol from dialdehyde starch. Industrial Eng Chem 53:267–268, 1961.
13. FJR Bornet. Undigestible sugars in food products. Am J Clin Nutr 59S:763S–769S, 1994.
14. J Goosens, H Röper. Erythritol: a new sweetener. Food Sci Tech Today 8:144–148, 1994.
15. J Goosens, H Röper. Erythritol: a new sweetener. Confec Prod 62:4, 6–7, 1996.
16. H Röper, J Goossens. Erythritol, a new raw material for food and non-food applications. Starch/Stärke 45:400–405, 1993.
17. J Goossens, M Gonze. Nutritional and application properties of erythritol: a unique combination? Part I: Nutritional and functional properties. Agro-Food-Industry Hi-Tech July/Aug:3–9, 1997.
18. T Kasumi. Fermentative production of polyols and utilization for food and other products in Japan. Jpn Agric Res Q 29:49–55, 1995.
19. MAY Aoki, GM Pastore, YK Park. Microbial transformation of sucrose and glucose to erythritol. Biotech Lett 15:383–388, 1993.
20. JB Park, C Yook, YK Park. Production of erythritol by newly isolated osmophilic *Trichosporon* sp. Starch/Stärke 50:120–123, 1998.
21. SY Kim, KH Lee, JH Kim, DK Oh. Erythritol production by controlling osmotic pressure in *Trigonopsis variabilis*. Biotech Lett 19:727–729, 1997.
22. L Slade, H Levine. Interpreting the behavior of low-moisture foods. In: TM Hardman, ed. Water and Food Quality. Essex: Elsevier Science Publishers Ltd., 1989, pp 71–134.
23. TB Carr, SD Pecore, KM Gibes. Sensory methods for sweetener evaluation. In: CT Ho, CH Manley, eds. Flavor Measurement. New York: Marcel Dekker, Inc., 1993, pp 219–237.
24. CL Austin, DJ Pierpoint. The role of starch-derived ingredients in beverage applications. Cereal Foods World 43:748–752, 1998.
25. G Vaamonde, G Scarmato, J Chirife, JL Parada. Inhibition of *Staphylococcus aureus* C-243 growth in laboratory media with water activity adjusted using less usual solutes. Lebensm-Wiss u-Technol 19:403–404, 1986.

26. M Hiele, Y Ghoos, P Rutgeerts, G Vantrappen. Metabolism of erythritol in humans: comparison with glucose and lactitol. Br J Nutr 69:169–176, 1993.
27. IC Munro, WO Bernt, JF Borzella, G Flamm, BS Lynch, E Kennepohl, EA Bär, J Modderman. Erythritol: an interpretive summary of biochemical, metabolic, toxicological and chemical data. Food Chem Toxicol 36:1139–1174, 1998.
28. FRJ Bornet, A Blayo, F Dauchy, G Slama. Plasma and urine kinetics of erythritol after oral ingestion by healthy humans. Regul Toxicol Pharmacol 24:S280–S286, 1996.
29. H Schiweck, SC Ziesenitz. Physiological properties of polyols in comparison with easily metabolisable saccharides. In: TH Grenby, ed. Advances in Sweeteners. Gaithersburg, MD: Aspen Publishers, 1995, pp 56–84.
30. J Kawanabe, M Hirasawa, T Takeuchi, T Oda, T Ikeda. Noncariogenicity of erythritol as a substrate. Caries Res 26:258–362, 1992.
31. WO Bernt, JF Borzelleca, G Flamm, IC Munro. Erythritol: a review of biological and toxicological studies. Regul Toxicol Pharmacol 24:S191–S197, 1996.
32. W Tetzloff, F Dauchy, S Medimagh, D Carr, A Bär. Tolerance to subchronic, high-dose ingestion of erythritol in human volunteers. Regul Toxicol Pharmacol 24: S286–S295, 1996.
33. Federal Register, 34(231) Dec 1997.
34. HP Til, J Modderman. Four-week oral toxicity study with erythritol in rats. Regul Toxicol Pharmacol 24:S214–S220, 1996.
35. HP Til, CF Kufer, HE Falke, A Bär. Subchronic oral toxicity studies with erythritol in mice and rats. Regul Toxicol Pharmacol 24:S221–S231, 1996.
36. AE Smits-van Prooije, DH Waalkens-Berendsen, A Bär. Embryotoxicity and teratogenicity study with erythritol in rats. Regul Toxicol Pharmacol 24:S232–S236, 1996.
37. DH Wallkens-Berendsen, AE Smits-van Prooije, MVM Wijnands, A Bär. Two-generation reproduction study of erythritol in rats. Regul Toxicol Pharmacol 24:S237–S246, 1996.
38. M Shimizu, M Katoh, M Imamura, J Modderman. Teratology study of erythritol in rabbits. Regul Toxicol Pharmacol 24:S247–S253, 1996.
39. WI Dean, F Jackson, RJ Greenough. Chronic (1-year) oral toxicity study of erythritol in dogs. Regul Toxicol Pharmacol 24:S254–S261, 1996.
40. Y Kawamura, Y Saito, M Imamura, JP Modderman. Mutagenicity studies on erythritol in bacterial reversion assay systems and in Chinese hamster fribroblast cells. Regul Toxicol Pharmacol 24:S261–S264, 1996.
41. BAR Lina, MHM Bos-Kuijpers, HP Til, A Bär. Chronic toxicity and carcinogenicity study of erythritol in rats. Regul Toxicol Pharmacol 24:S264–S279, 1996.
42. FRJ Bornet, A Blayo, F Dauchy, G Slama. Gastrointestinal response and plasma and urine determinations in human subjects given erythritol. Regul Toxicol Pharmacol 24:S296–S302, 1996.
43. J Goossens, M Gonze. Nutritional and application properties of erythritol: a unique combination? Part II: Application properties. Agro-Food-Industry Hi-Tech Sept/Oct:12–15, 1997.
44. M Ishikawa, M Miyashita, Y Kawashima, T Nakamura, N Saitou, J Modderman. Effects of oral administration of erythritol on patients with diabetes. Regul Toxicol Pharmacol 24:S303–S308, 1996.

14
Hydrogenated Starch Hydrolysates and Maltitol Syrups

Laura Eberhardt
Calorie Control Council, Atlanta, Georgia

I. INTRODUCTION

Hydrogenated starch hydrolysates (HSH), including maltitol syrups, sorbitol syrups, and hydrogenated glucose syrups, are a family of products found in a wide variety of foods. These food ingredients serve a number of functional roles, including use as bulk sweeteners, viscosity or bodying agents, humectants, crystallization modifiers, cryoprotectants, and rehydration aids. They also can serve as sugar-free carriers for flavors, colors, and enzymes. HSH were developed by a Swedish company in the 1960s and have been used by the food industry for many years, especially in confectionery products.

The term ''hydrogenated starch hydrolysate'' can correctly be applied to any polyol produced by the hydrogenation of the saccharide products of starch hydrolysis. In practice, however, certain polyols such as sorbitol, mannitol, and maltitol are referred to by their common chemical names. ''Hydrogenated starch hydrolysate'' is more commonly used to describe the broad group of polyols that contain substantial quantities of hydrogenated oligosaccharides and polysaccharides in addition to any monomeric or dimeric polyols (sorbitol/mannitol or maltitol, respectively).

The broad term HSH does not differentiate polyols having, for example, different levels of sweetness nor does it identify the principal polyol in the HSH. Common names for major HSH subgroups have, therefore, been developed. These common names are generally based on the most prevalent polyol comprising the HSH. For example, polyols containing maltitol as the major (50% or more) component are called maltitol syrups, maltitol solutions, or hydrogenated glucose syrups; those with sorbitol as the major component are called sorbitol syrups. Polyols that do not contain a specific polyol as the major component continue to be referred to by the general term ''hydrogenated starch hydrolysate.''

II. PRODUCTION

HSH are produced by the partial hydrolysis of corn, wheat, or potato starch and catalytic hydrogenation of the hydrolysate at high temperature under pressure. By varying the conditions and extent of hydrolysis, the relative occurrence of various mono-, di-, oligo- and polymeric hydrogenated saccharides in the resulting product can be obtained. The proportion of these hydrogenated species distinctly affects the functional properties and chemistry of the particular HSH.

In the United States, HSH are produced and marketed by two major manufacturers: SPI Polyols, New Castle, Delaware and Roquette America, Inc., Gurnee, Illinois (parent company Roquette Freres, Lestrem, France). The products produced by these manufacturers may differ slightly in viscosity, sweetness, and hygroscopicity, depending on their composition.

Product examples of maltitol syrups include SPI Polyols' Maltisweet™ B which is approximately 50% maltitol and Maltisweet™ 80 and Maltisweet™ 85, and Maltisweet™ 3145, which are approximately 65% maltitol; and Roquette's Lycasin® 80/55, which is 50–55 percent maltitol. Sorbitol syrups include Roquette's 70/100, and SPI Polyols' A-625, which contain 75 and 70% sorbitol, respectively. Examples of products called by the general term HSH include Roquette's 75/400 and SPI Polyols' Stabilite™ 1 and Stabilite™ 2.

III. PHYSICAL CHARACTERISTICS

HSH are colorless and odorless and generally available as a 75% solids syrup composed of variable relative amounts of hydrogenated saccharides, characterized by their degree of polymerization (DP). The components consist of sorbitol (DP-1), maltitol (DP-2), and higher hydrogenated saccharides—maltotriitol (DP-3) and others, continuing above DP-20. HSH structures, along with their precursors, are shown in Fig. 1.

IV. APPLICATIONS

The physical properties of HSH make them a valuable aid in a variety of applications. Because of the similarity to corn syrup, HSH can be used as a substitute in most applications in which corn syrup is used. Advantages over corn syrup

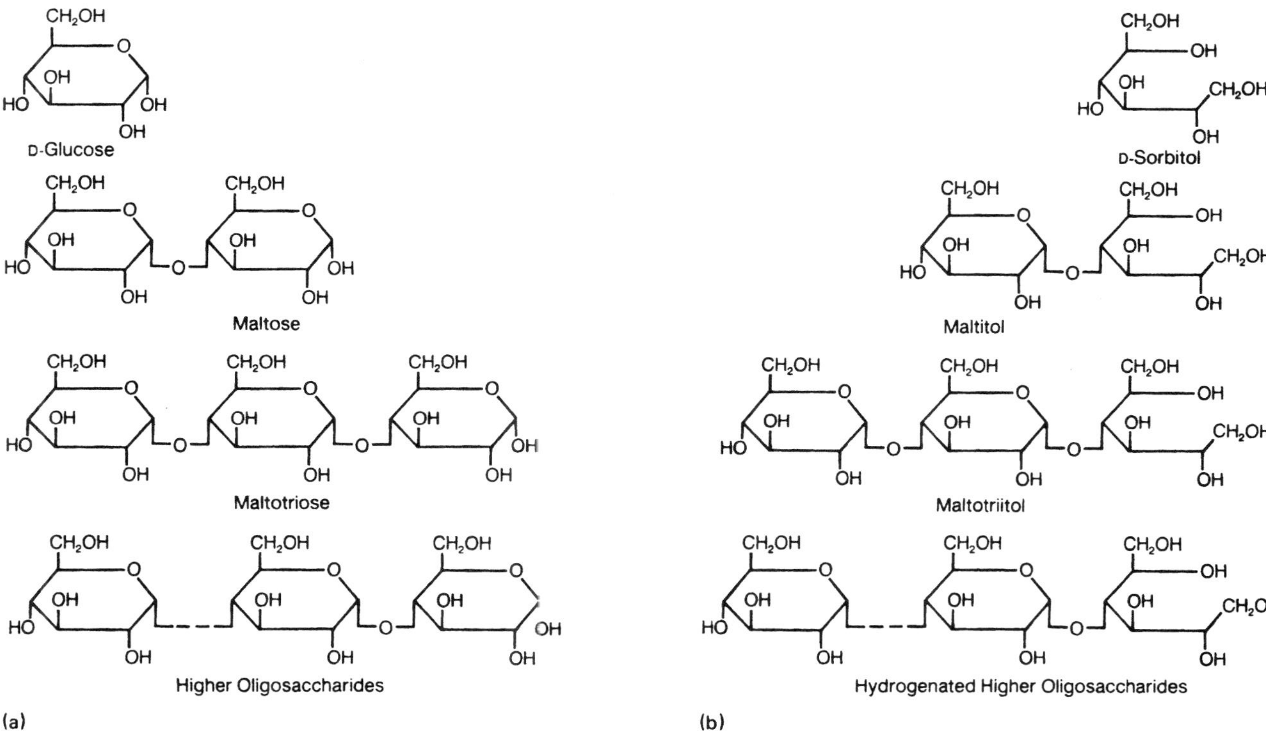

Figure 1 (a) Chemical structures for monosaccharide, disaccharide, and polysaccharide polyol precursors. (b) Chemical structures for the polyol components of hydrogenated starch hydrolysate.

include humectancy, nonreducing characteristics, cryoprotective properties, and resistance to heat and acid. HSH's noncariogenicity makes them a desirable alternative to sugar sweeteners.

A. Confectionery Products

Multiple uses are possible in confectionery products by replacing sugar and corn syrup with HSH. Detailed formulations for preparing caramels, gummy bears, jelly beans, hard candy, taffy, nougats, butterscotch, marshmallows, and chewing gum are available from HSH suppliers. HSH are outstanding humectants, which do not crystallize. This property enables the production of sugar-free confections with the same cooking and handling systems used to produce sugar candies and makes HSH advantageous over other polyols. Candies made with sorbitol, for example, require additional molding, curing, and demolding—thus requiring additional processing costs.

Hard candies made with HSH should be packaged while warm in a moisture-resistant container. During preparation, the boiling temperature must reach 160°C under a vacuum to reach a moisture level less than 1% (1). Also, flavors added to candies made with HSH must be nonaqueous. These moisture-minimizing procedures are essential regardless of the type of HSH used. In the case of sucrose or corn syrup candy, moisture pickup on the surface leads to crystallization of sucrose (graining), which insulates the candy against further moisture collection. However, because HSH do not crystallize, moisture collection at the surface causes a sticky layer of solubilized HSH that will eventually cause the candy to partly dissolve (2). HSH cannot be used as a replacement for sugar in chocolates or in pressed tablets, where moisture would be especially detrimental to the product (3).

The lack of a crystal structure makes HSH hard candies subject to "cold flow." Cold flow is a loss of shape caused by elevated moisture. This may occur even if the candy is protected from atmospheric water vapor. Products with longer chain polyols provide additional molecular structure and are more resistant to cold flow. However, this resistance to crystallization, even at low temperatures and high concentrations, makes HSH particularly advantageous for use in chewy candies. The crystallization of other components present in the formulation, such as sorbitol, mannitol, and xylitol, is also prevented.

Another advantage of HSH is that they do not have reducing groups, thus minimizing Maillard browning reactions. Because of this and HSH's resistance to heat and acid, it is possible to manufacture, at high temperatures with acidic ingredients, sweets that remain bright and colorless.

HSH are nutritive sweeteners that provide 40 to 90% of the sweetness of sugar. Unlike sugars, however, HSH are not readily fermented by oral bacteria so are used to formulate sugarless products that do not promote dental caries.

However, to make a candy noncariogenic, it is necessary to also remove other sugar-containing ingredients in the product and replace them with a suitable alternative. For example, this may include replacing milk powder with caseinates and milk fat. To manufacture a product matching the sweetness of sugar-sweetened candies may require the addition of a potent noncariogenic sweetener such as aspartame or acesulfame-K. HSH blend well with other sweeteners and can mask unpleasant off-flavors such as bitter notes. Using the multiple sweetener approach generally allows a better sweetness profile than using sweeteners alone.

B. Other Applications

HSH can be used as a replacement for sugar in a variety of frozen desserts, because they will not form crystals. They also act as mild freezing point depressants to increase freeze/thaw stability. These cryoprotective properties protect protein fibers from damage caused by ice crystal formation and thermal shock (e.g., freezer burn). For this reason, HSH can be used as a cryoprotectant glaze in seafood and other products to increase shelf life in frozen storage.

Because of their excellent humectancy, HSH products are also used extensively to partially replace sugar in baked goods, a broad range of other foods, medicinal syrups (e.g., pediatric medicines, cough syrups), dentifrices, and mouthwashes. Formulations for the preparation of these products may be obtained from the HSH manufacturer.

V. GENERAL COST AND ECONOMICS

Because of the need for additional processing, polyols generally are more expensive than the common carbohydrate sweeteners corn syrup and sucrose. The cost of HSH is two to three times more than that of liquid sorbitol. Purchasing in very large quantities generally lessens the cost.

VI. METABOLIC ASPECTS

On ingestion, HSH is enzymatically hydrolyzed to sorbitol, glucose, and maltitol. Only 10% of maltitol may be converted into monosaccharides that are absorbed through the intestinal mucosa (4). Maltitol is excreted mainly as gas, also in the feces and urine (5). The digestion of the oligosaccharides and polysaccharides is close to 90% (6), although the rate of enzymatic hydrolysis might be a function of the component polyol's chain length. The unabsorbed products of HSH hydrolyzation reach the lower digestive tract where they are metabolized by naturally occurring colonic bacteria. Evidence of colonic fermentation has been shown using the hydrogen breath test (7). The absorbed glucose and sorbitol are taken up into the bloodstream. Sorbitol is converted to fructose, which is then used through the glycolytic pathway. A review of the metabolism of sorbitol can be found in Chapter 18.

A. Cariogenicity

It is widely accepted that prolonged exposure of the teeth to acid produces dental caries. Sugars and starches are fermented by oral bacteria, producing organic acids that can solubilize tooth enamel and result in decay. Because of obvious ethical restrictions in conducting most tests of cariogenicity in human subjects, screening for proper conditions for cariogenic potential is conducted.

''Cariogenic potential'' was defined at the Scientific Consensus Conference on Methods for the Assessment of the Cariogenic Potential of Foods as, ''the ability of a food to foster caries in humans under conditions conducive to caries formation'' (8). This group also reached an agreement on a line of testing that could be used to establish that a food had either no cariogenic potential or low cariogenic potential. Consensus is that ''foods assessed by two recommended plaque acidity test methodologies that result in pH profiles statistically equivalent to those generated by sorbitol would be deemed as possessing no cariogenic potential'' (8). Demineralization of tooth enamel definitely occurs below a pH of 5.5; between 5.5 and 5.7 is a transition range, where some demineralization may begin.

The FDA has authorized the use of the ''does not promote tooth decay'' health claim for sugar-free food products sweetened with polyols (9). The regulation provides that ''when fermentable carbohydrates are present in the sugar alcohol–containing food, the food shall not lower plaque pH below 5.7 by bacterial fermentation either during consumption or up to 30 minutes after consumption, as measured by the indwelling plaque test found in 'Identification of Low Caries Risk Dietary Components,' T.N. Imfeld, Volume 11, *Monographs in Oral Science* (1983)'' (10).

The relative acidogenicity of test products may be predicted on the basis of the component makeup of the compounds. Theoretically, HSH with higher hydrogenated saccharides (molecules DP4 and larger) would be more likely to result in the release of free glucose after hydrolysis, which is readily fermentable by oral bacteria (11). Because HSH products differ in their saccharide profile, studies have been conducted to examine the cariogenicity of each HSH. Although HSH with higher concentrations of higher DP saccharides show a greater drop in pH than those with lower DP fractions, each has been shown to remain above the 5.7 ''safe for teeth'' value (3,11).

The American Dental Association (ADA) has recognized the usefulness of polyols as alternatives to sugars and as a part of a comprehensive program including proper dental hygiene. In October 1998, the ADA's House of Delegates approved a position statement acknowledging the ''Role of Sugar-Free Foods and Medications in Maintaining Good Oral Health'' (12)

B. Laxation

The ingestion of HSH several times a day on an empty stomach of unadapted subjects can result in a laxative effect. This is true of most polyols because of their incomplete absorption and resultant increased osmotic pressure. This effect increases with an increase in the relative amount of sorbitol in the HSH. In addition, the digestive system appears to adapt with a decrease in symptoms such as flatulence and diarrhea after repeated daily consumption (3). Persons in the 90th percentile for consumption of lycasin-containing products only consume 1.1–2.6 g per day (13). It is recommended that if the ingestion of 50 g or more is foreseeable, the statement ''excess consumption may have a laxative effect'' should be used.

C. Caloric Content

The components of HSH are slowly and incompletely absorbed, allowing a portion of HSH to reach the large intestine, thereby reducing the carbohydrate available for metabolism. Therefore, unlike sugar that contributes 4 calories per gram, the caloric contribution of HSH is not more than 3 calories per gram (14). For a product to qualify as ''reduced calorie'' in the United States, it must have at least a 25% reduction in calories. HSH may, therefore, be of use in formulating reduced-calorie food products.

The lower caloric value of HSH and other polyols is recognized in other countries. For example, the European Union has provided a Nutritional Labeling Directive stating that all polyols, including HSH, have a caloric value of 2.4 calories per gram (15).

D. Suitability in Diabetic Diets

Control of blood glucose, lipids, and weight are the three major goals in diabetes management today. Because of their slow and incomplete absorption, HSH have a reduced glycemic potential relative to glucose for individuals with and without diabetes (7). This property permits its use as a reduced-calorie alternative to sugar. Doses of 45–90 g day are well tolerated without glycemic effect in either diabetic or nondiabetic subjects (16). The reduced caloric value (75%, or less, that of sugar) of HSH is also consistent with the objective of weight control.

The American Diabetes Association acknowledges the lower caloric value of polyols but cautions their use may not contribute to a significant reduction in total calories or carbohydrate content of the daily diet. The calories and carbohydrate from HSH-sweetened products should be accounted for in the meal plan (17). Although studies have shown a reduced glycemic response compared with glucose, HSH and other ingredients in the food product may have the potential to affect blood glucose levels. Recognizing that diabetes is complex and requirements for its management may vary between individuals, the usefulness of HSH should be discussed between individuals and their physicians.

VII. TOXICITY

A broad range of safety studies in man and animals, including long-term feeding (18), multigeneration reproduction/development (19), and teratology studies (20) have shown no evidence of any adverse effects from HSH. The results of these studies have added to the substantial body of information establishing the safety of HSH (5).

The Joint Food and Agriculture Organization/World Health Organization Expert Committee on Food Additives (JECFA) has reviewed the safety information and concluded that maltitol syrups are safe (21). JECFA established an acceptable daily intake (ADI) for HSH (maltitol syrup) of ''not specified,'' meaning no limits are placed on its use. JECFA defines ''not specified'' as: ''on the basis of available scientific data, the total daily intake of a substance arising from its use at levels necessary to achieve the desired effect, does not, in the opinion of the Committee, represent a hazard to health.'' Many small countries that do not have their own agencies to review food additive safety often adopt JECFA's decisions. In 1984, the Scientific Committee for Food of the European Union evaluated maltitol syrups and also concluded it was not necessary to set an ADI for maltitol syrups (22).

VIII. REGULATORY STATUS

In the United States, generally recognized as safe (GRAS) petitions for HSH products have been accepted for filing by the Food and Drug Administration. Once a GRAS affirmation petition has been accepted for filing, manufacturers are allowed to produce and sell foods containing these sweeteners in the United

States. Products from the HSH family are approved in many other countries, including Canada, Japan, and Australia.

REFERENCES

1. MS Billaux, et al. Sugar alcohols. In: S Marie, JR Piggott, eds. Handbook of Sweeteners. Glasgow and London: Blackie and Son Publishers, 1991.
2. PJ Sicard, P Leroy, Mannitol, sorbitol and lycasin: Properties and food applications. In: TH Grenby, ed. Developments in Sweeteners—2. London and New York: Elsevier-Applied Science Publishers, 1983.
3. A Moskowitz. Maltitol and hydrogenated starch hydrolysate. In: L Nabors, R Gelardi, eds. Alternative Sweeteners. 2nd ed. New York: Marcel Dekker, 1991.
4. A Secchi, AE Pontiroli, L Cammelli, A Bizzi, M Cini, G Pozza. Effects of oral administration of maltitol on plasma glucose, plasma sorbitol, and serum insulin levels in man. Klin Wochenschr 64:265–269, 1986.
5. JP Modderman. Safety assessment of hydrogenated starch hydrolysates. Regul Toxicol Pharmacol 18:80–114, 1993.
6. L Beaugerie, B Flourié, C Franchisseur, P Pellier, H Dupas, JC Rambaud. Absorption intestinale et tolérance clinique au sorbitol, maltitol, lactitol et isomalt. (abstr) Gastroenterol Clin Biol 13:102, 1989.
7. ML Wheeler, et al. Metabolic response to oral challenge of hydrogenated starch hydrolysate versus glucose in diabetes. Diabetes Care, 13:733–740, 1990.
8. Working Group Consensus Report. Integration of methods. J Dent Res 1986; 65 (special issue):1537–1539.
9. U.S. Food and Drug Administration. Food labeling: Health claims; sugar alcohols and dental caries, final rule. Federal Register, Bol. 61 No. 165:43433, August 23, 1996.
10. T Imfeld. Identification of low caries risk dietary components. In: HM Myers, ed., Monographs in Oral Science. Vol. 11. Basel: Karger 1983, p 117.
11. D Abelson. The effect of hydrogenated starch hydrolysates on plaque pH in vivo. Clin Prev Dent 11(2):20–23, 1989.
12. American Dental Association. Position Statement on the Role of Sugar-Free Foods and Medications in Maintaining Good Oral Health. Adopted October 1998.
13. U.S. Food and Drug Administration. Roquette Corp.; Filing of petition for affirmation of GRAS status (hydrogenated glucose syrup). Federal Register, Vol. 49, No. 39:7153, February 27, 1984.
14. Federation of American Societies for Experimental Biology. The evaluation of the energy of certain polyols used as food ingredients. June 1994. Unpublished.
15. European Economic Community Council (EEC). Directive on food labeling. Official Journal of the European Communities. No. L 276/40 (Oct. 6), 1990.
16. A Tacquet. Study on the Clinical and Biological Tolerance of Polysorb Lycasin 80/33 in the Human Being. (Calmette Hospital, d'Lille, France). Report available

from U.S. FDA, Dockets Management Branch; specify Docket No. 84G-0003, Report E-35.
17. Nutrition Recommendations and Principles for People with Diabetes Mellitus. American Diabetes Association: Clinical Practice Recommendations, 2000.
18. H Dupas, et al. 24-Month Safety Study of Lycasin® 80/55 on Rats. (Roquette Freres, Lestrem, France). Report available from U.S. FDA Dockets management Branch; specify Docket No. 84G-0003, op cit, Report E-52.
19. P Leroy, H Dupas. Lycasin 80/55: Three Generation Reproduction Toxicity Studies. (Roquette Freres, Lestrem, France). Report available from U.S. FDA Dockets Management Branch, specify Docket No. 84G-0003, op cit, Report E-33.
20. H Dupas, G Siou. Lycasin® 80/55: Teratogenic Potential Study in Rats. (Roquette Freres, Lestrem, France). Report available from U.S. FDA Dockets Management Branch; specify Docket No. 84G-0003, op cit, Report E-53.
21. Joint FAO/WHO Expert Committee on Food Additives. Evaluation of certain food additives and contaminants: maltitol and maltitol syrup. Forty-first report. WHO Technical Report Series 837, pp 16–17. Geneva, 1993.
22. Commission of the European Communities. Reports of the Scientific Committee for Food Concerning Sweeteners. Sixteenth Series. Report EUR 10210 EN. Office for Official Publications of the European Communities, Luxembourg, 1985.

15
Isomalt

Marie-Christel Wijers*
Palatinit of America, Inc., Morris Plains, New Jersey

Peter Jozef Sträter
Palatinit Süßungsmittel GmbH, Mannheim, Germany

I. INTRODUCTION

In the past two decades, the interest in sugar-free bulk sweeteners has grown in the field of "tooth-friendly" and calorie-reduced confectionery, baked goods, and pharmaceutical products. Sugar-free bulk sweeteners are sweeteners that give body and texture to a product, as well as a sweet flavor. Ideally, they do not cause any aftertaste and provide the same functions as sucrose and glucose. Their physiological and nutritional metabolism in the gastrointestinal tract and their biochemical changes in the mouth, however, differ from sucrose and glucose.

The ideal sweetener should be chemically and biologically stable for an indefinite period of time and provide the same properties to a product as sucrose or glucose. Its processing parameters should be similar to that of sucrose or glucose, so that existing equipment can be used without requiring major changes. In addition, the finished products should have practically the same taste and appearance as those of traditional products, have an excellent shelf-life, and be readily accepted by the consumer.

This chapter presents the polyol isomalt as a sugar-free bulk sweetener. Isomalt is a sweet, low-calorie, bulking agent with properties and characteristics similar to sucrose. It is odorless, crystalline, and nonhygroscopic (1, 2). It is a nonreducing sugar and optically active. Unlike sucrose, however, it is extremely stable with respect to chemical and enzymatic hydrolysis. It cannot be fermented by a large number of yeasts and other microorganisms found in nature. Isomalt is manufactured and marketed by Palatinit GmbH, a wholly owned subsidiary

---

*Retired.

of Südzucker AG (Germany). For certain applications, special types of isomalt have been developed such as isomalt-ST, isomalt-HB, isomalt-GS, isomalt-DC, and isomalt-LM. These different types will be discussed in this chapter in the section "Applications and Product Development." Because a lot of properties are similar for all isomalt types, isomalt-ST (i.e., isomalt standard, the best known type) has been selected to describe properties of isomalt in general.

II. PRODUCTION

Isomalt is the only bulk sweetener derived exclusively from sucrose. It is manufactured in a two-stage process in which sugar is first transformed by enzymatic transglucosidation into isomaltulose, a reducing disaccharide (6-*O*-α-D-glucopyranosyl-D-fructose). The isomaltulose is then hydrogenated into isomalt. Isomalt is composed of 6-α-D-glucopyranosyl-D-sorbitol (1,6-GPS) and 1-*O*-α-glucopyranosyl-D-mannitol dihydrate (1,1-GPM dihydrate) (3, 4). The ratio of GPS and

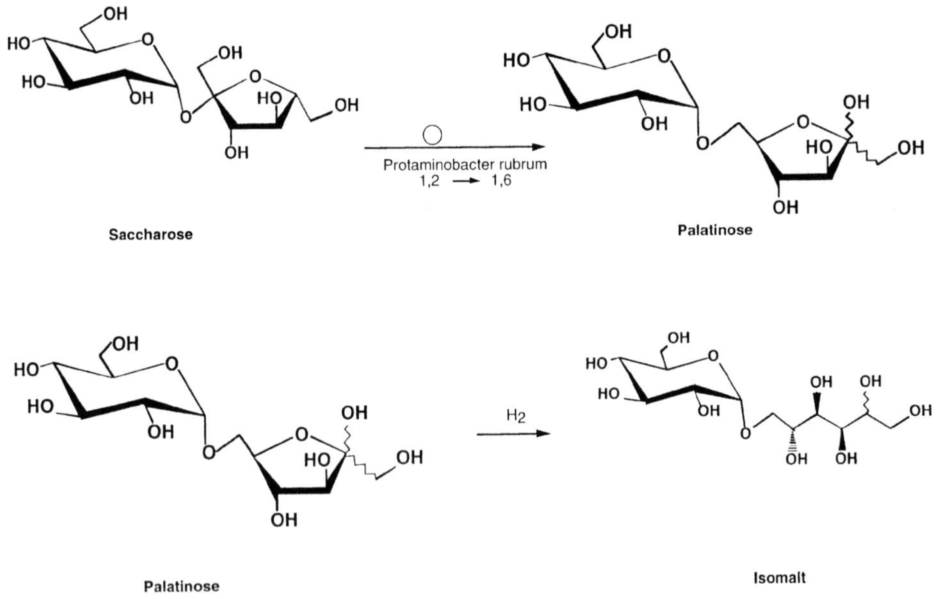

Figure 1 Production of isomalt from sucrose. The intermediate product is called isomaltulose or palatinose.

GPM depends on the isomalt type. In aqueous systems, GPS forms anhydrous crystals, whereas GPM has two molecules of water of crystallization. The production process is illustrated in Fig. 1.

III. SENSORY PROPERTIES

A. Sweetness and Taste

The sweetening power of isomalt lies between 0.45 and 0.6 compared with that of sucrose (= 1.0). Figure 2 shows that the sweetening power is a function of concentration (i.e., it increases with increasing isomalt concentration) (5). There is no difference between the sensorially tested and theoretically determined curves.

Isomalt has a pure sweet taste similar to sucrose without any aftertaste. Furthermore, it reinforces flavor transfer in foods. Synergistic effects occur when

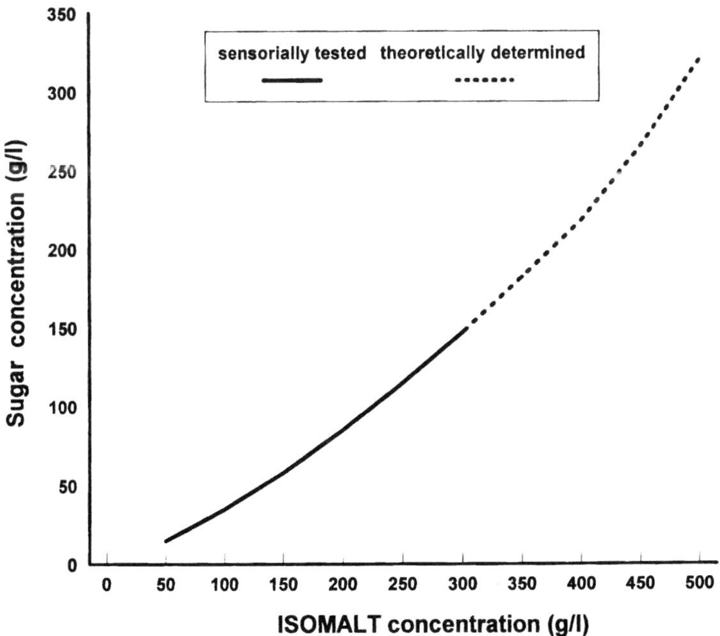

Figure 2 Isosweet aqueous solutions of isomalt-ST and sucrose, determined sensorially and theoretically.

isomalt is combined with other sugar alcohols (e.g., xylitol, sorbitol, mannitol, maltitol syrup, hydrogenated starch syrup) and with high-intensity sweeteners (e.g., acesulfame K, aspartame, sucralose, cyclamate, or saccharin). In addition, isomalt tends to mask the bitter metallic aftertaste of some intense sweeteners (4).

B. Cooling Effect

Several sugar alcohols, used as sugar substitutes, have a high negative heat of solution (6–8). This results in a cooling sensation in the mouth when these sugar alcohols are consumed in a crystalline or solid state. Although this mouth-cooling effect is a desirable feature for peppermint or menthol products, it is considered atypical in many other products, such as baked products and chocolate. The negative heat of solution of isomalt-ST lies between the values of GPS and GPM. Figure 3 shows that, compared with other sugar alcohols, isomalt has a very low negative heat of solution, which is comparable to that of sucrose. Isomalt, therefore, does not produce a cooling effect.

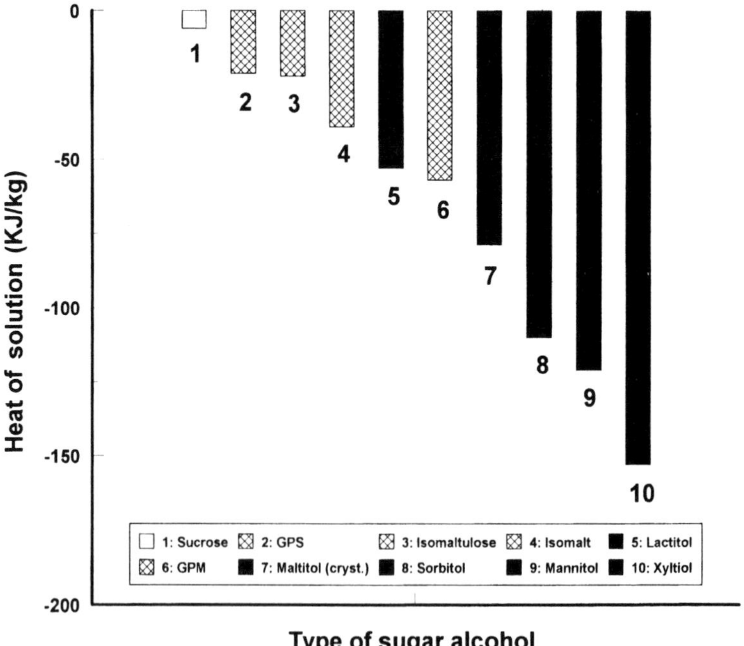

Figure 3 Negative heat of solution (kJ/kg) of sugar and sugar alcohols.

IV. PHYSICO-CHEMICAL PROPERTIES

A. Physical Properties

The sorption isotherm (see Fig. 4) shows that isomalt has a very low water activity. At 25°C isomalt absorbs virtually no water up to a relative humidity of 85%. Furthermore, isomalt does not start absorbing water until the temperature reaches 60°C at 75% relative humidity (rh) or 80°C at 65% rh (9). This very low hygroscopicity means that isomalt can be stored easily and distributed without much special care. Furthermore, this explains why products exclusively or mainly based on isomalt (e.g., hard candies) tend to be not sticky and have a long shelf-life.

As depicted in Fig. 5, the solubility of isomalt-ST is much lower than that of sugar at 20°C, namely 24.5 g/100 g solution compared with a value of 66.7 g/100 g solution for sugar (4, 9). Because of its low solubility, oversaturated solutions can be obtained with a relatively low concentration of isomalt. The solubility of isomalt increases with increasing temperature and is comparable to sucrose at most processing temperatures (>85°C).

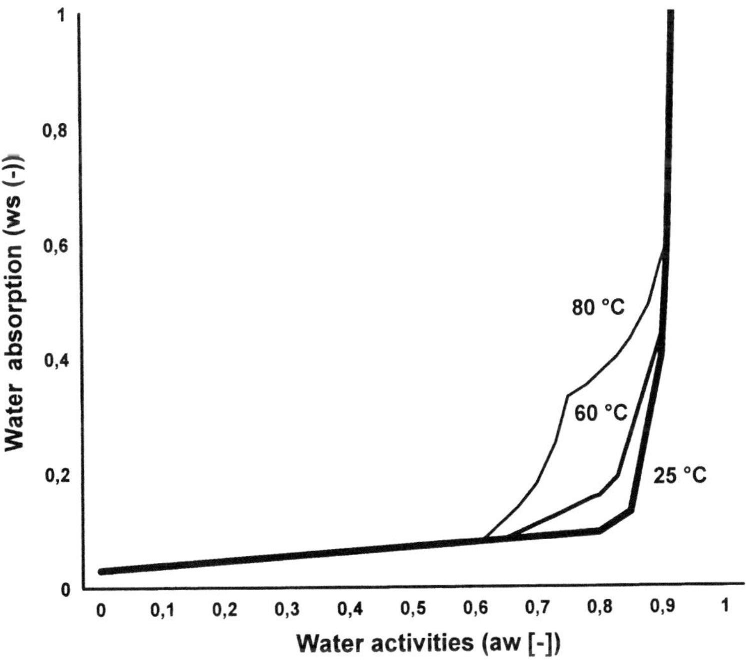

Figure 4 Sorption isotherm for isomalt-ST at three temperatures (25, 60, and 80°C); w_s is the amount of water absorbed by isomalt (kg water/kg isomalt).

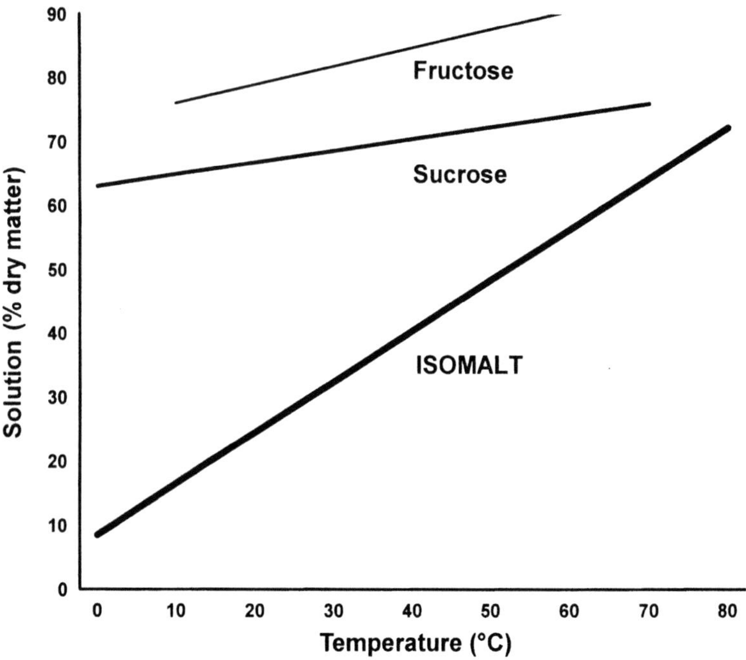

Figure 5 Solubility of isomalt in water compared with sucrose and fructose as a function of temperature.

The high melting temperature range of isomalt-ST (between 145 and 150°C) demonstrates its high heat resistance (4, 9). Because its chemical structure is not altered at normal cooking temperatures, isomalt is well suited for cooking, baking, and extrusion processes. Another important property of isomalt is that it can easily be ground to obtain 100-μm granules. These granules are needed to give the right texture for chocolates and chewing gum. Finely ground isomalt granules can be formed into larger agglomerates that can be used for instant products and tablets.

The viscosity of aqueous isomalt solutions does not differ significantly from that of corresponding sucrose solutions in a temperature range between 5 and 90°C (4, 9). No special engineering requirements need to be taken into consideration. The viscosity of a cooked isomalt mass depends on the shear velocity and can be higher or lower than a sucrose/corn syrup melt. For hard candy production for example, the cooked isomalt mass shows a lower viscosity than a sucrose/ corn syrup melt at the same temperature. Isomalt is usually boiled to a higher temperature than sucrose/corn syrup to obtain the required water content in the melt.

B. Chemical Properties

Isomalt is extremely resistant to chemical degradation because of its very stable 1–6 bond between the mannitol or sorbitol moiety and the glucose moiety. When the crystalline substance is heated above the melting point or the aqueous solution above the boiling point, no changes in the molecular structure are observed. No caramelization or other discoloration develops during the melting, extrusion, or baking processes. In general, isomalt does not react with other ingredients in the formulation (e.g., with amino acids to produce Maillard reactions) (melanoids). The stability of isomalt during acid hydrolysis was measured at 100°C in 1% hydrochloric acid (4, 9). Figure 6 illustrates that under these conditions sucrose is hydrolyzed in less than 5 minutes. In contrast, isomalt is only completely split after 5 hours. In an alkaline milieu, isomalt also shows a much higher resistance

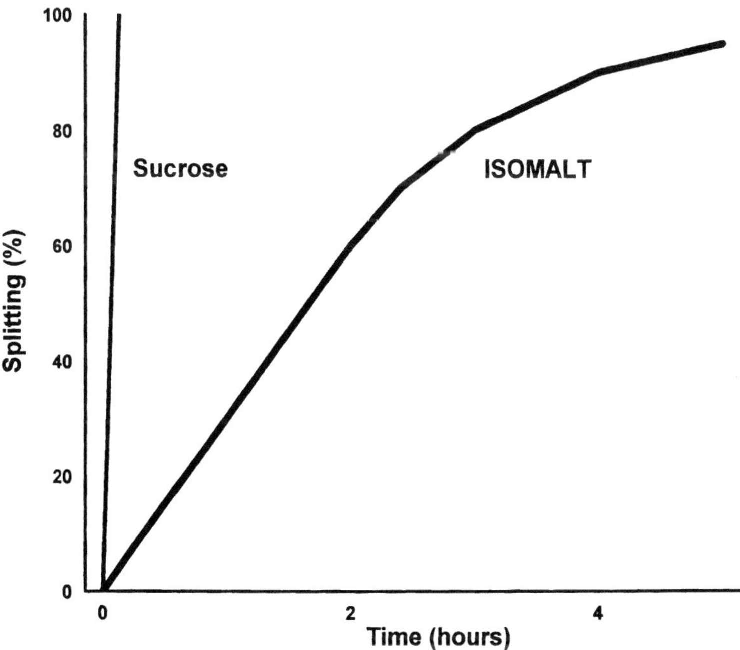

Figure 6 Hydrolysis of isomalt and sucrose versus time in a 1% HCl solution at 100°C.

than sucrose. Furthermore, isomalt is resistant to enzymatic hydrolysis. Most microorganisms found in foods are unable to use isomalt as a substrate (4, 9, 10).

V. PHYSIOLOGICAL PROPERTIES

As mentioned in the preceding section, the remarkable feature of isomalt is the very stable glycoside bond in GPS and GPM. Compared with sucrose, this difficult-to-split glycoside bond gives isomalt the following properties after oral intake (11):

- Only about 50% of isomalt is converted into available energy.
- Isomalt is noncariogenic.
- Isomalt is suitable for diabetics.

A. Caloric Value

Humans are only able to absorb and, subsequently, to metabolize monosaccharides from the digestive tract by means of the wall of the small intestine. Because of the stable glycoside bond, isomalt is hardly metabolized in the small intestine. It is mainly fermented in the colon. This results in a lower energy conversion (11–15).

Because of individual nutrition labeling regulations, the energy value used for food labeling purposes can vary from country to country. For example, the energy value for isomalt used for food labeling in Japan is 2.0 kcal/g (food) or 1.2–1.5 kcal/g (''special food for patient''); in the United States and Canada, it is 2.0 kcal/g; in Australia it is 2.15 kcal/g; and in the European Union it is 2.4 kcal/g.

B. Cariogenicity

Tooth decay (caries) is the result of acid affecting the tooth enamel. The acids dissolve calcium from the surface and holes occur. The acid is produced by micro-organisms adhering to the surface of the tooth, forming plaque. Although these micro-organisms easily ferment carbohydrates like sugar or glucose into decay-causing acids, they cannot convert isomalt. The critical pH value in the plaque at which the dental enamel may dissolve is 5.7 or less. Studies have shown that consumption of isomalt and products containing only isomalt as the bulk sweetener does not result in a plaque pH less than 5.7 (16–19).

C. Diabetics

A number of scientific studies have shown that insulin and blood glucose levels in humans increase only slightly after oral intake of isomalt—if at all (14, 15, 20). This means sharp increases in blood glucose, as occur after sucrose intake (especially when between-meal snacks and sweets are eaten), can be avoided with isomalt intake. Isomalt is, therefore, a suitable sugar substitute for diabetics.

D. Gastrointestinal Tolerance

Sugar alcohols and some other carbohydrates (e.g., polydextrose, dietary fiber, lactose) have low digestibility. Consumption of large amounts may result in a laxative effect. This is due to the slow degradation process in the gastrointestinal tract. It is not possible to give a useful tolerance limit for sugar alcohols such as isomalt because this tolerance depends on (21–24):

- The form in which it is ingested. In a liquid food, the intolerance is higher than in a solid food.
- The individual sensitivity. The tolerance varies from person to person.
- The moment of consumption. Differences in tolerances even exist for the same person. A person's tolerance is affected by diet (low or high in low digestible carbohydrates) and psychological well-being (emotions, prejudices).
- The adaptation to sugar alcohols. Frequent consumption results in a higher tolerance.

E. Toxicological Evaluations

Extensive toxicological and metabolic studies have been conducted that prove conclusively the safety of isomalt (25–28). The results of these studies have been summarized in a World Health Organization report prepared by the Joint FAO/WHO Expert Committee on Food Additives (JECFA) (29). This report concludes by assigning isomalt an ADI "not specified," the safest rating assigned to any evaluated food substance.

VI. APPLICATIONS

On the basis of its physical and chemical properties, isomalt is a suitable sugar replacer in many areas of the food and pharmaceutical industries. Existing processing equipment can be used for all applications without requiring major changes. Only formula and process parameter modifications are recommended to optimize processes and products. The spectrum of isomalt applications is broad and listed below:

- High-boiled candies
- Pan-coated products
- Chewing gum
- Low-boiled candies
- Compressed tablets/lozenges
- Chocolate
- Baked goods and baking mixes
- Ice cream
- Jams and preserves
- Fillings, fondant
- Marzipan, croquant, and nougat
- Cereal extrudates
- Pharmaceutical applications
- Technical applications

All these products can be made with isomalt-ST. The new isomalt types, however, are specifically designed to improve certain characteristics in finished products and/or to simplify production processes. The new isomalt types will be discussed together with their main applications.

A. High-boiled Candies

High-boiled candies made with isomalt can be stamped, filled, pulled, combed, or molded. Hard candies with a very good shelf-life will be obtained if the water content in the finished product is less than 2%. These candies are very stable against water absorption. Compared with candies based on sucrose/corn syrup, only minor changes in the production process (batch or continuous) are required. The minor changes are necessary because the following characteristics of isomalt differ from sucrose/corn syrup (9, 30–34):

- Lower solubility
- Higher boiling point
- Lower viscosity of the melt
- Higher specific heat capacity

Recently, a new type, isomalt-HB, has been specially designed, for example, for the manufacture of hard candies. Like isomalt-ST, isomalt-HB is an odorless, white, crystalline powder. The storage conditions of both isomalt types are the same, namely at 20 ± 10°C and 40–70% rh. They differ slightly in composition, for example, in content of mannitol and sorbitol.

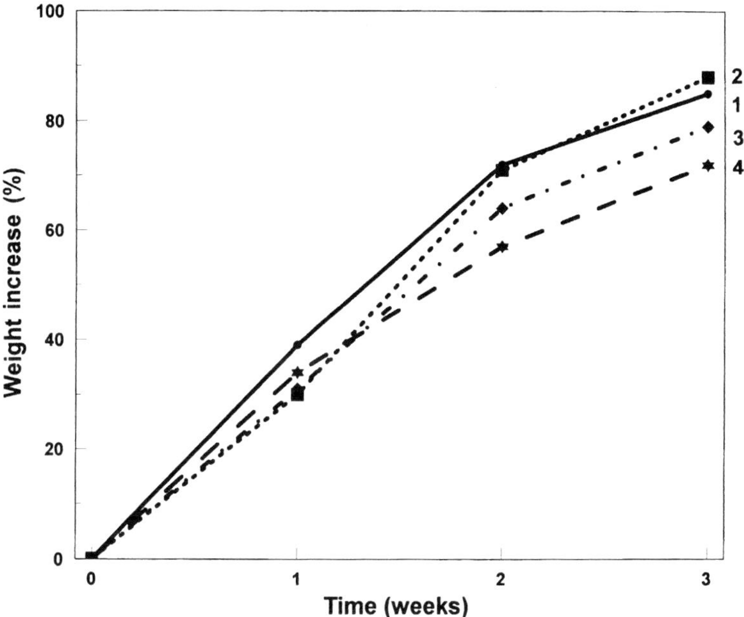

Figure 7 Changes in weight of hard candies with isomalt-ST and HB, with and without acid. The candies are stored at 25°C/80% rh in closed packaging; 1 = isomalt-ST without acid; 2 = isomalt-ST with crystalline acid; 3 = isomalt-ST with dissolved acid; 4 = isomalt-HB with dissolved acid.

The slightly greater amounts of mannitol and sorbitol in isomalt-HB result in a higher elasticity of the mass below 80°C. This allows a better processibility during rope-sizing. The candies, made with isomalt-HB, have the same sensorial profile as candies with isomalt-ST and are less brittle during sucking. The shelf-life of isomalt-HB and isomalt-ST candies is similar. Figure 7 shows the changes in weight of different isomalt hard candies during storage. These changes in weight, caused by moisture absorption, are under most climatic conditions not significantly different for isomalt-HB and isomalt-ST candies.

B. Pan-Coated Products

Panned goods consist of a center (e.g., chewing gum, nut, raisin, candy) coated with sugar, sugar substitutes, or chocolate. The coating procedure depends on the type of coating equipment and on the type of center. Isomalt can replace sucrose in all sorts of pan-coating (30, 31):

- Hard coating with isomalt suspension or solution, possibly in combination with isomalt powder
- Soft coating with isomalt in combination with another nonrecrystallizing sugar substitute
- Chocolate coating with isomalt chocolate

The low hygroscopicity of isomalt leads to products with an excellent shelf-life. Isomalt-ST and isomalt-GS can be used for coating purposes. These two types differ in their ratio of GPS and GPM. For isomalt-ST, this ratio is about 1 (wt%), whereas for isomalt-GS this ratio is approximately 4 (wt%). The greater amount of GPS in isomalt-GS improves the crunchiness of a coating made with isomalt-GS caused by differences in crystal structure between GPM and GPS. Because the solubility of GPS is higher than the solubility of GPM, isomalt-GS has a higher aqueous solubility than isomalt-ST. Therefore, a coating with isomalt-GS dissolves faster, which improves the sweetness impact and the flavor release. Figure 8 compares the dissolution kinetics of isomalt-GS, isomalt-ST, and xylitol.

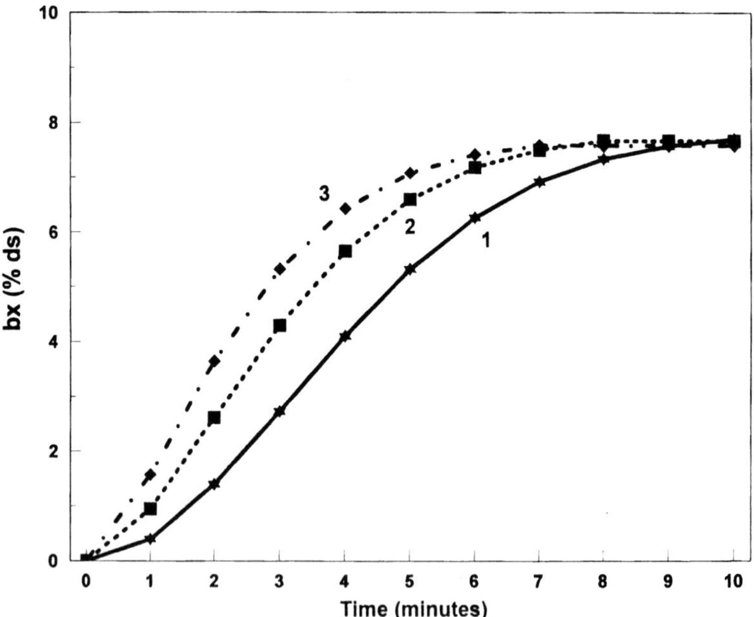

Figure 8 Dissolution kinetics of chewing gum coatings based on isomalt-ST, isomalt-GS, and xylitol, determined by brix measurements (bx); 1 = isomalt-ST; 2 = isomalt-GS; 3 = xylitol.

C. Chewing Gum

Sticks, pellets, or gum balls can be manufactured with powder-fine isomalt. Its low solubility causes isomalt to remain crystalline in the chewing gum mass, which leads to a softer/smoother textured product. Shelf-life tests indicate that chewing gums containing 10 to 15% isomalt remain flexible longer than other sugar-free chewing gums (30). Furthermore, sensorial tests have shown that the slow solubility of isomalt gives a longer lasting flavor release and sweetness. The addition of flavored coarse isomalt crystals to the gum base is an interesting new approach in sugar-free products to further extend the long-lasting flavor release.

D. Low-boiled Candies

Low-boiled candies often have a crystallizing and noncrystallizing phase. Isomalt can replace sugar in the noncrystallizing phase. The addition of gelatin and fat gives the candy its chewable consistency. The water content of the low boiling affects the crystallization and must be between 6 and 10%. Form, stability and stickiness are often problems in low-boiled candies. If the stickiness of the low-boiled candy has to be reduced, the level of the crystallizing phase has to be increased. Isomalt's low hygroscopicity minimizes stickiness during cooling, forming, cutting, and wrapping. In addition, it prevents the soft-boiled mass from sticking to the wrapper. Seeding the low boiling mass with powdered isomalt after the boiling process also results in an increased form stability and a reduced stickiness.

E. Compressed Tablets/Lozenges

Chewable tablets and lozenges can be made with isomalt by direct compressing without the use of excipients. The major advantage of compressed tablets based on isomalt is their sugar-comparable taste and slow solubility. Because of the low hygroscopicity, isomalt tablets have a long shelf-life. The newly developed isomalt-DC gives an even better compressibility (30, 34).

F. Chocolate

Isomalt is used in reduced-calorie and diabetic chocolate products because of its sugarlike neutral taste, snap, melting behavior, and negligible cooling effect. The normal formulation for a sucrose chocolate can be used for an isomalt chocolate (milk and bittersweet). Isomalt-LM has a lower water content (<1%) than isomalt-ST (5–7%). This allows higher temperatures during conching. Isomalt chocolate is both sugar free and calorie reduced. Because the major source of energy in chocolate is fat and not sugar, a combination of isomalt and fat substitutes and/or bulking agents is recommended to further lower the calorie content of chocolate (30, 31, 35–37).

G. Baked Products

Isomalt is a useful ingredient in the formulation of ''light'' or sugar-free baked goods (38–41). The low solubility of isomalt, its low hygroscopicity, and its browning reaction have to be considered during formulation development. Only minimal modifications in formulations and processing methods are required for baked goods made with isomalt. The final baked products made with isomalt have a sugarlike taste and a long shelf-life. Isomalt wafers and cookies absorb a lower amount of water than sugar formulations. Therefore, they stay crispier during storage, even if the sugar is only partly replaced. The texture of wafers and extruded food products can be improved by adding isomalt to the mixture (42). For texture improvement, isomalt can be added to sweet, and to salted products. Figure 9 illustrates that the moisture pickup in wafer sheets with only 5%

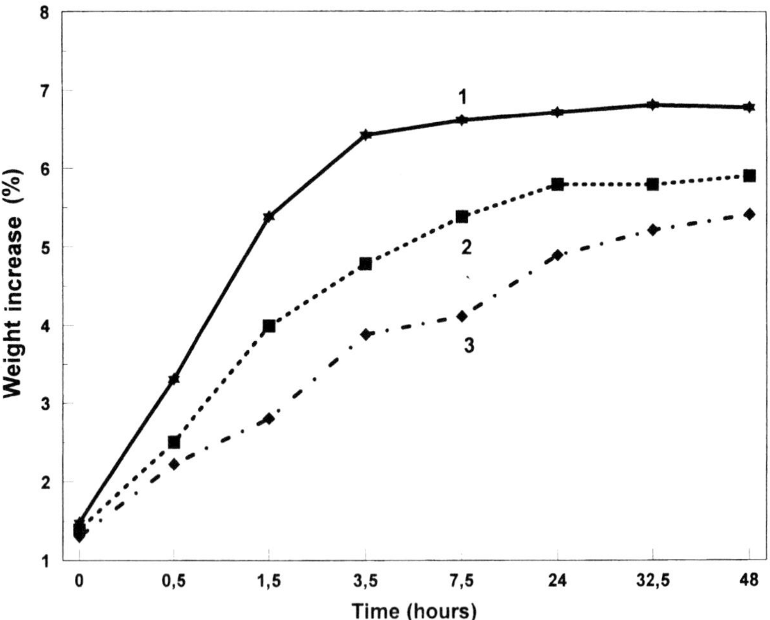

Figure 9 Moisture pickup in wafer sheets during conditioning at 30°C, 35% rh. The standard formulation does not contain sugar, or isomalt; 1 =standard with no sucrose; 2 = 5 parts isomalt-ST; 3 = 10 parts isomalt-ST.

isomalt is significantly reduced. A lower moisture pickup leads to an improved shelf-life.

H. Pharmaceutical Applications

Isomalt is of particular interest for pharmaceutical applications because of its:

- Chemical stability
- Slow dissolution kinetics
- Low hygroscopicity
- Heat stability

Isomalt can be used as a neutral vehicle in a wide range of pharmaceutical applications, for example, high-boilings, tablets, and instant drink powders. The low solubility of isomalt is an advantage in high-boilings and tablets with active ingredients (34). A high-boiling made with isomalt dissolves approximately one-third slower than a sugar high-boiling, and an isomalt tablet dissolves almost twice as slow as its sugar counterpart. Active ingredients are therefore released more slowly and remain in the oral cavity for a longer time. The low hygroscopicity of isomalt ensures that pharmaceutical products remain stable for several years. This is especially important when hydrolysis-prone ingredients are involved.

VII. REGULATORY STATUS

Isomalt is regulated in countries worldwide, including NAFTA, EU, Japan, Australia, New Zealand, East European countries, and a number of Asian and South-American countries. In most European countries, isomalt is approved as a food additive. A Generally Recognized As Safe affirmation petition was accepted for filing by the U.S. Food and Drug Administration in 1990. Since then, isomalt has been marketed in the United States as a GRAS substance. In Japan isomalt is accepted as a food.

REFERENCES

1. PJ Sträter. Palatinit (isomalt), an energy-reduced bulk sweetener derived from saccharose. In: GG Birch, MG Lindley, eds. Low-Calorie Products. London: Elsevier Applied Science, 1988, p 36.
2. PJ Sträter. Palatinit-technological and processing characteristics. In: LO Nabors, RC Gelardi, eds. Alternative Sweeteners. New York: 1986, p 217.

3. W Gau, J Kurz, L Müller, E Fischer, G Steinle, U Grupp, G Siebert. Analytische Charakterisierung von Palatinit. A Lebensmittel-Untersuchung und Forschung 168: 125, 1979.
4. H Schiweck. Palatinit-Production, technological characteristics, and analytical study of foods containing Palatinit. Translated from German. Alimenta 19:5, 1980.
5. K Paulus, A Fricker. Zucker-Ersatzstoffe: Anforderungen und Eigenschaften am Beispiel von Palatinit. Z Lebensmitteltechnologie und Verfahrenstechnik 31:128, 1980.
6. F Voirol. Xylitol: Food applications of a noncariogenic sugar substitute. In: Health and Sugar Substitutes. Basel: Verlag S Karger, 1979, p 130.
7. JN Counsel. Xylitol. London: Applied Science Publishers, 1978, p 7.
8. FR Benson. Alcohols, polyhydric (sugar). In: Kirk-Othmer Encyclopedia of Chemical Technology, 3rd ed. Vol. 1., 1978, p 754.
9. H Bollinger. Palatinit (Isomalt)-a calorie-reduced sugar substitute. Translated from German. Gordian (May):92, 1987.
10. CC Emeis, S Windisch. Palatinosevergärung durch Hefen. Z. Zuckerindustrie 10: 248, 1960.
11. U Nilsson, MJ Jägerstad. Hydrolysis of lactitol, maltitol and Palatinit by human intestinal biopsies. Br J Nutr 58:1999, 1987.
12. PJ Sträter, A Sentko. Energy utilization of isomalt. Presented at Sweeteners: Carbohydrate and Low Calorie International Conference, 196th meeting of American Chemical Society, Los Angeles, CA, Sept 22–25, 1988.
13. F Berschauer, M Spengler. Energetische Nutzung von Palatinit. Stsch Zahnärztl Z 42:145, 1987.
14. D Thiebaud, E Jacot, H Schmitz, M Spengler, JP Felber. Comparative study of isomalt and sucrose by means of continous indirect calorimetry. Metabolism 33(9):808, 1984.
15. U Grupp, G Siebert. Metabolism of hydrogenated Palatinose, an equimolar mixture of α-D-glucopyranosido-1,6 sorbitol and α-D-glycopyranosido-1,6 mannitol. Res Exp Med (Berl) 173:261, 1978.
16. EJ Karle, F Gehring. Kariogenitätsuntersuchungen von Zuckeraustauschstoffen an xerostomierten Ratten. Dtsch Zahnärztl Z 34:551, 1979.
17. J Ciardi, WH Bowen, G Rolla, K Nagorski. Effect of sugar substitutes on bacterial growth, acid production and glucan synthesis. J Dental Res 62:182, 1983.
18. F Gehring, EJ Karle. Der Saccharoseaustauschstoff Palatinit unter besonderer Berücksichtigung mikrobiologischer und kariesprophylaktischer Aspekte. Z Ernährungswiss 20:96, 1983.
19. TN Imfeld. Identification of Low Caries Risk Dietary Components. Basel: Verlag S Karger, 1983, p 117.
20. H Drost, M Spengler, P Gierlich, K Jahnke. Palatinit ein neuer Zuckeraustauschstoff: Untersuchungen bei Diabetikern vom Erwachsentyp. Aktuelle Endokrinolgie (Vol 2). Stuttgart, New York: Georg Thieme Verlag, 1980, p 171.
21. M Spengler, J Sommer, H Schmitz. Vergleich der Palatinit-Tolerance gegenüber der Sorbit-Toleranz gesunder Erwachsener nach einmaliger oraler Gabe in aufsteigender Dosierung. Bayer AG, Pharma Bericht Nr 8457, 1979.
22. M Spengler, J Sommer, H Schmitz. Vergleich der Palatinit-Tolerance gegenüber

der Sorbit-Toleranz gesunder Erwachsener nach 14-tägiger oraler Gabe. Bayer AG, Pharma Bericht Nr 8449, 1979.

23. DM Paige, TM Bayless, LR Davis. The evaluation of Palatinit digestibility—revised report. Johns Hopkins University, Baltimore, MD, May 1986.
24. M Spengler, JC Somogyi, E Pletcher, K Boehme. Tolerability, acceptance and energetic conversion of isomalt (Palatinit) in comparison with sucrose. Akt. Ernährung 12:210, 1987.
25. DH Waalkens-Berendsen, HBWM Koeter, G Schlüter, M Renhof. Developmental toxicity of isomalt in rats. Food Chem Toxicology 27(10):631, 1989.
26. DH Waalkens-Berendsen, HBWM Koeter, MW van Marwijk. Embryotoxicity/teratogenicity of isomalt in rats and rabbits. Food Chem Toxicology 28(1):1, 1990.
27. DH Waalkens-Berendsen, HBWM Koeter, EJ Sinkeldam. Multigeneration reproduction study with isomalt in rats. Food Chem Toxicology 28(1):11, 1990.
28. AE Smits-Van Prooije, AP De Groot, HC Dreef-Van Der Meulen, EJ Sinkeldam. Chronic toxicity and carcinogenicity of isomalt in rats and mice. Food Chem Toxicol 28(4):243, 1990.
29. Anonymous. Isomalt. In: Toxicological evaluation of certain food additives and contaminants. WHO Food Additive Series: 20. Cambridge, UK: Cambridge University Press, 1987, p 207.
30. I Willibald-Ettle, T Keme. Recent findings concerning Isomalt in different applications. Int Food Ingred Magazine 1:17–21, 1992.
31. B Fritzsching. Isomalt, a sugar substitute ideal for the manufacture of sugar-free and calorie-reduced confectionery. Proceedings of Food Ingredients Europe, Paris, 1993, pp 371–377.
32. B Frizsching. Isomalt in hard candy applications. The Manufacturing Confectioner (November):65–73, 1995.
33. WE Irwin. Reduced calorie bulk ingredients: isomalt. The Manufacturing Confectioner 70(11):55, 1990.
34. T Dörr, I Willibald-Ettle. Evaluation of the kinetics of dissolution of tablets and lozenges consisting of saccharides and sugar substitutes. Pharm Ind 58:947–952, 1996.
35. H Bollinger, T Keme. How to make sucrose-free chocolate: The use of the reduced calorie sugar substitute Palatinit. Zucker-und Süsswaren Wirtschaft 41(1), 1988.
36. WE Irwin. The use of Palatinit in the production of reduced calorie chocolate ingredients. Presented at 197th meeting of American Chemical Society, Dallas, TX, April 9–14, 1989.
37. WE Irwin. Sugar substitute in chocolate. Manufacturing Confectioner 70(5):150, 1990.
38. W Seibel, G Brack, F Bretschneider. Backtechnische Wirkung des Zuckeraustauschstoffes Palatinit (isomalt): Feinteige mit Hefe. Getreide, Mehl und Brot 40(8):239, 1986.
39. G Brack, W Seibel, F Bretschneider. Backtechnische Wirkung des Zuckeraustauschstoffes Palatinit (isomalt): Feinteige ohne Hefe. Getreide, Mehl und Brot 40(9):269, 1986.
40. G Brack, W Seibel, F Bretschneider. Backtechnische Wirkung des Zuckeraustausch-

stoffes Palatinit (isomalt): Massen mit Aufschlag. Getreide, Mehl und Brot 40(10): 302, 1986.

41. I Willibald-Ettle. Isomalt-a sugar substitute. Its properties and applications in confectionery and baked goods. Presented at Practical Confectionery Course, William Angliss College, Melbourne, Australia, 1992.
42. H Bollinger, H Steinhage. The production of low-calorie cereal extrudates. Zucker- und Süsswaren Wirtschaft 42(2), 1989.

16
Maltitol

Kazuaki Kato
Towa Chemical Industry Co., Ltd., Tokyo, Japan

Alan H. Moskowitz
Adams Division of Pfizer, Parsippany, New Jersey

I. INTRODUCTION

Maltitol, or hydrogenated maltose, is a member of a family of bulk sweeteners known as polyols or sugar alcohols. Early grades of maltitol were low in purity and were only available in syrup form. In 1940 solid maltitol, an amorphous, white substance, was developed (1). Grades of higher purity were then developed and have been in use in Japan (2). Because these products were very hygroscopic and difficult to prepare as a powder, maltitol was almost always handled in the form of aqueous solution; thus, its use was extremely restricted.

Maltitols of substantially nonhygroscopic crystalline form (hereinafter referred as maltitol), now called crystalline maltitol or crystalline mixture solids of maltitol, were discovered, developed, and have been in use since 1981 (3).

II. PRODUCTION

As with other polyhydric alcohols (such as sorbitol, mannitol, and xylitol), maltitol is produced by means of the catalytic hydrogenation of a precursor, in this case, maltose. A starch solution is liquefied to a paste of low dextrose equivalent (DE) and then hydrolyzed enzymatically into maltose syrup, which, after purification and concentration, is hydrogenated by means of a nickel or other transition metal catalyst. After additional purification steps to remove the starting materials and catalyst, the solution is concentrated to syrup, then crystallized (3). Maltitol may also contain trace or quite low levels of sorbitol, maltotriitol, and higher hydrogenated oligosaccharides. Currently, there are two kinds of crystalline maltitol powders on the market with minimum maltitol contents of 92.5% and 98% that are produced and marketed worldwide.

III. PHYSICAL CHARACTERISTICS

The structure of maltitol (alpha(1-4)-glucosylsorbibtol) has been confirmed by nuclear magnetic resonance and infrared absorption studies as a crystalline polyhydric alcohol obtained by the catalytic hydrogenation of maltose, a disaccharide consisting of two glucose units linked by means of an alpha(1-4) bond (Fig. 1). Crystallographic studies have also demonstrated that the molecular structure is a fully extended conformation with no intramolecular hydrogen bonding; all nine hydroxyl groups are involved in intermolecular hydrogen bonds and in bifurcated, finite chains (3).

Maltitol exhibits a negligible cooling effect in the mouth given its negative heat of solution (−23 kJ/kg), which is close to that of sucrose (−18 kJ/kg), and much less than that of other carbohydrates: xylitol (−153 kJ/kg), mannitol (−121 kJ/kg), sorbitol (−111 kJ/kg), isomalt (−39 kJ/kg), and dextrose (−104.6 kJ/kg).

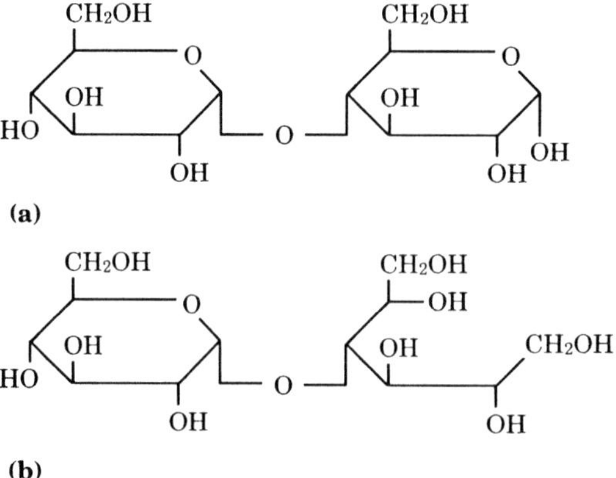

Figure 1 (a) Chemical structure for maltose. (b) Chemical structure for maltitol.

Like the other polyols, maltitol has no reducing groups and will not undergo Maillard reactions to produce caramel-colored species in the presence of amino acids or proteins. Maltitol has a sweetness of about 85–95% that of sucrose, making it sweeter than all other polyols except xylitol with a taste remarkably similar to sucrose.

Towa Chemical Industry, Co., Ltd., an affiliate of Mitsubishi Corporation, Tokyo, Japan, produces and markets both crystalline maltitol and crystalline mixture solids of maltitol under various brand names, including Amalty MR and Lesys. Towa received a license for the patent of the process from Hayashibara Biochemical Laboratories, Okayama, Japan, which pioneered the process. Towa's license is by geographical region and includes Japan, the United States, and several other countries in Asia, Oceania, and America. Hayashibara also licenses the patent to Cerestar and Roquette Freres who both market the product in Europe under the brand Malbit and Maltisorb. Roquette America, Inc., an affiliate of Roquette Freres, also received a license in America and markets the product in the United States.

The specifications in Table 1 show the varying chemical compositions of several commercialized brands of maltitol supplied by Towa Chemical Industry and Roquette Freres (Lestrem, France). Although both Towa (3) and Roquette have the technology for producing high-purity crystalline maltitol, this type of product may not necessarily be available in all countries. Patent constraints often allow only one supplier in a given country.

For most applications, crystalline maltitol and crystalline mixture solids of maltitol (i.e., min. 98% and min 92.5%) can be used interchangeably; exceptions include pan coating in which slight differences in crystallinity might be evident. Because sorbitol and other less sweet hydrogenated components are found in the lower maltitol product, one may expect that the high maltitol product would be

Table 1 Specifications and Analysis for Commercially Available Crystalline Maltitol Products

	Roquette Maltisorb	Towa Lesys	Towa Amalty MR
Maltitol (min, %)	98.0	98.0	92.5
Sorbitol (max, %)	—	—	1.5
Maltotriitol (max, %)	—	—	3.0
Hydrogenated oligosaccharides (max %)	—	—	2.5
Reducing sugars (max %, as dextrose)	0.01	0.01	0.5
Melting point (°C)	149.4	149.3	140.3

sweeter. However, there is little difference in sweetness when comparing actual food products.

IV. APPLICATIONS

Maltitol has many attributes and properties that allow it to be used in a wide variety of food applications:

- White crystalline powder
- Sweetness profile similar to sugar
- Reduced calorie, bulk sweetener
- Noncariogenic
- Non-insulin-dependent and can be used in diabetic diet
- Substantially nonhygroscopic
- Thermostable
- Fat replacer

A. Confectionery Products

Considerable formulation and processing information is available for producing maltitol-based confections and chewing gums (4–6). Confectionery products include chocolates, hard-boiled candy, and caramels.

Before the development of maltitol, the production of ''sugar-free'' or ''no sugar added'' chocolate proved difficult because of the lack of a polyol with the physical, chemical, and organoleptic properties of sucrose. Maltitol's anhydrous crystalline form, low hygroscopicity, high melting point, and stability allow it to replace sucrose in high-quality chocolate coatings, confectionery, and bakery chocolate.

Maltitol may also be used to partially or totally replace the fat component in foods, because it gives a creamy mouth-feel to products such as brownies, cakes, and cookies (7). High-quality sugarless chocolates can be made by simply substituting maltitol for sucrose. Chocolate made with maltitol does not have the mouth-cooling property associated with sorbitol-based chocolate, and because the sweetness is greater than chocolate made with sorbitol, the addition of a potent sweetener is not required. Maltitol-based hard-boiled candies can be manufactured with conventional ''roping and forming'' equipment, unlike sorbitol boiled candy, which generally must be seeded with crystalline sorbitol, deposited into plastic or metal molds, and allowed to cure (8). Resultant candy must be securely packaged in a moistureproof wrapper to minimize crystallization. Of note, there may be slight differences in stability with respect to ''cold flow'' (a distortion of shape over time), hygroscopicity, and crystallization, depending on the purity of the maltitol used. Use of a higher purity maltitol would result in a boiled candy having lower hygroscopicity, for example.

B. Other Food Applications

Dietetic foods for those on sugar-restricted diets or fat-reduced diets can also be formulated with maltitol (4,6,7). Examples may include granola bars, jams with no added sugar, ice cream analogue, pie fillings, salad dressings, spreads, cookies and cakes. In most cases, the use of potent sweeteners is not required, because maltitol is nearly as sweet as sucrose when used at approximately a one-for-one replacement. In cases in which sugarless products are to be formulated with lower caloric density, maltitol will work well with other potent sweeteners and will have a sweetness profile very similar to that of sucrose. For example, a tabletop sweetener can be formulated with maltitol and aspartame.

V. GENERAL COST AND ECONOMICS

Polyols, as a general rule, are more expensive than the common carbohydrate sweeteners, sucrose and corn syrup, because of the additional processing they must undergo. Although maltitol is more expensive than crystalline sorbitol (maltitol is about twice the cost of crystalline sorbitol), maltitol's outstanding taste, sweetness, and functionality make it worth the added cost in many applications.

VI. METABOLIC ASPECTS

A. General Metabolism

Mammalian metabolism of maltitol generally begins with the hydrolysis of maltitol into the monosaccharides glucose and sorbitol. Although this hydrolysis may not be complete at high maltitol intakes, primarily glucose and sorbitol will be transported into the bloodstream. Sorbitol is converted to fructose, which then is used by way of the glycolytic pathway. Detailed reviews of the metabolism of sorbitol have appeared (5,9). (See Chapter 18.)

B. Cariogenicity

It is well known that dental caries are caused by the prolonged exposure of tooth enamel to acid. Dextran, the principal component of plaque, is produced by microorganisms such as *S. mutans* and is responsible for adherence of the organisms to the tooth. Metabolism of sugars and starches by these microorganisms in the mouth leads to the formation of acids that if in sufficient quantity, will erode tooth enamel during prolonged contact. Oral bacteria do not metabolize polyols such as sorbitol, mannitol, maltitol, and xylitol; therefore no acid production occurs (10).

Laboratory assessment of cariogenic or acidogenic potential had been somewhat problematic. Quick preliminary screening of various materials for cariogenic potential had been accomplished with various bioassays, using microorganisms (11) or laboratory animals (12). A rigorous test, now well accepted, is plaque pH telemetry in human subjects (13–15). Telemetry refers to the continuous electronic data-gathering process used in these experiments. In these tests, an indwelling pH electrode is secured in a denture appliance and implanted onto a tooth, replacing an existing appliance. Several days after sufficient plaque has built up on the appliance, the test subject is given a food and the pH is monitored as a function of time. If acid formation does not reach below pH 5.7 in 30 minutes, the food may be certified as safe for teeth (''zahnschonend'' by the Swiss authorities). Demineralization of tooth enamel definitely occurs below pH 5.5; between 5.5 and 5.7 is a transition range where some demineralization may begin. The advantage of this in vivo experimental approach over other in vitro assay methods is that it takes into account salivary flow, the buffering capacity of saliva, and the removal of dissolved sugars from the mouth by swallowing.

More recently, however, another simplified in vitro test has been developed that claims results comparable to that of the pH telemetry method (16). For a comprehensive discussion of cariogenicity in foods, the reader is referred to a special issue of the *Journal of Dental Research*, December 1986 (pp. 1473–1543), that is based on the Scientific Consensus Conference on Methods for Assessment of the Cariogenic Potential of Foods that was held at the University of Texas Dental School, San Antonio, November 17–21, 1985. Among the many

Table 2 Plaque pH Minima After Rinsing with Test Solutions

Product	pH minimum
Crystalline maltitol	6.4 ± 0.4
Sorbitol	6.5 ± 0.4
Sucrose	5.5 ± 0.5

Source: Adapted from Ref. 21.

results of this conference was the promulgation of a set of criteria for assessment of cariogenic potential of sugarless foods.

There are extensive cariogenicity studies for maltitol. Several studies have shown that maltitol is noncariogenic (17–20), the plaque pH curve not dropping below pH 5.7 (17).

The U.S. Food and Drug Administration has authorized a ''does not promote tooth decay'' health claim for sugar-free foods sweetened with polyols, including maltitol. The regulation provides that ''when fermentable carbohydrates are present in the sugar alcohol containing food, the food shall not lower plaque pH below 5.7 by bacterial fermentation either during consumption or up to 30 minutes after consumption, as measured by the indwelling plaque test found in 'Identification of Low Caries Risk Dietary Components,' T.N. Imfeld, Volume 11, *Monographs in Oral Science* (1983).'' (22)

C. Laxation

As is the case for many polyols, particularly sorbitol and mannitol, excessive consumption of maltitol may have a laxative effect. The reason for this may be viewed as a direct consequence of the noncariogenicity of these materials. If these polyols are not readily fermented by oral microorganisms, it is because the microbes do not possess adequate quantities of the necessary enzymes, or their existing enzymes are not active on polyol substrates. Understandably, digestion by human enzyme systems will be similarly impaired.

Hydrolysis of maltitol to glucose and sorbitol is significantly slower than is the hydrolysis of maltose (10). Therefore, the intestinal brush border maltase enzymes may not have adequate time to complete hydrolysis before passage of the intestinal contents into the large intestine, where the delicate water balance is thereby upset. Sorbitol is absorbed slowly by means of passive diffusion, whereas glucose is readily absorbed by active transport. Not only do carbohydrates carry osmotically bound water into the large intestine, but also microbiological fermentation of these polyols will create still a large number of osmotically active components, as well as flatus. Excessive water absorption into the large intestine causes laxation.

Maltitol is less laxating compared with sorbitol; the maximum noneffective dose of maltitol (the dosage where no laxation or other discomfort occurs) is about twice that of sorbitol. This is consistent with the facts that disaccharides are less osmotically active and also that half the maltitol molecule is glucose, which, after hydrolysis, is readily absorbed (23–25). The U.S. Food and Drug Administration requires a food to be labeled with the warning ''Excess consumption may have a laxative effect'' if foreseeable daily consumption may result in ingestion of greater than 50 of sorbitol (21 CFR 184.1835 (e)). Towa Chemical

Industries therefore asserts (22) that a warning statement may apply to maltitol only when greater than 100 g/day is likely to be ingested.

A study by Koizumi (24) concluded that the maximum noneffective laxative dose for both men and women was 0.3 g maltitol/kg body weight, and that the 50% single laxative dose (i.e., producing laxation in 50% of the subjects) was 0.8 g/kg body weight. Data cited by Towa (23) show that the estimated daily intake of maltitol is only about half the 50% single laxative dose, supporting their contention that significant laxation problems will not occur in most individuals.

D. Suitability in Diabetic Diets

Because maltitol is hydrolyzed and absorbed slowly compared with sucrose or glucose in humans, it would appear that maltitol might be useful in diabetic diets. The goal in dietary management for persons with diabetes is to minimize fluctuations in blood glucose by not overloading the insulin requirements at any one time. Indeed, many studies (24,26–30) have shown that ingestion of maltitol does not produce marked elevations in blood glucose levels in both diabetic and healthy individuals. It appears that any sorbitol produced from the hydrolysis of maltitol is absorbed very slowly and, to a certain extent, inhibits absorption of the glucose also released (31). Thorough reviews of the literature are available (23,31). Diabetic patients should consult their physician to determine suitability for their particular circumstances.

E. Caloric Content of Maltitol

In 1990, the European Economic Community Council reviewed the net energy (NE) value of polyols, including maltitol, and adopted a single interim value for all polyols of 10 kJ/g (2.4 kcal/g) (32,33).

In 1991 the Japanese Ministry of Health and Welfare (JMHW) evaluated the NE value of low-digestible saccharides, including polyols, adopting 1.8 kcal/g for maltitol (34). After reviewing data in 1996, the JMHW adopted the NE value of 2 kcal/g for maltitol and most of other polyols except erythritol (0 kcal/g), xylitol, sorbitol, and some hydrogenated oligosaccharides (3 kcal/g).

The FDA nutrition labeling regulations set the energy value for all carbohydrates, including maltitol, at 17 kJ/g (4 kcal/g) (21 CFR 101.9). The Code of Federal Regulations provides for exceptions to this energy value when a request is supported by sufficient data to justify an alternative value.

Thus, the net energy value of maltitol and other polyols was of great concern to producers and manufacturers of foods worldwide, including those in the United States (35). Consequently, the members of the Calorie Control Council's Polyol Committee sponsored a study by the Life Sciences Research Office (LSRO) that would provide an objective assessment of the scientific information available on certain polyols, including maltitol. A report on the energy available from these polyols was prepared in 1994. The LSRO report (36), prepared in consultation with an expert panel, determined that the appropriate energy to be evaluated was net energy. The panel reviewed both published and unpublished studies relating to the energy determination of these polyols. They evaluated the methods used to determine energy values and set forth a set of basic premises with which net energy values could be evaluated. Last, they determined the best estimates of the metabolizable and net energy values for these polyols (36). Estimated net energy value for maltitol in the report was 2.8–3.2 kcal/g (11.7–13.4 kJ/g). But it is also stated in the report that in view of the more recent data of Tsuji et al (36), the NE available for maltitol might be lower than 2.8–3.2 kcal/g. LSRO also concluded that additional studies were needed to confirm the lower range of values.

In 1999, LSRO evaluated again the NE value for maltitol. Summarizing all the available ADME (absorption/distribution/metabolism/excretion) data: 10% of the energy available from maltitol is used at 4 kcal/g during the first 2 hours after ingestion; 85% of the energy available is used by bacterial fermentation in the large intestine at 2 kcal/g; and 5% of the energy available is excreted in the feces.

On the basis of the ADME data, one may estimate the factors in Equation (1) as A, 0.1; B, 1; C, 0.05; and R_e, 1.

$$E_w = [(A \times B) + (1 - (A + C)) \times 0.5] \times 16.5 \times R_e \text{ kJ/g} \tag{1}$$

Using these terms, Equation (1) leads to an estimate of the net energy value for maltitol of 8.8 kJ/g (2.1 kcal/g). Previous estimates of the NE value have ranged from 6.7 to 16.4 kJ/g. The current estimate for the NE value of maltitol is based on the most recent evidence of the ADME of maltitol in humans. The value of 8.8 kJ/g is in good agreement with the values estimated by Oku (37), Tsuji et al. (37), and the mid-range of Felber et al. (38) and the Japanese Ministry of Health and Welfare (34).

The FDA issued a no objection letter on May 28, 1999, in response to a letter proposing a self-determined energy value for maltitol of 2.1 kcal/g for the purpose of nutrition labeling.

VII. CARCINOGENICITY AND OTHER TOXICITY

Numerous studies examining the acute, subacute, chronic, and subchronic toxicities, and carcinogenic, mutagenic, and teratogenic activities of maltitol have shown no evidence of any adverse effects (23,31). In fact, a recent study has indicated that dietary maltitol decreases the incidence of cecum and proximal tumors in rats induced by a hydrazine derivative (40).

The Joint FAO/WHO Expert Committee on Food Additives (JECFA) first reviewed maltitol in conjunction with hydrogenated starch hydrolysate in 1980, when it concluded that the allocation of an ''acceptable daily intake (ADI) need not be specified'' (39). A second review in 1983 resulted in the allocation of a temporary ADI of 0–25 mg/kg body weight for hydrogenated glucose syrups containing 50–90% maltitol, mainly because the panel did not have, but requested, the results of chronic feeding studies (41).

In 1985, after the acquisition of lifetime feeding studies, JECFA reevaluated the toxicity of maltitol and HSH and reverted to an ADI of ''not specified'' (31). As defined by JECFA, ADI ''not specified'' means that, ''on the basis of available scientific data, the total daily intake of a substance arising from its use at levels necessary to achieve the desired effect, does not, in the opinion of the Committee, represent a hazard to health.'' Therefore, establishment of an ADI in numerical terms was not necessary.

Another group of scientists, the Scientific Committee for Food of the European Economic Community (EEC), in its December 1984 report on sweeteners, found maltitol to be ''acceptable.'' The committee found it inappropriate to establish an ADI for maltitol and other polyols (42).

VIII. REGULATORY STATUS

A. United States

Towa Chemical Industry Co., Ltd., filed a petition with the FDA on August 26, 1986, to affirm the generally recognized as safe (GRAS) status of maltitol (23). The petition was accepted for filing on September 23, 1986, and was announced in the *Federal Register* on December 23, 1986.

While the FDA is conducting their assessment of the petition, maltitol may be used in the United States, although potential customers should make their own assessment of the risk; that is, if the FDA rules adversely, products containing maltitol may need to be recalled from the marketplace. In this case, the risk is minimal, because maltitol is a principal component in hydrogenated starch hydrolysate (hydrogenated glucose syrup), which has been used in the United States since around 1977, and also the subject of a longstanding, prior FDA GRAS petition.

The Towa petition asks for use of maltitol in a variety of applications, including as a flavoring agent, formulation aid, humectant, nutritive sweetener, processing aid, sequestrant, stabilizer and thickener, surface-finishing agent, and texturizer. The ingredient shall be used in food at levels not to exceed current good manufacturing practices: a maximum level of 99.5% in hard candy and cough drops, 99% in sugar substitutes, 85% in soft candy, 75% in chewing gum, 55% in nonstandardized jams and jellies, and 30% in cookies and sponge cake.

Table 3 Worldwide Status of Maltitol

Country	Status
France	Approved
Switzerland	Approved
The Netherlands	Approved
Belgium	Approved
Denmark	Approved
Finland	Approved
Norway	Approved
Mexico	Approved
UK	Approved
Japan	Approved
Sweden	Approved
Germany	Approved
Austria	Approved
Spain	Approved
Australia	Approved
Canada	Approved

B. Worldwide

Table 3 shows the status of maltitol in a number of countries. Where approval is shown, it signifies approval in at least one category of food even if restrictions such as preregistration may apply.

REFERENCES

1. MJ Wolfrom et al. Melibiotol and maltitol. J Am Chem Soc 62:2553–2555, 1940.
2. T Hiraiwa. Free-flowing maltitol, Japan, Kokai 74-110,620. Chem Abstr 82–171360, 1975.
3. M Hirao, H Hijiya, T Miyaka, Anhydrous crystals of maltitol and the whole crystalline hydrogenated starch hydrolysate mixture solid containing the crystals, and process for the production and use thereof, U.S. Patent 4, 408,041, 1983.
4. Towa Chemical Industry. technical literature.
5. Anonymous. Industrial applications of malbit. Confectionery Production (Dec.): 741–743, 1988.
6. I Fabry. Malbit and its applications in the food industry. In: TH Grenby, ed. Developments in Sweeteners—3. London and New York: Elsevier-Applied Science Publishers, 1987, pp 83–108.
7. AI Bakal, S Nanbu, T Muraoka. Foodstuffs containing maltitol as sweetener or fat replacement. European Patent 039299 B1, 1993.

8. KJ Klacik, WV Vink, PR Fronczkowski. Sorbitol containing hard candy. U.S. Patent 4,452,825, 1984.
9. BK Dwivedi. Food products for special dietary needs. In: BK Dwivedi, ed. Low Calorie and Special Dietary Foods. CRC Press, West Palm Beach, FL, 1978, pp 1–22.
10. SC Ziesentiz, G Siebert. Polyols and other bulk sweeteners. In: TH Grenby, ed. Developments in Sweeteners–3. London and New York: Elsevier-Applied Science Publishers, 1987, pp 109–149.
11. HAB Linke. Sweeteners and dental health: The influence of sugar substitutes on oral microorganisms. In TH Grenby, ed. Developments in Sweeteners—3. London and New York: Elsevier-Applied Science Publishers, 1987, pp 151–188.
12. R Havenaar. Dental advantages of some bulk sweeteners in laboratory animal trials. In: TH Grenby, ed. Developments in Sweeteners—3. London and New York: Elsevier-Applied Science Publishers, 1987, pp 189–211.
13. T Imfeld. Identification of low caries risk dietary components. In: HM Myers, ed. Monographs in Oral Science. Vol. 11. Basel: Karger, 1983, pp 117.
14. T Imfeld, HR Muhlemann. Evaluation of sugar substitutes in preventing cariology. J Prev Dent 4:8, 1977.
15. T Imfeld, HR Muhlemann. Cariogenicity and acidogenicity of food, confectionery and beverages. Pharm Ther Dent 3:53, 1978.
16. BG Bibby, J Fu. Changes in plaque pH in vitro by sweeteners. J Dent Res 64(9): 1130, 1985.
17. K Matsuoka. On the possibility of maltitol and lactitol and SE-58 as non-cariogenic polysaccharides. Nihon Univ Dent J 49:334, 1975.
18. R Firestone, R Schmid, HR Muhlemann. The effects of topical applications of sugar substitutes on the incidence of caries and bacterial agglomerate formation in rats. Caries Res 14:324, 1980.
19. T Imfeld, F Lutz. Malbit, ein zahnfreundlicher Zuckeraustauschstoff. Swiss Dent 6: 44, 1985.
20. J Rundgen, T Koulourides, T Encson. Contribution of maltitol and Lycasin to experimental enamel demineralization in the human mouth. Caries Res 14:67, 1980.
21. DC Abelson. The effects of hydrogenated starch hydrolysate on plaque pH in vivo. Clin Prev Dent 11(2):21, 1989.
22. U.S. Food and Drug Administration. Food Labeling: Health Claims; Sugar Alcohols and Dental Caries, Final Rule. Federal Register, Vol. 61 No. 165:43433, August 23, 1996.
23. Towa Chemical Industry GRAS Affirmation Petition to the U.S. Food and Drug Administration for maltitol, 1986.
24. N Koizumi, M Fujii, R Ninomiya, Y Inoue, T Kagawa, T Tsukamoto. Studies on transient laxation effects of sorbitol and maltitol I: Estimation of 50% effective dose and maximum non-effective dose. Chemosphere 12(1):45, 1983.
25. PJ Sicard, and Y Lebot. From Lycasin to crystalline maltitol: A new series of versatile sweeteners. Presented at the International Conference on Sweeteners-Carbohydrate and Low Calorie, September 22–25, 1988, Los Angeles, California. Sponsored by the American Chemical Society.
26. H Atsuji, S Asano, S Hayashi. A study on the metabolism of maltitol, a disaccharide

alcohol. Unpublished report from Keio University, Japan. Submitted to WHO by ANIC S.p.A., 1982.

27. K Nishikawa. Variations in blood and urine sugar values with oral administration of maltitol, unpublished report from Nishikawa Hospital, Tadera, Japan, Submitted to WHO by ANIC S.p.A, 1982.
28. Takeuchi, M Yamashita. Clinical test results of a disaccharide alcohol (maltitol). Eiyo to Shokuryo (Nutrition and Food) 25:170, 1972.
29. T Yamakubi. Malbit, a new sweetener and its applications for diabetics. Proceedings of the 14th Congress of the Japanese Society of Diabetics, April 3–4, 1971.
30. M Kamoi, Y Shimiza, M Kawauchi, Y Fuji, Y Kikuchi, S Mizukawa, H Yoshioka, M Kibata, A study on the metabolism of maltitol (clinical study). J Jap Diab Soc 18:451, 1975.
31. Toxicological evaluation of certain food additives and contaminants. Prepared by the 29th Meeting of the Joint FAO/WHO Expert Committee on Food Additives. June 3–12, 1985, Geneva, Switzerland. Published by Cambridge University Press on behalf of WHO, 1987, pp 179–206.
32. KC Ellwood. Methods available to estimate the energy values of sugar alcohols. Am J Clin Nutr 62:1169S–1174S, 1995.
33. European Economic Community Council. Council Directive: Nutrition labeling for foodstuffs. Off. J. Eur. Communities No. L276/41, 1990.
34. Japanese Ministry of Health and Welfare, Evaluation of the energy value of indigestible carbohydrates in special nutritive foods. Official Notice of Japanese Ministry of Health and Welfare, no. Ei-shin 71. Tokyo, 1991.
35. RC Gelardi. The evaluation of the metabolizable energy of polyols used as food ingredients. Presentation for the LSRO Expert Panel on the metabolizable energy of polyols, June 16, 1992.
36. Life Sciences Research Office. In: JM Talbot, SA Anderson, KD Fisher, eds. The Evaluation of the Energy of Certain Sugar Alcohols Use as Food Ingredients. Prepared for the Calorie Control Council, Atlanta, Georgia, Federation of American Societies for Experimental Biology, Bethesda, MD, 1994.
37. T Oku. Caloric evaluation of reduced-energy bulking sweeteners. In: DR Romsos, J Himms-Hagen, M Suzuki, eds. Obesity: Dietary Factors and Control. Tokyo, Japan: Japan Sci Soc Press, 1991, pp 169–180.
38. JP Felber, L Tappy, D Vouillamoz, JP Randin, E Jequier. Comparative study of maltitol and sucrose by means of continuous indirect calorimetry. J Parenter Enteral Nutr 11:250–254, 1987.
39. Twenty-fourth report of Joint FAO/WHO Expert Committee on Food Additives. Rome, 1980. WHO Technical Report Series, No. 653, 1980.
40. Y Shimomura, et al. Dietary maltitol decreases the incidence of 1,2-dimethylhydrazine-induced cecum and proximal colon tumors in rats. J Nutr 128(3):536–540, 1998.
41. Twenty-seventh report of Joint FAO/WHO Expert Committee on Food Additives. Geneva, 1983. WHO Technical Report Series, No. 696, 1983.
42. Commission of the European Communities. Reports of the Scientific Committee for Food Concerning Sweeteners. Sixteenth Series. Report EUR 10210 EN. Office of Official Publications of the European Communities, Luxembourg, 1985.

17
Lactitol: A New Reduced-Calorie Sweetener

Paul H. J. Mesters, John A. van Velthuijsen, and Saskia Brokx
PURAC biochem bv., Gorinchem, The Netherlands

I. HISTORY

Lactitol (LACTY®) is a bulk sweetener. It is a sweet-tasting sugar alcohol derived from lactose (milk sugar) by reduction of the glucose part of this disaccharide. It has not been found in nature and was described in the literature for the first time by Senderens (1) in 1920. The first useful preparation was made by Karrer and Büchi in 1937 (2). Lactitol is also called lactit, lactositol, and lactobiosit. The CAS registry number is 585-86-4. The molecular weight of lactitol monohydrate ($C_{12}H_{24}O_{11}.H_2O$) is 362.34. The chemical structure of lactitol, 4-0-(β-D-galactopyranosyl)-D-glucitol, is shown in Fig. 1.

II. PREPARATION

The principles of the preparation of lactitol are general knowledge. The industrial process is a hydrogenation of a 30–40% lactose solution at about 100°C with a Raney nickel catalyst (3). The reaction is carried out in an autoclave under a hydrogen pressure of 40 bar or more. On sedimentation of the catalyst, the hydrogenated solution is filtered and purified by means of ion-exchange resins and activated carbon. The purified lactitol solution is then concentrated and crystallized. The monohydrate, as well as the anhydrate and the dihydrate, can be prepared depending on the conditions of crystallization (4).

Hydrogenation under more severe conditions (130°C, 90 bar) results in partial epimerization to lactulose and partial hydrolization to galactose and glu-

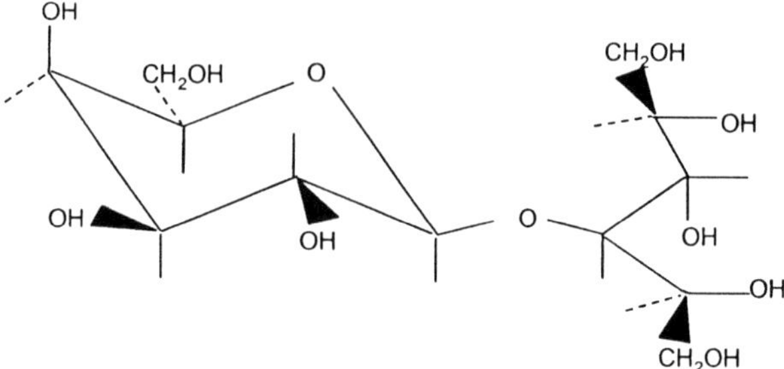

Figure 1 The chemical structure of lactitol.

cose and hydrogenation to the corresponding sugar alcohols lactitol, lactulitol, sorbitol, and galactitol (dulcitol) (5).

III. TOXICOLOGY

All the required studies for a food additive have been carried out (CIVO-TNO, The Netherlands). There are no deleterious effects of feeding lactitol at levels up to 10% of the diet. The Joint FAO/WHO Expert Committee of Food Additives (JECFA) approved lactitol in April 1983. The Committee allocated an acceptable daily intake (ADI) ''not specified'' to lactitol (6).

The Scientific Committee on Food of the EC (SCF-EC) evaluated lactitol and considered it to be a safe product; they stated that ''consumption of the order of 20 g per person per day of polyols is unlikely to cause undesirable laxative symptoms'' (7). The same was stated for isomalt, maltitol, mannitol, sorbitol, and xylitol.

Safety studies in experimental animals included long-term feeding studies at high dietary levels for 2½ years in rats and for 2 years in mice. The safety data are not summarized here. Full data are described in CIVO-TNO (Zeist, The Netherlands) reports, which are published in the *Journal of the American College of Toxicology* (7a). These reports have been evaluated by JECFA, SCF-EC, and the U.S. Food and Drug Administration.

IV. USE AND PURPOSE OF LACTITOL BASED ON BIOLOGICAL PROPERTIES

Lactitol is a sweet-tasting sugar alcohol with interesting nutritional, physiological, and pharmaceutical properties. Lactitol is neither hydrolyzed nor absorbed in the small intestine. It is metabolized by bacteria in the large intestine, where it is converted into biomass, short chain fatty acids, lactic acid, CO_2, and a small amount of H_2. Beneficial bacteria in the large intestine, such as *Bifidobacteria* and *Lactobacillus* spp., use lactitol as a substrate.

Lactitol consumption does not induce an increase of blood glucose or insulin levels, making it a desirable sugar substitute for diabetic patients. Furthermore, it has been demonstrated that its nutritional caloric use is half that of carbohydrates, with a metabolic energy value maximum of 2 kcal/g or maximum 8.4 kJ/g.

Lactitol is very slowly converted to organic acids (lactic acid) by tooth plaque bacteria, so that lactitol does not cause dental caries and satisfies the Swiss regulations as safe for teeth and U.S. health claim regulations for "does not promote tooth decay." These properties suggest important dietary applications e.g., products for people with diabetes, noncariogenic confections, low-calorie foods, and products with a prebiotic effect.

A. Caloric Value

1. Biochemical Aspects

Karrer and Büchi (2) have studied the action of galactosidase-containing enzyme preparations on the splitting of lactitol into galactose and sorbitol. They found that lactitol is only hydrolyzed very slowly by these enzyme preparations. Later studies reported in a German patent (8) have confirmed that lactitol is only slowly split by enzymes at about a tenth the speed at which lactose is split. On the basis of both these in vitro studies, a reduced calorie value can be expected.

In patents of Hayashibara (9, 10), it is claimed that lactitol has no caloric value because it is not digested or absorbed. The intestines of test rabbits not fed for 24 h beforehand were closed at both ends and were injected with a 20% aqueous solution of lactitol or with an equimolar amount of a sucrose solution. After a lapse of several hours, the lactitol or sucrose left in the intestines was determined. Although 85% of the sucrose intake had been lost because of absorption and digestion, all the lactitol was still present.

2. Metabolic Energy

At the Agricultural University of Wageningen in The Netherlands, a study was carried out (11) determining the energy balance of eight volunteers on diets supplemented with either lactitol or sucrose. In this study, volunteers were kept for 4 days in a respiratory room; this was done twice: once with 49 g of sugar a day in a diet; the other time, this sugar was replaced by lactitol. The dosage of 50 g lactitol monohydrate was ingested in four to six portions during the day. Intakes of metabolizable energy (ME) were corrected, within subjects, to energy equilibrium and equal metabolic body weight. Further correction of ME intake was made toward equal actometer activity. With regard to the value of ME to supply energy for maintaining the body in energy equilibrium, the energy contribution to the body of lactitol monohydrate was 60% less than for sucrose.

Because the lactitol monohydrate contains 5% water and the results had an inaccuracy of standard error (SE) 10%, the final conclusion is justified that the metabolic energy of lactitol, on the basis of the dry substance, is at most 50% sucrose. Rounding off upward, a caloric value of lactitol in man at 2 kcal/g seems fully justified.

B. Prebiotic Effects of Lactitol

Lactitol reaches the colon untouched, where it is used as an energy source by the intestinal microflora. The fermentation of lactitol favors the growth of saccharolytic bacteria and decreases the amount of proteolytic bacteria that are responsible for the production of ammonia, carcinogenic compounds, and endotoxins. This can positively influence the health of humans and animals.

In vitro studies show that lactitol stimulates the growth of *Lactobacillus* spp. and *Bifidobacteria*. The growth of proteolytic bacteria such as *Enterobacteria* and *Enterococci* is inhibited (see Table 1). Inhibition of these microorganisms is caused by the production of short-chain fatty acids, which reduce the pH, and inhibition of the adhesion of these bacteria to the epithelial cell walls (12–14).

In vivo lactitol is tested both in humans and animals. Stool samples and cecal material was retrieved from the tested subjects to determine the presence of micro-organisms, short-chain fatty acid compounds, fecal pH, moisture, and activity of certain enzymes (15–17). These studies confirmed the results of the in vitro tests. This prebiotic effect can be used to develop new products.

C. Lactitol and Diabetes

Because lactitol does not have any significant effect on blood glucose or insulin levels, it is suitable for insulin-dependent diabetic patients (type 1). It can also be consumed by noninsulin-dependent diabetic patients (type II). This group can limit its diabetic status by taking dietetic measures. Lactitol fits in their diets because of its reduced calorie value compared with sucrose and sorbitol and its similar metabolism to dietary fiber.

D. Tooth-Protective Properties

Sugars are a major factor in the pathogenesis of dental caries. Oral bacteria convert sugars into polysaccharides that are deposited on the teeth; these plaque

Table 1 Fermentability of Lactitol, and Glucose by Intestinal Microorganisms[a]

Species	Lactitol			Glucose			Sugar-free control
	No. positive[b]	pH[c]	Gas	No. positive[b]	pH[c]	Gas	pH[c]
Obligate anaerobic microorganisms							
Bacteriodes thetaiomicron	7/8	5.6 ± 0.5	0	8	5.3 ± 0.7	0	6.6
Bacteriodes fragilis	5/8	5.6 ± 0.4	0	8	5.1 ± 0.4	0	6.4 ± 0.4
Bacteriodes distasonis	8/8	6.1 ± 0.6	0	8	5.4 ± 0.6	0	6.6 ± 0.3
Bifidobacterium bifidum	1/1	5.5	0	1	4.0	0	6.5
Clostridium perfringens	6/8	5.7 ± 0.6	+++	8	5.1 ± 0.4	+++	6.8 ± 0.4
Lactobacillus acidophilus	7/9	4.7 ± 0.6	0	9	5.0 ± 0.9	0	6.4 ± 0.5
Lactobacillus species	2/2	5.3 ± 1.7	0	2	5.0 ± 1.4	0	6.8 ± 0.4
Streptococcus mutans	5/7	5.3 ± 0.4	0	7	4.0	0	6.6 ± 0.5
Clostridium glycolcium	0/1	—	0	1	5.0	+++	6.5
Propionibacterium species	1/2	6.5	0	1	5.5 ± 0.7	0	7.0
Lactobacillus fermentum	0/1	—	0	1	6.0	0	6.5
Facultative anaerobic and obligate anaerobic microorganisms							
Enterobacter	2/7	7.0	+	7/7	6.8 ± 0.4	+++	7.8 ± 0.4
Klebsiella	14/15	5.6 ± 0.4	+	15/15	5.4 ± 0.4	++	6.3 ± 0.3
E. coli	0/2	—	0	2/2	4.7 ± 0.4	++	7.5
Proteus mirabilis	0/1	—	0	1/1	4.5	+	7.5
Proteus morgani	0/2	—	0	2/2	4.5	++	7.7 ± 0.4
Proteus vulgaris	0/2	—	0	2/2	5.0	+	7.5
Serratia	0/2	—	0	2/2	6.2 ± 0.4	++	7.7 ± 0.4
Citrobacter	0/2	—	0	2/2	5.0 ± 0.7	+	7.7 ± 0.4
Hafnia	0/2	—	0	2/2	4.5	0	7.0
Pseudomonas aereeginosa	0/2	—	0	0/2	—	0	8.5
Candida albicans	0/2	—	0	2/2	6.2 ± 0.4	0	7.7 ± 0.4
Enterococci	8/9	5.7 ± 0.4	0	9/9	4.1 ± 0.2	0	7.0 ± 0.2
Staphylococcus epidermidis	5/10	5.5 ± 0.4	0	10/10	4.8 ± 0.3	0	7.8 ± 0.4
Staphylococcus aureus	4/8	6.3 ± 0.3	0	8/8	4.9 ± 0.5	0	6.5 ± 0.4

[a] All experiments done at least in triplicates.
[b] Number of positive strains per number of examined strains.
[c] pH is the mean on day 2.
Source: Ref. 12.

sugars are then fermented into acids. The acid demineralizes the enamel and causes cavities.

E. Dental Studies

Three main types of experiments are used to evaluate the cariogenicity of foods and food ingredients:

Studies in vitro
Experiments in laboratory animals
Clinical trials and investigations in human subjects

F. Studies In Vitro

Among the earliest reports are those of Havenaar (18, 19) on the use of lactitol by oral bacteria, with formation of acids. A number of plaque bacteria were found that could metabolize lactitol as a substrate, including *Streptococcus mutans*, which is known to possess cariogenic activity, and certain strains of *S. sanguis*, *Bifidobacterium*, and *Lactobacillus*. These first experiments did not establish the speed of fermentation, which because of the limited time that sugars and sweeteners remain in the mouth is a determinant of their cariogenicity. It was later shown that pH drop was slow, leading to the conclusion that lactitol could be fermented slowly by these microorganisms.

In addition to acid production, another important property of cariogenic bacteria is the capacity to synthesize extracellular polysaccharide from carbohydrate substrates, making dental plaque. No evidence for extracellular polysaccharide synthesis was found.

A more recent study on lactitol in vitro comparing five other bulk sweeteners, on incubation with mixed cultures of human dental plaque microorganisms, measuring acid development, insoluble polysaccharide synthesis, and the attack of the acid on enamel mineral, was done by Grenby and Phillips (20). The six different sweeteners fell into three groups, with the highest acid generation, polysaccharide production, and enamel demineralization from glucose and sucrose, less from sorbitol and mannitol, and least of all from lactitol and xylitol.

G. Experiments in Laboratory Animals

The cariogenicity of lactitol was also tested in rats (21). For this purpose, lactitol was incorporated in a powdered diet, consisting of 50% of a basal diet (SPP, Trouw & Co, Putten, The Netherlands), 25% wheat flour, and 25% of test substance. Lactitol was compared with sorbitol, xylitol, sucrose, and a control con-

Table 2 Average of Carious Fissure Lesions and Weight Gains in Five Dietary Groups of 12 Rats

Diet	Number of fissure lesions[a] B	C	Weight gain (g)
Sucrose	5.75	2.58	82.0
Sorbitol	1.33[b]	0.42[b]	46.0
Lactitol	1.17[b]	0.33[b]	72.2[b]
Wheat flour	0.67[b]	0.25[b]	80.8[b]
Xylitol	0.42[b]	0[b]	48.6[c]
Pooled standard error	0.56	0.43	5.4

[a] Twelve fissures at risk.
[b] $P < 0.01$, significantly different from the sucrose group.
[c] $P < 0.0001$ (Turkey's test).
Source: Ref. 15.

sisting of 50% wheat flour and 50% basal diet. The rats were program fed. None of the animals had diarrhea.

Especially in the lactitol group, no significant adverse effects on general health were observed. The results are shown in Table 2. Obviously, the substitution of sucrose by lactitol reduces caries significantly. The caries results for sorbitol and xylitol were in accordance with other studies in which these polyols were used.

In the second experiment, human food was fed to rats. Shortbread biscuits containing 16.6% sucrose or lactitol were incorporated at 66% in pulverized, blended diets fed to two groups of 21 rats, so that the final level of lactitol in the test diet was 11%. Again, after a period of 8 weeks, caries attack showed highly significant reduction when replacing the sucrose in the biscuits with lactitol.

The most recent studies on the dental properties of lactitol in rats tested it at lower levels in the diet and in the form of a finished human food product rather than as a raw ingredient in a blended animal diet (22). At a level of 16% lactitol or 16% xylitol in the blended diet, the carious scores, lesion counts, and severity of the lesions were so close on the two polyol regimens that they were indistinguishable but significantly beneficial compared with the 16% sucrose group.

H. Investigations in Man

At the university of Zunch, a special method has been developed for the in vivo determination of the plaque pH in humans. After consumption of chocolate or

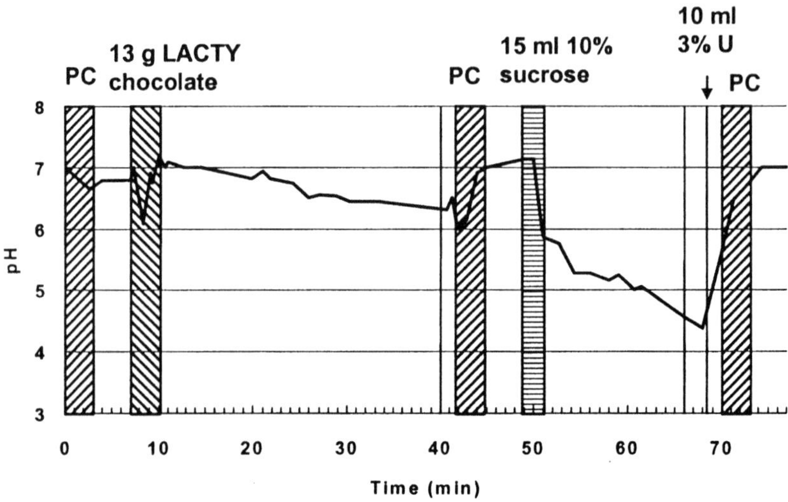

Figure 2 Telemetrically recorded pH of 5 days interdental plaque in subject H. H. during and after consumption of 13 g LACTY plain chocolate. A 10% sucrose solution was used as a positive control. PC, 3-min paraffin chewing gum; u, rinsing with 3% carbamide.

confections in which sucrose is replaced with lactitol, changes in plaque acidification are detected by the electrodes and transmitted electronically to a graph recorder.

The term ''zahnschonend'' (safe for teeth/friendly to teeth) is used officially in Switzerland when the pH of dental plaque does not drop below 5.7 during a 30-minute period. Professor Muhlemann demonstrated in chocolates that lactitol is ''safe for teeth'' (23, 24).

In Fig. 2, it is shown that during and after eating 13 g LACTY chocolate (plain lactitol-containing chocolate), the plaque pH is not affected, whereas with a 15-ml 10% sucrose solution, the plaque-pH is reduced to about 4.5, which indicates that sucrose is being fermented into acids by the oral bacteria.

V. PHYSICAL PROPERTIES

Lactitol is a white crystalline powder available as a monohydrate or anhydrate. In the following sections the most important characteristics of lactitol will be described.

Table 3 Relative Sweetness of Lactitol Solutions Compared with Sucrose Solutions at 25°C

Sucrose solution concentration (w/w)	Relative sweetness of lactitol (sucrose = 1.0)
2%	0.30
4%	0.35
6%	0.37
8%	0.39
10%	0.42

A. Sweetness

Lactitol is less sweet than sucrose. Lactitol increases in relative sweetness when its concentration is increased (see Table 3). In most applications, a sweetness of 0.4 times that of sucrose will be found. Lactitol has a clean sweet taste, without an aftertaste. In some products, the lower sweetness will allow the flavor of other ingredients to develop better. In other products, the sweetness will need to be increased. This is possible by using an intense sweetener. To obtain a sweetness equivalent to sucrose, approximately 0.3% aspartame or acesulfame K of lactitol, or approximately 0.15% saccharin needs to be added to lactitol.

B. Crystal Forms

Crystallization from aqueous solutions results in nonhygroscopic white monohydrate or dihydrate crystals or hygroscopic anhydrate crystals. The dihydrate has a melting point of 76–78°C, and the monohydrate has a more complicated melting behavior. Under the melting microscope, most of the crystals melt at 121–123°C, but part of the crystal-bound water evaporates below 100°C, converting this monohydrate into the anhydrous form. The melting point determination of the monohydrate by differential scanning calorimetry shows the main endothermal peak in the range of 96–107°C (water evaporation). The melting point of lactitol anhydrate is 145–150°C, determined by differential scanning calorimetry.

C. Hygroscopicity

Hygroscopicity is related to the amount of water absorbed by a product. Sucrose is not hygroscopic. Only under severe circumstances will sucrose absorb water.

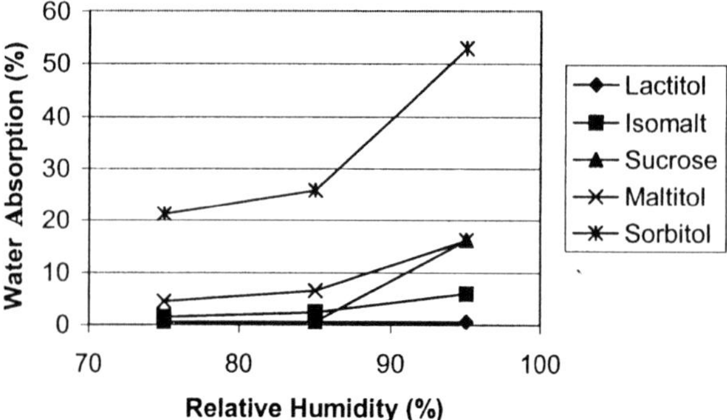

Figure 3 Moisture absorption of sucrose and polyols at different relative humidities after 7 days at 25°C.

The same applies to lactitol monohydrate. This can be seen in Fig. 3. When we look at sorbitol, we notice a high level of hygroscopicity. This means that, for example, crisp bakery products will easily become soggy because they absorb water from the air. Lactitol monohydrate is one of the least hygroscopic polyols together with mannitol. This makes lactitol suitable for all applications in which water absorption is a critical parameter, like bakery products, tablets, and panned confection.

D. Water Activity

Water activity influences enzymatic activities, Maillard reaction, fat oxidation, microbial stability, and texture. These elements combined influence the shelf-life. The influence of lactitol on water activity is similar to that of sucrose (on a dry solids basis) because the molecular weight of lactitol is almost identical to that of sucrose. Sorbitol, being a much smaller molecule, has a much larger effect on the water activity (Fig. 4).

E. Solubility

Solubility of the ingredients is important for many products (e.g., ice cream, hard-boiled sweets). A low solubility can make production processes more difficult (e.g., in hard-boiled sweets, most of the water will have to be evaporated because more water will have to be added when the solubility is low). The solubility of lactitol at lower temperatures is less than that of sucrose but still good enough not to cause inconveniences during processing (Fig. 5). The solubility of maltitol

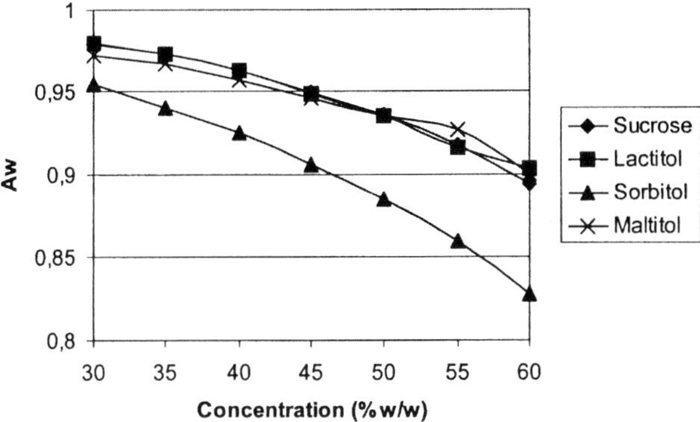

Figure 4 Water activity of lactitol, sucrose, and sorbitol at 25°C.

is comparable with lactitol. Mannitol and isomalt have a lower solubility, which influences processing, mouth-feel, and flavor release.

F. Viscosity

The viscosity of lactitol in aqueous solution is slightly higher than that of sucrose (Fig. 6). In the case of hard-boiled sweets, the viscosity of the lactitol-melt is more important. In this case, a combination of lactitol and hydrogenated starch hydrolysate is used (comparable to sucrose and glucose syrup). For this combination a slightly lower viscosity is found than for its sucrose counterpart, but on cooling down, the same viscosity (or plasticity) will be reached, so lactitol will also be suitable for molded sweets.

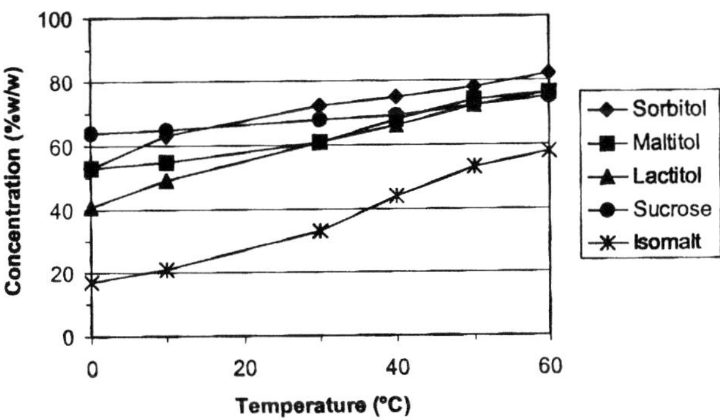

Figure 5 Solubility of lactitol and other polyols at different temperatures.

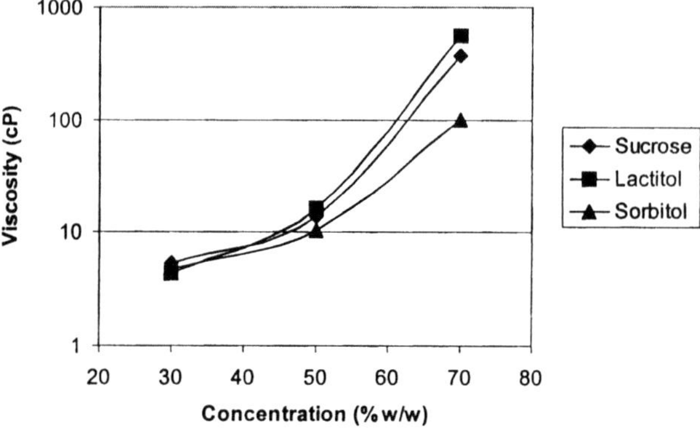

Figure 6 Viscosity of polyol solutions at different concentrations at 25°C.

G. Heat of Solution

Lactitol has a cooling effect that is slightly stronger than that of sucrose but far less than that of sorbitol and xylitol (Table 4).

H. Freezing-Point-Depressing Effect

If sugars are replaced in ice cream, a replacer causing a similar freezing-point decrease is required. Lactitol is such a product, as shown in Fig. 7. This effect is closely related to the molecular weight because this is virtually the same for lactitol and sucrose; the effect is very similar.

Table 4 Heat of Solution of Various Sweeteners

Sweetener	Cal/g
Sorbitol	−26.6
Mannitol	−28.9
Xylitol	−36.6
Lactitol monohydrate	−12.4
Lactitol anhydrate	−6.0
Maltitol	−18.9
Isomalt	−9.4
Sugar	−4.3

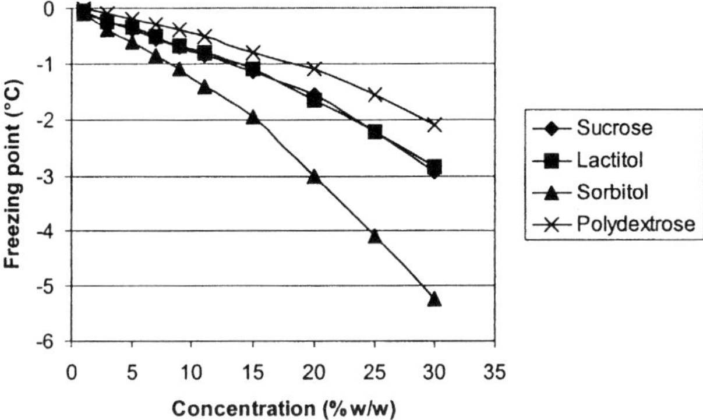

Figure 7 Freezing-point-depressing effect.

I. Chemical Properties

Lactitol is a polyol with nine OH-groups that can be esterified with fatty acids resulting in emulsifiers (3) or that can react with propylene oxide (to produce polyurethanes). Because of the absence of a carbonyl group, lactitol is chemically more stable than related disaccharides like lactose. Its stability in the presence of alkali is much higher than that of lactose. The stability of lactitol in the presence of acid is similar to that of lactose. The absence of a carbonyl group also means that lactitol does not take part in nonenzymatic browning (Maillard) reactions.

Lactitol solutions have an excellent storage stability. A 10% lactitol solution in the pH range 3.0–7.5 at 60°C shows no decomposition after 1 month. After 2 months at pH 3.0, some decomposition (15%) is detected. At a higher pH this does not occur. With increasing temperature and especially with increasing acidity, hydrolytic decomposition of lactitol is observed. Sorbitol and galac-

Table 5 Decomposition of a 10% Lactitol Solution under Extreme Conditions of pH and Temperature

	% Weight recovery at 24 hr at 105°C			
pH	Lactitol	Galactose	Sorbitol	Not identified
2.0	42.9	28.1	21.4	7.6
10.0	98.0	—	—	2.0 (acids)
12.0	98.3	—	—	1.7 (acids)

tose are the main decomposition products. At high pH lactitol is stable even at 105°C (Table 5).

When heated to temperatures of 179–240°C, lactitol is partly converted into anhydrous derivatives (lactitan), sorbitol, and lower polyols.

VI. APPLICATIONS

Any application in which carbohydrates are used could be discussed because lactitol could play a role. However, some general guidelines for specific application areas are offered.

Lactitol is used in products for its bulking properties. Therefore on most occasions a one-to-one replacement will be required. Because of the lower sweetness of lactitol, most products contain a combination of lactitol and intense sweeteners.

A. Bakery Products

In this area there are many products in which sugar can be replaced. For a large number of bakery products the crispiness of the product is important. Most bulk sweeteners have the tendency to be hygroscopic, but if lactitol products are handled like sucrose products, crispiness can be assured.

In the soft bakery products (e.g., cakes), we find that lactitol (being nonhygroscopic) often benefits by a small addition of a humectant (e.g., sorbitol). This gives a smooth, moist feel without affecting the good structure and texture obtained by lactitol.

Next to sugar-free bakery products, products low in calories or with functional claims, high fiber, are developed. In these products lactitol is a suitable bulk sweetener. In low-calorie products, lactitol is often combined with polydextrose, which is an optimal combination to reduce the calorie content and maintain a high-quality product. High-fiber products claim to have a positive effect on the colon. Because of the beneficial effects that lactitol has on the colon, it will be a suitable sweetener in such products.

B. Chewing Gum

Sugar-free chewing gum is becoming very popular. The advantage of the sugar-free chewing gum is the tooth-friendliness. Lactitol, like all other polyolsis tooth-friendly and suitable for use in chewing gum. Lactitol can be used as a sweetener in the gum base, as rolling compound/dusting powder, and in the panning layer.

In the gum base, lactitol gives a more flexible structure, and there is no need to increase the gum base level as is required in sorbitol containing chewing gum.

Lactitol and mannitol are the polyol, with the lowest hygroscopicity. This makes lactitol an excellent rolling compound. The low hygroscopicity will increase shelf-life, especially when stored at high temperatures and humidities. Furthermore, lactitol will improve mouth-feel and, compared to mannitol, lactitol has a better solubility that prevents sandiness of the chewing gum.

A lactitol panned chewing gum is very stable because of the low hygroscopicity.

C. Chocolate

Lactitol can successfully be applied when producing a sugar-free chocolate. Both lactitol monohydrate and anhydrate can be used for the production of chocolate. The difference between these two crystal forms during processing is the conching temperature. As discussed before, lactitol monohydrate is not hygroscopic; however, it does contain crystal water. When lactitol monohydrate is used, the crystal water will be bound sufficiently to use temperatures of up to about 60°C during the conching stage of chocolate. Above this temperature, the water will come free and the viscosity of the chocolate mass increases. With anhydrous lactitol the conching temperature can be 80°C. Another difference between the two forms is the cooling effect. The anhydrous lactitol has a lower cooling effect than the monohydrate. This will improve the taste of the chocolate, which should have a warm taste.

D. Confectionery

In confectionery a large range of products is imaginable. A combination of different sugars is often found when looking at sugar confectionery. This is necessary to obtain the desired effect and results. In general, lactitol replaces the sucrose part of recipes, whereas glucose syrup needs to be replaced by hydrogenated starch hydrolysate or maltitol syrup. These latter products are used as crystallization inhibitors. The right ratio differs from product to product, (e.g., hard-boiled sweets need a 70/30 ratio [lactitol/hydrogenated starch hydrolysate], chewies, fruit gums, and pastilles a 40/60 ratio).

E. Ice Cream

Lactitol shows a combination of physical properties (e.g., freezing-point-depressing effect, hygroscopicity, solubility) that are suitable for ice cream. Not

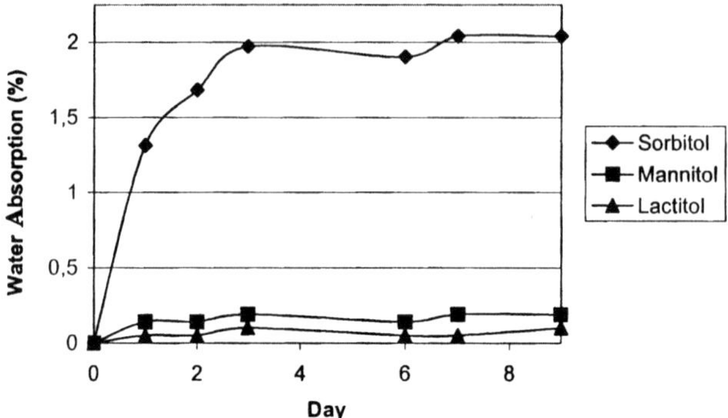

Figure 8 The water absorption of tablets made with direct compressible polyols. (Tablet hardness, 150N; temperature, 20°C; RH, 70%).

only will the use of lactitol reduce the caloric content of the ice cream but it will also permit a low fat percentage, allowing even further reduction in calories.

F. Tablets

Granulated lactitol is suitable for the use in tabletted confections, such as breath fresheners, and as an excipient in pharmaceutical preparations like vitamin tablets. In this application, the low hygroscopicity of lactitol is important. This low

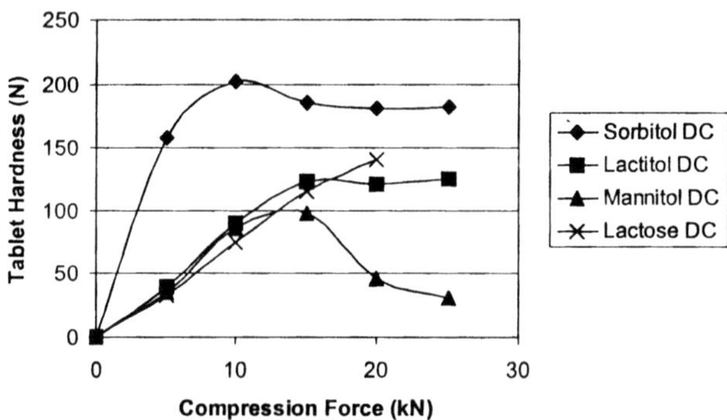

Figure 9 The compression profile of direct compressible polyols on a rotary press (30,000 rpm, 0.5% Mg-stearate, ∅ 9 mm, 650 mg).

hygroscopicity prolongs shelf-life and protects active agents against moisture. Figure 8 shows that lactitol tablets absorb hardly any water.

In pharmaceutical tablets, direct compressible sorbitol and mannitol are most often used. The advantage of lactitol in this application is the low hygroscopicity compared with sorbitol and the better solubility compared the mannitol. With respect to tablet hardness lactitol performs in between (Fig. 9). This makes direct compressible lactitol a good alternative.

G. Preserves

The solution to low-calorie preserves is often found in a low dry solids preserve. To combine this with lactitol will result in an even larger reduction in calories, while maintaining the good flavor and texture.

VII. REGULATORY STATUS FOR LACTITOL

Argentina	Allowed
Australia	Allowed
Brazil	Allowed
Canada	Allowed
European Union	Lactitol is allowed as a sweetener in all EU countries regulated by the sweeteners in food directive No. 94/35/EC.
	Lactitol can also be used as additive; this is regulated in the food additives other than colors and sweeteners No. 95/2/EC
	The E-number for lactitol is E 966
Israel	Allowed
Japan	Allowed
Norway	Allowed
Sweden	Allowed
Switzerland	Allowed
United States	Lactitol is self-affirmed GRAS. The petition to affirm lactitol as GRAS was accepted for filing by the Food and Drug Administration on May 17, 1993.

VIII. CONCLUSION

Lactitol is a unique, commercially available bulk sweetener. Its physical properties guarantee optimal product performance during processing and storage of the food containing lactitol. Being so similar to sucrose, it is possible to develop new sugar-free and light food products in a short period of time.

REFERENCES

1. JB Senderens. Catalytic hydrogenation of lactose. Comptes Rendus 170:47–50, 1920.
2. J Karrer, Büchi. Reduktionsprodukte von Disacchariden: Maltit, Lactit, Cellobit. Helvetica Chim Acta 20:86–90, 1937.
3. JA van Velthuijsen. Food additives derived from lactose: lactitol and lactitolpalmitate. J Agric Food Chem 27:680–686, 1979.
4. CF Wijnman, JA van Velthuijsen, H van den Berg. Lactitol monohydrate and a method for the production of crystalline lactitol. European Patent 39981, 1983.
5. T Saijonmaa, M Heikonen, M Kreula, et al. Preparation and characterization of milk sugar alcohol dihydrate (abstract). Sixth European Chrystallograph Meeting, Barcelona, 1980, p 26.
6. Joint FAO/WHO Expert Committee on Food Additives. IPCS Toxicological evaluation of certain food additives and contaminants: Lactitol. 27th report Geneva, WHO Food Additives Series 1983, pp 82–94.
7. The Scientific Committee for Food. Report on sweeteners, EC-Document III/1316/84/CS/EDUL/27 rev., 1984.

7a. RM Diener. J Am Coll Toxico 11(2):165–257, 1992.

8. Maizena GmbH, Verwendung von Lactit als Zuckeraustauschstoff. German Patent 2,133,428, 1974.
9. K Hayashibara. Improvements in and relating to the preparation of foodstuffs. U. K. Patent 1,253,300, 1971.
10. K Hayashibara, K Sugimoto. Containing lactitol as a sweetener. U. S. Patent 3,973,050, 1975.
11. AJH Van Es, L de Groot, JF Vogt. Energy balances of eight volunteers fed on diets supplemented with either lactitol or saccharose. Br J Nutr 56:545–554, 1986.
12. G Lebek, SP Luginbühl. Effects of lactulose and lactitol on human intestinal flora, In: HO Conn, J Bircher, eds. Hepatic Encephalopathy: Management with Lactulose and Related Compounds. Michigan:1989, pp 271–282.
13. D Scevola et al. Intestinal bacterial toxins and alcohol liver damage: effects of lactitol, a synthetic disaccharide. La Clinica Dietologica 20:297–314, 1993.
14. D Scevola, et al. The role of lactitol in the regulation of the intestinal microflora in liver disease. Giornale di malattie infettive e parassitaire 45:906–918, 1993.
15. YF Felix, et al. Effect of dietary lactitol on the composition and metabolic activity of the intestinal microflora in the pig and in humans. Microbial Ecology Health Dis 3:259–267, 1990.
16. Kitler, et al. Lactitol and lactulose an in vivo and in vitro comparison of their effects on the human intestinal flora. Drug Invest 4:73–82, 1992.
17. GP Ravelli, et al. The effect of lactitol intake upon stool parameters and the faecal

bacterial flora in chronically constipated women. Acta Therapeutica 21:243–255, 1995.

18. R Havenaar. Microbial investigations on the canogenicity of the sugar substitute lactitol. Unpublished report from the University of Utrecht (Netherlands), Department of Preventive Dentistry and Oral Microbiology; June, 1976.
19. R Havenaar, JHJ Huis in 't Veld, O Backer Dirks, et al. Some bacteriological aspects of sugar substitutes. In: B Guggenheim ed. Health and Sugar Substitutes. Proceedings of the ERGOB Conference. Geneva: Basel, Karger, 1978, pp 192–198.
20. TH Grenby, A Phillips. Studies of the dental properties of lactitol compared with five other bulksweeteners in vitro. Caries Res 23:315–319, 1989.
21. JS van der Hoeven. Cariogenicity of lactitol in program-fed rats. Caries Res 20: 441–443, 1986.
22. TH Grenby, A Phillips. Dental and metabolic observations on lactitol in laboratory rats. Br J Nutr 61:17–24, 1989.
23. HR Muhlemann. Gutachten uber die zahnschonenden Eigenschaften von Lacty Schokolade und Lacty-Milch-schokolade der Firma CCA. Holland, unpublished report from Zahnarztliches Institut der Universitat Zunch, July 1977.
24. TH Grenby, T Desai. A trial of lactitol in sweets and its effects on human dental plaque. Br Dent J 164:383–387, 1988.

18
Sorbitol and Mannitol

Anh S. Le and Kathleen Bowe Mulderrig
SPI Polyols, Inc., New Castle, Delaware

I. INTRODUCTION

Sorbitol and mannitol have existed as commercial products for more than 60 years. Today, sorbitol and mannitol are used in food, confectionery, oral care, pharmaceutical, and industrial applications because of their unique physical and chemical properties. Sorbitol and mannitol are classified as sugar alcohols or polyols.

Sorbitol was first discovered by a French Chemist named Joseph Boussingault in 1872. He isolated it from the fresh juice of the mountain ash berries. Mannitol is found in the extrudates of trees, manna ash, marine algae, and fresh mushroom. Although present in natural sources, sorbitol and mannitol were not commercially available until 1937, when they were first manufactured in full production scale by the Atlas Powder Company, Wilmington, Delaware (now SPI Polyols, Inc.) (1).

II. PRODUCTION OF SORBITOL AND MANNITOL

Glucose syrups, invert sugar, and other hydrolyzed starches are important raw materials for the manufacture of sorbitol and mannitol. Sorbitol is produced from the catalytic hydrogenation of glucose. The hydrogenation reaction is driven by a catalyst, such as nickel. After the reaction is complete, the catalyst is filtered out and the solution is purified. It is then evaporated to 70% solids and sold as sorbitol solution (1).

Crystalline sorbitol is made by further evaporating the sorbitol solution into molten syrup containing at least 99% solids. The molten syrup is crystallized into a stable crystalline polymorph that has one single melting point (99–101°C) and heat of fusion (42 cal/g, assuming 44 cal/g represents a fully crystallized crystalline sorbitol). The stable polymorph of sorbitol is known as gamma (γ) sorbitol and was determined from the work of Atlas Powder Company and University of Pittsburgh researchers in the early 1960s. Most commercially available crystalline sorbitol is the gamma polymorph (2, 3).

When hydrogenated invert sugar is the raw material, both sorbitol and mannitol are produced in the solution because they are isomers. The mannitol and sorbitol are separated on the basis of their different solubilities. Mannitol is crystallized from the solution because of its lower solubility than sorbitol in water. The resulting product is the most stable polymorph, the β-polymorph (4). Mannitol is then filtered, dried, and sold as white powder or free-flowing granules (directly compressible material).

III. PHYSICAL AND CHEMICAL PROPERTIES OF SORBITOL AND MANNITOL

Sorbitol and mannitol are six-carbon, straight-chain polyhydric alcohols, meaning they have more than one hydroxyl group. Both sorbitol and mannitol have six hydroxyl groups and the same molecular formula, $C_6H_{14}O_6$. They are isomers of one another and have different molecular configurations. The difference between sorbitol and mannitol occurs in the planar orientation of the hydroxyl group on the second carbon atom (Figs. 1 and 2). This dissimilarity has a powerful influence and results in an individual set of properties for each isomer.

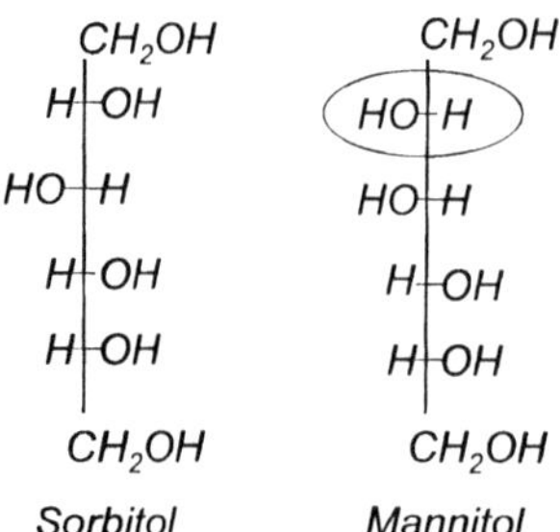

Figure 1 The chemical structures of sorbitol and mannitol.

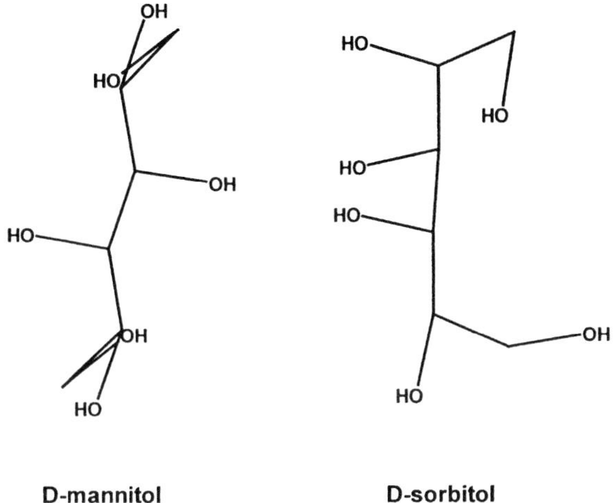

Figure 2 The planar configurations of sorbitol and mannitol.

The major difference between the two isomers is that sorbitol is hygroscopic and mannitol is nonhygroscopic. Therefore, sorbitol is used as a humectant because of its affinity for moisture, and mannitol is used as a pharmaceutical and nutritional tablet excipient because of its inertness and stability against moisture.

Sorbitol solution is hygroscopic, attracting and releasing moisture under varying humidity conditions, but it does so very slowly. Polyols of lower molecular weight, such as glycerin, tend to gain and lose water more rapidly. Sorbitol provides improved moisture control and is more likely to maintain equilibrium in the surrounding environment. This slower rate of change in moisture content protects the food products in which sorbitol is used, thus maintaining the as-made quality of the product and extending the shelf-life (Fig. 3).

There are many different polymorphs of both sorbitol and mannitol. The most stable polymorph of sorbitol is (γ)-sorbitol. The properties of γ-sorbitol and the other polymorphs are contained in Table 1 (2). The most stable polymorph of mannitol is the (β)-polymorph. Working with the most stable polymorph is important because one needs to be certain the product being used will not change during processing, (i.e., an unstable polymorph converting to the stable polymorph). Use of the most stable polymorph ensures that the properties of the finished product will not change (5). Most of the commercially available sorbitol and mannitol products are of the most stable polymorphs. One of the characteristics affected by polymorph is solubility. The solubility in water of γ-sorbitol is very high (235 g/100 ml at 25°C). The solubility in water of β-mannitol is lower (22 g/100 ml at 25°C).

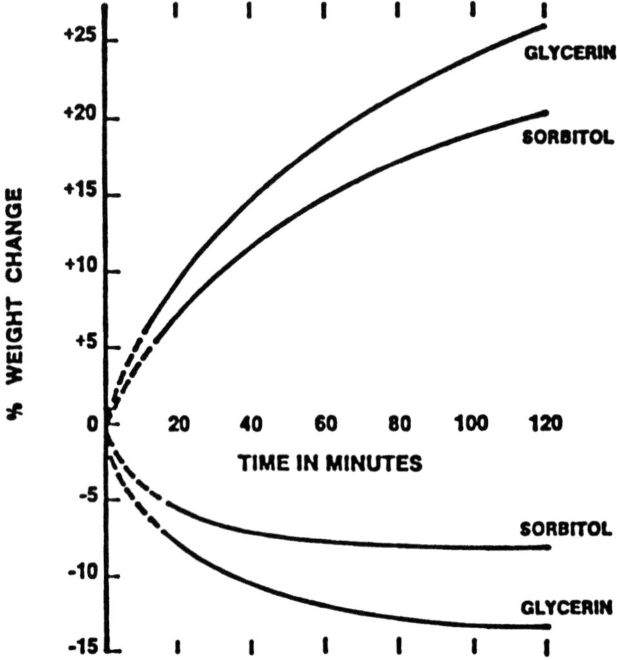

Figure 3 Rates of moisture exchange between air and sorbitol solutions and glycerin. These curves show the rate of moisture gain occurring when solutions of sorbitol or glycerin, which are in hygroscopic equilibrium at 58% relative humidity, are transferred to an atmosphere of 79% relative humidity; also the rate of moisture loss when such solutions are transferred to an atmosphere of 32% relative humidity. Note that the sorbitol solutions neither gain nor lose water as rapidly as do the glycerin solutions. (Courtesy of SPI Polyols, Inc., New Castle, Delaware.)

Sorbitol and mannitol are sweet, pleasant-tasting polyols. This characteristic makes them popular in confectionery and pharmaceutical taste-masking applications. Sorbitol is approximately 60% as sweet as sucrose, and mannitol is approximately 50% as sweet. They both have a negative heat of solution, which gives them a cooling sensation in the mouth. Sorbitol's heat of solution is −26.5 cal/g (at 25°C), and mannitol's heat of solution is −28.9 cal/g (at 25°C).

Like many polyols, sorbitol and mannitol have a laxation threshold associated with them. The average adult can consume up to 50 g of sorbitol and 20 g of mannitol in a day without a laxation problem. The general characteristics of sorbitol and mannitol are summarized in Table 2.

Table 1 Physical Properties of the Polymorphic Forms of Crystalline Sorbitol

	Form									
Property	E		α		Δ		β		γ	
Peak[a] (°C)	75/83		88		89		97.5		101	
Onset of peak[a] (°C)	68		85		86		94.5		99	
Melting range (°C)	77–		90.0–91.5		86.8–88.5		94.2–95.0		99.0–101.6	
Density (g/cm^{-3}) flotation	n.d.[d]		1.550		1.460		1.481		1.527	
H_2O solubility[b]	252		n.c.§		252		216		211	
X-ray interplanar spacings[c] (Å) (6 strongest)	d	I/I_0	d	I/I_0	d	I/I_0	d	I/I_0	d	I/I_0
	4.73	100	4.22	100	4.23	100	3.95	100	4.73	100
	3.50	67	3.93	75	3.71	66	5.95	80	3.49	67
	4.04	60	4.33	70	4.77	64	4.66	74	4.04	61
	6.00	45	2.65	51	2.12	53	4.98	50	3.91	53
	7.57	40	3.78	41	2.64	44	4.24	42	4.33	50
	5.07	40	2.77	41	8.85	27	3.48	38	3.50	50

[a] As determined by differential thermal analysis (DTA) and differential scanning calorimetry (DSC).
[b] H_2O at 20°C (parts per 100 parts H_2O).
[c] Space group: $P2_12_12_1$; Axes (Å): a = 8.677, b = 9.311, c = 9.727.
[d] n.d. = not determined.
Source: Reproduced with permission from Pharmaceutical Technology, Vol. 8, Number 9, September 1984, pages 50–56.

Table 2 Characteristics of Sorbitol and Mannitol

	Sorbitol	Mannitol
Chemical formula	$C_6H_{14}O_6$	$C_6H_{14}O_6$
Form	White powder or 70% solution	White powder
Sweetness	60% of sucrose	50% of sucrose
Taste	Sweet/cool	Sweet/cool
Odor	None	None
Noncariogenic	Yes	Yes
Moisture sensitivity	Hygroscopic	Nonhygroscopic
Solubility in H_2O (at 25°C)	235 g/100 g H_2O	23 g/100 g H_2O
Caloric value	2.6 cal/g	1.6 cal/g
Melting point	100°C	164°C
Molecular weight	182	182
Heat of solution (at 25°C)	−26.5 cal/g	−28.9 cal/g
Chemical stability	Stable in air in the absence of catalysts and in cold, dilute acids and alkalis. Sorbitol does not darken or decompose at elevated temperatures or in the presence of amines. It is nonflammable, noncorrosive, and nonvolatile.	Stable in dry state or in sterile aqueous solutions. In solutions, it is not attacked by cold dilute acids or alkalis, or by the atmospheric oxygen in the absence of catalysts. Mannitol does not undergo Maillard reactions.

IV. ABSORPTION AND METABOLISM

Sorbitol and mannitol are widely accepted by the food and pharmaceutical industries as nutritive ingredients because of their ability to improve the taste and shelf-life of regular foods and special dietary products.

Sorbitol and mannitol are slowly absorbed into the body from the gastrointestinal tract and are metabolized by the liver, largely as fructose, a carbohydrate that is highly tolerated by people with diabetes. Sorbitol is absorbed and metabolized in the liver by way of a pathway located entirely in the cytoplasmic compartment; this pathway is illustrated in Fig. 4 (6).

Ingestion of sorbitol and mannitol does not usually cause an immediate demand for extra insulin. The initial steps in the metabolism of sorbitol and mannitol in the liver, their uptake by the liver cells, and their conversion to glucose are independent of insulin, but the subsequent use of glucose by the muscle and

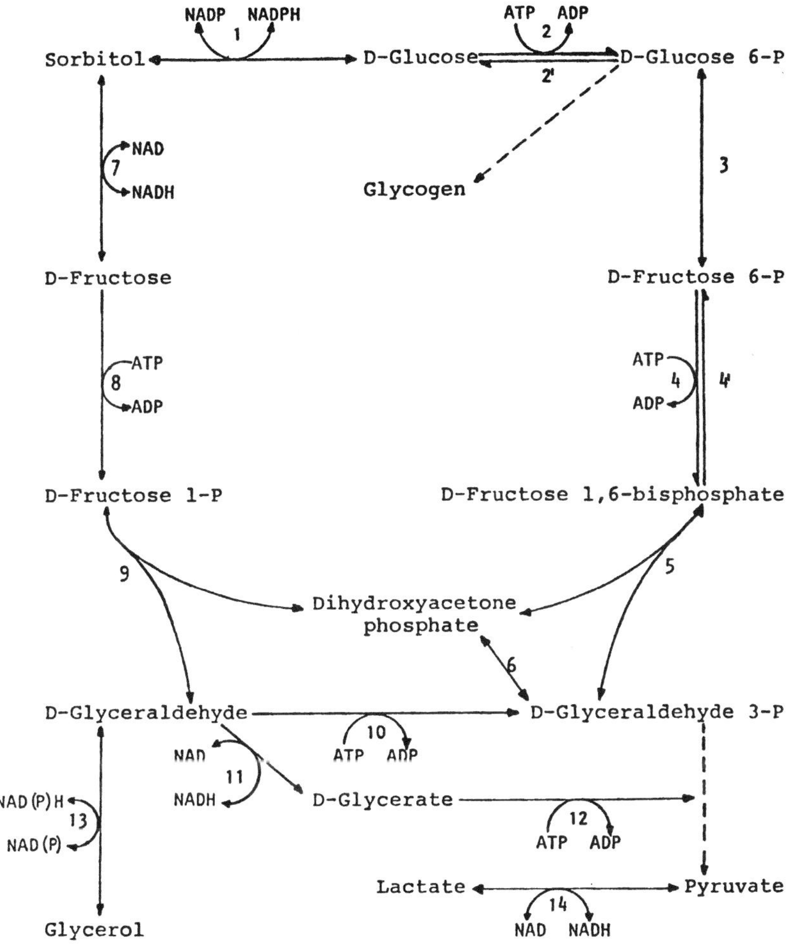

Figure 4 Sorbitol and mannitol metabolism. (From Ref. 6.)

adipose tissues is influenced by insulin (6). The person's condition at the time of ingestion will determine the amount of insulin required. In the calculation of nutrient intake, sorbitol and mannitol should be considered as carbohydrates; however, they have lower glycemic indices than other carbohydrates so can usually be used safely by people with diabetes.

The caloric value of sorbitol is 2.6 cal/g and the caloric value of mannitol is 1.6 cal/g in the United States. These are lower than the caloric value of sucrose, which is 4 cal/g.

Sorbitol and mannitol do not increase the incidence of dental caries, a condition started and promoted by acid conditions that develop in the mouth after eating carbohydrates and proteins. Results from pH telemetry tests show that sorbitol and mannitol do not increase the acidity or lower the pH of the mouth after ingestion. This means that they will not promote tooth decay. For this reason, sorbitol and mannitol are used in oral care and pediatric applications. The usefulness of polyols (e.g., including sorbitol and mannitol) as alternatives to sugars and as part of a comprehensive program including proper dental hygiene has been recognized by a numerous authorities, including the American Dental Association.

V. APPLICATIONS

Sorbitol and mannitol are used in the food, confectionery, oral care, and pharmaceutical industries because of their unique physical and chemical properties. Because of these properties, sorbitol and mannitol perform certain functions that are beneficial within the final product. As mentioned previously, sorbitol and mannitol have different properties from one another because of their different planar orientations. For this reason, they serve different functions in food and pharmaceutical products.

A. Sorbitol Applications

1. Chewing Gum

Crystalline sorbitol is widely used as a bulking agent in sugar-free chewing gum. Crystalline sorbitol does not promote dental caries. It provides sweetness and a pleasant cooling effect, which are synergistic with other flavoring agents such as spearmint, peppermint, cinnamon, wintergreen, and fruit flavors. The level of crystalline sorbitol used in a sugar-free chewing gum is typically between 50–55% by weight. High-intensity sweeteners such as aspartame and acesulfame potassium can be used with crystalline sorbitol to improve the sweetness and flavor release. The size of the crystalline sorbitol granules is very important in regard to flexibility, chew, cohesion, and smoothness in chewing gum. The particle size distribution of crystalline sorbitol used in sugar-free chewing gum is typically 94.5% through the U. S. Standard Testing Sieve Number 40.

The process of making sugar-free chewing gum is basically a blending operation that uses a horizontal sigma blade mixer. The mixer is equipped with a circulating water-heated jacket. The temperature of the water is heated to between 50–55°C. Crystalline sorbitol is mixed with the preheated gum base (the temperature of the gum base is 50–55°C) and the humectant solution until the mixture is homogenous. The mixing time is typically about 8 to 9 minutes. Gum bases are derived from natural sources and are synthetically produced to provide desired chewing properties. The humectant solution is usually a sorbitol solution, glycerin, and/or a maltitol solution. The humectant is used to prevent the sugar-free chewing gum from becoming too dry or stale during storage. The flavoring agent is then added to the homogenous mixture and mixed for an additional 3 to 4 minutes. The gum is removed from the sigma blade mixer and shaped into thin sheets by running it through a set of rollers several times. A dusting agent, mannitol powder, is usually used to dust the surfaces of the gum sheets to reduce the tendency of the gum to stick to the rolling, cutting, and wrapping equipment. The thin sheets of gum are then stored in a constant temperature room (25°C and less than 40% relative humidity) for 24 hours. They are then cut and wrapped into individual sticks of chewing gum.

2. Sugar-Free Hard Candy

To make a candy sugar free, sorbitol solution is used as the primary ingredient. In hard sugar candies, the corn syrup serves this purpose. Sorbitol is used for its sweet, cool taste and its ability to crystallize and form a hard candy. Sorbitol is noncariogenic, so sugar-free hard candies made with it will not promote tooth decay.

Sorbitol is used to make sugar-free candies by the batch depositing method. In this method, sorbitol solution is cooked at high temperatures until most of the water is driven off. The molten sorbitol is slowly and evenly cooled to a certain temperature, at which time flavor and a small amount of crystalline sorbitol are added. The crystalline sorbitol is used as a seed that nucleates the melt and starts the crystallization of sorbitol. The melt is deposited into molds and allowed to crystallize further. The sorbitol continues to crystallize and sets up into a hard candy.

The type of sorbitol solution used can control the rate of sorbitol crystallization. In the manufacture of sorbitol solution, mannitol is also produced because it is an isomer. Sorbitol solution 70%, USP, contains a small amount of mannitol (approximately 3%). Adjusting the mannitol levels up or down can increase or decrease the candy's set time. If the candy manufacturer has an automated batch depositing process and wants the candy to set up quickly, the mannitol level in sorbitol solution can be adjusted down. Alternatively, if the candy manufacturer has a manual process and does not want the candy to set up too quickly for fear of it setting up in the depositor, the mannitol level in sorbitol solution can be adjusted up. The mannitol inhibits or controls the rate of sorbitol crystallization.

3. Pressed Mints and Pharmaceutical and Nutritional Tablets

Crystalline sorbitol can be used in various types of tablets. It is most often used in a confectionery tablet such as a pressed mint. Because of its negative heat of solution, sorbitol has a cool taste that almost gives the perception of mint. Used with a mint flavor, this attribute of sorbitol enhances the tablet's flavor. Sorbitol is 60% as sweet as sucrose, so a mint tablet will not be too sweet, which would detract from its breath-freshening properties. (γ)-Sorbitol crystallized in a certain way can be used to improve flavor holding. This is due to its porous crystalline structure (7).

Sorbitol is very compressible and binds other tablet ingredients effectively. Grades of sorbitol that are made for tableting applications are available. These have special particle size distributions that make the product free flowing and suitable for direct compression. The particle size is controlled to optimize the flow characteristics of the granulations in modern high-speed tablet presses (8).

Sorbitol is also used in pharmaceutical and nutritional tablets. The hygroscopicity of sorbitol can pose a problem to some tablet formulations because many drug substances are moisture sensitive and degrade if moisture is present. A small amount of sorbitol can be used in the tablet without greatly affecting the moisture pickup. By adding a small amount of sorbitol, a pharmaceutical formulator can make a pleasant tasting, strong tablet without jeopardizing the potency of the active drug substance. In some pharmaceutical tablets, moisture uptake is not a great problem. Sugar-free antacid tablets use sorbitol as the major inactive ingredient to bind the active ingredients effectively.

4. Confections

Sorbitol is related to sugars but has a different carbohydrate structure. Despite the chemical relationship, sorbitol modifies the crystallization of sugar by complexing. As a result, it influences the rate of crystallization, crystal size, and crystal-syrup balance in sugar-based confections.

In grained confections and similar candies, sorbitol functions as one of the doctors used to modify crystal structures to extend shelf-life. In the production of any confection,there is a point at which sucrose crystallization should take place to achieve maximum shelf-life. If crystallization takes place on either side of this "optimal point," the crystals will be either too large or too small to achieve the stability required for maximum shelf-life. Including sorbitol in the doctor system of a confection will complex the total sucrose/doctor system. This provides the confectioner a broader area in which to crystallize the confectioner and still be at the optimum point as measured by quality and shelf-life. Sorbitol also improves as-made texture and moisture retention because of its humectant properties. It is unique because, unlike other doctors, enough sorbitol can be used to extend shelf-life without adversely affecting as-made texture or taste.

5. Baked Goods

Sorbitol can be used in both sugar-free and sugar-based baked good products. In sugar-free cakes, cookies, muffins, etc., sorbitol is used to replace sucrose. It is used as a bulking agent and sweetener. When only sorbitol is used as the bulking agent, a high-intensity sweetener may be needed because sorbitol is only approximately 60% as sweet as sucrose.

Sorbitol can also be used in sugar-based baked goods to extend the product's shelf-life. Sorbitol works as a humectant, attracting moisture from the environment. It holds onto this moisture, maintaining the proper balance and not allowing the baked good to dry out.

6. Surimi

Sorbitol is used to make Surimi products such as imitation crabmeat, shrimp, and lobster. Surimi is fish (pollack) that is processed and frozen. It is then shipped to the imitation seafood manufacturers who color it and mold it into the shape of the seafood they wish to imitate. This product has enjoyed tremendous growth in the U.S. market and has always been extremely popular in the Asian markets.

Because Surimi is frozen, shipped, and subjected to freeze-thaw cycling in transit, the primary muscle protein of the fish becomes denatured, resulting in a reduction of gel-forming ability and gel strength. As deterioration of Surimi quality progresses, the Surimi becomes unsatisfactory for the production of high-quality imitation seafood products. A cryoprotectant is needed to protect the Surimi from the freeze-thaw cycle (9).

Initially, the Japanese used sucrose as the primary cryoprotectant in Surimi at levels of 6–8%. The resulting product had good freeze-thaw stability, but the sucrose had imparted an undesirable sweetness to the Surimi analog and in some cases caused a significant color change in the Surimi paste. These two problems led to the evaluation of sorbitol in blends with sucrose.

Sorbitol proved effective. It lowered the sweetness (60% that of sucrose) and improved the flavor of the Surimi analog. It served as an effective cryoprotectant because of its ability to lower the freezing point of water, and it prevented the deterioration of the Surimi's texture because of its ability to maintain proper moisture balance. Crystalline sorbitol can be used alone or in combination with sucrose. Sorbitol solution is also effective as a flavor enhancer and cryoprotectant and can be used in place of crystalline sorbitol/sucrose blends (10).

7. Cooked Sausages

Sorbitol can be used in cooked sausages to improve the flavor and method of cooking. The term "cooked sausages" is defined in the USDA Regulation 9 CFR 318.7 and includes products labeled as "franks," "frankfurters," and "knockwurst." It does not include raw sausage products such as "brown and serve" products.

When frankfurters are cooked commercially, such as at a sports stadium or convenience market, the sausage is cooked continuously and can spend a great deal of time on the rotary grill. This can lead to burning or charring of the sausage casing, which makes its appearance unappealing to the customer. It is the sugar or corn syrup used in the cooked sausage that is caramelizing and causing the charring. Sorbitol does not caramelize as do the sugars or corn syrups, so the cooked sausage will not char on the grill.

Sorbitol enhances the flavor of the cooked sausage. It improves the reddish color of the sausage and the color stability. It has also been shown to reduce the potential for white slime bacterial growth that occurs with older frankfurters (11).

The previously mentioned benefits make the cooked sausage more appealing to the consumer. The addition of sorbitol to the cooked sausage also benefits the manufacturer in that it increases the ease of peeling off the sausage casing after the sausage has been made, such as is done in the skinless frankfurter manufacturing process.

8. Shredded Coconut

The pleasant, sweet taste and hygroscopicity of sorbitol protects shredded coconut from loss of moisture. Because it is nonvolatile, sorbitol has a more permanent conditioning effect than other humectants.

9. Beverages and Liquid Products

Sorbitol has many advantages for a wide range of liquid products. It can be used in carbonated and noncarbonated beverages, nutritional drinks, and pharmaceutical liquids (antacid and antibiotic suspensions, cough syrups, etc).

Sorbitol solution acts as a bodying agent because of its viscosity of approximately 110 centipoise (cps). It improves the mouth-feel of the finished product by eliminating a watery or thin organoleptic sensation. Sorbitol enhances flavor and imparts a characteristic sweet, cool taste that is not cloying.

Sorbitol exhibits stability and chemical inertness when in contact with many chemical combinations of ingredients. Sorbitol solution can be used alone or with other liquid vehicle components, if necessary, for enhanced solvent power or special taste and mouth-feel effects. With sorbitol solution as the vehicle, almost any medicament can be used. Sensitive, insoluble active ingredients such as vitamins or antibiotics retain their potency in oral liquids containing sorbitol solution, USP. Sorbitol can also act as a cryoprotectant and therefore lowers the freezing point of water. This ability gives the formulators the means to protect liquid pharmaceutical preparations from low-temperature damage in storage or transit (12).

Sorbitol can serve as a crystallization inhibitor for liquid sugar systems. In any concentrated solution of pure solids or suspension systems, crystallization can occur. When crystallization is localized on the threads of a bottle, the cap may become difficult or impossible to remove. The condition is known as "caplocking." Sorbitol helps eliminate crystallization and its undesirable effects by developing a more complex solids system in syrups that makes them less readily crystallizable.

10. Oral Care and Toothpaste

Sorbitol is used in toothpaste and other oral care products as the primary sweetener. Sorbitol solution and noncrystallizing sorbitol solution are the primary types of sorbitol used. The role of oral care products is to prevent tooth decay. As mentioned previously, sorbitol does not lower the pH of the mouth so will not promote the growth of bacteria that can lead to tooth decay. The fact that sorbitol is only 60% as sweet as sucrose is a benefit because the oral care product will not be too sweet and cloying. The cool taste of sorbitol caused by its negative heat of solution is also helpful because it enhances the mint flavor of the mouthwash or toothpaste.

Sorbitol is also used because of its humectant properties. Sorbitol is a static humectant, meaning it picks up moisture and retains that moisture over time. It does not have the rapid moisture gain and loss cycle of other polyols, such as glycerin. For this reason, sorbitol is used by itself or in combination with other polyols to maintain the appropriate moisture balance in the toothpaste or mouthwash (Fig. 3).

Sorbitol is used as a bodying agent to give oral care products the appropriate mouth-feel. The viscosity of sorbitol solution or noncrystallizing solution gives the mouthwash or toothpaste a smooth feel without making them too thick (Fig. 5).

11. Special Dietary Foods

Flavor, palatability, and texture are often lost to a noticeable extent when foods are modified for dietary reasons. Sorbitol has been used as an adjunct in these products to make them more palatable. Its sweet, cool taste and humectant properties lend it to application in many product types. The variety of foods is almost endless and includes dried fruits, granola bars, ice cream, jams, roasted nuts, and pancake syrup in addition to the products discussed in the previous sections.

B. Mannitol Applications

1. Chocolate-Flavored Compound Coatings

Mannitol can be used in chocolate-flavored compound coatings that are used to enrobe ice creams and confections such as marshmallows and butter creams. Mannitol replaces the sucrose in this application to make a sugar-free compound

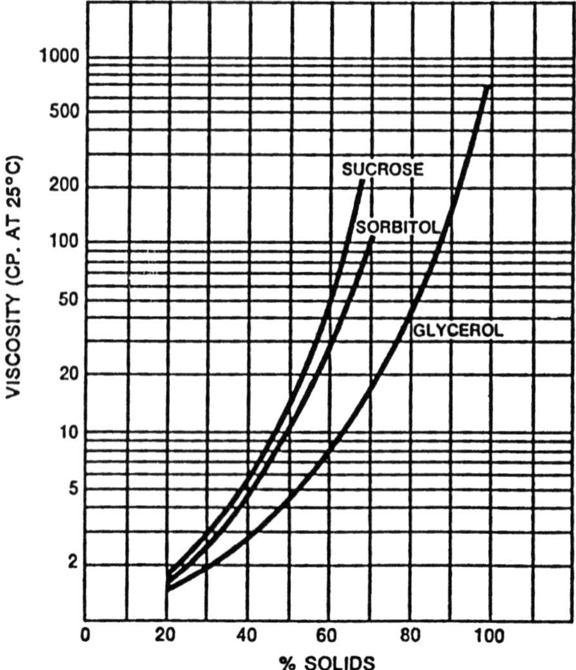

Figure 5 Viscosity of solutions of sucrose, sorbitol, and glycerol at various concentrations. (Courtesy of SPI Polyols, Inc., New Castle, Delaware.)

coating. Sugar-free chocolates are popular for consumers with dietary restrictions, such as people with diabetes. Mannitol sweetens the compound coating formulation.

Mannitol is used in compound coating manufacture because it is nonhygroscopic. When sugar-free chocolate compound coatings were first manufactured, sorbitol was used as the replacement for sucrose. Sorbitol is hygroscopic and picks up moisture, which proved to be a problem. In the early stages of manufacture, the chocolate is refined and the particle size reduced so that it will have a smooth mouth-feel in the end product. During this step, if too much moisture is introduced from the equipment or environment, the chocolate viscosity will be adversely affected. Because the viscosity is too high, the end product will require more fat than usual to achieve the appropriate texture. Fat is one of the most expensive ingredients in chocolate manufacture. The chocolate manufacturer wants to use the least amount of fat possible to make the best chocolate at the lowest cost; this is possible with mannitol. It is widely known in the industry that mannitol compound coatings perform better than other sugar-free coatings because less chocolate is required to successfully enrobe a confection.

2. Chewing Gum

In chewing gum, mannitol is used in a small percentage. It is used in two areas, both in the gum and in the dusting on top of the stick of gum. In the gum, mannitol is used as part of the plasticizer system to help maintain the soft texture of the gum. In the dusting powder, mannitol is used as the sole or primary ingredient. The surface of stick gum is dusted so that the gum does not stick to the manufacturing equipment or gum wrapper. Mannitol is used because it is nonhygroscopic. It protects the gum from picking up moisture and becoming too sticky, thus preventing it from adhering to the equipment or gum wrapper.

3. Pharmaceutical and Nutritional Tablets

Mannitol is used in tableting applications as a diluent or filler. A diluent or filler is an inactive ingredient that binds with the active drug substance and helps to compress it into a tablet. As a diluent, mannitol often makes up the bulk of the tablet imparting its properties to the rest of the granulation. Mannitol comes in two forms for tablet applications. There is powdered mannitol for wet granulation tablet manufacture and granular mannitol for direct compression tablet manufacture. The only difference between these products is their particle size distribution. The granular mannitols contain larger particles and less fine particles so they are free flowing. This is important for the direct compression method of tablet manufacture because the tablet ingredients are simply dry blended before they are introduced into the tablet press hopper.

Mannitol is most often used in chewable tablets because of its pleasant sweet taste and good mouth-feel. Mannitol is 50% as sweet as sucrose and has a cooling effect because of its negative heat of solution. These attributes lend themselves to masking the bitter tastes of vitamins and minerals, herbs, or pharmaceutical actives.

One of the major benefits of mannitol for tableting is its chemical inertness. Mannitol is one of the most stable tablet diluents available. It will not react with other tablet ingredients under any conditions. Many other tablet ingredients, such as lactose or dextrose, undergo the Maillard reaction when combined with active drug substances that contain a free amine group. This reaction causes the tablet to discolor, turning a brownish color. Mannitol does not undergo the Maillard reaction so therefore will not result in tablet discoloration. Mannitol has a high melting point and will not discolor at higher temperatures, as do some other tablet ingredients (13).

Another extremely important benefit of mannitol is that it is nonhygroscopic. Mannitol is very stable and will not pick up moisture, even under condi-

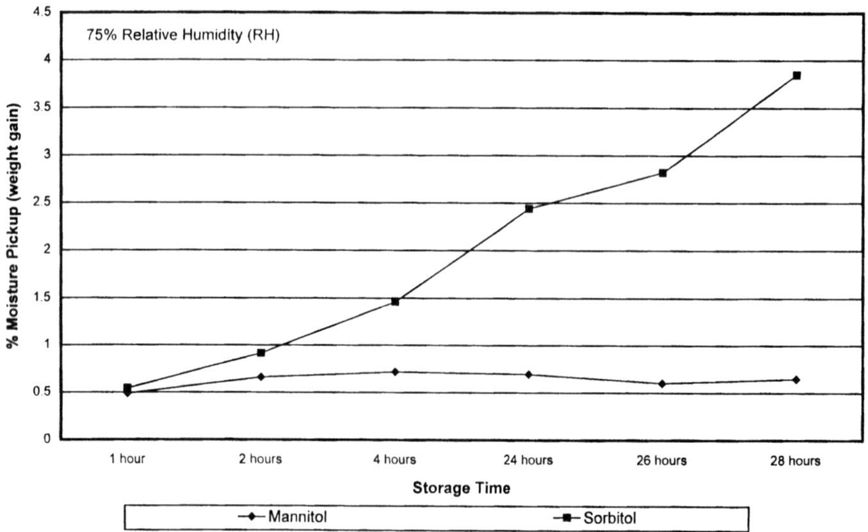

Figure 6 Moisture pickup of sorbitol and mannitol.

tions of high temperature and high humidity. Many pharmaceutical active substances are moisture sensitive and will degrade when moisture is present. Mannitol protects the active substance from moisture in the environment so that its potency will not be lost (Fig. 6).

4. Special Dietary Foods and Candy

Mannitol can be used in a variety of foods and candies, but its actual use is limited because of its price and laxation potential. It is popular in special dietary foods because of its inert nature and the fact that it is considered safe for people with diabetes. It is used as a flavor enhancer because of its sweet and pleasantly cool taste. Because of its nonhygroscopic nature, mannitol can be used to maintain the proper moisture balance in foods to increase their shelf-life and stability. It is noncariogenic and can be used in pediatric and geriatric food products because it will not contribute to tooth decay.

VI. REGULATORY STATUS

Sorbitol is generally recognized as safe (GRAS) for use as a direct additive to human food according to the FDA regulation 21 CFR 184.1835 (sorbitol) (14).

Mannitol is permitted in food on an interim basis, 21 CFR 180.25 (mannitol) (15). Sugar-free foods sweetened with sorbitol and mannitol may bear the ''does not promote tooth decay'' health claim in accordance with U.S. regulations (16).

The Joint Food and Agriculture Organization/World Health Organization Expert Committee on Food Additives (JECFA) has reviewed the safety data for both sorbitol and mannitol and determined they are safe. JECFA has established an acceptable daily intake (ADI) for sorbitol of ''not specified,'' meaning no limits are placed on its use (17). JECFA has allocated a temporary ADI of 0–50 mg/kg for mannitol (18).

Sorbitol solution 70%, noncrystallizing sorbitol solution 70%, crystalline sorbitol, and mannitol all have monographs in the United States Pharmacopoeia/National Formulary (USP/NF), as well as the various pharmacopoeias around the world (European, British, Japanese, etc.). Sorbitol and mannitol are also included in the Food Chemical Codex (FCC).

VII. SUMMARY

Sorbitol and mannitol are naturally occurring sugar alcohols found in animals and plants. They are present in small quantities in almost all vegetables. They are widely used in specialty foods and pharmaceuticals. Sorbitol and mannitol are important ingredients in sugar-free baked goods, candies, chewing gum, and tablets. The sweet, cool taste of sorbitol and mannitol makes them useful for many taste-masking or sweetening applications. Their unique moisture retention properties make them useful in improving the shelf-life of food products. The nonhygroscopic nature and chemical inertness of mannitol are attractive benefits for pharmaceutical tablets. Sorbitol and mannitol have a wide variety of uses as evidenced by the tremendous number of applications in which they are found.

REFERENCES

1. I Mellan. Polyhydric Alcohols. Washington, DC: McGregor & Werner, 1962, pp 185–202.
2. JW DuRoss. Modification of the crystalline structure of sorbitol and its effects on tableting characteristics. Pharm Technol 8:50–56, 1984.
3. JW DuRoss. Modified crystalline sorbitol. Manuf Confect 35–41, 1982.
4. B Debord, C Lefebvre, AM Guyot-Hermann, J Hubert, R Bouche, JC Guyot. Study of different crystalline forms of mannitol: comparative behaviour under compression. Drug Devl Indust Pharm 13:1533–1546, 1987.
5. DA Wadke, ATM Serajuddin, H Jacobson. Preformulation testing. In: HA Lieberman, L Lachman, JB Schwartz, ed. Pharmaceutical Dosage Forms: Tablets. Vol. 1. New York: Marcel Dekker, 1989, pp 1–69.

6. RG Allison. Dietary Sugars in Health and Disease III. Sorbitol. Bethesda, MD: Life Sciences Research Office Federation of American Societies for Experimental Biology, March 1979, pp 1–51.
7. SPI Polyols, Inc. Essential ingredients for confectionery/food, oral care, and industrial applications. Product guide. New Castle, DE: SPI Polyols, 1995.
8. RF Shangraw, JW Wallace, FM Bowers. Morphology and functionality in tablet excipients for direct compression. Pharm Technol 5:68–78, 1981.
9. CM Lee. Surimi process technology. Food Technol 38:69, 1984.
10. KS Yoon, CM Lee. Assessment of the cryoprotectability of liquid sorbitol for its application in surimi and extruded products. The 31st AFTC Meeting, Halifax, Nova Scotia, 1986.
11. AS Geisler, GC Papalexis. Method of preparing a frankfurter product and compositon for use therein. US Patent no. 3,561,978, February 9, 1971.
12. ICI Americas, Inc. ICI Americas products for cosmetics and pharmaceuticals. Applications brochure. Wilmington, DE: ICI Americas, 1977.
13. NA Armstrong, GE Reier, DA Wadke. Mannitol. In: A Wade, PJ Weller, eds. Handbook of Pharmaceutical Excipients. Washington, DC: American Pharmaceutical Association, 1994, pp 294–298.
14. Office of the Federal Register, General Services Administration. Code of Federal Regulations, Title 21, Section 184.1835. Washington, DC, US Government Printing Office, 1999.
15. Office of the Federal Register, General Services Administration. Code of Federal Regulations, Title 21, Section 180.25. Washington, DC, US Government Printing Office, 1999.
16. Office of the Federal Register, General Services Administration. Code of Federal Regulations, Title 21, Section 101.80. Washington, DC, US Government Printing Office, 1999.
17. Joint FAO/WHO Expert Committee on Food Additives. Toxicological evaluation of certain food additives: sorbitol. Twenty-sixth report. WHO Technical Report Series 683, p 27. Geneva, 1982.
18. Joint FAO/WHO Expert Committee on Food Additives. Toxicological evaluation of certain food additives: mannitol. Twenty-ninth report. WHO Technical Report Series 733, p 35. Geneva, 1982.

19
Xylitol

Philip M. Olinger
Polyol Innovations, Inc., Reno, Nevada

Tammy Pepper
Danisco Sweeteners, Redhill, Surrey, United Kingdom

I. INTRODUCTION

Xylitol is a five-carbon polyol with a sweetness similar to sucrose (1,2). It is found in small amounts in a variety of fruits and vegetables (3,4) and is formed as a normal intermediate in the human body during glucose metabolism (5). Xylitol has been shown to be of value in the prevention of dental caries because it is not an effective substrate for plaque bacteria (6–10,23,24). Because of its largely insulin-independent metabolic utilization, it may also be used as a sweetener in the diabetic diet and as an energy source in parenteral nutrition (11–13). As a sweetening agent, xylitol has been used in human food since the 1960s. Different aspects of its applications, metabolic properties, and safety evaluation have been the subject of several previous reviews (14–20). Its physicochemical properties are summarized in Table 1.

II. PRODUCTION

Xylitol was first synthesized and described in 1891 by Emil Fischer and his associate (21). On a commercial scale it is produced by chemical conversion of xylan (22,25). Potential sources of xylan include birch wood and other hardwoods, almond shells, straw, corn cobs, and by-products from the paper and pulp industries. The practical suitability of these raw materials depends, among other factors, on their xylan content, which may vary considerably, and the presence of

Table 1 Some Physical and Chemical Properties of Xylitol

Formula	$C_5H_{12}O_5$ (molecular weight, 152.15)
Appearance	White, crystalline powder
Odor	None
Specific rotation	Optically inactive
Melting range	92–96°C
Boiling point	216°C (760 mmHg)
Solubility at 20°C	169 g/100 g H_2O; 63 g/100 g solution; sparingly soluble in ethanol and methanol
pH in water (100 g/liter)	5–7
Density of solution	10%, 1.03 g/ml; 60%, 1.23 g/ml
Viscosity (20°C)	10%, 1.23 cP; 60%, 20.63 cP
Heat of solution	+34.8 cal/g (endothermic)
Heat of combustion	16.96 kJ/g
Refractive index (25°C)	10%, 1.3471; 50%, 1.4132
Moisture absorption (4 days, RT)	60% RH, 0.051% H_2O, 92% RH, 90% H_2O
Relative sweetness	Equal to sucrose; greater than sorbitol and mannitol
Stability	Stable at 120°C, no caramelization; stable also under usual conditions in food processing; carmelization occurs if heated for several minutes near the boiling point

cP = centipoise.
RH = relative humidity.
RT = room temperature.

by-products (e.g., poly- or oligosaccharides), which have to be removed during the production process. In principle, the commercial synthesis of xylitol involves four steps. The first step is the disintegration of natural xylan-rich material and the hydrolysis of the recovered xylan to xylose. The second step is the isolation of xylose from the hydrolysate by means of chromatographic processes to yield a pure xylose solution. Third, xylose is hydrogenated to xylitol in the presence of a nickel catalyst. Alternatively, hydrogenation of the impure xylose solution may be conducted first, followed by purification of the xylitol syrup. Ultimately, xylitol is crystallized in orthorhombic form (26).

The synthesis of xylitol by fermentativeor enzymatic processes is possible in principle (27–30). However, such procedures have so far not been used on a commercial scale. Other approaches to the synthesis of xylitol are of merely scientific interest (31,32).

Xylitol is supplied commercially in crystalline, milled, and granulated forms. Crystalline xylitol has a mean particle size of about 400 to 600 microns.

Milled forms of xylitol range in mean particle sizes from approximately 50 to 200 microns. The granulated forms of xylitol are suitable for direct compression.

III. PROPERTIES AND APPLICATIONS

At present, xylitol is used as a sweetener mainly in noncariogenic confectionery (chewing gum, candies, chocolates, gumdrops), and less frequently in dietetic foods (food products for people with diabetes), in pharmaceutical preparations (tablets, throat lozenges, multivitamin tablets, cough syrup), and in cosmetics (toothpaste and mouthwash) (33–39). Xylitol is used at low levels in selected low-calorie soft drink applications to improve product mouthfeel and sweetness profile. In principle, the manufacture of various baked goods with xylitol is possible. However, if crust formation, caramelization, or nonenzymatic browning is required, the addition of a reducing sugar is necessary. Because xylitol inhibits the growth and fermentative activity of yeast, it is not a suitable sweetener for products containing yeast as a leavening agent (40,41).

A. Sweetness

Xylitol is the sweetest polyol (Fig. 1) (42–46). At 10% solids (w/w) xylitol is isosweet to sucrose, whereas at 20% solids (w/w) xylitol is about 20% sweeter

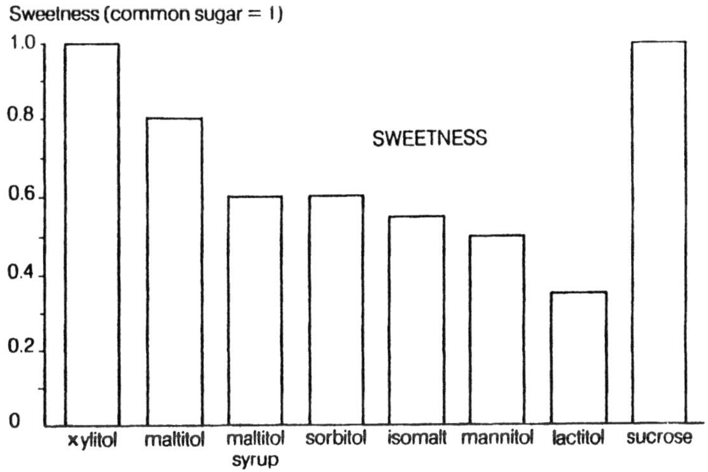

Figure 1 Relative sweetness of polyols.

than sucrose (2). Combinations of xylitol and other polyols, such as sorbitol, create significant sweetness synergisms.

B. The Cooling Effect

The heats of solution of crystalline polyols and sucrose are shown in Fig. 2. The loss of heat when dissolving polyols in water is much greater than with sucrose. Crystalline xylitol provides a significant cooling effect. This interesting organoleptic property is most notable in sugar-free chewing gum, compressed candies, and chewable vitamins. The cooling effect enhances mint flavor perception, and the presence of xylitol contributes a refreshing coolness.

A cooling effect is obviously not perceived from products in which xylitol is already dissolved (e.g., toothpaste, mouth rinse) or in which it exists in an amorphous form (jellies; boiled, transparent candies).

C. Solubility of Polyols

The solubility of a bulk sweetener has a critical impact on the mouthfeel and texture of the final product. Bulk sweetener solubility also affects the release or onset of flavor and sweetness perception and the release and bioavailability of the active ingredients of pharmaceuticals. Table 2 shows the solubility of selected polyols at 20°C. The solubility of xylitol is 2% higher than sucrose at body temperature.

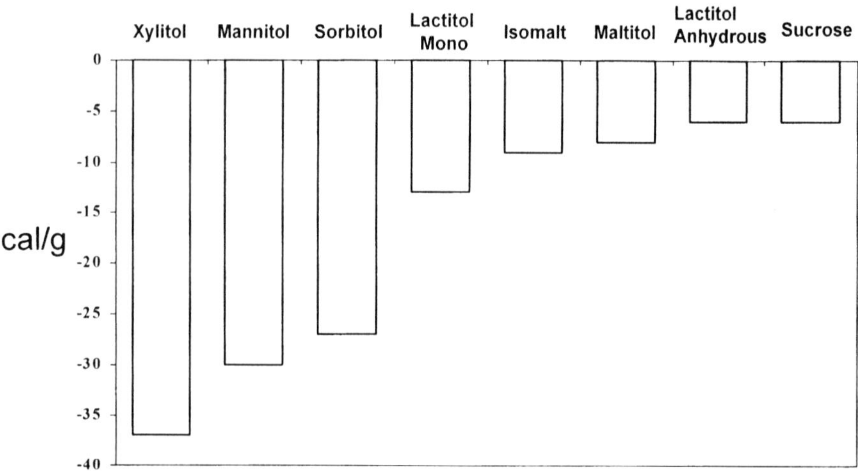

Figure 2 Cooling effect of polyols.

Table 2 Solubility of Sugar Alcohols in Water at 20°C

Sweetener	Solubility (% w/w)
Sucrose	66
Xylitol	63
Sorbitol	75
Maltitol	62
Lactitol	55
Isomalt	28
Mannitol	18

D. Other Characteristics

The viscosity of polyol solutions depends on their molecular weight. Monosaccharide alcohols such as sorbitol and xylitol have a relatively low viscosity, whereas maltitol syrups with a high content of hydrogenated oligosaccharides are fairly viscous.

Xylitol is chemically quite inert because of the lack of an active carbonyl group. It cannot participate in browning reactions. This means that there is no caramelization during heating, as is typical of sugars. Because xylitol does not form Maillard reactants, it is ideally suited to use as an excipient in conjunction with amino active ingredients.

E. The Use of Xylitol in Chewing Gum and Confectionery

Sugar-free chewing gum is the leading worldwide application of xylitol. Xylitol is used to sweeten both stick and pellet (dragee) forms of chewing gum. In addition to the dental benefits attributed to it, xylitol provides a pleasant cooling effect and a rapid onset of both sweetness and flavor. Because of its rapid drying and crystallization character, xylitol is often used to coat pellet forms of sugar-free chewing gum.

Sugar-free chewing gums usually contain mixtures of polyols and intense sweeteners to ensure a satisfactory sweetness profile over time. Although a few brands are sweetened solely with xylitol, most products contain less than 50% xylitol. The use of xylitol-containing chewing gum in either stick or pellet form as a between-meal supplement has been associated with improvements in oral health. Among the reported benefits are reductions in plaque levels, plaque bacteria levels, and caries increment (47–51).

Xylitol is commercially used in many countries either alone or in combination with other sugar substitutes in the manufacture of noncariogenic sugar-free or no sugar added confectionery. Although it is in principle possible to produce gum arabic pastilles, chocolate, hard candies, ice cream fillings, and other confections with xylitol alone, other sugar substitutes are normally added to optimize sweetness, texture, or shelf-life. Sugar-free pectin jellies, which are indistinguishable from conventional sugar/corn syrup jellies, can be produced with a combination of xylitol and hydrogenated starch hydrolysates. Crystalline xylitol, which has an appearance similar to sucrose, is an excellent sugar-free sanding material in conjunction with pectin jellies and other confection forms.

Xylitol can be used either with sorbitol or hydrogenated starch hydrolysate to produce an exceptional sugar-free fondant. In each case, xylitol is applied as the crystalline phase in the production. The resulting fondants exhibit a fine texture and a pleasant cooling effect.

IV. PHARMACEUTICAL AND ORAL HYGIENE PREPARATIONS

Xylitol can be used as an excipient or as a sweetener in many pharmaceutical preparations. As in foods, the advantages are suitability for diabetic patients, noncariogenic properties, and nonfermentability. Cough syrups, tonics, and vitamin preparations made with xylitol can neither ferment nor mold. Because xylitol is chemically inert, it does not undergo Maillard reactions or react with other excipients or active ingredients of pharmaceuticals. Xylitol-sweetened medications can be given to children at night after toothbrushing without any harm to the teeth (38,52–56).

In tablets, xylitol can be used as a carrier and/or as a sweetener (57,58). In addition to sweetness, nonreactivity, and microbial stability, xylitol offers the advantages of high solubility at body temperature and a pleasing, cooling effect. As a carrier, xylitol has also been tested in solid dispersions because of its low melting point and stability up to 180°C. In these studies, it was found that solid dispersions of hydrochlorothiazide or *p*-aminobenzoic acid with xylitol showed a faster release than the micronized drugs (59,60). Milled xylitol may be granulated and pressed into tablets after flavoring (50). Alternatively, directly compressible grades of xylitol can be used to facilitate tablet production (61,62). Coatings with xylitol or mixtures of xylitol and sorbitol (up to 20%) can be made by conventional pan coating and by sintering the surface of compressed tablets in a hot-air stream (63).

In toothpaste, xylitol may partially or completely replace sorbitol as a humectant. Because of its greater sweetness, xylitol improves the taste of the dentifrice, and in the manufacture of transparent gels, it has been said to exhibit properties slightly superior to those of sorbitol. Furthermore, xylitol can enhance the anticaries effect of a fluoride toothpaste. For example, the inclusion of 10% xylitol in a fluoride toothpaste resulted in a 12% reduction in decayed/filled surfaces after 3 years compared with a fluoride-only toothpaste (64). In addition, there is evidence that xylitol exerts a plaque-reducing effect (65,66) and that it interferes with bacterial metabolism particularly in the presence of fluoride and zinc ions (67). An inhibitory effect on enamel demineralization has been postulated as well (68,69). There is also evidence that use of a xylitol-containing dentifice can result in a significant reduction of *Streptococcus mutans* in saliva (70). Because of its overall favorable effects on dental health, xylitol has also been applied in other oral care products, such as in mouth rinses and in artificial saliva (71–74).

V. METABOLISM

In principle, two different metabolic pathways are available for the use of xylitol: (a) direct metabolism of absorbed xylitol in the mammalian organism, mainly in the liver, or (b) indirect metabolism by means of fermentative degradation of unabsorbed xylitol by the intestinal flora.

A. Indirect Metabolic Utilization

All polyols, including xylitol, are slowly absorbed from the digestive tract because their transport through the intestinal mucosa is not facilitated by a specific transport system. Therefore, after ingestion of large amounts, only a certain proportion of the ingested xylitol is absorbed and enters the hepatic metabolic system through the portal vein blood. A comparatively larger amount of the ingested xylitol reaches the distal parts of the gut, where extensive fermentation by the intestinal flora takes place. Besides minor amounts of gas (H_2, CH_4, CO_2), the end-products of the bacterial metabolism of xylitol are mainly short-chain, volatile fatty acids, (i.e., acetate, propionate, and butyrate) (75–77). These products are subsequently absorbed from the gut and enter the mammalian metabolic pathways (78). Acetate and butyrate are efficiently taken up by the liver and used in mitochondria for production of acetyl-CoA. Propionate is also almost quantitatively removed by the liver and yields propionyl-CoA (79–81).

The production of volatile fatty acids (VFA) is a normal process associated with the consumption of polyols and dietary fibers (cellulose, hemicelluloses, pectins, gums) for which hydrolyzing enzymes are lacking or poorly efficient in the small intestine. Under normal conditions, most of the generated VFA are absorbed from the gut and are further used by established metabolic pathways in animals and man (77,82). For the energetic use of slowly digestible materials, this fermentative, indirect route of metabolism plays an important role, and evidence has been presented that, even under normal dietary conditions (i.e., in the absence of polyols), the contribution of intestinal fermentations to the overall energy balance is significant (83,84).

B. Direct Metabolism Via the Glucuronic Acid–Pentose Phosphate Shunt

In addition to the indirect route of use, a direct metabolic pathway is available for the portion of xylitol that is absorbed unchanged from the gastrointestinal tract (5,85,86).

The metabolism of xylitol and its general relationship to the carbohydrate metabolism by means of the pentose phosphate pathway is shown in Fig. 3. This scheme illustrates how the transformation of L-xylulose by way of xylitol to D-xylulose links the oxidative branch of the glucuronate pentose phosphate pathway with the nonoxidative branch, which yields glyceraldehyde-3-phosphate and fructose-6-phophate, as well as ribose-5-phosphate for ribonucleotide biosynthesis (87–89). Thus, xylitol can be converted by means of the pentose phosphate pathway to intermediates of the glycolytic pathway, which may either undergo further degradation or transformation to glucose-1-phosphate, a precursor of glycogen (5,85–87). Because gluconeogenesis from exogenous xylitol is associated with the generation of NADH in the cytosol, reoxidation of cytosolic NADH is a necessary step for xylitol use.

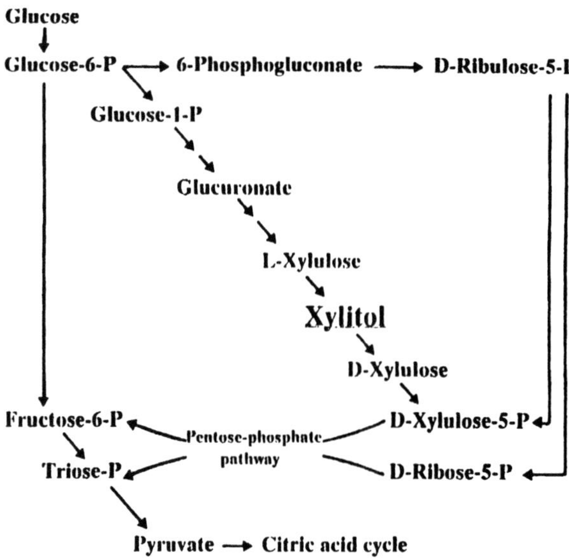

Figure 3 Metabolism of xylitol.

The recognition of xylitol as a normal endogenous metabolite in humans originates from observations in patients with a genetic abnormality, essential pentosuria. These persons excrete considerable quantities of L-xylulose in the urine. When incubated in tissue preparations, L-xylulose was found to be metabolized to xylitol by a specific NADPH-linked enzyme, L-xylulose-reductase. This enzyme has subsequently been demonstrated to be deficient in essential pentosuria. Because pentosuric patients excrete 2–15 g of L-xylulose per day, the daily production of xylitol was estimated to be of a similar order of magnitude (5,90,91).

C. Estimation of the Caloric Value of Xylitol

If the caloric value of polyols is to be estimated on the basis of their metabolic fate, precise knowledge about the different steps of their digestion and metabolic use is required. In particular, it is, for example, essential to know (a) how much of an ingested dose is absorbed directly from the gut, (b) by which pathways the absorbed portion is metabolized in the human organism, (c) which proportion of an ingested dose is fermented by the gut microflora, (d) to what extent the resulting fermentation products are absorbed and metabolically used by the host, and (e) how much of an ingested dose leaves the intestinal tract unchanged with the feces or urine. Although the experimental database on xylitol does not allow one to answer all these questions with sufficient precision, it is possible to obtain reasonable estimates by extrapolation from existing data and studies on other polyols that are also incompletely absorbed and are subject to the same fermentative degradation in the gut.

On the basis of a thorough assessment of numerous in vitro and in vivo experiments with xylitol and other polyols, it has been estimated that approximately one fourth of an ingested xylitol dose is absorbed from the gastrointestinal tract (92). This portion of xylitol, which is effectively metabolized by means of the glucuronate-pentose phosphate shunt, is energetically fully available and provides about 4 kcal/g. The nonabsorbed three-fourths of the ingested load, however, are almost completely fermented by the intestinal flora. Combining a 78% retention of energy in bacterial fermentation products and a growth yield of 20 g bacteria per 100 g substrate, it has been estimated that about 42% of the energy provided with unabsorbed xylitol is consumed by bacterial metabolism and growth, whereas about 58% of the energy becomes available to the host after absorption of the fermentation end-products (VFA) (93). This value is well in line with an estimated 50% energy salvage proposed by a Dutch expert group (94).

On the basis of these estimates, a metabolizable energy value of about 2.8–2.9 kcal/g may be calculated for xylitol (92). This value is in line with the results of an in vivo study in which xylitol was found to be about 60% as effective as glucose in promoting growth (95).

Besides laborious and expensive energy balance trials, only indirect calorimetry allows one to determine the energetic use of substrates in humans. By use of this noninvasive technique, the energetic use of xylitol was examined in 10 healthy volunteers (96). The results of this study revealed that, during a 2.5-hour period, the cumulated increase in carbohydrate oxidation amounted to only one fourth of that caused by glucose. The total increase in metabolic rate was 52% lower after the xylitol load than after the glucose treatment. This result suggests that xylitol has a caloric value of only about 50% that of glucose. Considering, however, that the absorption (and hence metabolic use) of the bacterial fermentation products may not have been complete within the 2.5-hour experimental period, it is likely that the true caloric value for xylitol is somewhat higher than the proposed 2 kcal/g. The net energy value of 2.4 kcal/g has been assigned to xylitol by FASEB. The net energy is defined as that portion of gross energy intake that is deposited or mobilized in the body to do physical, mental, and metabolic work (92).

Food regulatory authorities are increasingly taking note of the reduced caloric value of polyols, which is by now well supported by a still growing volume of scientific literature. In the EU, a caloric value of 2.4 kcal/g has been allocated for all polyols including xylitol (Council Directive No. 90/496/EEC of 24 September 1990 on nutrition labelling for foodstuffs (Off. J. European Communities 1990, 33 (L276), 42–46). On the basis of the 1994 FASEB review, the Food and Drug Administration acknowledged the xylitol caloric value of 2.4 kcal/g (letter to American Xyrofin Inc from FDA concerning use of a self-determinated energy-value for xylitol; 1994).

VI. DENTAL BENEFITS

A. Caries Formation

According to current knowledge, dental caries is caused by bacteria that accumulate in large masses, known as dental plaque, on the teeth in the absence of adequate oral hygiene. Fermentation of common dietary carbohydrates by plaque bacteria leads to the formation of acid end-products. Acid accumulation and a decrease in plaque pH will follow. A decrease in plaque pH caused by bacterial fermentation of carbohydrates may lead to undersaturation of the plaque with respect to calcium and phosphate ions, to demineralization of the tooth enamel, and eventually to formation of a cavity.

Approaches aimed at the prevention or elimination of dental caries include reduction of acid dissolution of tooth mineral and stimulation of enamel remineralization by fluoride, removal of dental plaque by brushing the teeth, and reduction of the availability of fermentable carbohydrates by appropriate dietary habits (97).

For obvious reasons, fermentable sugars cannot be completely eliminated from our daily food supply. From the point of view of caries prevention, this is not even necessary, because a number of studies have indicated that eating sucrose-sweetened foods in moderation and with regular meals results in little or no caries (98). On the other hand, experiments with rats, as well as some studies in humans, indicate a strong relationship between the frequency of consumption of sucrose between meals and caries activity (99,100). The reason for this observation is that under conditions of frequent consumption of sugary snacks, the plaque pH drops and remains below the critical value of 5.7 for prolonged periods of time (101). As a result, periods with demineralization overwhelm the recovery phases, and dental caries may subsequently develop. This suggest that substitution of sucrose and other fermentable sugars by nonfermentable sugar substitutes has considerable impact on caries activity, even if this substitution is limited to snacks and beverages that are consumed between the main meals. Because of its noncariogenic or even cariostatic properties, xylitol is a particularly suitable substance for this purpose (51,102,103).

B. Acidogenicity

Considerable evidence documents the fact that xylitol is not fermented by most oral microorganisms and that exposure of dental plaque to xylitol does not result in a reduction of plaque pH (6,104–109). Even after chronic exposure for 2 years, no adaptation occurred with respect to the ability of the dental plaque to ferment xylitol (110,111).

The *noncariogenicity* of xylitol was demonstrated in several rat caries studies in which xylitol was fed in the absence of other readily fermentable dietary sugars. Under these conditions, xylitol did not exhibit any cariogenic potential (6,112). The *cariostatic properties* of xylitol were investigated in numerous studies in which rats received xylitol in admixture to a cariogenic, (i.e., sucrose-containing diet). The results of these studies indicate that xylitol tends to reduce the cariogenicity of such diets and that this effect is greater for moderately cariogenic diets with ≤25% fermentable sugars than for diets containing ≥50% sucrose (6,113–115). More pronounced and more consistent reductions of caries formation were observed in rat studies in which xylitol was administered alternately with sucrose-containing meals (i.e., between the cariogenic challenges) (6,116–118). Under these conditions, which mimic the human meal pattern more closely, the caries scores of the xylitol-treated animals were on average 35% less than those of the corresponding controls. In one study it was even found that xylitol may promote the remineralization of early caries lesions of rats (118,119).

In summary, the results obtained in the rat model suggest that xylitol exhibits cariostatic properties.

C. Efficacy of Xylitol in Human Caries Studies

The clinical benefit of xylitol for caries prevention has been demonstrated in several studies with children and adult human volunteers in which consumption of xylitol was consistently associated with a significant reduction of the caries increment. In these studies, which were conducted by independent investigators at different locations, the efficacy of xylitol was tested under carefully controlled conditions with randomized assignment of the subjects to the different treatment groups (120–124) and under the less standardized but ''real-life''–oriented conditions of field studies involving large numbers of caries-prone children (125–127). Regardless of the differences in study design, significant reductions of the caries increment were observed in all xylitol-treated groups. Depending on the dose (128) and frequency of the xylitol administration (129), this reduction varied between 45 and >90%. The dental benefits attributed to xylitol have been shown to extend beyond the study period. When subjects from a Finnish chewing gum study (129) were evaluated up to 5 years after the initial study, subjects from the xylitol group continued to demonstrate a reduction in caries rate compared with the control group (130).

In a further study involving 10-year-old children, xylitol, xylitol/sorbitol, and sorbitol-sweetened gums were evaluated in terms of caries risk and their impact on remineralization. Although each chewing gum group exhibited a reduction in caries risk and an increase in remineralization, the xylitol group had a significantly lower caries risk and higher remineralization than the other gum groups (131–133). The results support the view that xylitol has an active anticaries effect above that of a mere sugar substitute (51).

It is particularly noteworthy that daily fluoride brushing plus xylitol consumption was found to be more efficient than daily fluoride brushing alone (125–128). Considering the different modes of the cariostatic action of fluoride and xylitol, this is not surprising, but it is certainly relevant in terms of a most efficient combination of caries preventive measures (102).

D. Discrimination Against *Streptococcus mutans*

Microbiological investigations in caries-active humans, as well as studies in experimental animals, have demonstrated that certain types of bacteria are particularly active in initiating and promoting dental caries. Culprit number one in this respect is *Streptococcus mutans*. This micro-organism has the ability to adhere strongly to the crowns of the teeth by means of sticky extracellular glucans, to grow and metabolize optimally in a relatively acidic environment, and to live under microaerobic or strictly anaerobic conditions, as met in the depth of pits and fissures (134–137).

Several studies have suggested that the predominance of *S. mutans* in cariogenic plaque depends on the ability of this bacterium to remain metabolically active even in a relatively acidic environment. In fact, *S. mutans* is more active at pH 5 than at pH 7, whereas many other members of the plaque flora become metabolically inactive under such conditions. This observation suggests that the frequent ingestion of surcrose gives a competitive edge to *S. mutans* over the other plaque microbes. In this way, a circulus vitiosus for the formation of a more cariogenic plaque is formed (138–140).

In this regard, the use of xylitol as a sucrose substitute becomes an extremely attractive means to control and prevent dental caries for two different reasons. First, no acid is formed from xylitol by the dental plaque. In fact, during and after chewing of a xylitol-sweetened gum, an elevation rather than a decrease of the plaque pH is observed (141,142). Under such conditions, however, the metabolism of *S. mutans* is, as mentioned previously, not optimally active, and other bacteria may successfully compete with *S. mutans*. Second, in addition to this indirect, pH-mediated effect, xylitol appears to inhibit the growth and metabolism of *S. mutans* in a more direct way. Several experiments have shown that the addition of xylitol to a glucose-containing medium reduced the growth of *S. mutans* (143–150). This inhibition by xylitol appears to be related to the accumulation of xylitol-5-phosphate and xylulose-5-phosphate within the cells (151–156). Probably as a result of the intracellular accumulation of these metabolites, the ability of *S. mutans* to adhere to surfaces is decreased (157,158), and disintegration of the ultrastructure of the cells may occur (159).

In line with these results, lower *S. mutans* counts were found in the plaque and/or saliva of xylitol-treated human volunteers (160–166). These observations suggest that the formation of xylitol-tolerant strains of *S. mutans* after chronic xylitol exposure does not annihilate its inhibitory effect (167,168).

E. General Effects on Dental Plaque

Because dental plaque plays a crucial role in the formation of tooth decay, many investigators have examined the effects of xylitol on dental plaque. It is well established that frequent consumption of sucrose promotes the growth of a voluminous, sticky dental plaque. Xylitol is not a substrate for oral microorganisms, and it therefore does not favor plaque formation. Consequently, studies comparing the effects of sucrose and xylitol on dental plaque have demonstrated consistently that the plaque weights are lower in the xylitol-treated subjects than in the positive controls consuming sucrose (6). Studies suggest that xylitol inhibits the formation of insoluble glucans and lipoteichoic acid, two products of bacterial metabolism that play an important role in the adhesion and cohesion of dental plaque (143,157).

Chewing gums sweetened with xylitol or sorbitol have been investigated regarding their impact on the formation of plaque. The results of these studies suggest that chewing gums sweetened with xylitol have a greater impact on plaque reduction (169,170).

F. Dental Endorsements and Claims

The dental benefits provided by xylitol have been recognized by a number of dental associations.

VII. EVOLVING APPLICATIONS

New, beneficial and potentially existing applications of xylitol continue to be either discovered or suggested. These new applications include the following.

1. The use of xylitol as a bulk satiety sweetener
2. The use of xylitol to reduce the potential occurrence of acute otitis media

The effects of xylitol on gastric emptying and food intake when measured in 10 healthy male volunteers suggests that ingestion of a 25 g xylitol preload may be associated with an approximate 25% reduction of caloric intake during a subsequent test meal (171). Other bulk sweeteners did not suppress caloric intake when tested at the same 25 g level. The study suggests that xylitol could be applied at efficacious levels in meal replacement or diet-supporting products to facilitate weight reduction programs.

Acute otitis media, an infection of the middle ear, is a common illness that affects a significant number of young children. During 1990, for example, an estimated 24.5 million visits were made to office-based physicians in the United States at which the principal diagnosis was otitis media. A study of the effect of xylitol ingestion on acute otitis media with either a xylitol-sweetened chewing gum, syrup, or lozenge, suggests that each xylitol-sweetened material has a positive influence in the reduction of acute otitis media occurrence. Reductions of 40%, 30%, and 20% were reported for the chewing gum, syrup, and lozenge, respectively (172,173). The observed benefit is believed to be associated with the ability of xylitol to reduce the growth of *Streptococcus pneumonia* and thus minimize the attacks of acute otitis media caused by pneumococci (174).

VIII. USE OF XYLITOL AS A SWEETENER IN DIABETIC DIETS

Historically, the first proposed application of xylitol concerned its use as a sugar substitute for diabetic patients (175). Diabetes mellitus is a chronic metabolic disorder characterized by fasting hyperglycemia and/or plasma glucose levels above defined limits during oral glucose tolerance testing. It is caused either by a total lack of insulin (type I, insulin-dependent diabetes mellitus) or by insulin resistance in the presence of normal or even elevated plasma insulin levels (i.e., by a decreased tissue sensitivity or responsiveness to insulin [type II, non-insulin dependent diabetes mellitus]).

The major goals of dietetic and drug-based management of diabetes mellitus are to achieve normal control of glucose metabolism and glycemia, and thereby to prevent macro- and microvascular complications. The recommended treatment modalities are dietary modification, increased physical activity, and pharmacological intervention with either an oral hypoglycemic agent or insulin.

Modification of the diet is the most important element in the therapeutic plan for diabetic patients, and for some patients with type II diabetes, it is the only intervention needed to control the metabolic abnormalities associated with the disease.

A specific goal of medical nutrition therapy for people with diabetes is the maintenance of as near-normal blood glucose levels as possible. This includes balancing food intake with either endogenous or exogenous or oral glucose-lowering medications and physical activity levels. It is the current position of the American Diabetes Association that first priority be given to the total amount of carbohydrates consumed rather than the source of the carbohydrate. According to ADA, the calories and carbohydrate content of all nutritive sweeteners must be taken into account in a meal plan and that all have the potential to affect blood glucose levels. ADA recognizes, however, that polyols produce a lower glycemic response than sucrose and other carbohydrates and have approximately 2 calories per gram compared with 4 calories per gram from other carbohydrates (176).

Even an "ideal" diet plan is worthless if patients do not adhere to it. To increase compliance, it seems therefore appropriate to use low glycemic or nonglycemic sweeteners for the preparation of special diabetic products. Traditionally, one may consider using nonnutritive, intense sweeteners for this purpose. These compounds are noncaloric and have no deleterious effect on diabetic control. Undoubtedly, they are most useful for the sweetening of beverages (soft drinks, coffee, tea). However, incorporation of nonnutritive sweeteners into solid foods causes a major technological problem. In normal products, sucrose represents a considerable portion of bulk, and its replacement by nonnutritive sweeteners in special products for diabetic patients requires the addition of bulking agents (i.e., typically fat and/or starch). Because diabetic patients are predisposed to macrovascular disease and are required to restrict their fat intake, products with increased fat content, although carbohydrate-modified, are not recommended. The addition of starch, on the other hand, increases the glycemic index of the food and eliminates the glycemic and caloric advantage that one hoped to achieve by using an intense sweetener. To avoid these problems, bulk sugar substitutes like fructose, sorbitol, and xylitol may be used as an appropriate alternative. These sweeteners produce only a slight increase in blood glucose concentration and require only small amounts of insulin for their metabolism in both healthy and diabetic individuals.

The effects of xylitol on blood glucose and insulin levels and its general suitability for inclusion in foods for diabetic patients have been examined in several acute and subchronic studies with healthy and diabetic volunteers. When xylitol is given orally, no increase in blood glucose levels is observed even in diabetic patients (96,177–179). Similarly, plasma insulin concentrations do not rise at all (177) or only moderately (96,180) after oral xylitol application in normal and diabetic subjects. These observations indicate that the conversion of xylitol to glucose—which, in principle, is possible—is apparently too slow to raise the blood glucose concentration to a significant extent.

In a recent study involving eight healthy nonobese men, it was observed that ingestion of xylitol caused significantly lower increases in plasma glucose and insulin concentrations compared with the ingestion of glucose. The glycemic index of xylitol was determined to be 7 (181).

It has been suspected that the obvious advantage of xylitol in terms of blood glucose and insulin requirement may disappear when it is incorporated into a meal. Therefore, a study was conducted in which 30 g xylitol or sucrose was substituted for an equal amount of starch in a meal of a diabetic diet regimen. The results of this investigation demonstrate that the insulin requirement after sucrose is significantly higher than after starch or xylitol (182).

In an early subchronic study, the application of 45–60 g/day of xylitol had no adverse effects on the metabolic condition of 20 diabetic patients (175). These results were confirmed in a subsequent study in which the effect of 30 g xylitol/day on the carbohydrate and lipid metabolism of 12 well-controlled diabetic patients (type II) was examined for a duration of 2–6 weeks (183). The urinary glucose excretion disappeared with the xylitol diet at least in some of the participating patients. The good tolerance of xylitol was also established in a study with 18 diabetic children (type I) who received 30 g/day of xylitol for 4 weeks (184).

Even at a dose of 70 g/day administered over a period of 6 weeks, xylitol was well tolerated by type I and type II diabetic individuals and by healthy nondiabetic controls. Contrary to expectations, however, no significant differences were noted in plasma glycosylated Hb, in fasting or postprandial glucose, and in total urinary glucose (185). In conclusion, these investigations demonstrate that xylitol can safely be incorporated as a sweetening agent in the diabetic diet without any negative effects on their metabolic condition.

IX. USE OF XYLITOL IN PARENTERAL NUTRITION

The aim of parenteral nutrition is to optimize fluid, energy, nitrogen, and electrolyte balance by the infusion of adequate solutions providing energy (e.g., glucose or lipids), amino acids, and mineral salts (186,187). However, a stable metabolic condition is often difficult to reestablish in severely injured patients because the posttrauma metabolism may be seriously disturbed in various respects. In particular, the use of glucose, which is the body's normal energy source, is often impaired, especially in patients with shock, severe trauma, burns, sepsis, and diabetes.

Under the conditions of such a general disturbance of metabolism and of the hormone-regulated control mechanisms, insulin must be administered concomitantly with the glucose to ensure sufficient use of this energy source. However, this procedure requires careful regular monitoring of the blood glucose levels to avoid severe complications. In addition, the infusion of glucose in excess of 0.12–0.24 g/kg/hr may be generally questioned because higher infusion rates lead neither to a further suppression of the endogenous gluconeogenesis at the expense of physiologically important proteins nor to an increase of the peripheral glucose use (188). Because with this recommended maximum rate of glucose infusion, the energy requirement of the injured or stressed organism is not completely covered, the infusion of fructose and xylitol in addition to glucose has been advocated (12,13,189,190). In several studies, the positive influence of such infusion regimens on nitrogen balance and visceral protein synthesis has been demonstrated in stressed or post-trauma patients (13,189), as well as in an animal model (190). However, these effects were not always reproducible (191,192).

More recently, it has been proposed to supplement lipid emulsions with xylitol. In this combination, xylitol is preferred over glucose because it only marginally stimulates insulin secretion and therefore does not suppress lipolysis, as demonstrated by a positive adenosine triphosphate (ATP) balance. In this way, xylitol could improve the metabolic situation (193,194).

At present, xylitol is used mainly as a glucose substitute in parenteral amino acid solutions, as well as in combination with fructose and glucose in so-called trisugar solutions. These products are frequently applied, particularly in Germany. Solutions of xylitol alone are nowadays rarely used, except in Japan.

Most clinical investigations on the use of xylitol in parenteral solutions have concentrated on the general energetic and amino acid–sparing effects. Other potentially beneficial, metabolic effects have not yet been explored in detail. Because xylitol is metabolized by way of the pentose phosphate pathway, it is conceivable that the levels of phosphorylated nucleotides (ATP) might be enhanced. This could be relevant in reperfused myocardial tissue, as has been suggested by studies with ribose, which is subject to the same catabolic pathway (195–197). Other potential advantageous effects include an increase of the 2,3-diphosphoglycerate levels and a NADH-mediated reduction of oxidized glutathione in erythrocytes (198–202). This latter effect may be particularly relevant in the tens of millions of people that are affected by different degrees of glucose-6-phosphate dehydrogenase deficiency, which as such causes little concern but which predisposes to radical-induced pathology under certain conditions (drug therapy, sepsis, shock). Whether a reported effective treatment of cardiac arrhythmias with xylitol is also related to a shift to a more reduced status is not known (203,204).

Although a large number of studies indicate the absence of adverse side effects with parenteral xylitol administration, some authors advise against the use of xylitol because of the possibility of increased serum lactic acid, uric acid, and bilirubin concentrations (205,206) and because of the possible deposition of calcium oxalate crystals in the kidney and in the brain (207–210). However, regarding the occurrence of renocerebral oxalosis in association with xylitol infusions, it is noteworthy that in most of these cases the recommended maximum daily dose and/or infusion rate was surpassed. If xylitol is infused at recommended rates ($\leq$0.25 g/kg/hr and $\leq$3g/kg/day) (211) and in combination with glucose and fructose, such complications are not likely to occur more frequently than in patients not treated with xylitol (212–214). In general, no adverse side effects are observed if these recommendations are observed. Contrary to fructose, sorbitol, and glucose, a metabolic intolerance to xylitol is not known.

X. TOXICITY AND TOLERANCE

The results of animal tests for acute toxicity have indicated that xylitol is of very low toxicity by all routes of administration. Conventional tests for embryotoxicity and teratogenicity and for adverse effects on reproduction have given entirely negative results. Similarly, both in vitro and in vivo tests for mutagenicity and clastogenicity have given uniformly negative results (215).

Long-term studies in animals for safety evaluation have included 2-year treatment of rats, mice, and dogs. In these studies, xylitol was tested at a maximum dose level of 20% of the diet. Although the findings of these animal studies generally supported the safety of oral xylitol, some observations required further investigation. These observations were urinary tract calculi in mice and a slight increase in the incidence of adrenal medullary pheochromocytomas in male rats (216).

The results of subsequent studies and additional data from experiments with other polyols and lactose demonstrated, however, that the adverse effects observed in mice and rats are generic in nature and lack significance for safety evaluation in humans because of the species specificity of the underlying mechanisms (19,20,217–220).

Human tolerance of high oral doses of xylitol has been investigated in numerous studies with healthy and diabetic volunteers. The results of these studies invariably demonstrate the good tolerance even of extremely high intakes (up to 200 g/day) of xylitol. Adverse changes of clinical parameters were not observed. The only side effect that was occasionally noted was transient laxation and gastrointestinal discomfort (16,215,221–224). Such effects are generally observed after consumption of high doses of polyols and slowly digestible carbohydrates (e.g., lactose). The slow absorption of these compounds from the gut and the resulting osmotic imbalance are considered to be the cause of these effects, which are readily reversible on cessation or reduction of the amounts consumed. With continued exposure, tolerance usually develops (225,226).

XI. REGULATORY STATUS

On a supranational level, the Joint FAO/WHO Expert Committee on Food Additives (JECFA) has allocated an acceptable daily intake (ADI) "not specified," the most favorable ADI possible, for xylitol (227) and the Scientific Committee for Food for the European Economic Community (EEC) proposed "acceptance" of this polyol in 1984. On a national level, xylitol is approved for foods, cosmetics, and pharmaceuticals in many countries. Claims such as "noncariogenic" or "safe for teeth" may be applied where permitted.

REFERENCES

1. L Hyvoenen, R Kurkela, P Koivistoinen, P Merimaa. Effects of temperature and concentration on the relative sweetness of fructose, glucose and xylitol. Lebensm Wiss Technol 10:316–320, 1977.
2. SL Munton, and GG Birch. Accession of sweet stimuli to receptors. I. Absolute dominance of one molecular species in binary mixtures. J Theor Biol 112:539–551, 1985.
3. KK Maekinen, E Soederling. A quantitative study of mannitol, sorbitol, xylitol, and xylose in wild berries and commercial fruits. J Food Sci 45:367–374, 1980.
4. J Washuett, P Riederer, E Bancher. A qualitative and quantitative study of sugar-alcohols in several foods. J Food Sci 38:1262–1263, 1973.
5. O Touster. The metabolism of polyols. In HL Sipple, KW McNutt, eds. Sugars in Nutrition. New York: Academic Press, 1974, pp 229–239.
6. A Baer. Caries prevention with xylitol, a review of the scientific evidence. Wld Rev Nutr Diet 55:183–209, 1988.
7. KK Maekinen. Sweeteners and prevention of dental caries. Oral Health 78:57–66, 1988.

8. HAB Linke. Sugar alcohols and dental health. Wld Rev Nutr Diet 47:134–162, 1986.
9. WJ Loesche. The rationale for caries prevention through the use of sugar substitutes. Int Dent J 35:1–8, 1985.
10. I Kleinberg. Oral effects of sugars and sweeteners. Int Dent J 35:180–189, 1985.
11. H Foerster, H Mehnert. Die orale Anwendung von Xylit als Zucker-austauschstoff in der Diät des Diabetes mellitus. Akt Ernährungsmedizin 4:296–314, 1979.
12. M Georgieff, RH Ackermann, KH Baessler, H Lutz. Die Vorteile von Xylit gegenüber Glucose als Energiedonator in Rahmen der frühen postoperativen, parenteralen Nährstoffzufuhr. Z. Ernährungswiss 21:27–42, 1982.
13. M Georgieff, LL Moldawer, BR Bistrian, GL Blackburn. Xylitol, an energy source of intravenous nutrition after trauma. J Enteral Parenteral Nutr 9:199–209, 1985.
14. JN Counsell. In JN Counsell, ed., Xylitol. London, Applied Science Publishers Ltd., 1978, p 191.
15. L Hyvoenen, P Koivistoinen. Food technological evaluation of xylitol. Adv Food Res 28:373–403, 1982.
16. KK Maekinen. Biochemical principles of the use of xylitol in medicine and nutrition with special consideration of dental aspects. Experientia, 30:1–160, 1978.
17. R Ylikahri. Metabolic and nutritional aspects of xylitol. Adv Food Res 25:159–180, 1979.
18. KK Maekinen, A Scheinin. Xylitol and dental caries. Ann Rev Nutr 2:133–150, 1982.
19. A Baer. Safety assessment of polyol sweeteners—some aspects of toxicology. Food Chem 16:231–241, 1985.
20. A Baer. Toxicological aspects of sugar alcohols—studies with xylitol. In: Low digestibility carbohydrates, Proceedings of a workshop held at the TNO-CIVO Institutes Zeist, The Netherlands, November 27–28, 1986, pp 42–50.
21. E Fischer, R Stahel. Zur Kenntnis der Xylose. Berichte Dtsch Chem Gesellschaft 24:528–539, 1891.
22. GM Jaffe. Xylitol—a specialty sweetener. Sugar y Azucar 73:36–42, 1978.
23. KK Makinen. Latest dental studies on xylitol and mechanisms of action of xylitol in caries limitation. In: Grenby TH, ed. Progress in Sweeteners. New York, Elsevier, 1989, pp 331–362.
24. KK Makinen. Sugar alcohols. In: Goldberg I, ed. Functional Foods, Designer Foods, Pharmafeeds Nutraceuticals. New York, Chapman & Hall, Chap. 11, 1994, pp 219–241.
25. C Aminoff, E Vanninen, TE Doty. The occurrence, manufacture and properties of xylitol. In JN Counsell, ed. Xylitol. London, Applied Science Publishers Ltd., 1978, pp 1–9.
26. JF Carson, SW Waisbrot, FT Jones. A new form of crystalline xylitol. J Am Chem Soc 65:1777–1778, 1943.
27. H Onishi, T Suzuki. Microbial production of xylitol from glucose. Appl Microbiol 15:1031–1035, 1969.
28. LF Chen, CS Gong. Fermentation of sugarcane bagasse hemicellulose hydrolysate to xylitol by a hydrolysate-acclimatized yeast. J Food Sci 50:226–228, 1985.

29. CS Gong, TA Claypool, LD McCracken, CM Maun, PP Ueng, GT Tsao. Conversion of pentoses by yeasts. Biotechnol Bioeng 25:85–102, 1983.
30. V Kitpreechavanich, N Nishio, M Hayashi, S Nagai. Regeneration and retention of NADP (H) for xylitol production in an ionized membrane reactor. Biotechnol Lett 7:657–662, 1985.
31. D Holland, JF Stoddart. A stereoselective synthesis of xylitol. Tetrahed Lett 23: 5367–5370, 1982.
32. J Kiss, R D'Souza, P Taschner. Präparative Herstellung von 5-Desoxy-L-arabinose, Xylit und D-Ribose aus ''Diacetonglucose. Helv Chim Acta 58:311–317, 1975.
33. T Pepper, PM Olinger. Xylitol in sugar-free confections. Food Techn 42:98–106, 1988.
34. C Krueger. Zuckerfreie Pralinen—zuckerfreie Füllungen und Schokoladen-massen. Süsswaren 11:506–516, 1987.
35. T Pepper. Sugar substitutes, their use in chocolate and chocolate fillings. Manuf Conf 67(6):83, 1987.
36. C Krueger. Zuckeraustauschstoffe, Arten, technologische, sensorische und ernährungsphysiologische Eigenschaften und Synergie-Effekte, Zucker-und Süsswaren Wirtschaft. 41(11):360–365, 1988.
37. AG Dodson, T Pepper. Confectionery technology and the pros and cons of using non-sucrose sweeteners. Fd Chem 16:271–280, 1985.
38. T Imfeld. Zahnschonende Halswehlutschtabletten. Swiss Dent 5:19–22, 1984.
39. F Bray. Zuccheri alternativi e loro influenza nel gelato. Ind Aliment 24:895–898, 1985.
40. A Askar, MG Abd El-Fadeel, MA Sadek, AMA El Rakaybi, GA Mostafa. Studies on the production of dietetic cake using sweeteners and sugar substitutes. Dtsch Lebensm Rundschau 83:389–394, 1987.
41. P Varo, C Westermarck-Rosendahl, L Hyvoenen, P Koivistoinen. The baking behavior of different sugars and sugar alcohols as determined by high pressure liquid chromatography. Lebensm. Wiss Technol 12:153–156, 1979.
42. HR Moskowitz. The sweetness and pleasantness of sugars. Am J Psychol 84:387–405, 1971.
43. J Gutschmidt, G Ordynsky. Bestimmung des Süssungsgrades von Xylit. Dtsch Lebensm Rdsch 57:321–324, 1961.
44. MG Lindley, GG Birch, R Khan. Sweetness of sucrose and xylitol, structural considerations. J Sci Food Agric 27:140–144, 1976.
45. S Yamaguchi, T Yoshikawa, S Ikeda, T Ninomiya. Studies on the taste of some sweet substances. Agric Biol Chem 34:181–197, 1970.
46. CK Lee. Structural functions of taste in the sugar series: Taste properties of sugar alcohols and related compounds. Fd Chem 2:95–105, 1977.
47. E Soederling, KK Maekinen, CY Chen, HR Pape, PL Maekinen. Effect of sorbitol, xylitol or xylitol/sorbitol gum on plaque. J Dent Res 67:279 (Abstr. 1334), 1988.
48. P Soparkar, P DePaola, S Vrasanos, I Mandel. The effect of xylitol chewing gums on plaque and gingivitis in humans. J Dent Res 68: Abstr. 833, 1989.
49. D Birkhed. Cariogenic aspects of xylitol and its use in chewing gums: A review. Acta Odontal Scand 25:116–127, 1994.

50. J Gordon, M Cronin, R Reardon. Ability of a xylitol chewing gum to reduce plaque accumulation. J Dent Res 69:136.
51. W Edgar. Sugar substitutes, chewing gum and dental caries—a review. Br Dent J 184 (1):29–31, 1998.
52. RJ Feigal, ME Jensen, CA Mensing. Dental caries potential of liquid medications. Pediatrics 68:416–419, 1981.
53. PE Souney, DA Cyr, JT Chang, AE Kaul. Sugar content of selected pharmaceuticals. Diabetes Care 6:231–240 (1983).
54. IF Roberts, GJ Roberts. Relation between medicines sweetened with sucrose and dental disease. BMJ 2:14–16, 1979.
55. PJM Crawford. Sweetened medicine and caries. Pharm J 226:668–669, 1981.
56. HR Muehlemann, A Firestone. Drei neue "zahnscshonende" Präparate. Swiss Dent 3:25–30, 1982.
57. E Kristoffersson, S Halme. Xylitol as an excipient in oral lozenges. Acta Pharm Fenn 87:61–73, 1978.
58. R Laakso, K Sneck, E Kristoffersson. Xylitol and Avicel pH 102 as excipients in tablets made by compression and from granulate. Acta Pharm Fenn 91:47–54, 1982.
59. DW Bloch, MA El Egakey, PP Speiser. Verbesserung der Auflösungsgeschwindigkeit von Hydrochlorothiazid durch feste Dispersionen mit Xylitol. Acta Pharm Technol 18:177–183, 1982.
60. I Sirenius, VE Krogerus, T Leppaenen. Dissolution rate of *p*-aminobenzoates from solid xylitol dispersions. J Pharm Sci 68:791–792, 1979.
61. L Morris, J Moore, J Schwartz. Characterisation and Performance of a New Direct Compression Excipient for Chewable Tablets: Xylitab, American Association of Pharmaceutical Scientist Annual Meeting, 1993.
62. JSM Garr, MH Rubinstein. Direct compression characteristics of Xylitol. Inst Pharm 64:223–226, 1990.
63. F Voirol, R. Etter. Verfahren zur Erzeugung einer Deckschicht auf Tablettenkernen. German Patent Appl. 29 13 555, 1979.
64. J Sintes, C Escalante, B Stewart, J McCod, L Garcia, Volpe, C Triol. Enhanced anticaries efficacy of a 0.243% sodium fluoride/10% xylitol/silica dentifrice: 3-year clinical results. Am J Dent 8(5):231–235, 1995.
65. KK Maekinen, E Soderling, I Laeikkoe. Zuckeralkohole (Polyole) als "aktive" Zahnpastenbestandteile. Oralprophylaxe, 9:115–120, 1987.
66. KK Maekinen, E Soederling, H Hurttia, OP Lehtonen, E Luukkala. Biochemical, microbiologic and clinical comparisons between two dentifrices that contain different mixtures of sugar alcohols. J Am Dent Assoc 111:745–751, 1985.
67. AA Scheie, O Fejerskov, S Assev, G Roella. Ultrastructural changes in *Streptococcus sobrinus* induced by xylitol, NaF and $ZnCl_2$. Caries Res 23:320–327, 1989.
68. J Arends, J Christoffersen, J Schuthof, MT Smits. Influence of xylitol on demineralization of enamel. Caries Res 18:296–301, 1984.
69. MT Smits, J Arends. Influence of extraoral xylitol and sucrose dippings on enamel demineralization in vivo. Caries Res 22:160–165, 1988.
70. M Svanberg, D Birkhed. Effects of dentrifices containing either xylitol and glycerin or sorbitol on *Streptococcus mutans* in saliva. Caries Res 25:74–79, 1995.

71. JDB Featherstone, TW Cutress, BE Rodgers, PJ Dennison. Remineralization of artificial caries-like lesions in vivo by a self-administered mouthrinse or paste. Caries Res 16:235–242, 1982.
72. A Vissink, EJ's-Gravenmade, TBFM Gelhard, AK Panders, MH Franken. Rehardening properties of mucin- or CMC-containing saliva substitutes on softened human enamel. Caries Res 19:212–218, 1985.
73. A Cobanera, E Mopasso, P White, Cuevas, R Espinosa. Xylitol-sodium fluoride: Effect on plaque. J Dent Res 66:Abstr. 56, 1987.
74. LG Petersson, M Johansson, G Joensson, D Birkhed, A Gleerup. Caries preventive effect of toothpastes containing different concentration and mixture of fluorides and sugar alcohols. Caries Res 23:109, 1989.
75. G Grimble. Fibre, fermentation, flora, and flatus. Gut 30:6–13, 1989.
76. AA Salyers, JAZ Leedle. Carbohydrate metabolism in the human colon. In DJ Hentges, ed. Human Intestinal Flora in Health and Disease. New York, Academic Press, 1983, pp 129–146.
77. JH Cummings. Short chain fatty acids in the human colon. Gut 22:763–779, 1981.
78. H Ruppin, S Bar-Meir, KH Soergel, CM Wood, MG Schmitt Jr. Absorption of short-chain fatty acids by the colon. Gastroenterol 78:1500–1507, 1980.
79. JH Cummings, EW Pomare, WJ Branch, CPE Naylor, GT Mac Farlane. Short chain fatty acids in human large intestine, portal, hepatic and venous blood. Gut 28:1221–1227, 1987.
80. BM Buckley, DH Williamson. Origin of blood acetate in the rat. Biochem J 166: 539–545, 1977.
81. CL Skutches, CP Holroyde, RN Myers, P Paul, GA Reichard. Plasma acetate turnover and oxidation. J Clin Invest 64:708–713, 1979.
82. JH Cummings, HN Englyst, HS Wiggins. The role of carbohydrates in lower gut function. Nutr Rev 44:50–54, 1986.
83. NI McNeil. The contribution of the large intestine to energy supplies in man. Am J Clin Nutr 39:338–342, 1984.
84. SE Fleming, DS Arce. Volatile fatty acids: Their production, absorption, utilization, and roles in human health. Clin Gastroenterol 15:787–814, 1986.
85. GE Demetrakopoulos, H Amos. Xylose and xylitol. World Rev Nutr Diet 32:96–122, 1978.
86. KH Baessler. Absorption, metabolism, and tolerance of polyol sugar substitutes. Pharmacol Ther Dent 3:85–93, 1978.
87. JF Williams, KK Arora, JP Longenecker. The pentose pathway: A random harvest. Int J BIochem 19:749–817, 1987.
88. T Wood. Physiological functions of the pentose phosphate pathway. Cell Biochemistry and Function 4:241–247, 1986.
89. NZ Baquer, JS Hothersall, P McLean. Function and regulation of the pentose phosphate pathway in brain. Curr Top Cell Requl 29:265–289, 1988.
90. AB Lane. On the nature of L-xylulose reductase deficiency in essential pentosuria. Biochem Genet 23:61–72, 1985.
91. K Soyama, N Furukawa. A Japanese case of pentosuria—case report. J Inher Metab Dis 8:37, 1985.
92. The Evaluation of the Energy of Certain Sugar Alcohols Used as Food Ingredients,

Federation of American Societies for Experimental Biology. FASEB, 1994. Unpublished.
93. A Baer. Factorial calculation model for the estimation of the physiological caloric value of polyols. (1991).
94. Anonymous. De energetische waarde van suikeralcoholen. Voeding 48:357–365, 1987.
95. E Karimzadegan, AJ Clifford, FW Hill. A rat bioassay for measuring the comparative availability of carbohydrates and its application to legume foods, pure carbohydrates and polyols. J Nutr 109:2247–2259, 1979.
96. R Mueller-Hess, CA Geser, JP Bonjour, E Jequier, JP Felber. Effects of oral xylitol administration on carbohydrate and lipid metabolism in normal subjects. Infusionstherapie 2:247–252, 1975.
97. JH Shaw. Medical progress—causes and control of dental caries. N Engl J Med 317:996–1004, 1987.
98. E Newburn. Sugar and dental caries: A review of human studies. Science 217:418–423, 1982.
99. HG Huxley. The effect of feeding frequency on rat caries. J Dent Res 56:976, 1977.
100. B Gustaffson, CE Quensel, L Lanke, C Lundqvist, W Grahnen, BE Gonow, B Krasse. The Vipeholm dental caries study: The effect of different levels of carbohydrate intake on caries activity in 436 individuals observed for 5 years. Acta Odont Scand 11:232–363, 1953.
101. T Imfeld, HR Muehlemann. Cariogenicity and acido-genicity of food, confectionery and beverages. Pharmacol Ther Dent 3:53–68, 1978.
102. A Baer. Significance and promotion of sugar substitution for the prevention of dental caries. Lebensm-Wiss u-Technol 22:46–53, 1989.
103. WJ Loesche. The rationale for caries prevention through the use of sugar substitutes. Int Dent J 35:1–8, 1985.
104. HJ Guelzow. Ueber den anaeroben Umsatz von Palatinit durch Mikro-organismen der menschlichen Mundhöhle. Dtsch zalmärztl Z 37:669–672, 1982.
105. BG Bibby, J Fu. Changes in plaque pH in vitro by sweeteners. J Dent Res 64: 1130–1133, 1985.
106. AR Firestone, JM Navia. In vivo measurements of sulcal plaque pH in rats after topical applications of xylitol, sorbitol, glucose, sucrose and sucrose plus 53 mM sodium fluoride, J Dent Res 65:44–48 1986.
107. M Rekola. Acid production from xylitol products in vivo and in vitro. Proc Finn Dent Soc 84:39–44, 1988.
108. TN Imfeld. Identification of low caries risk dietary components. Monogr Oral Sci 11:1983.
109. Y Maki, K Ohta, I Takazoe, et al. Acid production from isomaltulose, sucrose, sorbitol, and xylitol in suspensions of human dental plaque (short communication). Carries Res 17:335–339, 1983.
110. KK Maekinen, E Soederling, M Haemaelaeinen, P Antonen. Effect of long-term use of xylitol on dental plaque. Proc Finn Dent Soc 81:28–35, 1985.
111. F Gehring. Mikrobiologische Untersuchungen im Rahmen der "Turku Sugar Studies." Dtsch. Zahnärztl Z 32:84–88, 1977.

112. JS Van der Hoeven. Cariogenicity of lactitol in program-fed rats (short communication). Caries Res 20:441–443, 1986.
113. R Havenaar, JHJ Huis in't Veld, JD de Stoppelaar, O Backer Dirks. Anti-cariogenic and remineralizing properties of xylitol in combination with sucrose in rats inoculated with *Streptoccus mutans*. Caries Res 18:269–277, 1984.
114. R Havenaar. The anti-cariogenic potential of xylitol in comparison with sodium fluoride in rat caries experiments. J Dent Res 63:120–123, 1984.
115. EJ Karle, F Gehring. Zur Kariogenität von Mischungen aus Kohlenhydraten und Xylit im konventionellen und gnotobiotischen Tierversuch. Dtsch Zahnärztl Z 42: 835–840, 1987.
116. R Havenaar. Effects of intermittent feeding of sugar and sugar substitutes on experimental caries and the colonization of *Streptococcus mutans* in rats. In: R Havenaar, ed. Sugar Substitutes and Dental Caries. Utrecht Drukkerij Elinkwijk BV, 1984, pp 73–79.
117. A Hefti. Cariogenicity of topically applied sugar substitutes in rats under restricted feeding conditions. Caries Res 14:136–140, 1980.
118. SA Leach, RM Green. Effect of xylitol-supplemented diets on the progressional and regression of fissure caries in the albino rat. Caries Res 14:16–23, 1980.
119. SA Leach, EA Agalamanyi, RM Green. Remineralisation of the teeth by dietary means. In: (SA Leach, WM Edgar, eds.) Demineralisation and Remineralisation of the Teeth. Oxford, IRL Press Ltd., 1983, pp 51–73.
120. A Scheinin, KK Maekinen, K Ylitalo. Turku sugar studies V. Final report on the effect of sucrose, fructose and xylitol diets on the caries incidence in man. Acta Odontol Scand 33(Suppl. 70): 67–104, 1975.
121. M Rekola. Changes in buccal white spots during 2-year consumption of dietary sucrose or xylitol. Acta Odontol Scand 44:285–290, 1986.
122. M Rekola. Approximal caries development during 2-year total substitution of dietary sucrose with xylitol. Caries Res 21:87–94, 1987.
123. A Scheinin, KK Maekinen, E Tammisalo, M Rekola. Turku sugar studies XVIII. Incidence of dental caries in relation to 1-year consumption of xylitol chewing gum. Acta Odontol Scand 33(Suppl. 70):307–316, 1975.
124. M Rekola. A planimetric evaluation of approximal caries progression during one year of consuming sucrose and xylitol chewing gums. Proc Finu Dent Soc 82:213–218, 1986.
125. A Scheinin, J Banoczy, J Szoeke, I Esztari, K Pienihaekkinen, U Scheinin, J Tiekso, P Zimmermann, E Hadas. Collaborative WHO xylitol field studies in Hungary. I. Three-year caries activity in institutionalized children. Acta Odontol Scand 43: 327–347, 1985.
126. D Kandelman, A Baer, A Hefti. Collaborative WHO xylitol field study in French Polynesia. I. Baseline prevalence and 32 months caries increment. Caries Res 22: 55–62, 1988.
127. D Kandelman, G Gagnon. Clinical results after 12 months from a study of the incidence and progression of dental caries in relation to consumption of chewing-gum containing xylitol in school preventive programs. J Dent Res 66:1407–1411, 1987.

128. D Kandelman, G Gagnon. Effect on dental caries of xylitol chewing gum; two year results. J Dent Res 67:172 (abstr. no. 472), 1988.
129. P Isokangas. Xylitol chewing gum in caries prevention. A longitudinal study on Finnish school children. Proceedings of the Finnish Dental Society, 83:1–117, 1987.
130. P Isokangas, K Makinen, J Tieko, et al. Long-term effect of xylitol chewing gum in the prevention of dental caries: a follow-up five years after termination of a prevention program. Caries Res 27:495–498, 1993.
131. K Makinen, C Bennett, P Hujoel, et al. Xylitol chewing gums and caries rate: as 40-month cohort study. J Dent Res 74:1904–1913, 1995.
132. K Makinen, P Hujoel, C Bennett, et al. Polyol chewing gums and caries rates in primary dentition: a 24-month cohort study. Caries Res 30:408–417, 1996.
133. PL Makinen, P Allen, C Bennett, et al., Stabilization of rampant caries: polyol gums and arrest of dentin caries in two long-term cohort studies in young subjects. Int Dent J 45:93–107, 1995.
134. S Hamada, HD Slade. Biology, immunology and cariogenicity of *Streptococcus mutans*. Microbiol Rev 44:331–384, 1980.
135. J van Houte. Bacterial specificity in the etiology of dental caries. Int Dent J 30: 305–326, 1980.
136. S Alaluusua, M Nystroem, L Groenroos, L Peck. Caries-related microbiological findings in a group of teenagers and their parents. Caries Res 23:49–54, 1989.
137. C Stecksen-Blicks. Lactobacilli and *Streptococcus mutans* in saliva, diet and caries increment in 8- and 13-year-old children. Scand J Dent Res 95:18–26, 1988.
138. AA Scheie, P Arneberg, D Orstavik, J Afseth. Microbial composition, pH-depressing capacity and acidogenicity of 3-week smooth surface plaque developed on sucrose regulated diets in man. Caries Res 18:74–86, 1984.
139. DS Harper, WJ Loesche. Growth and acid tolerance of human dental plaque bacteria. Arch Oral Biol 29:843–838, 1984.
140. HD Sgan-Cohen, E Newbrun, R Huber, MN Sela. The effect of low-carbohydrate diet on plaque pH response to various foods. J Dent Res 66:911 (Abstr. 22), 1987.
141. HJ Maiwald, J Banoczy, W Tietze, Z Toth, A Vegh. Die Beeinflussung des Plaque-pH durch zuckerhaltigen und zuckerfreien Kaugummi. Zahn- Mund-Kieferheilk 70: 598–604, 1982.
142. TM Graber, TP Muller, VD Bhatia. The effect of xylitol gum and rinses on plaque acidogenesis in patients with fixed orthodontic appliances. Swed Dent J 15:41–55, 1982.
143. S Assev, SM Waler, G Roella. Further studies on the growth inhibition of some oral bacteria by xylitol. Acta Path Microbiol Immunol Scand Sect B 91:261–265, 1983.
144. C Vadeboncoeur, L Trahan, C Mouton, D Mayrand. Effect of xylitol on the growth and glycolysis of acidogenic oral bacteria. J Dent Res 62:882–884, 1983.
145. S Assev, G Vegarud, G Roella. Addition of xylitol to the growth medium of *Streptococcus mutans* omz 176—effect on the synthesis of extractable glycerol-phosphate polymers. Acta Path Microbiol Immunol Scand Sect B 93:145–149, 1985.
146. S Assev, G Roella. Further studies on the growth inhibition of *Streptococcus mu-*

tans omz 176 by xylitol. Acta Path Microbiol Immunol Scand Sect B 94:97–102, 1986.

147. SC Ziesenitz, G Siebert. Nonnutritive sweeteners as inhibitors of acid formation by oral microorganisms. Caries Res 20:498–502, 1986.
148. HJA Beckers. Influence of xylitol on growth, establishment, and cariogenicity of *Streptococcus mutans* in dental plaque of rats. Caries Res 22:166–173, 1988.
149. G Siebert, P Thim, HP Brenner, SC Ziesenitz. Antiacidogenic effects of non-cariogenic bulk sweeteners on fermentation of sucrose by *S. mutans* NCTC 10449. J Dent Res 67:696 (Abstr. 110), 1988.
150. B Forbord, H Osmundsen. On the mechanism of xylitol-dependent inhibition of glycolysis in *Streptococcus sobrinus* OMZ 126. Inst J Biochem 24:509–514, 1992.
151. S Assev, AA Scheie. Xylitol metabolism in xylitol-sensitive and xylitol-resistant strains of *Streptococci*. Acta Path Microbiol Immunol Scand Sect B 94:239–243, 1986.
152. L Trahan, M Bareil, L Gauthier, C Vadeboncoeur. Transport and phosphorylation of xylitol by a fructose phosphotransferase system in *Streptococcus mutans*, Caries Res 19:53–63, 1985.
153. S Assev, G Roella. Sorbitol increases the growth inhibition of xylitol on *Strep. mutans* OMZ 176. Acta Path Microbiol Immunol Scand Sect B, 94:231–237, 1986.
154. SM Waaler, S Assev, G Roella. Metabolism of xylitol in dental plaque. Scand J Dent Res 93:218–221, 1985.
155. A Pihanto-Leppala, E Soderling, K Makinen. Expulsion mechanism of xylitol-5-phosphate in *Streptococcus mutans*.
156. S Waher. Evidence for xylitol 5-P production in human dental plaque. Scand J Dent Res 100:204–206, 1992.
157. E Soederling, L Alaraeisaenen, A Scheinin, KK Maekinen. Effect of xylitol and sorbitol on polysaccharide production by and adhesive properties of *Streptococcus mutans*. Caries Res 21:109–116, 1987.
158. J Verran, DB Drucker. Effects of two potential sucrose-substitute sweetening agents on deposition of an oral Streptococcus on glass in the presence of sucrose. Archs Oral Biol 27:693–695, 1982.
159. H Tuompo, JH Meurman, K Lounatmaa, J Linkola. Effect of xylitol and other carbon sources on the cell wall of *Streptococcus mutans*. Scand J Dent Res 91:17–25, 1983.
160. CA Overmyer, E Soederling, P Isokangas, J Tenovuo, KK Maekinen. Oral biochemical status of children given xylitol gums. J Dent Res 66: (Spec. Issue) Abstr. 1551, 1987.
161. E Soederling, P Isokangas, J Tenovuo, KK Maekinen, S Mustakallio, H Maennistoe. Habitual xylitol consumption and *Streptococcus mutans* in plaque and saliva. J Dent Res 68: (Spec. Issue) Abstr. 953, 1989.
162. WJ Loesche. The effect of sugar alcohols on plaque and saliva level of *Streptococcus mutans*. Swed Dent J 8:125–135, 1984.
163. KK Maekinen, E Soederling, P Isokangas, J Tenovuo, J Tiekso. Oral biochemical status and depression of *Streptococcus mutans* in children during 24- to 36-month use of xylitol chewing gum. Caries Res 23:261–267, 1989.
164. J Banoczy, M Orsos, K, Pienihaekkinen, A Scheinin. Collaborative WHO xylitol

field studies in Hungary. IV. Saliva levels of *Streptococcus mutans*. Acta Odontol Scand 43:367–370, 1985.
165. F Gehring. Mikrobiologische Untersuchungen im Rahmen der "Turku Sugar Studies." Dtsch Zahnärztl Z 32:84–88, 1977.
166. E Soderling, P Isoksangas, J Tenoureo, S Mustakallio, K Makinen. Long-term xylitol consumption and mutans streptococci in plaque and saliva. Caries Res 25:153–157, 1991.
167. L Trahan, C Mouton. Selection for *Streptococcus mutans* with an altered xylitol transport capacity in chronic xylitol consumers. J Dent Res 66:982–988, 1987.
168. L Trahan. Xylitol: a review of its action on mutans streptococci and dental plaque—the clinical significance. Int Dent J 45(Suppl 1): 77–92, 1995.
169. E Soederling, KK Maekinen, CY Chen, HR Pape, PL Maekinen. Effect of sorbitol, xylitol or xylitol/sorbitol gum on plaque. J Dent Res 67:279 (Abstr 1334) (1988).
170. P Soparkar, P DePaola, S Vratsanos, I Mandel. The effect of xylitol chewing gums on plaque and gingivitis in humans. J Dent Res 68:Abstr 833 (1989).
171. R Shafer, A Levine, J Marlette, J Morley. Effects of xylitol on gastric emptying and food intake. Am J Clin Nutr 45:744–747, 1987.
172. M Uhari, T Kontiokari, M Koskela, M Niemela. Xylitol chewing gum in prevention of acute otitis media: double blind randomized trial. BMJ 313:1180–1184, 1996.
173. M Uhari, T Kontiokari, M Niemala. A novel use of xylitol in preventing acute otitis media. Pediatrics 102(4):879–884, 1998.
174. T Kontiokari, M Uhari, M Koskela. Effect of xylitol on growth of nasoharyngal bacteria in vitro. Antimicrob Agents Chemother 39:1820–1823, 1995.
175. CH Mellinghoff. Ueber die Verwendbarkeit des Xylit als Ersatzzucker bei Diabetikern. Klin Wochenschr 39:447, 1961.
176. American Diabetes Association. Nutrition recommendations and principles for people with diabetes mellitus. Diabetes Care 22(S1):S42–S45, 1999.
177. F Manenti, L Della Casa. Effetti dello xylitolo sull'equilibrio glicidico del diabetico. Boll Soc Medico-Chirurgica di Modena 65:1–8, 1965.
178. JK Huttunen, KK Maekinen, A Scheinin. Turku Sugar Studies XI. Effects of sucrose, fructose, and xylitol diets on glucose, lipid and urate metabolism. Acta Odontol Scand 33:(Suppl 70):239–245, 1975.
179. S Yamagata, Y Goto, A Ohneda, M Anzai, S Kawashima, M Shiba, Y Maruhama, Y Yamauchi. Clinical effects of xylitol on carbohydrate and lipid metabolism in diabetes. Lancet ii:918–921, 1965.
180. W Berger, H Goeschke, J Moppert, H Kuenzli. Insulin concentrations in portal venous and peripheral venous blood in man following administration of glucose, galactose, xylitol and tolbutamide. Horm Metab Res 5:4–8, 1973.
181. S Natah, K Hussein, J Tuominen, V Koivisto. Metabolic response to lactitol and xylitol in healthy men. Am J Clin Nutr 65:947–950, 1997.
182. W Hassinger, G Sauer, U Cordes, J Beyer, KH. Baessler. The effects of equal caloric amounts of xylitol, sucrose and starch on insulin requirements and blood glucose levels in insulin-dependent diabetics. Diabetologia 2:37–40, 1981.
183. S Yamagata, Y Goto, A Ohneda, M Anzai, S Kawashima, J Kikuchi, M Chiba, Y Maruhama, Y Yamauchi, T Toyota. Clinical application of xylitol in diabetics. In:

BL Horecker, K Lang, Y Takagi, eds. Pentoses and Pentitols. Berlin, Springer-Verlag, 1969, pp 316–325.

184. H Foerster, S Boecker, A Walther. Verwendung von Xylit als Zuckeraustauschstoff bei diabetischen Kindern. Fortschr Med 95:99–102, 1977.
185. RA Bonner, DC Laine, BJ Hoogwerf, FC Goetz. Effects of xylitol, fructose and sucrose in types I and II diabetics: A controlled, cross-over diet study on a clinial research center. Diabetes 31 (Suppl. 2): 80A (Abstr.), 1982.
186. LKR Shanbhogue, WJ Chwals, M Weintraub, GL Blackburn, BR Bistrian. Parenteral nutrition in the surgical patient. Br J Surg 74:172–180, 1987.
187. JE Schmitz, R Doelp, KH Altemeyer, A Gruenert, FW Ahnefeld. Parenterale Ernährung: Stoffwechsel und Substrate. Arzneimit-teltherapie 3:162–172, 1985.
188. RR Wolfe, F O'Donneil Jr, MD Stone, DA Richmand, JF Burke. Investigation of factors determining the optimal glucose infusion rate in total parenteral nutrition. Metabolism 29:892–900, 1980.
189. R Doelp, A Gruenert, E Schmitz, FW Ahnefeld. Klinische Untersuchungen zur periphervenösen parenteralen Ernährung. Beitr Infusionstherapie klin Ernähr 16: 64–76, 1986.
190. M Georgieff, LL Moldawer, BR Bistrian, GL Blackburn. Mechanisms for the protein-sparing action of xylitol during partial parenteral feeding after trauma. Surg Forum 35:105–108, 1984.
191. W Behrendt, C Minale, G Giani. Parenterale Ernährung nach herzchirurgischen Operationen. Infusionstherapie 11:316–322, 1984.
192. M Semsroth, K Steinbereithner. Stickstoffmetabolismus und renale Aminosäurenausscheidung während totaler parenteraler Ernährung hypermetaboler Patienten mit verschiedenen Kohlenhydratregimen. Infusionstherapie 12:136–148, 1985.
193. M Georgieff. Zur Stoffwechselwirkung von intravenös verabreichter Glukose, Xylit oder Glyzerin während unterschiedlicher Dosierung und Kombination. Beitr Infusionstherapie klin Ernähr 16:103–119, 1986.
194. C Dillier, A Leutenegger. Vereinfachung der parenteralen Ernährungstherapie durch eine neue, gebrauchsfertige all-in-one-Lösung, presented at the Annual Meeting of the Schweizerische Gesellschaft für Ernährungsforschung, March 5, 1988.
195. HG Zimmer, E Gerlach. Stimulation of myocardial adenine nucleotide biosynthesis by pentoses and pentitols. Pflügers Arch 376:223–227, 1978.
196. HG Zimmer, et al. Combination of ribose with calcium antagonist and β-blocker treatment in closed-chest rats. J Mol Cell Cardiol 19:635–639, 1987.
197. JF McAnulty, JH Southard, FO Belzer. Improved maintenance of adenosine triphosphate in five-day perfused kidneys with adenine and ribose. Transplant Proc 19:1376–1379, 1987.
198. V Stigge, D Strauss, H Roigas, HJ Raderecht. Verbesserte Erhaltung der zellulären Konzentration von 2,3-Diphosphoglyzerat im Konservenblut durch Zusatz von Xylit. Dtsch Ges Wesen 28:1805–1810, 1973.
199. S Vora. Metabolic manipulation of key glycolytic enzymes: A novel proposal for the maintenance of red cell 2,3-DPG and ATP levels during storage. Biomed Biochim Acta 46:285–289, 1987.
200. T Asakura, K Adachi, H Yoshikawa. Reduction of oxidized glutathione by xylitol. J Biochem 67:731–733, 1970.

201. JL Pousset, B Boum, A Cavé. Action antihémolytique du xylitol isolé des écorces de carica papaya. J Med Plant Res 41:40–47, 1981.
202. WA Ukab, J Sato, YM Wang, J van Eys. Xylitol mediated amelioration of acetylphenylhydrazine-induced hemolysis in rabbits. Metabolism 30:1053–1059, 1981.
203. T Tanaka, H Arimura. Validity of xylitol for management of cardiac arrhythmia during anesthesia. J Uoeh 1:365–368, 1979.
204. N Yoshimura, H Yamada, M Haraguchi. Anti-arrhythmic effect of xylitol during anesthesia. Jap J Anesthesiol 28:841–848, 1979.
205. ER Froesch. Parenterale Ernährung: Glucose oder Glucoseersatzstoffe? Schweiz Med Wochenschr 108:813–815, 1978.
206. K Korttila, MAK Mattila. Increased serum concentrations of lactic, pyruvic and uric acid and bilirubin after postoperative xylitol infusion. Acta Anesthesiol Scand 23:273–277, 1979.
207. RAJ Conyers, AM Rofe, R Bais, HM James, JB Edwards, DW Thomas, RG Edwards. The metabolic production of oxalate from xylitol. Int Z Vitam Ernährungsforsch (Beih.) 28:9–28, 1985.
208. B Ludwig, E Schindler, J Bohl, J Pfeiffer, G Kremer. Reno-cerebral oxalosis induced by xylitol. Neuroradiology 26:517–521, 1984.
209. RG Galaske, M Burdelski, J Brodehl. Primär polyurisches Nierenversagen und akute gelbe Leberdystrophie nach Infusion von Zuckeraustauschstoffen im Kindesalter. Dtsch med Wschr 111:978–983, 1986.
210. K Rosenstiel. Nierenversagen nach Infusion von Zuckeraustauschstoffen. Dtsch med Wschr 111:1340–1341, 1986.
211. R Schroeder. Beachting der Dosierungsgrenzen für Xylit bei der parenteralen Ernährung. Dtsch med Wschr 109:1047–1048, 1984.
212. J Pfeiffer, E Danner, PF Schmidt. Oxalate-induced encephalitic reactions to polyol-containing infusions during intensive care. Clin Neuropathol 3:76–87, 1984.
213. S Mori, T Beppu. Secondary renal oxalosis: A statistical analysis of its possible causes. Acta Pathol Jpn 33:661–669, 1983.
214. A Baer, D Loehlein. Limited role of endogenous glycolate as oxalate precursor in xylitol-infused patients. Contr Nephrol 58:160–163, 1987.
215. WHO/FAO. Summary of toxicological data of certain food additives, Twenty-first Report of the Joint FAO/WHO Expert Committee on Food Additives. Geneva, WHO Technical Report Series No. 617, 1977, pp 124–147.
216. WHO/FAO. Summary of toxicological data of certain food additives and contaminants, Twenty-second Report of the Joint FAO/WHO Expert Committee on Food Additives. Geneva, WHO Technical Report Series No. 631, 1978, pp 28–34.
217. A Baer. Sugars and adrenomedullary proliferative lesions: The effects of lactose and various polyalcohols. J Am Coll Toxicol 7:71–81, 1988.
218. Life Sciences Research Office. Health aspects of sugar alcohols and lactose. Report prepared for the Bureau of Foods, Food and Drug Administration, Washington, DC under Contract No. FDA 223-2020 by the Life Sciences Research Office, Federation of American Societies for Experimental Biology, Bethesda, MD, 1986, p 85.
219. 46th JECFA. Toxicological significance of proliferative lesions of the adrenal medulla in rats fed polyols and other poorly digestible carbohydrates, In: Evaluation

of Certain Food Additives and Contaminants. World Health Organization Technical Report Series 868, 8–12, 1997.
220. Cantox, Inc. Low digestible carbohydrates (polyols and lactose): significance of adrenal medullary proliferative lesions in the rat. Report for the Project Committee on Polyols, ILSI North America, 1995.
221. H Foerster, R Quadbeck, U Gottstein. Metabolic tolerance to high doses of oral xylitol in human volunteers not previously adapted to xylitol. Int J Vit Nutr Res (Suppl. 20): 67–88, 1981.
222. KK Maekinen, R Ylikahri, PL Maekinen, E Soederling, M Haemaelaeinen. Turku sugar studies XXIII. Comparison of metabolic tolerance in human volunteers to high oral doses of xylitol and sucrose after long-term regular consumption of xylitol. Int J Vit Nutr Res (Suppl. 22):29–51, 1981.
223. SJ Culbert, YM Wang, HA Fritsche Jr, D Carr, E Lantin, J vanEys. Oral xylitol in American adults. Nutr Res 6:913–922, 1986.
224. HK Akerblom, T Koivukangas, R Punka, M Mononen. The tolerance of increasing amounts of dietary xylitol in children. Int J Vit Nutr Res (Suppl. 22): 53–66, 1981.
225. H Foerster. Tolerance in the human adults and children. In JN Counsell, ed. Xylitol. London, Applied Science Publishers, Ltd., 1978, pp 43–66.
226. KK Maekinen, A Scheinin. Turku sugar studies VI. The administration of the trial and the control of the dietary regimen. Acta Odontol Scand 33:105–127, 1975.
227. WHO/FAO. Evaluation of certain food additives and contaminants, Twenty-seventh Report of the Joint FAO/WHO Expert Committee on Food Additives, Geneva, WHO Technical Report Series No. 696, 1983, pp 23–24.

20
Crystalline Fructose

John S. White
White Technical Research Group, Argenta, Illinois

Thomas F. Osberger
Food Industry Consultant, Upland, California

I. INTRODUCTION

Crystalline fructose first became widely available for industrial food and pharmaceutical applications nearly 25 years ago. It is physically and functionally distinct from other carbohydrates in solubility, freezing point depression, boiling point elevation, water activity, osmotic pressure, Maillard browning and flavor development, flavor enhancement, starch synergy, and metabolism. Most important are its high relative sweetness, unique sweetness intensity profile, and synergy with other sweeteners.

Students of organic chemistry learn very early that sucrose is a disaccharide that can be readily hydrolyzed (inverted) to provide equimolar amounts of two monosaccharides, fructose and dextrose. Although it would seem simple to obtain both fructose and dextrose from the inversion of sucrose, it was only in the mid-1970s that sucrochemical advances enabled commercial quantities of pure fructose to become available in the United States.

It was certainly not the case that food scientists and members of the medical profession were unaware of the increased sweetening power and unique metabolic properties of pure fructose. However, it made little sense to expend research resources to formulate foods and diets that used crystalline fructose without assurance that bulk industrial quantities (i.e., truckloads and railcars) would be available. The late 1980s and early 1990s saw the convergence of a plentiful raw material, proven refining technologies, economies of scale, and reduced prices that ensured the integration of crystalline fructose into mainstream food and beverage applications.

II. MANUFACTURE OF CRYSTALLINE FRUCTOSE

Producers have tested many processes and raw materials, seeking an economical manufacturing process for crystalline fructose. In the 1960s, Finnish, French, and German manufacturers, expanding on the pioneering work of the Frenchman Dubrufant (circa 1850), began producing industrial quantities of crystalline fructose from sucrose. Their processes involved the straightforward hydrolysis of sucrose with release of the simple sugars fructose and dextrose. However, the subsequent separation of the sugars by ion-exclusion chromatography and isolation of the fructose by carefully controlled crystallization extended the total manufacturing time impracticably to more than a week.

The use of inulin—a polyfructose storage polymer isolated from tuberous plants such as dahlias, Jerusalem artichokes, and the Hawaiian Ti plant—has also been examined as a raw material for fructose production. Through the controlled hydrolysis of inulin's β $(2 \rightarrow 1)$ fructosidic linkages, fructofuranose units are freed and converted to the more stable fructopyranose anomer. Despite repeated attempts, however, inulin processes have never been developed to produce crystalline fructose at competitive prices or in quantities required for ingredient use in the food industry.

As the worldwide demand for fructose grew, the need to increase capacity, find cheaper raw materials, and develop less energy-intensive processes became critical. In 1981, the world's first facility for making crystalline fructose from corn came on stream in Thomson, Illinois. American Xyrofin used liquid dextrose derived from cornstarch as the starting material. Finnish refining technology was used to produce a particularly high-quality crystalline fructose with production time reduced to about 5 days.

The solubility of fructose in methanol or ethanol is very low—approximately 0.07 g per gram of alcohol—compared with water at 4 g per gram of water. For this reason alcohol became the solvent of choice for many early European crystallization processes. However, removal and recovery of the alcohol, coupled with disposal of mixed waste streams, has caused alcohol crystallization to largely fall out of favor.

Finnish manufacturers were among the first to experiment with aqueous crystallization. Although aqueous crystallizations are the most successful processes in use today, the Finns and their successors have learned that these processes hold unique challenges beyond the high solubility of fructose in water. Unfavorable crystallization conditions result in the formation of fructose-hemihydrate and fructose-dihydrate crystalline forms, which are more hygroscopic (moisture absorbing), deliquesce in humid environments, and melt at lower temperatures than pure fructose. Difructose dianhydrides also form under certain crystallization conditions, reducing yields and altering physical properties of the crystalline product.

Table 1 Typical Physical Properties

Appearance	White crystalline powder, forming anhydrous needle-shaped crystals
Empirical formula	$C_6H_{12}O_6$
Molecular weight	180.16
Melting point	102°–105°C
Density	1.60 g/cm^3
Bulk density (loose)	0.8 g/cm^3
Caloric value	3.7 cal/g
Loss on drying (70°C for 4 hr in vacuum oven)	Less than 0.2%
Residue on ignition	Less than 0.5%
Heavy metals	Less than 5 ppm
Arsenic	Less than 1 ppm
Chloride	Less than 0.018%
Sulfate	Less than 0.025%
Calcium and magnesium	Less than 0.005%
Hydroxymethylfurfural	Less than 50 ppm
Glucose	Less than 0.1%
Assay (dry basis)	98.0–102.0%
pH in aqueous solution (1 g/10 ml)	5.0–7.0

State-of-the-art instrumentation is now available to permit rapid, precise measurements of fructose purity utilizing liquid chromatography. The reader is referred to the Sweetener Group of A. E. Staley Manufacturing Company of Decatur, IL, for this methodology.

Over time, the following general processing scheme has emerged for the successful manufacture of crystalline fructose (1):

1. Preparation of concentrated fructose feed; generally HFCS with >90% fructose and >90% solids
2. Seeding with anhydrous crystalline fructose or undried crystals
3. Crystallization by means of batch or continuous process, and aqueous, alcohol, or aqueous alcohol solvents
4. Centrifugation and washing of crystals to remove surface impurities
5. Drying by means of fluidized bed, tray, rotary, box, or belt particulate dryer
6. Crystal conditioning
7. Screening
8. Packaging

The Finnish company, Suomen Sakeri Osakeyhtio, received one of the first U. S. patents issued for an aqueous process in 1975 (2). Subsequent patents describe improved processes for fructose concentration under atmospheric or reduced-pressure conditions; seeding to initiate crystallization; cooling to allow for crystal growth; crystal washing and drying steps; and batch and continuous processes with computer-controlled crystallization cycles (1). Table 1 shows the composition of a typical lot of fructose produced to *United States Pharmacopoeia* and *Food Chemicals Codex* specifications.

III. EXISTENCE IN NATURE

Fructose, levulose, and fruit sugar are synonyms for this sweetest of all naturally occurring sugars. Fructose is the chief constituent in honey, historically the most available form of naturally occurring fructose that ''until the end of the Middle Ages . . . was everywhere the sweetener par excellence'' (3). The earliest recorded mentions of sweet-tasting substances reference honey thousands of years before its principal component was isolated and characterized as fructose. Honey-sweetened foods were popular in Rome, where honey was referred to as ''chief among all sweet things.'' Biblical references to honey are found throughout both the Old and New Testaments. Although no single list of constituents can define the wide variety of honeys harvested worldwide, Table 2 gives representative values approximating those found in most commercial honeys.

The characterization of fructose as ''fruit sugar'' stems from its significant presence in fruits and berries. Table 3 shows the percentages of fructose in a variety of fruits, based on the entire fruit and in relation to the total fruit solids (4).

IV. PHYSICAL AND FUNCTIONAL PROPERTIES

Fructose possesses physical and functional properties that distinguish it from sucrose, dextrose, corn syrups, starches, high-intensity sweeteners, and the myriad other ingredients used in food formulation. The following discussion reviews

Table 2 Constituents of Honey

Fructose	40%
Glucose	35%
Water	18%
Other saccharides	4%
Other substances	3%

Table 3 Fructose Contents of Fruits

Fruit	Percent fructose in fruit	Percent fructose in total solids
Apple	6.04	37.8
Blackberry	2.15	14.1
Blueberry	3.82	24.0
Currant	3.68	20.8
Gooseberry	3.90	26.3
Grape	7.84	41.0
Pear	6.77	49.9
Raspberry	4.84	17.2
Sweet cherry	7.38	32.9
Strawberry	2.40	25.4

these unique differences. ''Practical Applications of Crystalline Fructose,'' found later in this chapter, illustrates ways in which these properties may be used to advantage in food products.

A. Relative Sweetness

The functional property that most distinguishes fructose from other nutritive carbohydrates is its high relative sweetness. Relative sweetness is a subjective comparison of the peak organoleptic perception of sweetness of a substance, usually in relation to a sucrose reference. Reported relative sweetness values fall in the range of 1.8 times that of sucrose for crystalline fructose and 1.2 times that of sucrose for liquid fructose (5). However, it must be emphasized that the relative sweetness of fructose depends on the anomeric state of fructose at the time the sweetness comparison is made.

Only the sweetest, β-D-fructopyranose, anomer exists in crystalline fructose. Fructose rapidly mutarotates on dissolution in water, forming three additional tautomers possessing lower sweetness (6). The extent of mutarotation can be determined using optical rotation, gas-liquid chromatography, and nuclear magnetic resonance (NMR). These techniques have been used to determine that at 22°C the tautomeric equilibrium concentrations of 20% solids fructose in D_2O are as illustrated in Fig. 1 (7).

Temperature, pH, concentration of the solution, and presence of other sweeteners are factors that most influence sweetness intensity. Of these, only temperature exerts a significant effect on the mutarotational behavior of fructose in solution and on the transformation from the sweetest β-D-fructopyranose form to an equilibrium state in which less sweet tautomeric forms are present (6, 8).

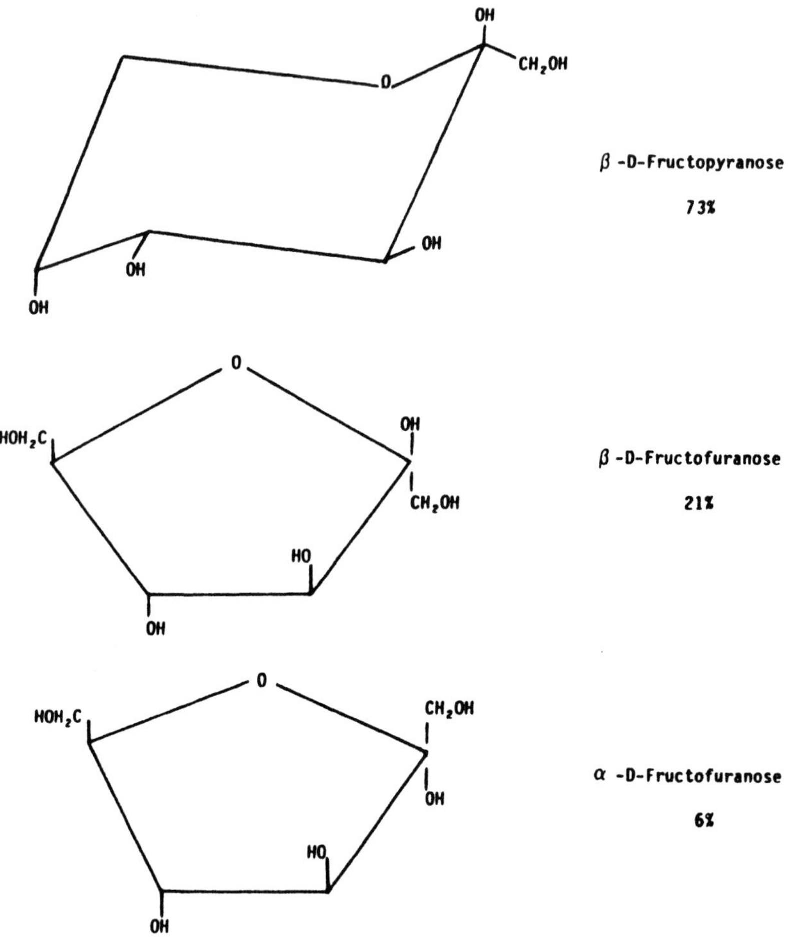

Figure 1 Cyclic tautomers of fructose. (From Ref. 7.)

NMR measurements at different temperatures demonstrate that the change in relative sweetness is directly related to the shift in tautomeric equilibrium as temperature is increased, as illustrated in Table 4 (9). On the basis of the correlation between fructose sweetness and mutarotational behavior, Shallenberger deduced that the furanoses are nearly devoid of sweet taste (6). Although this deduction is based primarily on conformational considerations, it is evident that an increase in the furanose anomer at the expense of the pyranose form will cause a reduction in perceived sweetness.

Table 4 The Tautomeric Equilibrium of Fructose at Different Temperatures

Temperature (°C)	α-D-Fructo-furanose (%)	β-D-Fructo-furanose (%)	β-D-Fructo-pyranose (%)
20	7	24	69
40	7	31	62
60	9	33	58
80	11	38	51

In practice, the degree of sweetness loss caused by this partial change to other, less sweet furanose tautomers can be minimized through the use of cold solutions and slightly acid conditions. Experience has shown that citrus-flavored beverage bases, sweetened with pure crystalline fructose and containing the usual amounts of acidulents, can realize a reduction of up to 50% of usual sweetener calories. Conversely, one of the least efficient uses of fructose is in hot coffee, in which mutarotation to furanose forms diminishes sweetness to the point that fructose is isosweet with sucrose. One further note in regard to fructose sweeteners is that the equilibrium state of the anomers is determined by the temperature at the time of consumption. Thus, cakes made with pure crystalline fructose will taste sweeter after they have been allowed to cool than they will if tasted just out of the oven.

B. Sweetness Intensity Profile and Flavor Enhancement

The sweetness of fructose is perceived more quickly, peaks more sharply and with greater intensity, and dissipates sooner from the palate than either sucrose or dextrose. It is this early sweetness intensity profile that accounts for the flavor enhancement so often observed in fructose formulations. Many fruit, spice, and acid flavors come through with greater clarity and identity after the sweetness of fructose has dissipated; they are not masked by the lingering sweetness of sucrose. The use of fructose thus makes possible the formulation of a more flavorful product or, alternatively, offers an opportunity for cost savings through lower flavor use.

C. Sweetness Synergy

Fructose exhibits sweetness synergy when used in combination with other caloric or high-intensity sweeteners. The relative sweetness of fructose blended with sucrose, aspartame, saccharin, or sucralose is perceived to be greater than the sweetness calculated from individual components in the blend (10–13).

When aspartame received clearance for use in food products in 1981, it became the task of food scientists to determine precisely how to use this sugar substitute. Contrary to oft-expressed opinion, aspartame does not exhibit a sweetness profile exactly identical to that of sucrose. Some subtle organoleptic differences can be perceived, with the significance of these differences being dependent on the specific conditions under which the aspartame is used. Hyvonen in Finland (14) and Johnson in the United States (15) discovered that combinations of aspartame and crystalline fructose could be used to achieve a synergistic sweetening effect and to minimize any lingering, nonsweet flavors of aspartame.

Thus, the sweetness synergy between fructose and other sweeteners offers formulators the choice of accepting finished products with greater sweetness or reducing sweetener levels and accepting ingredient cost savings. Blending fructose with reduced levels of high-intensity sweeteners has the added benefit of eliminating their bitter, metallic, or lingering aftertastes that are disagreeable to some consumers.

D. Colligative Properties

Fructose is a monosaccharide molecule with very different colligative properties than dextrose, another monosaccharide, or sucrose, a disaccharide. Colligative properties are those physical properties that depend solely on the concentration of particles in the specific system of interest. The concentration of particles, in turn, depends on the solubility and molecular weight of the particles.

The solubility of fructose, sucrose, and dextrose versus temperature is illustrated in Fig. 2. Fructose is more soluble at all temperatures than either sucrose or dextrose. Because it is half the molecular weight of sucrose and possesses greater solubility, it is easy to understand how the concentration of particles and concomitant colligative properties are accentuated with fructose.

Osmotic pressure, water activity, and freezing point depression are three colligative properties that are accentuated with fructose. Fructose creates higher osmotic pressure and lower water activity than sucrose, dextrose, or higher saccharides, resulting in greater product microbial stability. Fructose depresses the freezing point more than sucrose. When used in dairy desserts (both soft-serve and hard-frozen), the tendency of fructose to depress the freezing point can be countered with the addition of higher molecular weight corn sweeteners, gums, or stabilizers. The depressed freezing point is actually an asset when fructose is substituted for sucrose in frozen juice concentrates.

E. Hygroscopicity and Humectancy

Fructose is quicker to absorb moisture (hygroscopicity) and slower to release it to the environment (humectancy) than sucrose, dextrose, or other nutritive sweet-

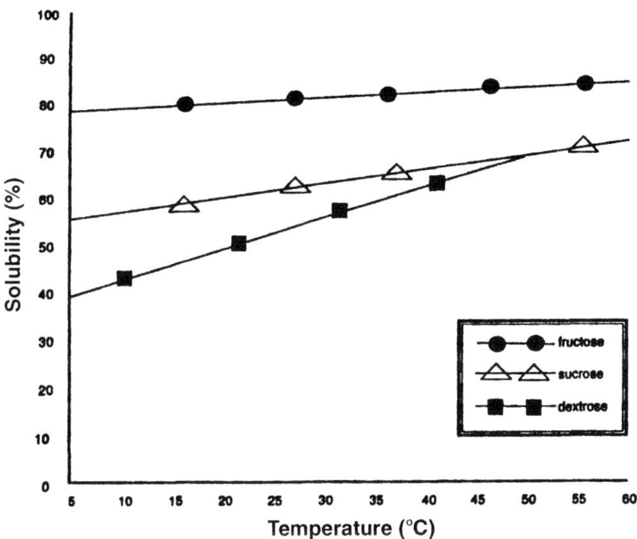

Figure 2 Solubility of sucrose, dextrose, and fructose at different temperatures.

eners. At approximately 55% relative humidity (RH), fructose begins absorbing moisture from the environment; sucrose does not absorb appreciable moisture until the RH exceeds 65%.

Although the "yin" of fructose hygroscopicity can present challenges in ingredient handling and storage (see later), the "yang" of fructose humectancy is a valuable functional attribute. Fructose is useful in sustaining product moistness at low RH, retarding sweetener recrystallization at high sweetener solids, delaying product staling, improving product eating qualities, and prolonging product shelf-life.

F. Browning and Flavor Development

Fructose is a reducing sugar—sucrose is not. If conditions of temperature and pH are favorable, reducing sugars can undergo a series of chemical condensation and degradation reactions with proteins and amino acids that produce flavored compounds and are responsible for product browning. In baked goods, the golden crust and delectable flavors and aromas of breads and cakes are the highly anticipated fruits of reducing sugars like fructose. If taken to extreme, however, this valuable attribute can result in surface burning and the development of undesirable off-flavors.

It should be noted that sucrose, a nonreducing sugar, may only significantly

Table 5 Temperature Range for Loss of Birefringence by Starch in Bleached Commercial Cake Flour in Several Sugar Solutions and % Loss Birefringence

Sugar Solutions (% w/w)	Sucrose			Glucose			Fructose		
	2% (°C)	50% (°C)	98% (°C)	2% (°C)	50% (°C)	98% (°C)	2% (°C)	50% (°C)	98% (°C)
None	55	58.5	62.5	55	58.5	62.5	55	58.5	62.5
10	58	61	65	58.5	61	64.5			
20	59	63.5	69.5	61	64	67.5	59	63	67
30	66	70	74	66	68	72			
40	73	76	79	68.5	73	76	67	71	74
50	82	84	87	77	79	83			
57	90	91.5	94	82	84	86.5	78	80	83
60	93.5	94.5	96.5	85	86.5	90.0	81	83	85
62				88	89.5	91.5	84	86	88.5
65	98	101	104	90	91.5	94	85	87	90
70	104.5	106	Boiled	95	97.5	100.5	89	91	94.5
73							91.5	94.5	97
80							99	103	105

contribute to browning and flavor development after inversion to its constituent reducing sugars, fructose and dextrose.

G. Starch-Fructose Synergy

It has been observed that the gelatinization of starch in heated foods is altered in the presence of carbohydrates. Carbohydrate sweeteners delay gelatinization of the starch or cause the starch to gelatinize at a higher temperature. Bean et al. (16) tabulated differences in the loss of birefringence in three different sugars over a range of concentrations and temperatures (Table 5). In a comparison of the effects of fructose versus sucrose, White and Lauer used differential scanning calorimetry to demonstrate that fructose causes starch to gelatinize at a lower temperature than sucrose (17). The delay in starch gelatinization appears to be due to the change in water mobility induced by the starch-fructose combination. Practical implications of substituting fructose for sucrose in heated starch systems will be discussed later.

V. METABOLISM OF FRUCTOSE

For a comprehensive examination of fructose intake and metabolism, readers are referred to the highly regarded monograph edited by Forbes and Bowman (18).

A. Absorption

Fructose in the diet occurs either bound covalently to dextrose in the disaccharide, sucrose, or as the free monosaccharide. Free monosaccharide fructose can originate from dietary fruits and berries or from sweeteners like honey, high fructose corn syrup, and crystalline or liquid fructose. Ingested sucrose is hydrolyzed to fructose and glucose by sucrase enzymes associated with the brush border of the intestinal epithelium. The resulting monosaccharides are immediately transported through the brush border membrane by the disaccharidase-related transport system, without being released to the lumen (19). Free dietary glucose is transported through the brush border membrane by one or more specific carrier systems. It is generally considered that fructose is absorbed across the intestinal mucosa of humans by facilitated diffusion.

Riby et al (20) reported that the capacity for fructose absorption is small compared with sucrose and glucose. They found that the simultaneous ingestion of glucose could prevent fructose malabsorption by increasing intestinal absorptive capacity for fructose. This suggests that the pair of monosaccharides might be absorbed by the disaccharidase-related transport system as if they were products of the enzymatic hydrolysis of sucrose. The human intestinal capacity for fructose absorption in the absence of glucose appears to vary significantly from one individual to another. It is important to note U. S. dietary consumption data, which indicate that typical glucose/fructose ratios are more than adequate to support fructose absorption in the general population (21). Those individuals who experience symptoms of malabsorption are advised to avoid consumption of products in which fructose is the sole carbohydrate (20).

B. Metabolism

All dietary fructose is absorbed and transported by the intestinal epithelium into the hepatic portal vein. The active hepatic enzyme system for metabolizing fructose efficiently extracts this sugar into the liver, leaving relatively low fructose concentrations in systemic blood vessels (22). Significantly, its entry into liver cells and subsequent phosphorylation by fructokinase is insulin independent (23). After cleavage by liver aldolase, the resulting trioses can be used for gluconeogenesis and glycogenesis or the synthesis of triglycerides, or they can enter the glycolytic pathway. The ultimate fate of these trioses depends on the metabolic state of the individual. Figure 3 illustrates the pathways involved in fructose metabolism.

C. Fructose and Diabetes

Studies by Crapo, Kolterman, and Olefsky in the early 1980s showed that acute administration of fructose results in lower glycemic and insulin responses in nor-

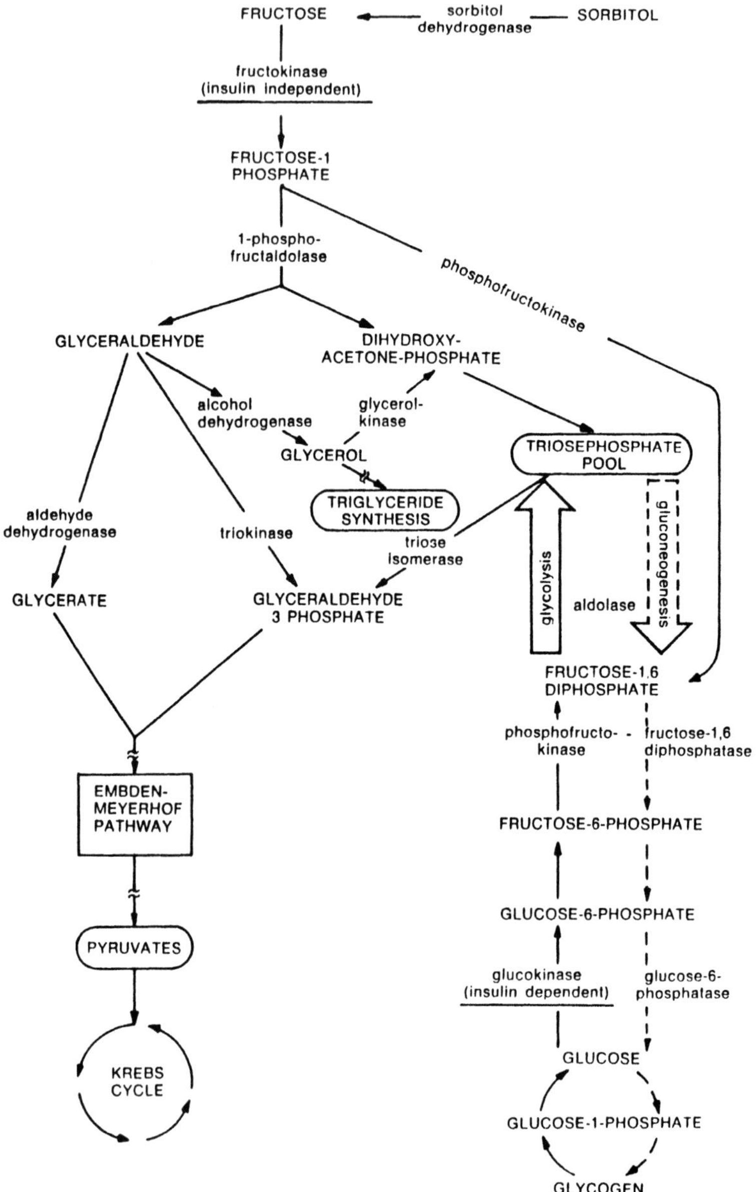

Figure 3 Metabolic pathways of fructose in the liver.

mal subjects, individuals with impaired glucose tolerance, and patients with non-insulin-dependent diabetes mellitus (24). Figures 4 and 5 illustrate these comparative effects.

Increased awareness of the harmful side effects of diabetes and the necessity for keeping blood sugar levels of diabetic patients close to normal without inducing severe hypoglycemia have led diabetologists and food technologists to renew investigations into the use of fructose as a preferred sweetener for diabetic

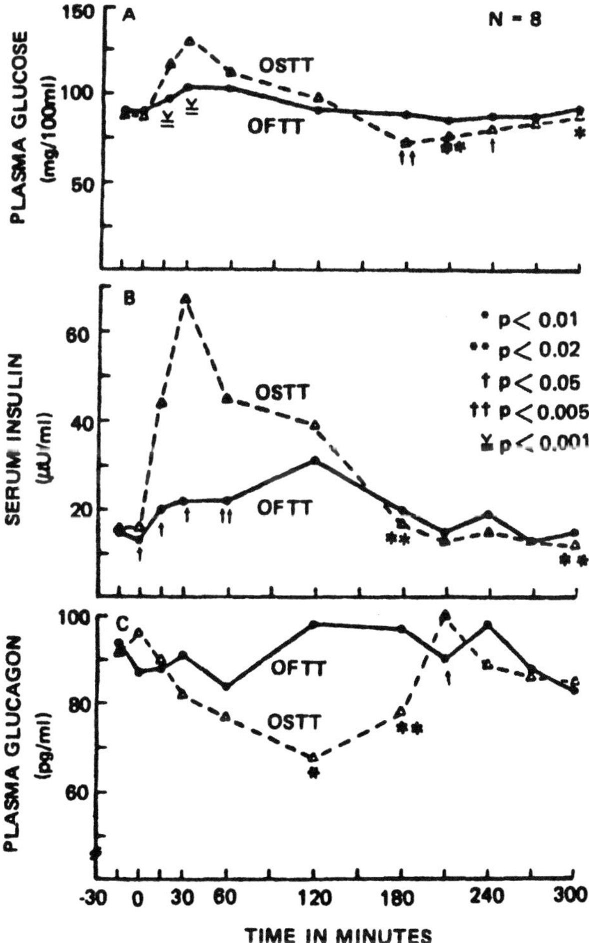

Figure 4 Comparison of glucose, insulin, and glucagon responses to oral sucrose versus oral fructose (OSTT = oral sucrose tolerance test; OFTT = oral fructose tolerance test).

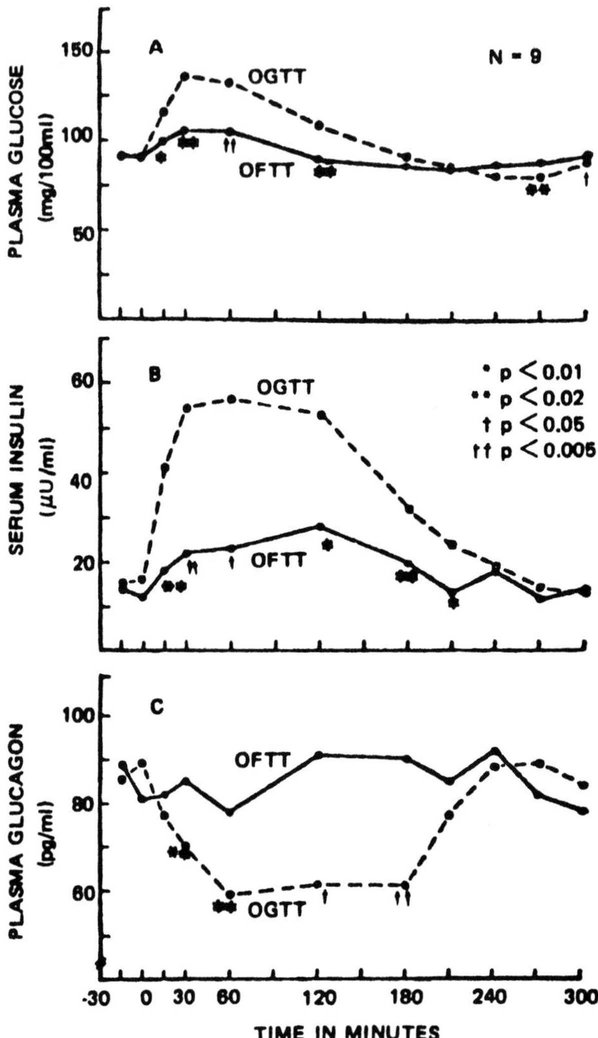

Figure 5 Comparison of glucose, insulin, glucagon responses to oral glucose versus oral fructose (OGTT = oral glucose tolerance test; OFTT = oral fructose tolerance test).

individuals (25). Short-term studies have now shown that substitution of fructose for sucrose in the diets of individuals with diabetes improves glycemic control; long-term effects are still inconclusive (26).

D. Food Intake

To test the effect of fructose on appetite, Moyer and Rodin fed subjects a pudding meal containing a test sugar (fructose or dextrose), milk protein, and lipid. Subjects were subsequently offered unrestricted access to a buffet meal and food choices were recorded. When the pudding was sweetened solely with fructose, subjects consumed significantly less energy in the buffet meal than when it was sweetened with glucose. When pudding was sweetened with more than one test sugar, however, responses to fructose and dextrose did not differ significantly. Thus, fructose shows promise in controlling food intake when used as the sole carbohydrate sweetener in a meal (27).

E. Physical Performance

Athletes have long used dietary supplements in an effort to sustain peak physical performance. Although much of the data concerning fructose supplementation is contradictory, fructose feeding before or during exercise can enhance performance under certain conditions. Fructose intake before exercise appears to spare muscle glycogen by elevating liver glycogen, thereby prolonging activity. In addition, good evidence suggests that the addition of fructose supplementation during ultraendurance events can improve performance by 126% (28). As indicated earlier, gastrointestinal discomfort created by intake of large amounts of fructose in the absence of glucose can hinder athletic performance. Because absorption capacity varies widely in the general population, the benefits of fructose will also vary from one athlete to another. Future research will refine the role fructose can play in enhancing athletic performance.

F. Glycemic Effect

Jenkins et al. determined the effect of 62 commonly eaten foods and sugars fed to groups of human volunteer subjects. They constructed a glycemic index, defined as "the area under the blood glucose response curve for each food expressed as a percentage of the area after taking the same amount of carbohydrate as glucose" (29). Table 6 reports the glycemic index for several sweeteners and foods relative to glucose, which is assigned a value of 100.

These glycemic values suggest that simple diabetic carbohydrate exchanges based on carbohydrate content may not, in fact, be an accurate predictor of physiological response. A clinical comparison of orally administered fructose, sucrose,

Table 6 Glycemic Index

Glucose	100
Sucrose	59 ± 10
Fructose	20 ± 5
Maltose	105 ± 12
Apples	39 ± 3
Raisins	64 ± 11
White bread	69 ± 5

Source: From Ref. 17.

and glucose in subjects with reactive hypoglycemia was reported in *Diabetes Care* (30). The investigators determined that the use of 100-g loads of pure fructose as the sweetener in cakes and beverages or by itself resulted in a significantly reduced glycemic effect, as indicated by markedly less severe glucose and insulin responses. The authors concluded, "fructose may thus prove useful as a sweetening agent in the dietary treatment of selected patients with reactive hypoglycemia."

VI. PRACTICAL APPLICATIONS OF CRYSTALLINE FRUCTOSE

Crystalline fructose was first promoted as a nutritionally beneficial sweetener by virtue of its unique metabolic disposition in the body. Early product formulations targeting diet and health-conscious consumers included powdered diet and sports beverages, nutritional candy bars, and specialty diabetic and dietetic food items. Breakthrough technology to crystallize fructose from HFCS feedstock was implemented in the mid-1980s, enabling large-scale and low-cost fructose production. Only then did formulation scientists attempt to integrate crystalline fructose into high-volume, mainstream food products.

Providing incentive to alert food scientists looking for applications for crystalline fructose was a new class of dietetic foods defined by the U. S. Food and Drug Administration. Products in this class could be labeled "reduced calorie" if they contained at least twenty-five percent fewer calories than their full-calorie counterpart (31). Crystalline fructose was immediately recognized as a critical ingredient in achieving the necessary reduction in sweetener calories (32). Substituting crystalline fructose for other sweeteners in existing formulas is not simply a case of plugging in a factor to determine the quantity of fructose required. Considerable food technology expertise is required to formulate reduced-calorie products because unwanted texture and body changes arise, in many instances, from the reduction in sweetener solids.

The availability of a new dry sweetener alternative to sucrose that combined high relative sweetness, dry physical form, and unique functional properties coincided with emerging opportunities to formulate food products for new regulatory classes. This collision of tool with opportunity provided all the impetus food scientists needed to formulate crystalline fructose into many food and beverage categories, the benefits of which follow.

A. Dry Mix Beverages, Puddings, and Gelatins

- Reduced sweetener content and concomitant lower calories are possible because of the intense sweetness and dry form of crystalline fructose.
- Fruit flavors are beneficially enhanced, permitting reduced levels of these expensive ingredients.
- Fructose-starch synergy sets puddings in about half the time required for all-sucrose products, permitting reduced starch use, cleaner flavor, and improved product performance.

The formulas in Tables 7, 8, and 9 illustrate these concepts.

B. Lite Pancake Syrups and Carbonated Beverages

- Reduced total sweetener use and ''Lite'' or reduced-calorie label claims are possible because of the sweetness synergy between fructose and sucrose, acesulfame potassium, saccharin, or aspartame.
- Fruit flavors are enhanced and high-intensity sweetener off-flavors may be eliminated as lower levels are used.

Table 7 Reduced-Calorie Lemonade Mix

Krystar® crystalline fructose	93.0 lb
Citric acid (anhydrous fine granular)	5.45 lb
Ascorbic acid	0.42 lb
Carrageenan (Viscarin #402)	0.10 lb
Riboflavin	0.0016 lb
Syloid #244	0.50 lb
Permaseal clouding agent FD-9208-B	0.20 lb
Naturalseal lemon flavor FD-8949-D	0.33 lb

Usage: Mix 8 oz of above product with sufficient water to make 1 gal of reduced-calorie lemonade. Chill and serve.

Table 8 Low-Calorie Fructose Lemonade Mix[a]

Fructose + saccharin blend (98:1)	89.1 lb
Citric acid (anhydrous fine granular)	9.2 lb
Ascorbic acid	0.7 lb
Carrageenan (Viscarin #402)	0.16 lb
Riboflavin	0.0025 lb
Permaseal clouding agent FD-9208-D	0.32 lb
Naturalseal lemon flavor FD-8949-D	0.53 lb

[a] Usage: Mix 5 oz of above product with sufficient water to make 1 gal of low-calorie lemonade. Chill and serve.

- Fructose provides lasting sweetness when paired with a sweetener like aspartame, which has a limited shelf-life because of thermal decomposition.

C. Breakfast Cereals

- Flavor enhancement and sweetness synergies improve product performance or allow cost reductions (e.g., cinnamon sugar coating).
- Simple sugars may be moved down the list of ingredients on the nutrition label, a strategy aimed at health-conscious consumers (5).

D. Baked Goods

- A reducing sugar like fructose used in place of a nonreducing sugar like sucrose improves product flavor and browning development in microwave and conventional ovens.
- The increased solubility of fructose makes it resistant to recrystallization. The development of soft, moist cookies was made possible by the partial substitution of fructose for sucrose in these formulations.

Table 9 Reduced-Calorie Gelatin Dessert

Crystalline fructose	79 lb, 8 oz
300 Bloom gelatin	14 lb
Fumaric acid	3 lb, 8 oz
Sodium citrate	2 lb
Syloid No. 244	1 lb
Spray-dried color and flavor	As required

Table 10 Reduced-Calorie Cake Mixes

Ingredient	White cake (%)	Yellow cake (%)	Chocolate cake (%)
Krystar® crystalline fructose	47.28	47.28	44.90
Hi-ratio bleached cake flour	38.75	38.75	37.72
N-Flate (National Starch)	7.71	7.71	7.52
Henningsen type P-20 egg white solids	3.44	3.44	3.35
Baking powder	2.24	2.24	1.49
Givaudan FD-9993 vanilla powder	0.58	0.58	0.57
De Zaan 11SB cocoa powder	—	—	4.17
Baking soda	—	—	0.28
Roche 1% CWS dry β-carotene	—	0.15	—

Directions for use: Blend 16 oz (454 g) of reduced-calorie cake mix with 10 oz of water for 30 sec at low speed, followed by 5 min whipping at high speed. Transfer the heavy batter into two 9-inch round cake pans (or one 9½ × 13½ rectangular pan) that have been sprayed with a low-calorie, nonstick spray. Bake at 325°F 30–32 min.

- Humectant properties of fructose can replace glycerin to improve product moisture and shelf-life in baked goods and bakery fillings. Multiple benefits gained include eliminating the off-flavor of glycerin, removing glycerin from the product label, and ingredient cost savings (33).
- It is possible to exploit the fructose-starch synergy to control the starch gelatinization temperature in cakes formulated with fat replacers. Reduced-calorie cake formulas like those illustrated in Table 10 allow bakers to produce cakes comparable in height and volume to full-fat products (34).
- Reduced fructose-starch cook temperatures may allow reduced product heat damage, faster line speeds, and lower energy costs.

E. Dairy Products

- Fructose may sweeten both the base and fruit in all-nutritively sweetened yogurt for enhanced fruit flavor and ingredient image and ingredient cost savings.
- Enhanced chocolate flavor allows reduced costs and fewer calories in chocolate milk.

F. Confections

- Fructose reduces the water activity of chocolate bar caramel fillings, resulting in greater microbial stability.
- It enhances hard candy flavors and improves product color.
- Fructose is more resistant to recrystallization than sucrose or dextrose. This quality can obviate the need for invertase when fructose replaces sucrose in liquid center confections like chocolate-covered cherries.
- Its lower relative viscosity offers faster production speeds.
- Use of a reducing sugar like fructose in place of a nonreducing sugar like sucrose allows shorter cook time to develop color and flavor.
- The fructose-starch synergy increases line efficiency by reducing starch levels and the viscosity of the deposit hot mix.
- Because fructose makes up 35% of the premium chocolate or carob coatings to enrobe energy bars, a thinner carob coating can be applied without reducing total sweetness (35). The authors warned, however, that temperatures must not exceed 120°F or agglomeration of crystals and severe glazing of pumps and pipelines will result.
- Most attempts at making hard candies with fructose have been unsuccessful, primarily because boiled candy made from it will not set properly. However, the addition of maltodextrin (44%) reportedly improved the set in a hard candy marketed in Italy (36).

G. Frozen Dairy Products and Novelties

- Calorie reduction and flavor enhancement are possible, but care must be taken to balance the freezing point depression with higher molecular weight compounds like corn syrups.

H. Frozen Fruit Packs and Juice Concentrates

- Greater sweetener solubility and lower osmotic pressure speed fruit infusion, preserve fruit integrity, and prolong storage stability.
- Improved solubility, resistance to recrystallization, and depressed freezing point hasten frozen juice concentrate reconstitution in water.
- Fructose enhances the natural flavors in fruit packs and juices.

I. Sports Drinks

- The increased solubility of fructose relative to sucrose and dextrose allows formulation to greater caloric density, particularly in chilled beverages.
- Flavor enhancement and compatible sweetness intensity profile help

mask unpalatable mineral and vitamin supplements apparent in sucrose-sweetened beverages.

VII. FRUCTOSE MARKETING

Crystalline fructose has been available as an industrial food ingredient in the United States since 1975. Fructose sales were initially strong as a result of intense marketing to educate the medical community, the health food industry and health-conscious consumers about the metabolic and sweetening advantages of crystalline fructose. Sales reached a plateau after several years, however, chiefly because of three contributing factors:

- Higher sales price relative to competing caloric and high-intensity sweeteners
- Sales efforts targeting the relatively small medical and health food markets
- The arrival of significant quantities from Europe and Asia at prices sharply below domestic prices

United States crystalline fructose manufacture would likely have failed early on had not A.E. Staley Manufacturing Company and Archer Daniels Midland, two leaders in the corn wet milling industry, decided in the mid 1980s that the time was ripe to begin large-scale production of crystalline fructose. Although long overdue in the view of many industry observers, this decision to enter the crystalline fructose market made a great deal of sense: corn wet millers had access to low-cost raw materials and were expert in refining the high fructose corn syrup feed stock to crystalline fructose. Furthermore, sales and distribution channels already in place to supply the food industry with corn sweeteners and starches enabled them to market crystalline fructose to this enormous industry far more aggressively than had been attempted to date.

Economies of scale, improved manufacturing processes, and increased awareness among food scientists of its unique physical and functional properties have spurred the growth of crystalline fructose manufacturing and formulation into finished foods and beverages. Although manufactured today in many countries throughout the world, crystalline fructose production is clearly dominated by two U. S. companies, A.E. Staley Manufacturing Company and Archer Daniels Midland. Actual production volumes are closely guarded and difficult to come by; however, it is roughly estimated that crystalline fructose manufacturing has grown from 12 million pounds in 1986 to approximately 185 million pounds in 1998 (37). Over the same period, the selling price of crystalline fructose has dropped in the United States from more than $1.00/lb. to approximately $0.35/lb. At this price, crystalline fructose is fully able to compete with sucrose and dextrose on a sweetness and functionality basis.

In 1993, Vuilleumier estimated the potential crystalline fructose market to be 200–300 million pounds (38). Viewed against projected 1998 market figures, his estimate appears to be well within reach of fructose manufacturers.

VIII. REGULATORY STATUS

Crystalline fructose, HFCS, and sucrose are generally recognized as safe (GRAS) in the United States. The *Food Chemicals Codex* and *United States Pharmacopoeia* define fructose as containing not less than 98% or more than 102% fructose and not more than 0.5% dextrose. Only crystalline fructose and crystalline fructose syrup meet this definition—HFCS and sucrose do not.

The *Codex Alimentarius Commission* defines fructose as "purified and crystallized D-α-fructose." These requirements, as well, are met only by crystalline fructose (39).

IX. STORAGE AND HANDLING

Relative humidity is the most important factor in crystalline fructose storage and handling. Table 11 illustrates the effect of RH on the moisture absorption of crystalline fructose (40). Product lumping is caused by the following.

- An initial increase in RH, causing surface water absorption and localized dissolving of fructose at crystal surfaces

Table 11 The Effect of Relative Humidity on the Moisture Absorption of Crystalline Fructose at 20°C[a]

Relative humidity (%)	Days exposed								
	1	2	5	7	9	12	16	20	26
33	0.02	0.01	0.02	0.02	0.01	0.01	0.01	0.01	0.01
48	—	0.03	0.03	0.03	0.03	0.03	—	—	—
55	—	0.06	0.06	0.06	0.06	0.05	—	—	—
60	0.10	0.12	0.13	0.13	0.13	0.13	0.13	0.13	0.13[b]
66	0.32	0.52	1.12	1.51	1.87	2.45	3.16	3.84	4.89
72	0.68	1.33	3.34	4.66	6.01	7.91	10.3	12.2	14.7
76	0.91	1.78	4.40	6.14	7.86	10.3	12.8	14.9	17.7
82	1.27	2.49	6.14	8.67	11.0	13.7	16.9	19.7	23.8

[a] The amount of moisture absorbed is indicated as a percent of dry weight.

[b] When moisture content exceeds 0.7%, fructose becomes increasingly wet and sticky. Processing conditions should be controlled to achieve values above this line.

- Subsequent lowering of RH, resulting in release of moisture, recrystallization, and sticking of adjacent crystals to one another

A conditioned air system with RH ≤55% and a maximum temperature of 24°C is recommended for bulk handling crystalline fructose. Crystalline fructose is typically packaged and shipped to customers in 25-kg multiwall (foil-lined) bags, 600- and 1000-kg tote bags, and bulk railcar. Recommended storage conditions are RH <50% and 21°C.

Fructose syrups, made by dissolving crystalline fructose in water, can be stored at 21–29°C. These pure fructose syrups may be held indefinitely with little likelihood of crystallization and are very stable microbially (1, 39).

REFERENCES

1. LM Hanover. In: FW Schenck, RE Hebeda, eds. Starch Hydrolysis Products. New York, VCH Publishers, Inc., 1992.
2. KH Forsberg, L Hamalainen, AJ Melaja, JJ Virtanen. Suomen Sakeri Osakeyhtio, US 3,883,365, 1975.
3. R Tannahill. In: Food in History. Briarcliff Manor, NY, Stein & Day, 1973.
4. H Moskowitz. In: H Sipple, K McNutt, eds. Sugars in Nutrition. New York, Academic Press, Inc., 1974, pp 37–64.
5. JS White, DW Parke. Fructose adds variety to breakfast. Cereal Foods World 34: 392, 1989.
6. RS Shallenberger. Pure and Applied Chemistry. London, Pergamon Press Ltd., 1978.
7. L Hyvonen, P Varo, P Koivistonen. Tautomeric equilibria of D-glucose and D-fructose. J Food Sci 42:657, 1977.
8. L Hyvonen. In: Varying Relative Sweetness. University of Helsinki EKT-546, 1980.
9. R Scott. Hoffmann-La Roche, Internal Report, Nutley, NJ, 1980, unpublished.
10. CK Batterman, ME Augustine, JR Dial. A. E. Staley Manufacturing Company, Sweetener composition, US 4,737,368, 1988.
11. CK Batterman, J Lambert. A. E. Staley Manufacturing Company, Synergistic sweetening composition, International Patent Publication WO88/08674, 1988.
12. P Van Tournout, J Pelgroms, J Van der Meerer. Sweetness evaluation of mixtures of fructose with saccharin, aspartame or acesulfame K. J Food Sci 50:469, 1985.
13. PK Beyts. Sweetening compositions. UK Patent Application GB2210545A, 1989.
14. L Hyvonen. Research Report. University of Helsinki, 1981.
15. L Johnson. unpublished report, Hoffmann-La Roche, Nutley, NJ, 1982.
16. MM Bean, WT Yamazak, DH Donelson. Wheat starch gelatinization in sugar solutions. II. Fructose, glucose and sucrose—cake performance. Cereal Chem 55(6): 945, 1978.
17. DC White, GN Lauer. Predicting gelatinization temperature of starch/sweetener systems for cake formulation by differential scanning calorimetry. I. Development of a model. Cereal Foods World, 35(8):728, 1990.

18. AL Forbes, BA Bowman, eds. Health effects of dietary fructose. Supplement to Am J Clin Nutr 58:721S, 1993.
19. AM Ugolev, BZ Zaripov, NN Iezuitova, et al. A revision of current data and views on membrane hydrolysis and transport in the mammalian small intestine based on a comparison of techniques of chronic and acute experiments: experimental re-investigation and critical review. Comp Biochem Physiol 85A:593, 1986.
20. JE Riby, T Fujisawa, N Kretchmer. Fructose absorption. Am J Clin Nutr 58:748S, 1993.
21. YK Park, EA Yetley. Intakes and food sources of fructose in the United States. Am J Clin Nutr 58:737S, 1993.
22. PA Mayes. Intermediary metabolism of fructose. Am J Clin Nutr 58:754S, 1993.
23. H Mehnert. In: Handbuch des Diabetes Mellitus. Munich, J.F. Lehmanns Verlag, 1971.
24. PA Crapo, OG Kolterman, J, Olefsky. Effects of oral fructose in normal, diabetic and impaired glucose tolerance subjects. Diabetes Care 3:575, 1980.
25. J Talbot, K Fisher. The need for special foods and sugar substitutes by individuals with diabetes mellitus. Diabetes Care 1:231, 1978.
26. PM Gerrits, E Tsalikian. Diabetes and fructose metabolism. Am J Clin Nutr 58: 796S, 1993.
27. AE Moyer, J Rodin. Fructose and behavior: does fructose influence food intake and macronutrient selection? Am J Clin Nutr 58:810S, 1993.
28. BW Craig. The influence of fructose feeding on physical performance. Am J Clin Nutr 58:815S, 1993.
29. DJA Jenkins, TMS Wolever, RH Taylor et al. Glycemic index of foods: A physiological basis for carbohydrate exchange. AM J Clin Nutr 34:362, 1981.
30. P Crapo, J Scarlett, O Koltermann, et al. The effects of oral fructose, sucrose and glucose in subjects with reactive hypoglycemia. Diabetes Care 5:512, 1982.
31. S Gardner. Special dietary foods. Federal Register 43:185, 1978.
32. T Osberger, H Linn. In: Low Calorie and Spec. Diet Foods. West Palm Beach, FL, CRC Press, 1978, pp. 115–123.
33. Anon. Fructose replaces glycerin in bakery fillings. Food Product Design 4(11):93, February, 1995.
34. SD Horton, GN Lauer, JS White. Predicting gelatinization temperature of starch/sweetener systems for cake formulation by differential scanning calorimetry. I. Evaluation and application of a model. Cereal Foods World 35(8):734, 1990.
35. M Zimmermann. Fructose in the manufacture of dietetic chocolate. Manufact Confectioner 39, 1974.
36. Sweet Topic No. 35, Xyrofin, Baar, Switzerland, 1982.
37. U.S. Economic Research Service. Personal communication with Steve Haley, March 1999.
38. S Vuilleumier. Worldwide production of high-fructose syrup and crystalline fructose. Am J Clin Nutr 58:733S, 1993.
39. LM Hanover, JS White. Manufacturing, composition and applications of crystalline fructose. Am J Clin Nutr 58:724S, 1993.
40. Research Report. Finnish State Technical Research Center #A-6759-73, Helsinki, 1973.

21
High Fructose Corn Syrup

Allan W. Buck
Archer Daniels Midland Company, Decatur, Illinois

I. INTRODUCTION

More than 30 years have elapsed since the first commercial shipment of modern high fructose corn syrup (HFCS) in the United States. Since then, the growth of this food ingredient has been tremendous, and production in the United States is predicted to expand to 20 billion pounds dry weight by the end of the century (1). Initially, one product was supplied with a concentration of 42% fructose on a dry basis. Because of its quality; sweetening properties; and availability as a bulk liquid, economical sweetener; that product, 42% HFCS, rapidly penetrated large markets formally held exclusively by liquid sucrose products. Not being quite as sweet as medium invert sugar used by the soft drink industry, technical demands from this industry led to the development of a 55% fructose product with higher sweetness levels. The development was brought about by the commercialization of a fractionation process that allows molecular separation of fructose from dextrose and concentration of the sweeter fructose fraction. This product, typically produced at levels of 90% fructose, was made available as a commercial product but was usually blended back with 42% HFCS to make 55% HFCS. The transition from sucrose to 100% HFCS in major soft drink brands occurred over about a 5 year period (1980–1985) as the corn wet milling industry increased production capacity to match needs and proved to the soft drink industry that quality and consistency in HFCS products were available. In the end, 55% HFCS had become the standard sweetener for carbonated beverages in this country, and the ratio between 42 and 55% HFCS has remained fairly constant since approval was given for the major soft drink brands. Table 1 shows the

Table 1 Market Division between 42 and 55% HFCS

Year	% 42% HFCS	% 55% HFCS
1978	88	12
1979	78	22
1980	67	33
1981	57	43
1982	50	50
1983	45	55
1984	40	60
1985	36	64
1986	35	65
1987	35	65
1988	38	62
1989	40	60
1990	41	59
1991	41	59
1992	42	58
1993	41	59
1994	40	60
1995	39	61
1996	38	62
1997	37	63
1998	36	64
1999	36	64

Source: USDA Economic Research Service.

market breakdown for the two major HFCS products. HFCS has matured to the point that growth is a function of increased population consumption of food products and not on penetration of new markets. Table 2 shows major events leading to the maturity of HFCS in the United States, and Fig. 1 depicts the growth in pounds.

Although corn is an economical and abundant source of starch for the manufacture of HFCS, the technology for the manufacture of high fructose syrups (HFS) is applicable to starch sources other than corn. For all practical purposes properties are identical, and the remainder of this chapter will refer to that product made from cornstarch, which is such an important part of the U.S. food chain. Estimated world consumption of HFS is shown in Table 3.

Most of this chapter will deal with the common commercial products 42 and 55% HFCS. Commercial products with higher levels of fructose (80, 90 and 95% HFCS) are available but in limited commercial supply compared with the widespread use of 42 and 55% HFCS. Because of the high level of fructose,

Table 2 Brief History of HFCS

1957	Patent by Marshall and Kooi disclosed a microbial enzyme capable of isomerizing dextrose to fructose.
1964	Clinton Corn Processing Company started work to commercialize a process.
1965	Clinton Corn initiated work with Japanese Agency of Industrial Science and Technology to commercialize process.
1967	First commercial shipment of HFCS by Clinton Corn was 15% fructose and was made by batch system using soluble glucose isomerase.
1968	First commercial shipment of 42% HFCS by Clinton Corn batch process and with both soluble and insoluble isomerase.
1972	Clinton Corn brought on stream first continuous plant using immobilized enzyme.
May 1978	First large-scale 90% HFCS plant by Archer Daniels Midland. Permitted volume production of 55% HFCS.
Jan. 1980	HFCS approved as 50% of the sweetener in brand Coca-Cola.
Apr. 1983	HFCS approved as 50% of the sweetener in brand Pepsi Cola.
Nov. 1984	100% level of 55% HFCS approved for both brands.

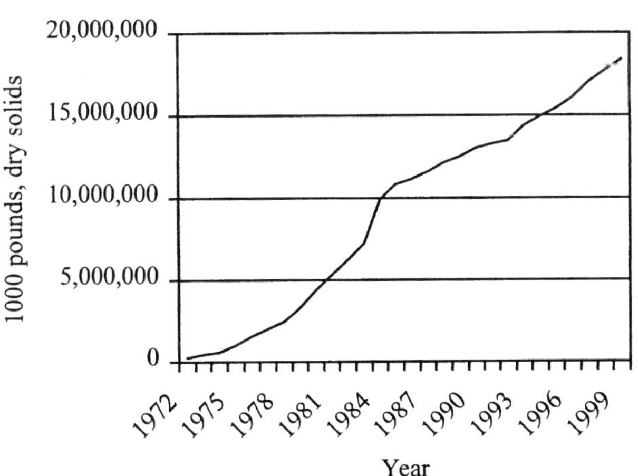

Figure 1 United States shipments of HFCS. (From USDA data.)

Table 3 World HFS Consumption (1000 metric tons dry)

	1989		1995	
	42% HFCS	55% HFCS	42% HFCS	55% HFCS
North America	1942	3563	2447	3947
South America	53	43	59	54
Europe-E.C.	293.2	—	293.5	—
Scandanavia	8	—	—	16
Eastern Europe	120	—	340	30
Africa	—	—	20	50
Asia	467	613	794	910
Oceania	10	—	12	—

Source: McKeany-Flavell Company, Oakland, CA.

many of the applications, functional, and nutritional properties of these other products are similar to pure crystalline fructose.

II. MANUFACTURE

Manufacture of HFCS is a highly sophisticated, automated operation starting with a continuous flow of corn and ending with a continuous output of finished sweetener. Manufacturing plants are relatively new and use the latest technology in separations, purification, and computer-automated control, resulting in the industry providing the highest quality food ingredient. The production of HFCS involves four major processing steps: (a) wet milling of corn to obtain starch, (b) hydrolysis of the starch to obtain dextrose, (c) conversion of a portion of the dextrose to fructose, and (d) enrichment of the dextrose-fructose stream to increase the fructose concentration.

A. Corn Wet Milling

Referring to Fig. 2, the objective of the corn wet milling process is to separate the starch from the other parts of the corn kernel (hull, gluten, and germ). Cleaned, shelled corn is soaked or steeped in a battery of tanks (steeps) in warm water containing 0.1–0.2% sulfur dioxide. The steepwater, which swells and softens the kernel to facilitate separation of the various components, flows countercurrently and is ultimately drawn off and replaced by fresh water. The steeped corn is degerminated in a water slurry by passing it first through a shearing mill that

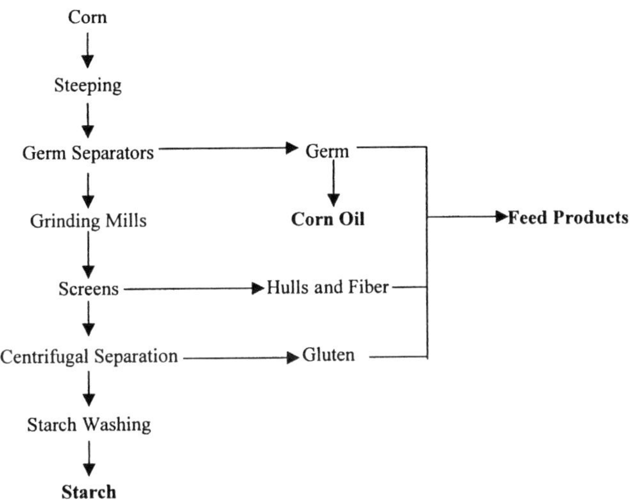

Figure 2 Corn wet milling process.

releases the germ and subsequently to a continuous liquid cyclone, which separates the germ for oil extraction. After fine grinding, the separated endosperm and hull are screened, and the resulting slurry is passed to a continuous centrifuge for starch and gluten separation. The separated starch fraction is filtered and redispersed in water repeatedly to reduce solubles (2) and to provide a pure starch slurry available for further processing into starch-based products such as HFCS, corn syrups, and ethanol.

B. 42% High Fructose Corn Syrup

The process for making 42% HFCS is shown in Fig. 3. The starch slurry made from the corn wet milling process is treated with alpha-amylase enzyme at high temperature to liquefy the starch to a dextrose equivalent (DE) of 10–20. This material is then treated with a saccharifying enzyme, glucoamylase, which hydrolyzes the liquefied starch to a dextrose content of about 95%. This high dextrose syrup is filtered, decolorized with carbon, and ion exchange refined to remove salts and other ionic compounds. It is then passed through a fixed-bed column of immobilized glucose isomerase enzyme. This isomerizing enzyme converts glucose to fructose (Fig. 4) and reaches equilibrium at about 45% fructose. This material is again filtered and refined by carbon and ion exchange treatment and evaporated to become the commercial 42% HFCS.

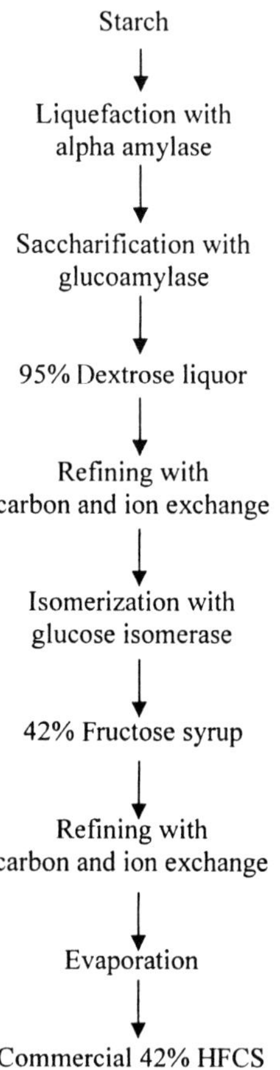

Figure 3 Production of 42% HFCS.

C. 55% High Fructose Corn Syrup

The refined glucose isomerase–treated liquor may be passed through fractionation units to enrich the fructose content to 80–95%. This material is filtered and refined by carbon and ion exchange treatment. It may be evaporated and provided as a commercial product with properties similar to crystalline fructose in liquid applications; however, more typically, this high fructose–containing syrup is blended with the 42% HFCS to a concentration of about 55% fructose and evaporated to become the commercial 55% HFCS. The dextrose from the fractionation unit is fed back to the isomerization columns in the 42% HFCS process. This process is shown in Fig. 5, and a comparison of the composition of these three commercially available products is shown in Table 4.

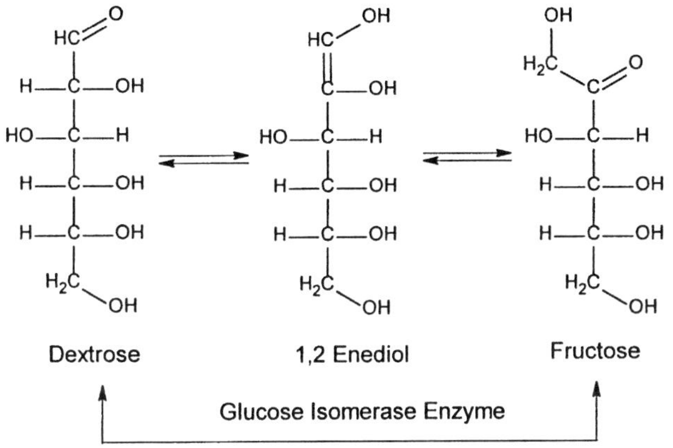

Dextrose ⇌ 1,2 Enediol ⇌ Fructose

Glucose Isomerase Enzyme

Figure 4 Isomerization of dextrose to fructose.

For a thorough description of the corn wet milling process, HFS manufacture, and special unit operations, see the chapter reference list (3–5).

III. CHARACTERISTICS

Typical chemical and physical characteristics are shown in Table 4. As the fructose level increases, so does sweetness. This is the major characteristic provided by HFCS.

Conventional corn syrups are characterized by their DE. DE is defined as the measurement of the total reducing sugars in the syrup calculated as dextrose

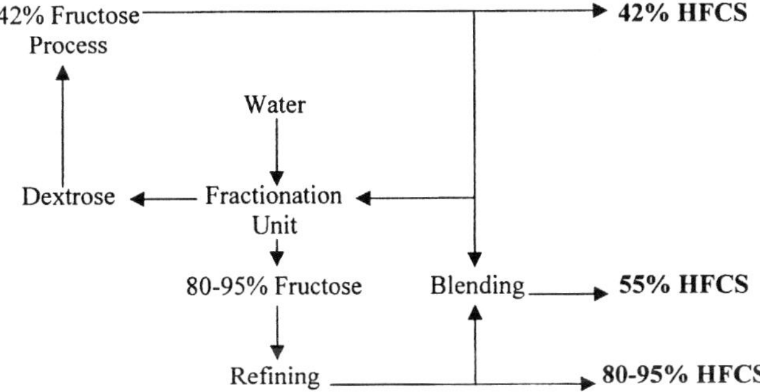

Figure 5 Production of 55% HFCS.

Table 4 Typical Characteristics of HFCS

	42% HFCS	55% HFCS	95% HFCS
Saccharides			
Fructose, %	42	55	80–95
Dextrose, %	52	41	4–19
Highers, %	6	4	1–3
Sweetness	1.0	1.0+	1.2
Dry Solids, %	71	77	77
Moisture, %	29	23	23
Color, RBU	<25	<25	<25
Ash, %	0.03	0.03	0.03
Viscosity (cps)			
80°F	150	700	575
90°F	100	400	360
100°F	70	250	220
pH	3.5	3.5	3.5
Lb./gal. (20°C)	11.23	11.55	11.56
Fermentable solids, %	96	98	99–100
Flavor	Sweet, bland	Sweet, bland	Sweet, bland

and expressed as a percentage of the total dry substance. HFCS is composed of a high amount of reducing sugars. Because the DE of HFCS is so high, HFCS is not characterized by DE but rather by the percent of fructose on a dry basis. Measurement of reducing sugars is based on the reducing action of aldose sugars on certain metallic salts (e.g., copper sulfate in Fehling's solution). Fructose, being a ketose sugar, is converted to an aldose under the alkaline conditions of the tests and behaves as dextrose. The reducing characteristic is one that particularly distinguishes HFCS from sucrose. Sucrose is not a reducing sugar and therefore does not participate in reducing reactions such as nonenzymatic browning until it has inverted into its constituent monosaccharides of fructose and dextrose.

pH for all products is low. Because these products are ion exchange refined, they have little buffering capacity and the pH tends to drift down with time. Because of the low buffering capacity and low acidity, HFCS has little effect on the pH of foods. Lower pH favors color stability for reducing sugars.

The treatment with carbon in the production of HFCS yields products with low color and flavor. Ion exchange treatment provides a product with extremely low ash, which helps to maintain the color and flavor stability of HFCS by removing color precursors and catalysts. Both treatments serve to reduce trace components that may promote or otherwise catalyze color and flavor development. The refining treatments and resulting lack of contaminant result in extremely bland products with few off-flavor notes.

HFCS products are extremely high in fermentable extract. This would be expected because of the high amount of monosaccharides and disaccharides (94–100%) and negligible oligosaccharide composition.

HFCS is microbiologically stable and meets the most stringent specifications requested by the food industry. The stability is a function of the high osmotic pressure created caused by the predominance of the monosaccharides in their composition and the high solids levels. Forty-two percent HFCS is supplied at lower solids than the others because its greater dextrose content renders it more susceptible to crystallization.

HFCS has lower viscosity compared with other nutritive sweeteners such as sucrose or conventional corn syrups but is equivalent to dextrose solutions and invert sugar because of the similar monosaccharide composition. At equal solids levels HFCS products have comparable viscosity; however, as supplied, 42% HFCS at lower solids has a lower viscosity.

IV. AVAILABILITY, TRANSPORT, STORAGE, HANDLING, AND SHELF-LIFE

The finished products are typically shipped from the manufacturing plants to the customer in bulk jumbo railcars equipped with food-grade epoxy lining or in bulk stainless steel tank trucks. Tank trucks normally have self contained positive displacement pumps for transferring the product to customer storage tanks. In addition to direct plant shipments, customers may be supplied the product by way of distributors or sweetener transfer facilities. Hundreds of strategically located facilities around the United States exist for the purpose of transferring product from railcar to storage tanks, trucks, drums, etc. for distribution to the customer.

Railcars are equipped with steam coils in the event the product cools sufficiently to promote crystallization of the more insoluble dextrose component. When this occurs, the syrup can be reliquefied with gentle heat. All three products are microbially stable and, under recommended storage conditions, the shelf-life is indefinite and depends on the final application's sensitivity to color development. HFCS is a clear, colorless liquid. It tends to develop a light straw color (color value of about 25 reference brain unit [RBU]) after prolonged time and/or higher temperatures. The product remains usable for many months under recommended storage conditions, maintaining sensory, physical, and chemical specifications. In general, storage temperatures of around 80 to 90°F will allow the product to be pumpable, prevent crystallization, and minimize color development. Cooler storage temperatures will increase shelf-life based on color development.

Table 5 Effect of Temperature on Density

	Pounds/gallon	
Temperature (°F)	42% HFCS	55% HFCS
60	11.258	11.573
70	11.230	11.542
80	11.201	11.511
90	11.172	11.479
100	11.142	11.448

Fourty-two percent will crystallize less than around 68°F, 55% at slightly cooler temperatures, and 95% will not crystallize at reasonable storage temperatures.

Stainless steel or mild steel tanks lined with food-grade epoxy lining are recommended for storage. Tanks should be equipped with sterile ultraviolet conditioning systems. These units are used to prevent condensation in the tank headspace that may dilute the surface of syrup and make it less microbially stable. The conditioning systems consist of blowers that pass filtered air past ultraviolet lights for sterilization before being introduced into the tank headspace.

Most users will take advantage of the convenience provided by bulk handling systems; however, these products may also be available in intermediate bulk containers, drums, or pails. Users of these containers should be equipped to warm the contents if necessary to facilitate pumping or reliquefy the contents.

HFCS is sold by the pound. It is commonly used by metered volume. The change in density with temperature may be sufficient for very large volume users to effect inventory and yield calculations so the effect of temperature must be taken into account when measured by volume. Table 5 demonstrates the change in weight on the basis of temperature.

V. SPECIFICATIONS AND ANALYSIS

HFCS is a consistent and high-quality bulk sweetener in the U.S. market. Because of refining to remove nonsweetener components and extensive microbiological controls in the process, it is able to meet very stringent specifications. Composition, identification, and testing requirements for HFCS have been published by the Committee on Food Chemicals Codex (FCC) of the National Academy of Sciences with the latest update in 1996 (Table 6). Because the beverage manufacturers are the single largest users of these products, stringent guidelines and test procedures (6) have been published by the International Society of Beverage Technologists (ISBT), which was known up until 1996 as the Society of Soft

Table 6 Food Chemical Codex IV—High Fructose Corn Syrup Specification

Description

High fructose corn syrup (HFCS) is a sweet, nutritive saccharide mixture prepared as a clear, aqueous solution form high dextrose-equivalent corn starch hydrolysate by the partial enzymatic conversion of glucose (dextrose) to fructose, using an insoluble glucose isomerase preparation that complies with 21 CFR 184.1372 and that has been grown in a pure culture fermentation that produces no antibiotics. It is a water-white to light yellow somewhat viscous liquid that darkens at high temperatures. It is miscible in all proportions with water.

Functional use in foods: nutritive sweetener

Requirements

Labeling: Indicate the color range and the presence of sulfur dioxide if the residual concentration is greater than 10 mg/kg.

Identification: To 5 ml of hot alkaline cupric tartrate TS add a few drops of a 1 in 10 solution of the samples. A copious red precipitate of cuprous oxide is formed.

Assay: 42% HFCS—Not less than 97.0% total saccharides, expressed as a percent of solids, of which not less than 42.0% consists of fructose, not less than 92.0% consists of monosaccharides, and not more than 8.0% consists of other saccharides; 55% HFCS—Not less than 95.0% total saccharides, expressed as a percent of solids, of which not less than 55.0% consists of fructose, not less than 95.0% consists of monosaccharides, and not more than 5.0% consists of other saccharides.

Arsenic (as As): Not more than 1 mg/kg.

Color: Within the range specified by the vendor.

Heavy metals (as Pb): Not more than 5 mg/kg.

Lead: Not more than 0.1 mg/kg.

Residue on ignition: Not more than 0.05%.

Sulfur dioxide: Not more than 0.003%.

Total solids: 42% HFCS—Not less than 70.5%; 55% HFCS—not less than 76.5%.

Drink Technologists (SSDT). Those guidelines published in November 1994 are shown in Table 7 and were the result of extensive collaboration between the soft drink manufacturers and the corn wet millers. Many of the test procedures incorporated were adapted from standard analytical methods developed and published by the Corn Refiners Association (CRA) (7).

Table 7 International Society of Beverage Technologists—Quality Guidelines

42% HFCS	Parameter	55% HFCS
71.0% ± 0.5%	Solids	77.0% ± 0.5%
0.05% Maximum	Ash (sulfated)	0.05% Maximum
42% Minimum	Fructose	55% Minimum
92% Minimum	Dextrose and fructose	95% Minimum
1.15 CRA (x100) max.	Color	1.15 CRA (x100) max.
10 Mold/10 g	Microbiological	10 Mold/10 g
10 Yeast/10 g		10 Yeast/10 g
200 Mesophiles/10 g		200 Mesophiles/10 g
6.0 ppm Maximum	Extraneous material	6.0 ppm Maximum
3.3–4.3	pH	3.3–4.3
Not objectionable	Flavor and odor	Not objectionable
4.0 ml Maximum of 0.05N sodium hydroxide to pH 6.0	Titratable acidity	Not objectionable
3 ppm Maximum	SO_2	3 ppm Maximum
97% T Minimum	Sulfonated polystyrene	97% T Minimum
80 ppb Maximum (11% ds)	Acetaldehyde	80 ppb Maximum (11% ds)

Because HFCS products are of high and consistent purity, useful properties such as solids and specific gravity can be determined by accurately measuring the refractive index. The CRA-sponsored studies by the Augustana Research Foundation in 1983, which resulted in a regression model being generated showing the refractive index–dry solids–density relationship of specific corn sweeteners. This information is used in the physical constants shown in Table 8. Complete tables covering all solids ranges are available from HFCS manufacturers.

The most critical characteristics of HFCS include the saccharide profile, dry solids, and color. Saccharide profile is best determined by the use of high-pressure liquid chromatography analysis. Polarimeters are often used to confirm acceptance of the correct product; however, because of the complex saccharide composition, they are not suitable for specification confirmation.

Color of HFCS is from virtually colorless to light straw. Because light straw is a subjective visual observation, spectrophotometric tests have been developed to measure this parameter. For years, soft drink bottlers used tests measuring RBUs on liquid sugars and set a specification of 35 RBU maximum. With the advent of HFCS with lower typical color, that specification was lowered to 25 RBU maximum. Since that time, the ISBT has adopted the CRA color test. A 25-RBU syrup is equivalent to a CRA color (times 100) of 1.15 maximum. Color

Table 8 HFCS Physical Constants

Solids %	Refractive index 20°C	Refractive index 45°C	Refractometer Brix 1966 ICUMSA	Specific gravity (in air) 20°C	Pounds per U.S. gal. at 20°C	Hydrometer Brix 20°C
42% HFCS						
11.0	1.3493	1.3457	10.92	1.04414	8.689	10.99
70.5	1.4632	1.4577	69.10	1.34717	11.211	69.54
71.0	1.4643	1.4589	69.55	1.35022	11.236	70.02
71.5	1.4655	1.4600	70.04	1.35327	11.261	70.50
72.0	1.4667	1.4612	70.54	1.35633	11.287	70.99
55% HFCS						
11.0	1.3492	1.3456	10.90	1.04418	8.689	11.00
76.5	1.4774	1.4716	74.84	1.38454	11.522	75.38
77.0	1.4786	1.4728	75.31	1.38769	11.548	75.87
77.5	1.4798	1.4740	75.78	1.39084	11.574	76.35

Source: Complied from CRA, ISBT, and ICUMSA.

values can be mathematically converted between the two units and, for the convenience of those more familiar with RBU color, that conversion equation is as follows:

$$RBU = \frac{CRA \times 100 + 0.0417}{0.0479}$$

The most commonly tested attributes are dry solids and pH, probably because many quality control laboratories are equipped to perform these tests. Dry solids are determined by refractometer. Many refractometers are calibrated with the Brix scale rather than refractive index. For sucrose solutions, Brix corresponds to percent solids in solution. The term *Brix* cannot be used when designating solids concentrations in other nutritive sweetener solutions such as invert sugars, HFCS, and corn syrups (8). From Table 8 it can be seen that for HFCS products the refractive index-Brix relationship is not the same as the refractive index-solids relationship. For 42% HFCS, the Brix value is roughly 1.4 less than the actual delivered solids, and for 55% HFCS, the difference is about 1.7. The difference is less at lower solids levels and at 10% solids the difference is only 0.1° Brix. Also, it is important to realize the effect of temperature on the converted dry solids value. There is a 0.1% solids error in the dry solids reading for every 1°C difference from 20°C. For example a refractive index or Brix reading taken at 25°C will result in a 77.0% solids value being reported as 76.5%.

Because of the high purity of HFCS, it has low ionic strength and buffering capacity. For this reason it is imperative that pure dilution water be used and that residual buffer solutions be removed from the probes to prevent erroneous pH results. Although pH is the most commonly used test, titratable acidity is a better indicator of the effect HFCS may have on the pH of a food system.

VI. PROPERTIES

Functionally, HFCS provides sweetness, flavor enhancement, viscosity, humectancy, nutritive solids, fermentability, and reducing characteristics such as flavor and color development. HFCS can be used to control crystallization, modify freezing point, raise osmotic pressure, and lower water activity.

By far, sweetness is the most important functional property of HFCS. Table 9 shows the relative sweetness of several sweeteners compared with sucrose. Although sweetness levels may be comparable, the actual sweetness may vary in a food system, depending on such factors as the temperature, pH, and solids. Sweetness is also affected by the flavor profile (9). HFCS has a rapid sweetness release. Compared with sucrose, HFCS is considered to enhance fruit flavors, spicy notes, and tartness. This is because the sweetness of HFCS is quickly perceived and does not linger. Sucrose on the other hand has a slower onset of sweetness but a longer lasting action. This can have a flavor-masking effect. The combination of the two can result in a synergism that results in a system perceived sweeter than the individual ingredients on their own.

Over the shelf-life of a product, acidic food systems (such as most beverages) sweetened with sucrose will invert, changing the sugar profile and consequently the flavor and sweetness of the product. HFCS has an advantage over sucrose in acidic food systems because it does not invert. Table 10 illustrates the effect of sucrose inversion.

Table 9 Relative Sweetness of Nutritive Sweeteners

Sweetener	Sweetness relative to sucrose
Sucrose	1.0
Crystalline fructose	1.2–1.8
90% HFCS	1.2–1.6
42% HFCS	1.0
55% HFCS	1.0+
Sucrose medium invert	1.0+
Dextrose	0.7

Table 10 Chemical Comparison between Sucrose and HFCS

Sucrose inverts in acidic foods or the body to dextrose and fructose.
100 g Sucrose $\xrightarrow{\text{Total inversion}}$ 52.6 g dextrose + 52.6 g fructose
100 g HFCS (dry basis) yields 94 to 100 grams dextrose and fructose

Although HFCS as supplied has a significantly lower viscosity than sucrose at the same level, the difference is minimal at lower solids levels. Viscosities of sweetener systems as delivered are shown in Table 11.

HFCS is an economical humectant and widely used for that purpose in the food industry. It has greater sweetening power than other common humectants such as glycerine and sorbitol.

Having a DE of more than 97, HFCS is composed predominantely of reducing sugars. Reducing sugars react with proteins in food systems to generate brown pigments through the Maillard or nonenzymatic browning reaction. The reaction is greatly enhanced at elevated temperatures such as in baking, giving rise to the brown crust color of items such as breads and rolls. At room temperature and below, the reaction is greatly reduced.

The monosaccharide makeup of HFCS results in changes in colligative properties versus the disaccharide sucrose. HFCS depresses the freezing point of food systems compared with sucrose, making a lower freezing, softer product. This property can be an advantage in some applications but a disadvantage in others. When identical freezing points are required, conventional corn syrups are often used to balance the freezing point.

The high osmotic pressure of HFCS is also a result of the monosaccharide composition. This property makes HFCS as supplied microbially stable and has been suggested as one reason monosaccharides are better for pickling and preserving processes such as in the manufacture of pickles and processed fruit.

The smaller molecules also result in lower water activity compared with the disaccharide sucrose. Low water activities allow greater microbial stability at higher moisture content, which often improves texture and shelf-life.

Table 11 Viscosity of Nutritive Sweeteners at 100°F

	Viscosity (centipoises)
42% HFCS	70
55% HFCS	250
67.5° Brix liquid sucrose	115
42 DE/43 Bé corn syrup	21,000

VII. APPLICATIONS

HFCS replaces sucrose on a solids basis in most foods except when a dry sweetener is required, such as dry mixes, and when hygroscopicity is a concern, as in some confections. Table 12 shows the major markets for HFCS. Some of the diverse product categories that find HFCS as an ingredient are bakery products, beverages of all types, processed fruits, condiments, frozen desserts, jams, jellies and preserves, pickles, wines, and liqueurs.

A. Beverages

By far the largest market segment is in beverages. Most 55% HFCS is used in this market. The major cola brands use that product because of its higher sweetness level. The availability of HFCS as a bulk liquid sweetener along with its consistent carbohydrate composition, narrow dry solids specification, and low ash level allows for highly automated bottling lines that use online instrumentation, which continuously meters and blends beverages to exacting beverage specifications at rates exceeding 1700 12-ounce cans per minute.

Typically beverages use 10–12.5% HFCS solids for desired sweetness. More acidic, stronger flavored beverages are at the high end. Sports drinks with isotonic requirements encouraging quick gastric emptying are typically at lower solids levels of about 2–9% (10). HFCS is also used as a fermentation adjunct in wines and as a sweetener for wine coolers and liqueurs.

B. Baking

The second single largest market for HFCS is the baking industry. Again, HFCS as a bulk liquid sweetener is ideal for large, automated bread and bun manufacturing lines. It is ideal in yeast-raised goods such as breads, buns, rolls, and yeast-

Table 12 HFCS Markets in United States—1995 Forecast

Industry	Percent of market
Beverages	72.3
Canning/processed foods	10.9
Baking	5.9
Dairy products	3.2
Other	7.6

Source: USDA Economic Research Service.

raised donuts because it is directly fermentable by yeast without the inverting action required before sucrose is used. In general, at equal solids levels, proof times are slightly shorter using HFCS compared with sucrose and volume, crust color, flavor, and grain are the same (11).

Besides its use as a sweetener and fermentable sugar, it is widely used in the baking industry as a humectant. It is incorporated into items such as cakes, cookies, and brownies as humectant and, compared with other commonly used humectants such as glycerine and sorbitol, it has increased sweetness.

High substitution levels for sucrose in chemically leavened products such as cakes and cookies result in different characteristics in the finished product. Cookies are plumper and softer, with less spread (12) but with lower water activity (13). Cakes are denser with less volume and flatter tops (14), which is reported to be due to starch gelatinization at a lower temperature (15). Under equal conditions, chemically leavened bakery products are browner. In general, cakes of equal quality can be achieved with a partial replacement (10–60%) of sugar with HFCS; however, changes in formulation can be made to achieve quality products at higher levels. Color may be controlled or minimized by lowering baking temperatures and lowering the pH slightly. The effect on leaving action may be modified by using higher protein flour and/or a highly emulsified shortening.

C. Processed Foods

The fruit canning industry uses a tremendous amount of HFCS, usually in combination with corn syrup and sucrose. HFCS is increasingly being used in vegetable canning such as peas and corn. Because of the improvements in refining of HFCS over the years, color development caused by retort temperatures and extended shelf-life is not the problem it once was. Water-soluble proteins in freshly cut and washed produce are not available to react with the reducing sugars, and modern refining techniques render HFCS virtually free of residual protein.

Reducing properties are important in maintaining the bright red color of tomato catsup and strawberry preserves (16), and HFCS as a sweetener is used in combination with conventional corn syrups in these items.

A large amount of HFCS is used in the production of jams, jellies, and preserves, usually in combination with conventional corn syrups to provide solids and sweetness. The HFCS/corn syrup formulation is an economical sweetener system that enhances fruit flavors, prevents crystallization, and preserves color.

D. Ice Cream and Dairy Products

HFCS is used in ice cream and other frozen desserts at levels from 2–8%. Removing the disaccharide sucrose from frozen desserts and replacing it with HFCS has an effect on the freezing point and texture of the product. In some cases, this is a benefit and in other cases both sucrose and HFCS are the preferred combination in these products. Formulators adjust the mixture with corn syrup, maltodextrin, or other bulking agents to fine-tune the sweetness, freezing point, and texture of the system.

Dairies incorporating HFCS in frozen desserts are likely to manufacture chocolate milk as well. HFCS works well in this beverage and has an ingredient-sparing effect on cocoa powder, allowing a 5–15% reduction (11).

VIII. REGULATORY STATUS

HFCS containing 42 or 55 percent fructose is affirmed GRAS with no limitations other than current good manufacturing practices per Chapter 21 of the U.S. Code of Federal Regulations Part 184.1866.

IX. NUTRITIONAL, SAFETY, AND REGULATORY STATUS

The nutritional value of HFCS is essentially that of 100% carbohydrate. Of the carbohydrate fraction, roughly 98% of HFCS is classified as sugars (monosaccharides and disaccharide) for nutritional labeling. The product is marketed as a nutritive sweetener and most manufacturers simply suggest using a caloric value of 4.0 kcal/g for the dry substance.

Because of the tremendous increase in HFCS consumption since the 80s with a corresponding decrease in sucrose consumption, from time to time we see references in the media of increased consumption of fructose. Commercially, much of the sucrose sold before the advent of HFCS was in the form of medium invert, which is a liquid and contains, at manufacture, about 50% sucrose, 25% dextrose, and 25% fructose. Referring again to Table 10, in acidic media such as soft drinks and in the body, sucrose will invert to approximately equal amounts of dextrose and fructose. Therefore, despite the many new applications for HFCS over the last several years, fructose per capita consumption is not much different because the increase in free fructose consumption was associated with a decrease in sucrose consumption, and the amount consumed has remained 8% of total daily energy intake (17). In addition, in the generally recommended as safe affirmation review by FDA, the agency concluded that HFCS "contains approximately the same glucose to fructose ratio as honey, invert sugar, and the disaccharide sucrose." Table 13 shows the changes in the per capita sweetener consumption since 1970.

Glycemic index is a measure of the rate at which a food substance raises blood sugar after ingestion and is often used by persons with diabetes as an indicator of the suitability of a particular food or ingredient in their meal plan.

Table 13 United States per Capita Sweetener Consumption[a]

Crop year	Refined sugar	HFCS	Glucose syrups	Dextrose	Total corn sweeteners	Honey and edible syrups	Total caloric sweeteners
1970	101.8	0.7	14.0	4.6	19.3	1.5	122.6
1971	102.1	0.9	14.9	5.0	20.8	1.4	124.3
1972	102.3	1.3	15.4	4.4	21.1	1.5	124.9
1973	100.8	2.1	16.5	4.8	23.4	1.4	125.6
1974	95.7	3.0	17.2	4.9	25.1	1.1	121.9
1975	89.2	4.9	17.5	5.0	27.4	1.4	118.0
1976	93.4	6.9	17.5	5.0	29.4	1.3	124.1
1977	94.2	9.1	17.6	4.1	30.8	1.4	126.4
1978	91.5	11.2	17.8	3.8	33.8	1.5	125.8
1979	89.3	14.4	17.9	3.6	35.9	1.4	126.6
1980	83.6	18.0	17.6	3.5	39.1	1.2	123.9
1981	79.4	22.2	17.8	3.5	43.5	1.2	124.9
1982	73.6	26.7	18.0	3.5	48.2	1.3	127.8
1983	71.0	31.0	18.0	3.5	52.6	1.3	130.4
1984	67.6	37.3	18.0	3.5	58.8	1.4	128.7
1985	63.2	44.6	15.9	3.5	63.9	1.5	128.6
1986	60.8	45.1	16.0	3.5	64.6	1.6	127.0
1987	63.1	47.1	16.2	3.6	66.8	1.7	131.6
1988	62.6	48.3	16.4	3.6	68.3	1.5	132.4
1989	62.8	47.5	16.7	3.7	68.0	1.6	132.4
1990	64.8	49.2	17.4	3.8	70.4	1.6	136.8
1991	64.4	50.0	18.2	3.8	72.0	1.6	138.0
1992	65.5	51.6	19.0	3.8	74.4	1.6	141.4
1993	65.2	54.4	19.6	3.8	77.8	1.6	144.6
1994	65.9	56.4	20.0	3.9	80.3	1.5	147.6
1995	66.2	58.2	15.1	4.0	77.3	1.5	144.9
1996	66.9	59.4	20.6	4.0	84.0	1.5	152.3
1997[b]	67.1	61.4	20.8	4.0	86.3	1.5	154.8

Source: U.S. Department of Agriculture, Economic Resource Service.
[a] Dry basis.
[b] Estimate.

Although pure fructose has a very low glycemic index of 20, compared with white pan bread at 100 and sucrose at 90 because of their high dextrose content, 42 and 55% HFCS would not have a low glycemic advantage and would be expected to be similar to sucrose in this area.

Fructose, in the form of HFCS used in the food supply, was included in a comprehensive evaluation of the health effects of sugars conducted by FDA in

1986. This evaluation concluded that, "Other than the contribution to dental caries, there is no conclusive evidence that demonstrates a hazard to the general public when sugars are consumed at the levels that are now current and in the manner now practiced."

X. ECONOMICS/COST

The cost of the raw material, corn, has a major influence on industry economics; however, other factors can seriously offset that parameter. Net corn costs are determined after subtracting the value of corn wet milling by-products of corn oil, corn gluten feed, corn gluten meal, and ethanol. Strong demand and pricing for these by-products has the net effect of reducing overall corn costs, whereas

Table 14 United States Spot Price for 42% HFCS, Midwest Markets

Calendar year	$ Per hundred pounds (dry weight)
1975	22.67
1976	14.03
1977	12.44
1978	12.12
1979	13.15
1980	23.64
1981	21.47
1982	14.30
1983	18.64
1984	19.94
1985	17.75
1986	18.07
1987	16.50
1988	16.47
1989	19.21
1990	19.69
1991	20.93
1992	20.70
1993	—
1994	18.77
1995	15.63
1996	14.46
1997	10.70

Source: Milling and Baking News.

weak markets drive the effective price of the raw material up. Major additional costs are for energy (steam and electricity), manpower (salaries, wages, and overhead), and supplies (chemicals, enzymes, packaging, etc.). These are in addition to water and waste treatment costs, depreciation, maintenance, etc. required to operate the process.

Early in the history of HFCS, price was related to the price of sugar and net corn costs. At present, supply and demand have a major influence as the industry expands production capacity past its current need. Table 14 shows the average free on board costs over the past several years.

REFERENCES

1. Milling and Baking News, March 12, 1996. p 23.
2. Corn Refiners Association. Nutritive Sweeteners from Corn. 5th ed. 1989.
3. Schenck FW, Hebeda RS, eds. Starch Hydrolysis Products: Worldwide Technology, Production, and Applications. New York: VCH Publishers, Inc., 1992.
4. May JB. Wet milling: Process and products. In: Watson SA, Ramstad PE, eds. Corn: Chemistry and Technology. St. Paul, MN: American Association of Cereal Chemists, Inc., 1994, pp 377–397.
5. Corn Refiners Association. Corn Refiners Association Symposium on Food Products of the Wet Milling Industry. September 25, 1986.
6. Society of Soft Drink Technologists. Quality Guidelines and Analytical Procedure Bibliography for ''Bottlers,'' High Fructose Corn Syrup 42 and 55. November, 1994.
7. Corn Refiners Association, Inc. Standard Analytical & Microbiological Methods. 1999.
8. Junk WR, Pancoast HM. Handbook of Sugars. 2nd ed. Westport, CT: AVI Publishing Company, 1980.
9. Shallenberger RS. Sweetness theory and its application in the food industry. Food Technol 52, (7):72–76, 1998.
10. Coleman E. Sports drink research. Food Technol March: 104–108, 1999.
11. Strickler AJ. Corn syrup selections in food applications. In: Lineback, DR, Inglett GE, eds. Food Carbohydrates. Westport, CT: AVI Publishing Company, 1982, pp 12–24.
12. Dubois D, ed. AIB Technical Bulletin, July 1986.
13. Dubois D, ed. AIB Technical Bulletin, July 1987.
14. Ohr LM. Prepared foods. March 1999.
15. White DC, Lauer DN. Predicting gelatinization temperature of starch/sweetener systems for cake formulation by differential scanning calorimetry. Cereal Foods World 35:728–731, 1990.
16. Palmer TJ. Nutritive sweeteners from starch. In: Birch GG, Parker KJ, eds. Nutritive Sweeteners. London: Applied Science Publishers, 1982, pp 83–108.
17. Park YK, Yetley EA. Intakes and food sources of fructose in the United States. Am J Clin Nutr 58:737S–747S, 1993.

22
Isomaltulose

William E. Irwin*
Palatinit Süßungsmittel GmbH, Elkhart, Indiana

Peter Jozef Sträter
Palatinit Süßungsmittel GmbH, Mannheim, Germany

I. INTRODUCTION

Isomaltulose is a reducing disaccharide. Its systematic name is 6-*O*-α-D-glucopyranosyl-D-fructofuranose (CAS Reg. No. 13718-94-0). Isomaltulose is a natural constituent of honey (1), occurring at levels up to 1%, and has also been found in sugar cane extract (2).

Isomaltulose was discovered in 1952 in a sucrose-containing substrate of *Leuconostoc mesenteroides* (3). Süddeutsche Zucker AG in 1957 first reported their interest in isomaltulose (4). This isomer of sucrose is produced commercially by Palatinit® Süßungsmittel GmbH using a sucrose–glucosylfructose–mutase enzyme. Isomaltulose (Palatinose®) is beginning to be used as a food ingredient and is the basic raw material for the production of isomalt (Palatinit®).

The chemical formula of isomaltulose is $C_{12}H_{22}O_{11} \cdot H_2O$. The molecular weight is 360.32. The chemical structure of isomaltulose has been confirmed by Loss (5) and is shown in Fig. 1. The crystalline structure of isomaltulose has been studied by Dreissig and Lugar (6).

II. PRODUCTION

The transglucosidation of sucrose to produce isomaltulose is accomplished using *Protaminobacter rubrum* (ATCC No. 8457) (7). In addition to this strain, other

*Retired.

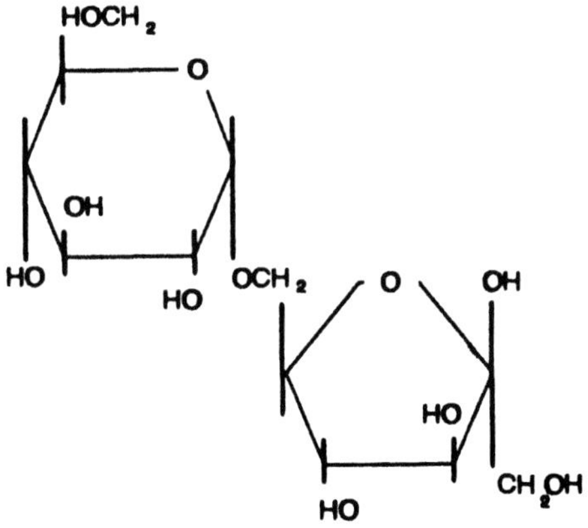

Figure 1 Chemical structure of isomaltulose.

microorganisms are capable of carrying out this transformation (e.g., *Serratia plymuthica* [8] and *Erwinia rhapontici* [9]).

The immobilized cells of the bacterium are used to pack a reactor column. A concentrated sucrose solution is pumped through the column, where the sucrose is transformed into isomaltulose.

Table 1 Food-Grade Isomaltulose: Typical Analysis

Test	Results	Method
Appearance	White crystalline substance	FCC
Melting Range	120–126°C	FCC
Water	5.0–6.5%	FCC
Specific rotation $[\alpha]^{20}$	±97–104° (C = 2, water)	FCC
Ash	0.02–0.10% (dry basis)	AOAC
Color of solution	max. 80 ICUMSA units	ICUMSA
Isomaltulose	NLT 98.0%	HPLC or GC

FCC = Food Chemicals Codex.
ICUMSA = International Commission for Uniform Methods of Sugar Analysis.
AOAC = Association of Official Analytical Chemists.

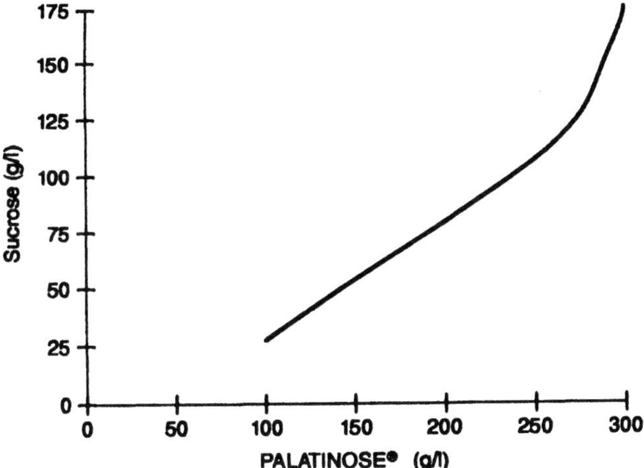

Figure 2 Isosweet aqueous solutions of Palatinose® and sucrose at 20°C.

The clarified solution from the fermenter is concentrated by evaporation and is then crystallized in two stages. The first crop is obtained by evaporation and the second by cooling. The crystals are dissolved in demineralized water and recrystallized. Isomaltulose crystallizes as a monohydrate.

The typical analysis of food grade isomaltulose is given in Table 1.

III. SENSORY PROPERTIES

The sweetness quality of isomaltulose is similar to sucrose. It is perceived quickly, it is refreshing and heavy without any aftertaste. Its sweetening power in comparison to a 10% sucrose solution is 0.48, but there is an increase in sweetening power as the concentration is increased, as seen in Fig. 2 (10).

Isomaltulose is reported to mask the off-flavors of some intense sweeteners (11).

IV. PHYSICAL AND CHEMICAL PROPERTIES

A. Physical Properties

The appearance of isomaltulose, a white crystalline material, is similar to that of sucrose. Its melting point (123–124°C) is lower than that of sucrose (190°C). It is practically nonhygroscopic. It is easily ground, an important property in many applications.

The solubility of isomaltulose, as seen in Fig. 3, although lower than that of sucrose, is adequate for most applications. The viscosity of isomaltulose solutions shown in Fig. 4 is very similar to that of sucrose solutions.

B. Chemical Properties

Isomaltulose is very acid stable compared with sucrose. At pH 2.0 with HCl, a 20% isomaltulose is stable for more than 60 minutes when heated at 100°C, whereas a 20% sucrose solution is completely hydrolyzed under these conditions. Isomaltulose is a reducing sugar and as such undergoes Maillard reactions. The microbial stability of products prepared with isomaltulose is very good.

V. METABOLISM

Isomaltulose has been shown to be hydrolyzed to glucose and fructose by intestinal α-glucosidases (12). The relative cleavage rate of isomaltulose by homogenates of jejunal mucosa is 44 compared with a sucrose rate of 100. Relative cleavage rates of 30 and 100 are found with rat tissue (13).

Experiments with rats have shown that the metabolism of isomaltulose is independent of the size of the dose and sex of the animals and compares in all respects to that of sucrose. It was concluded in the study by Macdonald and Daniel (14) that the V_{max} for the enzymatic hydrolysis of isomaltulose is about half that for sucrose.

In a study with pigs, one group of animals was provided with a reentrant fistula in the end of the ileum, which made possible the examination of the enzymatic gastrointestinal digestion process in the small intestine separately from the fermentative digestive processes in the large bowel. A second group of nonfistulated animals were used to study digestibility over the whole gastrointestinal tract. Study diets included either 20% sucrose or isomaltulose. Analysis of the ileal chyme from the fistulated pigs showed that there was no passage of isomaltulose (<0.3%) at the terminal end of the ileum. Both sugars were almost completely digested and absorbed in the small intestine; fermentation in the large intestine was not a significant factor (15).

A. Cariogenicity

Isomaltulose has been shown to be less cariogenic than sucrose. Ciardi has reported that isomaltulose does not support growth by oral streptococci and, in addition, has the added advantage of inhibiting bacterial growth and/or glucan

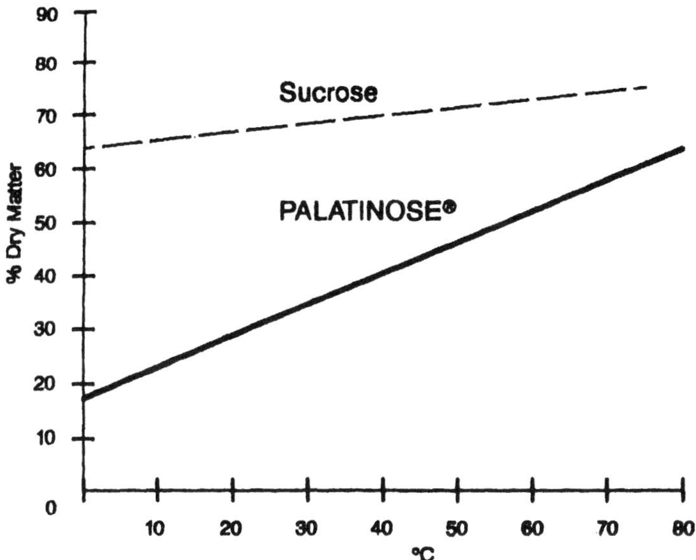

Figure 3 Solubility of Palatinose® and sucrose in water.

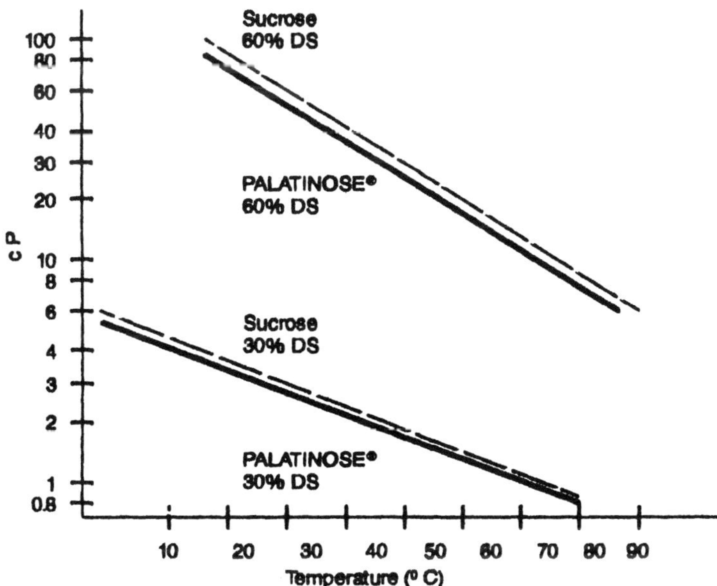

Figure 4 Viscosity of Palatinose® and sucrose solutions.

synthesis (16). A similar conclusion was reached by Gehring (17), who reported the final pH values of nutrient media containing different sugar substitutes after incubation with 100 *Streptococcus mutans* strains. These results have been confirmed in vitro by pH-telemetric studies according to Muhlmann with pure isomaltulose and isomaltulose-based chewing gum (18, 19).

It has been reported that when isomaltulose was added to reaction mixtures that contained various concentrations of sucrose, a significant reduction in the yield of insoluble glucan was noted. With the addition of 4% isomaltulose, insoluble glucan synthesis in a 1% sucrose solution was reduced from 100 to 4%. The addition of 0.125% isomaltulose reduced it to approximately 50% (20). The competition of isomaltulose in glucosyl reception is believed to cause the reduction in the amount of insoluble glucan produced from sucrose (21).

B. Effect on Blood Glucose and Insulin

Either isomaltulose or sucrose, in doses ranging from 0.25–1.0 g/kg of body weight, was given to healthy adult males. The insulin and fructose concentrations in the serum measured after isomaltulose ingestion were approximately half of those measured after sucrose ingestion. At all doses the rate of rise of blood glucose, insulin, and fructose levels and the peak values were lower for isomaltulose than for sucrose. When the area under the glucose response curve was calculated, there was no significant difference between sucrose and isomaltulose, except at the highest dose, whereas the area under the insulin response curve was lower in the case of isomaltulose at all dose levels. As a result of the slower cleavage rate and therefore slower absorption of the monosaccharides from isomaltulose, no rise in any intestinal discomfort was observed. This is compatible

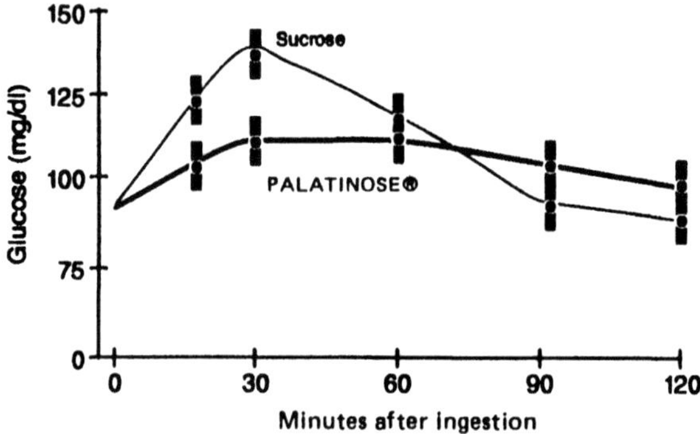

Figure 5 Plasma glucose levels after oral administration of 50 g sucrose or Palatinose® in man.

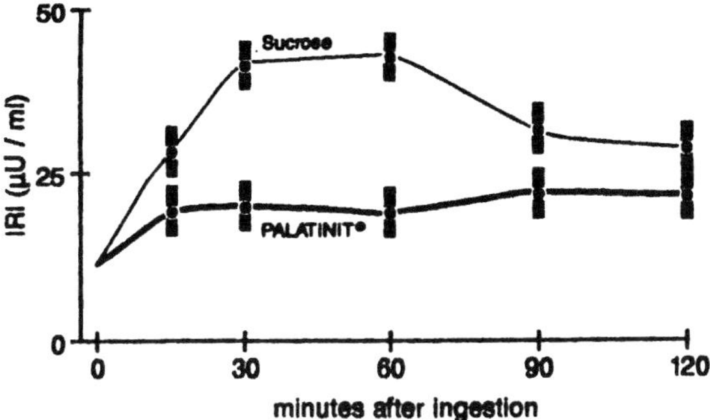

Figure 6 Plasma insulin levels after oral administration of 50 g sucrose or Palatinose® in man.

with the view that isomaltulose is completely absorbed, although more slowly than is sucrose (14). These results are similar to those reported by Kawai et al. (22) as seen in Figs. 5 and 6.

C. Tolerance

No evidence of intestinal discomfort has been observed in human studies with isomaltulose (14, 22).

D. Toxicological Evaluations

The monomers and chemical bonds comprising the chemical structure of isomaltulose are the same as those found in sucrose or starch. It is not surprising that only a limited number of toxicological studies have been conducted. No evidence of toxicity was noted in a 26-week toxicological study conducted with rats (23).

VI. APPLICATIONS AND PRODUCT DEVELOPMENT

In the manufacture of most foods, isomaltulose can be substituted for sucrose without any changes in the traditional manufacturing processes. Product development studies have been successful in many applications (24). The following applications are representative of these studies.

Anticaking powder
Baked goods
Canned fruits

Chewing gum
Chocolate products
Confections
Frostings
Hard and soft candy
Puddings
Sports and dairy drinks
Herbal tea drinks
Toothpaste

The solubility of isomaltulose is adequate for most applications. An exception is in the manufacture of traditional jams and jellies, in which the lower solubility of isomaltulose does not allow the production of high-solid products. Isomaltulose is suitable for use in the manufacture of the newer "light" fruit spreads.

Because of its higher bacterial and chemical stability than sucrose, isomaltulose is used in a major way as a sweetener in dairy products containing active cultures with acidophilus and bifidus bacteria. These bacteria cannot split isomaltulose as opposed to sucrose in the dairy product so that the sweetness level remains constant. Isomaltulose can also be found on the market in sugar-free drinks, such as teas, for children. The use of isomaltulose in such teas is of special interest because the tolerance of isomaltulose is very good compared with polyols and other slowly digestible carbohydrates.

VII. REGULATORY STATUS

In Japan and Korea isomaltulose is considered as a food and is marketed as such. Isomaltulose is used as a food additive in China. A regulatory status has not yet been established in other countries.

REFERENCES

1. Siddiqua R, Furhala B. Isolation and characterization of oligiosaccharides from honey. J Apicultural Res 6(3):139, 1967.
2. Takazoe I. New trends on sweeteners in Japan. Int Dent J 35:58, 1985.
3. Stodala FH, Hoepsell H, Sharpe ES. Ein neues von *Leuconostoc mesenteroides* gebildetes Disaccharid. J Am Chem Soc 74:3202, 1952.
4. Weidenhagen R, Lorenz S. Palatinose (6-α-glucopyranosido-fructo-furanose), ein neues bakterielles Umwandlungsprodukt der Saccharose, Zeitschr Zuckerindustrie 7:533, 1957.
5. Loss F. Über einige Derivate der Palatinose (Isomaltulose), zugleich ein Beitrag zur Frage der Konstitution. Zeitschr Zuckerindustrie 19:323, 1969.

6. Dreissig VW, Lugar P. Die Strukturbestimmung der Isomaltulose, $C_{12}H_{22}O_{11}H_2O$. Acta Cryst ect. B 29:514, 1973.
7. Schiweck H. Palatinit—Herstellung, technologische Eigenschaften und Analytik Palatinit haltiger Lebensmittel. Alimenta 19:5, 1980.
8. Schmidt-Berg-Lorenz S, Mauch W. Ein weiterer Isomaltulose bildender Bakterienstamm. Zeitschr Zuckerindustrie 14:625, 1964.
9. Cheetham PSJ, Imber CE, Isherwood J. The formation of isomaltulose by immobilized *Erwinia rhapontici*. Nature 299:628, 1982.
10. Paulus K. Abschlussbericht, sensorische Untersuchungen zur Süßkraft von Palatinose. Private communication, May 28, 1985.
11. Anonymous. Low cariogenic sugar with a bright future; Palatinose. Food Engineering 57(5):1985.
12. Dahlquist A, Auricchio S, Semenza G, Prader A. Human intestinal disaccharidases and hereditary disaccharide intolerance. The hydrolysis of sucrose, isomaltulose (Palatinose), and a 1,6-α-oligiosaccharide (isomalto-oligosaccharide) preparation. J Clin Invest 42:556, 1963.
13. Grupp U, Siebert G. Metabolism of hydrogenated Palatinose, an equimolar mixture of α-D-glucopyranosido-1,6-sorbitol and α-D-glucopyranosido-1,6-mannitol. Res Exp Med (Berl.) 173:261, 1978.
14. Macdonald I, Daniel JW. The bioavailability of isomaltulose in man and rat. Nutr Rep Intern 28:1083, 1983.
15. van Weerden IER, Huisman LJ, Leeuwen P. Digestion processes of Palatinose® and saccharose in the small and large intestine of the pig. ILOB-report 320, 20 June, 1983.
16. Ciardi J, Bowen WH, Rolla G, Nagorski K. Effect of sugar substitutes on bacterial growth, acid production and glucan synthesis. J Dent Res 62:182, 1983.
17. Gehring, Über sie Säurebildung kariesätiologisch wichtige Streptokokken aus Zuckern und Zuckeralkoholen unter besonderer Berücksichtigung von Isomaltit und Isomaltose. Ztschr. für Ernährungswiss Suppl. 15:16, 1973.
18. Mühlmann HR. Private communication. Zahnärztliches Institut der Universität Zürich, 1983.
19. Imfeld T. Private communication. Zahnärztliches Institut der Universität Zürich, 1983.
20. Takazoe I, Frostell G, Ohta K, Topitsoglou V, Saski N. Palatinose—a sucrose substitute. Swed Dent J 9:81, 1985.
21. Nakajima Y, Takaoka M, Ohta K, Takazoe I. Inhibitory activity of Palatinose on the synthesis of insoluble glucan from sucrose by *Streptococcus mutans*. Seito gijutsu kenkyukaishi 34:58, 1985.
22. Kawai K, Okuda Y, Yamashita K. Changes in blood glucose and insulin after an oral Palatinose administration in normal subjects. Endocrinol Japan 32(6):933, 1985.
23. Yamaguchi K, Yoshimura S, Inada H, Matsui E, Ohtaki T, Ono H. 26-week Palatinose toxicity study with rats through oral administration. Pharmacometrics 31(5): 1015, 1986.
24. Sträter PJ. Palatinose. Zuckerindustrie, 112(10):900, 1987.

23
Trehalose

Alan B. Richards
Hayashibara International Inc., Westminster, Colorado

Lee B. Dexter
Lee B. Dexter & Associates, Austin, Texas

I. INTRODUCTION

Trehalose is a unique disaccharide with important functional properties. Although these properties have been recognized for many years, it is only recently that the use of trehalose in a wide variety of products has been considered economically feasible. This change resulted from the development of a novel system of manufacture that dramatically reduces production costs, making it possible to integrate trehalose into a wide range of cost-sensitive products. The various physical properties as they relate to food applications are only now being explored. This chapter reviews various aspects of trehalose and provides preliminary information on its functional properties, potential use, regulatory status, and safety profile.

II. HISTORICAL BACKGROUND

Trehalose is believed to have been first isolated in 1832 from the ergot of rye by Wiggers (1). The name "trehalose" was coined years later when Berthelot extracted the same disaccharide as the primary sugar contained in a cocoonlike secretion of a beetle found in the Iraqi desert. The insect cocoon or shell was known as "trehala manna" (2). Other insects make similar structures using different sugars (2, 3). Trehala manna was described as a sticky raw material, and Leibowitz reported that about 30–45% of the dry weight of the manna was trehalose (2, 3). Bedouins have been known to gather manna, found on leaves of various plants, for use as a sweetening agent. Some believe that it may be similar to the manna mentioned in the Old Testament of the Bible, recorded as being eaten by the Israelites in the wilderness (4).

III. HISTORICAL CONSUMPTION OF TREHALOSE AND NATURAL OCCURRENCE

Trehalose appears to have been a part of the human diet since the beginning of mankind. It occurs in a wide range of fungi, insects, and other invertebrates such as lobster, crab, prawns, and brine shrimp (5,6, S Miyake, unpublished data, 1997). In addition, it can be isolated from certain plants such as sunflowers, particularly the seeds. Other common foods in which trehalose occurs are honey, wines, sherries, invert sugars, breads, and mirin (a sweet saké used for cooking) (6, S Miyake, unpublished data, 1997). Both brewer's and baker's yeast accumulate trehalose, the latter to concentrations of 15–20%, on a dry weight basis (7). Certain bacteria, including *Escherichia coli*, a common enteric bacteria of the human digestive tract, can synthesize trehalose (8, 9). Early man likely had a much greater exposure to trehalose than modern man because of greater dietary dependance on fungi, insects, and aquatic invertebrates. Although many of these substances are not common in our modern diet, several items are still consumed by a large portion of the population and provide a constant exposure to trehalose in the course of a normal diet. Modern food sources shown to contain significant quantities of trehalose are honey (0.1–1.9%), mirin (1.3–2.2%), brewer's (0.01–5.0%) and baker's yeasts (15–20%) (5–7, 10). Commercially grown mushrooms can contain 8–17% (w/w) trehalose. Reported values of trehalose from natural sources may vary substantially because of experimental conditions, analytical methods, and the life-cycle stage of the organism assayed.

Elbein provides an extensive review of the reported general occurrence of trehalose and the number of derivatives of trehalose that are found in nature (6). Further work will likely show the presence of trehalose in many additional organisms.

Since 1995, trehalose has been incorporated into hundreds of food products in Japan. Total consumption of trehalose added to food products in Japan will likely exceed 30,000 metric tons by the end of 1999. It is also sold in Taiwan and Korea for use in processed foods.

IV. CHEMICAL COMPOSITION

α,α-Trehalose (α-D-glucopyranosyl α-D-glucopyranoside) has commonly been called mushroom sugar or mycose. It is a conformationally stable, nonreducing

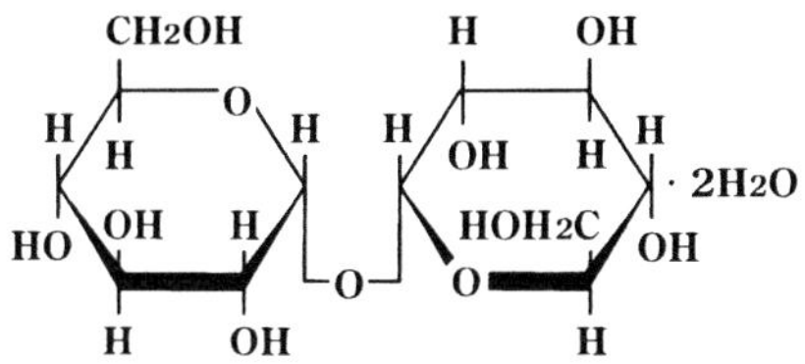

Figure 1 Structure of α,α-trehalose dihydrate.

sugar consisting of two glucose molecules bound by an α,α 1,1 glycosidic linkage (Fig. 1). After ethanol extraction or enzymatic synthesis, trehalose crystallizes to a stable dihydrate form (1, 5). Trehalose can also be found in nature as α,β (neotrehalose) and β,β (isotrehalose) isomers; however, these are uncommon, and they possess chemical and physical properties that are distinct from α,α-trehalose. In this review "trehalose" will be used to designate the α,α-trehalose form. The chemical formula and molecular weight of trehalose dihydrate are $C_{12}H_{22}O_{11} \cdot 2H_2O$ and 378.33. The Chemical Abstracts Service Registry Number is 6138-23-4.

V. PHYSICAL PROPERTIES

Physical properties reported in this section are, in large part, obtained from assays of trehalose produced enzymatically from starch by Hayashibara Company Ltd. (HBC) of Okayama, Japan; however, data from other sources are included (5).

Trehalose is available as a white crystalline powder that is colorless and of prismatic rhomboid form. X-ray diffraction reveals cell parameters of a = 12.33, b = 17.89, and c = 7.66 A (11, 12). The orthorhombic cell formula is given as $P2_12_12_1$, and the units per cell are 4. The theoretical density of 1.511 g/cm^3 agrees closely with the actual value of 1.512 g/cm^3. Reports of the melting point of trehalose dihydrate range from 94–100°C (1). Continued heating above 100°C vaporizes the water of hydration, resulting in resolidification; a second melting point is reached at 205°C (1). The melting point of trehalose dihydrate produced and assayed by HBC (97°C) equals the value listed in reference texts (13, HBC, unpublished data, 1997). The melting point of Hayashibara anhydrous trehalose is 210.5°C, approximately 7.5°C higher than reported in the literature (13, HBC, unpublished data, 1997).

Birch reports data from several studies in which the optical rotation of various preparations of trehalose ranged from +177° to +197.14°, but gives a value of +199° for his own anhydrous preparation (1, 14). Unpublished data from assays performed by HBC (1997) give a specific rotation equal to $[\alpha]^{20}{}_{D}$: +199° (c = 5).

VI. PRODUCTION OF TREHALOSE

A. Historical Processes for Trehalose Production

The isolation of trehalose from natural sources has been ongoing for several decades. Historically, the production of trehalose has been limited to small quantities obtained by extraction from microorganisms such as baker's yeast (1, 15, 16, 17). One complex purification method produced maximum yields of 0.7%, whereas later procedures resulted in yields of up to 16% of active dried yeast (92% solids) (15, 17).

Recently, investigators have focused on the use of bacterial synthesis, transgenic technology, or enzymatic conversion for the production of trehalose (5). Trehalose has been isolated during the fermentation of n-alkanes by *Arthrobacter* sp. in quantities of 5–6 g/l (18). Three Japanese patents, issued between 1975 and 1993, described the production of trehalose by bacteria using common carbon sources such as glucose and sucrose. Genetic recombination has been attempted by insertion of a gene that converts glucose into trehalose in a sugar-producing crop (5). Murao et al. reported on an enzymatic method for producing trehalose from maltose using trehalose- and maltose-phosphorylases, which provided a yield of 60% (19). Trehalose has also been produced enzymatically by a combination of acid reversion of D-glucose and trehalose (5, 20).

Although these newer methods have reduced the cost of trehalose, they have not been able to provide trehalose at a cost low enough for use in most food applications. This has effectively restricted its use to research, pharmaceuticals, and high-value cosmetics.

B. The Hayashibara Manufacturing Process

Hayashibara Biochemical Laboratories of Okayama, Japan, screened soil samples for bacteria that could produce trehalose from relatively inexpensive carbon sources such as starch or maltose.

One group of bacteria were shown to convert maltose to trehalose in a manner that was more efficient than previously reported enzyme systems (19, 21). A second bacterial group was found to use starch as the substrate for trehalose production (22, 23). Hayashibara scientists isolated and characterized the two different trehalose-producing enzyme systems from these organisms (21–23). Those species that produced trehalose directly from maltose required only one novel enzyme, trehalose-synthase (21). The bacteria that produced trehalose

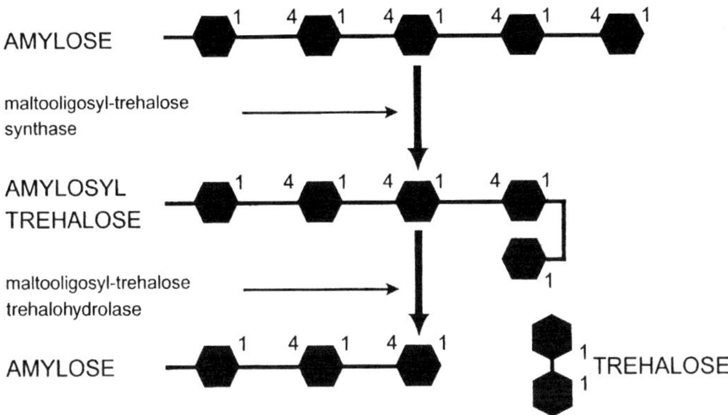

Figure 2 Enzymatic production of Hayashibara trehalose from starch. A patented debranching enzyme is used to produce the amylose substrate.

from starch possessed two novel enzymes, maltooligosyl-trehalose synthase and maltooligosyl-trehalose trehalohydrolase (22, 23).

HBC has commercialized the latter process using starch to manufacture trehalose. This new process has reduced the price of trehalose approximately 100-fold. Before this discovery, the cost of trehalose in Japan ranged from 20,000–30,000 Yen (¥) per kilogram ($170–250 USD/kg). Food-grade trehalose is now sold in Japan for about 300 ¥/kg ($2.50/kg).

The direct conversion of starch to trehalose on a commercial scale depends on a system in which several enzymes coreact in the same medium (5, 22–24). Isoamylase debranches the starch, producing amylose (Fig. 2). Amylose is then converted to amylosyl-trehalose by maltooligosyl-trehalose synthase, which identifies the reducing end glucose units and catalyzes the conversion of α-1,4 bonds into α,α-1,1 bonds. In the next step, maltooligosyl-trehalose trehalohydrolase interacts with amylosyl-trehalose molecules, hydrolyzing the α-1,4 bond between the amylosyl-moiety and the trehalose, releasing the trehalose into the media. This reaction terminates when the degree of polymerization of amylose becomes two or less; that is, when only glucose or maltose remains. This enzymatic system followed by conventional methods of saccharide production results in high-purity crystalline trehalose (5, 22–24).

VII. TECHNICAL QUALITIES

In nature, trehalose functions as one of the primary molecules responsible for maintaining the bioactivity of cell membranes during stress caused by freezing, desiccation, or high temperatures (25). It is present in relatively high concentrations in organisms that can be desiccated or frozen and on rehydration or thawing return to essentially full function. Examples include brine shrimp, baker's and brewer's yeast, insects, and certain plants (25). In nature, trehalose is thought to function through a number of mechanisms to assist such organisms as baker's yeast in resuming normal metabolic activity after freezing or drying. These mechanisms include the following:

1. Water replacement—Trehalose has been shown to replace structural water molecules that are hydrogen bonded to the end groups of proteins. By substituting for water, trehalose maintains the three-dimensional structure of protein molecules, and hence their ability to remain active after drying or freezing (25–28).
2. Formation of a glassy solid—Scientists have found that trehalose can form an amorphous glassy matrix around the vital tertiary structures of proteins and phospholipids, thus protecting them during drying and freezing (28–30).
3. Chemical inactivity—The chemical inertness of trehalose may be important in protecting biomembranes from destructive molecular reactions (31).
4. Formation of a tight-fitting complex with water—Trehalose has the ability to form a hydrogen-bonded complex with water. This complex retards ice crystal formation and allows trehalose to penetrate more deeply than other sugars into certain membrane structures (32, 33).

Because of these and other characteristics, trehalose possesses a number of technical qualities that indicate that it may provide significant beneficial properties to foods, cosmetics, and pharmaceuticals. Other properties of interest include mild sweetness; low hygroscopicity; good solubility in water; lack of color in solution; a high glass transition temperature; stability over a wide range of pH and temperatures; and chemical stability (25, 27, 29, 34, HBC, unpublished data, 1997). Because of its stability, it resists caramelization and doesn't participate in Maillard reactions in food systems (29, 31, 35, HBC, unpublished data, 1997). Trehalose is also known to stabilize proteins and starches (25–29, 36, HBC, unpublished data, 1997).

A. Sweetness and Organoleptic Properties

1. Intensity and Persistence

Portmann and Birch compared the intensity and persistence of various concentrations of trehalose to other sugars (32). Using a trained taste panel, the authors studied the perceived sweetness of trehalose versus other sugars at concentrations of 2.3, 4.6, 6.9, and 9.2%. The report showed that the perceived sweetness of trehalose increased in a nonlinear fashion as the concentration increased. The ratios of the intensity of sweetness at the various concentrations between sucrose and trehalose were 6.5 : 1, 5.0 : 1, 3.5 : 1, and 2.5 : 1, respectively. Comparisons of sucrose to glucose and fructose showed that the sweetness intensity ratio did not change with increasing concentration.

Increasing the concentration of trehalose also lead to a disproportionate increase in the perception of the persistence of sweetness as compared with sucrose (32). The ratio of the persistence of sweetness of sucrose to trehalose was 1.5 : 1 (2.3%), 0.8 : 1 (4.6%), 0.6 : 1 (6.9%), and 0.5 : 1 (9.2%), whereas glucose and fructose showed little change. The ratio of the intensity and persistence of sucrose to maltose had a similar pattern as trehalose but was less dramatic.

HBC performed a comparative sweetness test using a 22.2% solution (w/w) of trehalose and various concentrations of sucrose (HBC, unpublished data, 1997). The study showed that trehalose at 22.2% is about 45% as sweet as sucrose (Fig. 3). The studies suggest that although the sweetness of trehalose is less intense, it can be more persistent than sucrose (32).

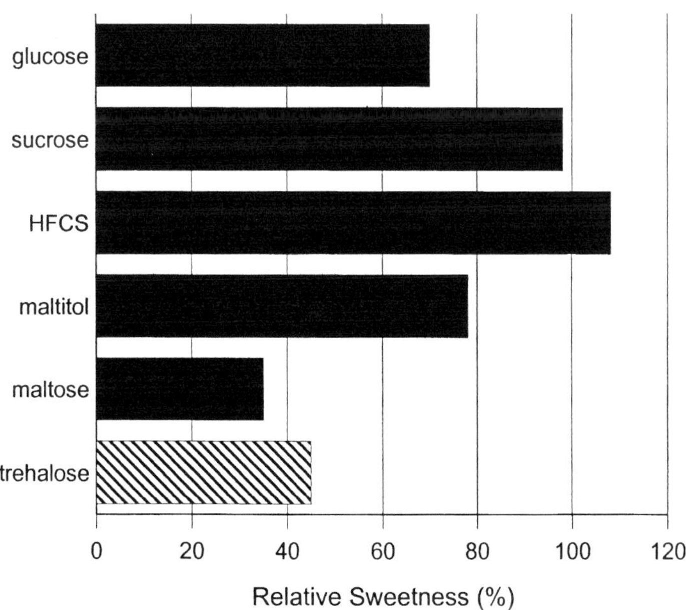

Figure 3 The relative sweetness of various sugar solutions compared with a 22.2% (w/w) solution of trehalose.

2. Taste Quality

Trehalose was also tested for taste quality by a Japanese taste panel. A 22.2% trehalose solution was compared with a 10% sucrose solution (HBC, unpublished data, 1997). Results showed that 17 of 20 subjects preferred the taste of trehalose to that of sucrose. Anecdotal responses to the quality of the taste of trehalose have described it as mild with a clean aftertaste that seems to clear the palate easily.

3. Taste Chemistry of Trehalose

Trehalose, a nonreducing disaccharide, is similar to other reducing disaccharides in that it does not produce as great a sensation of sweetness as does sucrose. Disaccharides are also generally accepted to be less sweet than the monosaccharides from which they are made, with the notable exception of sucrose. Trehalose is believed to have only one glucose molecule occupying the binding site on the sweet taste receptor (37). This is thought to result from steric hindrance around the C-1 site on the glucopyranose ring.

Physicochemical studies suggest that the shape and charge of trehalose are such that water molecules form a closer (tighter) association with trehalose than with other saccharides. Theoretically, these small molecular clusters can migrate deeper into the taste epithelium (34, 38). Trehalose and the associated water may concentrate in these areas more efficiently than other sugars, which would result in a more persistent sweet taste (32).

In contrast, the perception of sweetness intensity has been hypothesized to be enhanced by the ability of water molecules to dissociate or move in relation to an associated sugar (39). It has been suggested by Mathlouthi that free water molecules are required to allow the transport of Na^+/K^+ ions across the membrane of the sweet taste receptor. The more free water, the higher the possible membrane potential (Na^+/K^+ transport), resulting in the sensation of more intense sweetness (39). Trehalose holds the water of hydration tightly so that water is not available to facilitate the membrane potential. Both macroscopic and physiochemical results suggest that there is an optimal molecular volume for binding to sweet taste receptors, which corresponds to a particular packing of the water molecules around dissolved sugar molecules (40). The tight interaction of the trehalose/water complex could inhibit the bonding of trehalose to the receptor, further reducing the sweetness compared with other sugars (32).

B. Hygroscopicity

Low hygroscopicity is one of the hallmark qualities of trehalose. Figure 4 shows the moisture content of trehalose after storage at 25°C for 7 days at various humidity levels. The initial water content of trehalose dihydrate is 9.54%, whereas

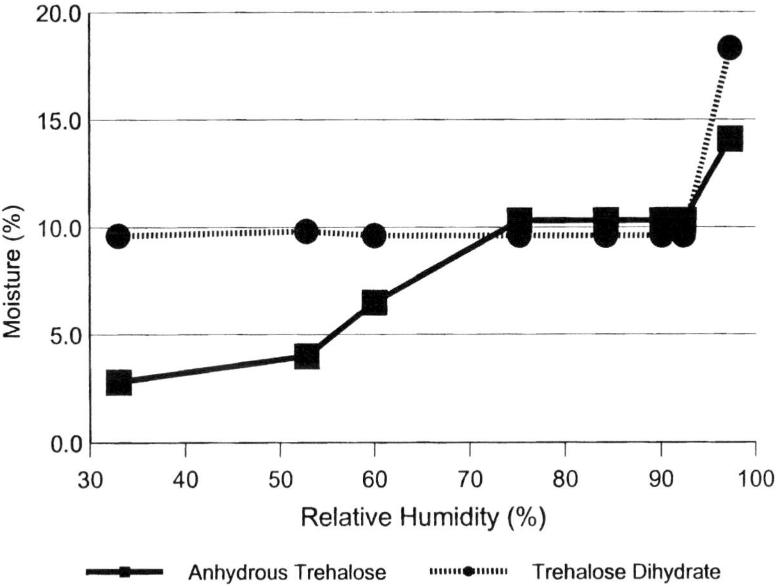

Figure 4 Hygroscopicity of anhydrous and dihydrous trehalose. Trehalose was stored at various relative humidity concentrations for 7 days at 25°C and then measured for the percent moisture.

anhydrous trehalose is 0.65%. The dihydrate form remained stable up to a relative humidity (RH) of approximately 92%. The moisture content of anhydrous trehalose increased as RH increased from 35–75% and then paralleled the moisture content of the dihydrate until 92% RH was reached. The maximum moisture content was nearly 14% (HBC, unpublished data, 1997). Table 1 provides com-

Table 1 Water Absorption Ratio in Percent of Various Sugars Stored at 90% Relative Humidity and 25°C for 10 Days

	Storage time in days			
Sugar	1.0	3.0	6.0	10.0
Trehalose	0.2	0.0	0.1	0.2
Lactose	0.2	0.3	0.3	0.7
Maltose	3.5	6.6	8.4	8.4
Glucose	4.0	7.7	10.8	11.3
Sucrose	5.1	15.0	33.4	50.0

parative data on the water absorption (%) of several sugars stored for 10 days at 90% RH and 25°C. It appears that trehalose could be of benefit compared with other sugars in dry blending operations and other applications in which low hygroscopicity is desired.

C. Solubility and Osmolarity

Trehalose is soluble in water and insoluble in absolute alcohol. Aqueous ethanol has been used to crystallize trehalose, which produces the dihydrate form (14). The solubility of trehalose in 70% aqueous ethanol has been reported as 1.8 g/100 ml (14). Table 2 provides data on the solubility of trehalose in water at various temperatures (HBC, unpublished data, 1997). Trehalose has a solubility similar to maltose up to about 80°C at which temperature the solubility increases at a rate greater than the other sweeteners. The osmolarity of trehalose is also essentially the same as maltose (Table 3).

D. Glass Transition Properties

Like most sugars, trehalose is capable of forming amorphous gels called glasses, which are characterized by high viscosity and low molecular mobility (34, 41). Glasses form as water is removed from sugar solutions during thermal- or freeze-drying (42). These vitreous solids can become sticky (or rubbery in the case of polymers) when their glass transition temperature is exceeded or there is an increase in moisture. In this viscous liquid state, many commonly used sugars, such as sucrose, become unstable and either crystallize or decompose chemically. Crystallization will usually result in an anhydrous sugar with a consequent increase in water activity in the residual phase. This, in turn, causes an increase in the rate of browning and other degradative reactions in foods (35).

Trehalose solutions, on the other hand, solidify to form heat-stable glasses that are capable of binding water and thus controlling water activity (25, 27, 41, 42). The glass transition temperature of these glasses is higher than for any other disaccharide (34). If crystallization of trehalose glasses does occur, the dihydrate forms, rather than the anhydrous crystal, resulting in a reduction of the water activity (43).

Because of its special glass-forming properties, trehalose can be used to stabilize proteins and lipids against the denaturation caused by desiccation and freezing (28, 36, 42, 44). The glass transition temperature for pure trehalose is the highest of any disaccharide (115°C) (42). This temperature is approximately 43°C higher than that of sucrose. The significance of a high glass transition temperature and the unique characteristics of trehalose glasses are discussed later.

Several studies have shown that proteins and other cell membrane components become embedded in the vitreous matrix formed by trehalose during drying

Table 2 Water Solubility at Various Temperatures of Trehalose Produced by the Hayashibara Method on an Anhydrous Basis

Temperature (°C)	10	20	30	40	50	60	70	80	90
g/100 g H_2O	55.3	68.9	86.3	109.1	140.1	184.1	251.4	365.9	602.9
% Saturation									
Concentration (w/w)	35.6	40.8	46.3	52.2	58.3	64.8	71.5	78.5	85.8

Table 3 Comparison of the Osmotic Pressure of Concentrations of Trehalose and Maltose

	Concentration of sugar (% w/w)			
Sugar	5	10	20	30
Trehalose	193[a]	298	690	1229
Maltose	195	299	676	1221

[a] Osmotic pressure measured in milliosmoles.

(28, 29, 36, 44, 45). Proteins dried in the presence of trehalose are physically constrained from undergoing degradative reactions (41). Several reports in the literature have cited instances in which enzymes preserved in trehalose have remained active, even after exposure and extended storage at high temperatures (36).

A review of the literature suggests that there may be several important reasons for the superiority of trehalose in preventing the destructive chemical and physical reactions that accompany drying and freezing. Trehalose glasses are purported to function by a number of mechanisms including the following:

1. Reduction in molecular mobility—The formation of a stable glass reduces molecular mobility and prevents destructive physical forces, like fusion, from taking place during freezing or drying.

2. Reduction in water activity—Water activity is necessary for degradative interactions such as the Maillard reaction, which occurs between proteins and other biocomponents (35). Trehalose glasses are hydrophilic and bind water tightly if excess water is introduced into the system (42). Trehalose glasses are almost unique in their ability to form a dihydrate compound, when excess water is present, effectively lowering the water activity (43). Ding et al. stated that the kinetics of most chemical reactions can be essentially shut down by the high viscosity of vitreous trehalose glasses, even at temperatures well above the glass transition temperature (42). They demonstrated that trehalose glasses can absorb up to 0.15 moles of water per mole of trehalose without loss of rigidity at ambient temperature (42). Others have reported that the freeze-drying of a 0.25 molar (9.45%) trehalose solution did not remove all the water. After storing the preparation for 48 hours under vacuum, the sample retained 0.02 g of water per gram of trehalose (41). The glass transition temperature determined for this preparation was 92°C. When the sample was stored at ambient temperature and high humidity for 66 hours, it absorbed water vapor, and 70% of the sample reverted to trehalose dihydrate (41).

3. Immobilization of active end-groups—Proteins are reported to contain between 0.25–0.75 g of water per gram of protein (27). Phospholipids also reside

in an aqueous environment. As water is removed during drying, the active end-groups, usually associated with hydrogen bonded water, are exposed and can bind to other molecules. This denaturation of both proteins and phospholipids, which causes distortion of the three-dimensional structure of the molecules, results in the loss of function. Trehalose can substitute for water by hydrogen bonding to the end-groups, thus minimizing changes and preserving the natural structure and function of the protein or phospholipid.

4. Nonhygroscopicity—Trehalose glasses are hydrophilic but not hygroscopic. It is reported that they are dry to the touch (27). This means that vapor pressure of water in equilibrium with trehalose glasses is high relative to other sugars. Therefore, trehalose glasses are able to lose water to the environment, rather than absorbing it (25, 27). This becomes important in maintaining a level of water activity below the threshold at which destructive interactions occur (35).

By contrast, the glass structure of other sugars commonly used in the food industry are hygroscopic, thus becoming tacky in the presence of water vapor. When moisture from the atmosphere is absorbed, the surface layers of the sugar dissolves, and crystallization can occur because this is the most stable state for most sugars. The onset of crystallization allows weeping, particularly with sucrose-based confections (27). Water, because of its plasticizing effect, creates a more mobile structure in these types of glasses, inducing degradative chemical reactions. These reactions can denature proteins and other food components (35).

The mobile state will also occur if the storage temperature exceeds the glass transition temperature of a particular sugar-containing matrix. If food products are stored above their glass transition temperature, crystallization is likely, with an accompanying increase in browning reactions (35). Because trehalose has a very high glass transition temperature and can control water vapor by forming the dihydrate, it can remain in the glass state at temperatures higher than other sugars. Therefore, glasses formed from trehalose have a greater capability for preserving or stabilizing proteins, lipids, or carbohydrates embedded in them (42).

Other interesting properties of trehalose glasses have been reported. Crowe et al. found that because trehalose glasses are permeable to water, products suspended in trehalose solutions can be dried to very low residual water contents without resorting to extreme dehydration methods (41). In another study, cytochrome c was dried with 27% trehalose (w/w) to a 6% moisture content under vacuum at ambient temperature. The compound was found to resist denaturation at temperatures up to 115°C, whereas the authors reported that the normal temperature at which cytochrome c denatures is 65°C (41). In general, proteins protected by trehalose retain full activity after exposure to high temperature (36).

It has been reported that the hydrophilic nature of trehalose glasses makes them impenetrable to the volatile lipid molecules responsible for food aromas and flavors. These molecules are trapped within the trehalose glass and cannot dissipate to the atmosphere. Therefore, the aromas and flavors are still present when the preparation is reconstituted (25, 46). It is speculated that trehalose might be particularly effective in stabilizing foods and biomaterials that are subject to high humidity and temperatures (41). Roser has reported that trehalose glasses do not attract significant quantities of water at a relative humidity of 90% or less (25).

Although a great deal of applications research remains to be done, HBC has demonstrated that trehalose can be added to foods that undergo phase transitions in preparation or storage (HBC, unpublished data, 1997). Such foods include frozen bakery products, frozen desserts, dried fruits and vegetables, glazes, and spray-dried products.

E. Storage Stability

The storage stability of a 10% solution of trehalose was tested by dividing aliquots of the solution and placing them in the dark at 25 and 37°C for 12 months (HBC, unpublished data, 1997). Samples were examined for pH, color development, turbidity, and residual sugar content monthly for the first 6 months and thereafter at 9 and 12 months. After 12 months, there was no development of color or turbidity, nor was there any degradation of trehalose at either temperature. Samples stored at 25 and 37°C for 12 months showed a drop in pH from an initial value of 6.80 to 5.27 and 5.15, respectively.

Four percent solutions of trehalose were prepared at nine different pH concentrations. The samples were heated to 100°C for 8 and 24 hours (HBC, unpublished data, 1997). The buffer systems (0.02 mM) used were acetate (pH 2–5), phosphate (pH 6–8), and ammonium (pH 9–10). After incubation, the samples were evaluated for pH and residual sugar content. The results of pH changes are presented in Table 4. Samples appeared stable for the first 8 hours and were relatively stable for the 24-hour incubation period. High-performance liquid chromatography analysis of the samples showed that even after 24 hours of incubation at 100°C, all samples retained greater than 99% of the original concentration of trehalose.

F. Heat Stability and Caramelization

The heat stability of trehalose was examined in a test system in which a 10% trehalose solution was made in a pH 6.0 buffer (HBC, unpublished data, 1997). The buffered solution was incubated in an oil bath at 120°C for up to 90 minutes. Maltose control samples were also included. Solutions were analyzed for absorbance at 480 nm (Table 5). The trehalose solution remained stable after a slight initial increase in absorbance; whereas the maltose solution increased at

Table 4 The pH of 4% Trehalose Solutions Prepared in 0.02 mM Buffers and Incubated for 8 and 24 Hours at 100°C

Incubation time (hr)		
0	8	24
3.46	3.41	3.83
3.71	3.64	4.02
4.25	4.22	4.50
5.14	5.10	5.25
6.69	6.67	6.52
7.46	7.44	7.13
8.25	8.14	7.64
8.88	8.91	8.42
9.86	9.85	9.16
Mean change ± SD	0.04 ± 0.03	0.37 ± 0.19

each sampling time. At 90 minutes the absorbance of the maltose sample was 15-fold greater than that of trehalose. The increase in absorbance is related to the production of colors associated with the caramelization process. Trehalose resisted the production of such color and may be an effective inhibitor of undesired caramelization.

G. The Maillard Reaction

The stability of 10% solutions of trehalose or maltose were tested in 0.05 M buffered solutions (pH 4.0, 5.0, 6.0, 6.5, and 7.0) containing 1% glycine (HBC,

Table 5 The Absorbance of 10% Trehalose and 10% Maltose Buffered Solutions (0.033 M Sodium Phosphate in Water) Incubated at 120°C for up to 90 Minutes

	Absorbance at 480 nm	
Incubation time (min)	Trehalose	Maltose
0	0.005	0.000
30	0.013	0.051
60	0.010	0.121
90	0.012	0.184

unpublished data, 1997). The solutions were placed in a oil bath at 100°C, and samples were taken at 0, 30, 60, and 90 minutes. After cooling, the samples were analyzed for changes in absorbance at 480 nm. After 90 minutes the absorbance values of maltose were 0.003 (pH 4.0), 0.014 (pH 5.0), 0.324 (pH 6.0), 0.610 (pH 6.5), and 0.926 (pH 7.0), whereas trehalose values were 0.000, 0.002, 0.010, 0.010, and 0.012, respectively. In another experiment, 10% solutions of trehalose and maltose were combined with a 5% (w/w) solution of polypeptone. Samples were placed in an oil bath at 120°C for up to 90 minutes. Color change was measured at 480 nm. Figure 5 displays the data from the experiment. These data suggest that trehalose may be an excellent sugar to use when inhibition or control of the Maillard reaction is desired.

H. Viscosity

Trehalose has a relatively low viscosity, even at higher concentrations (HBC, unpublished data, 1997). Figure 6 summarizes the viscosity of trehalose solutions

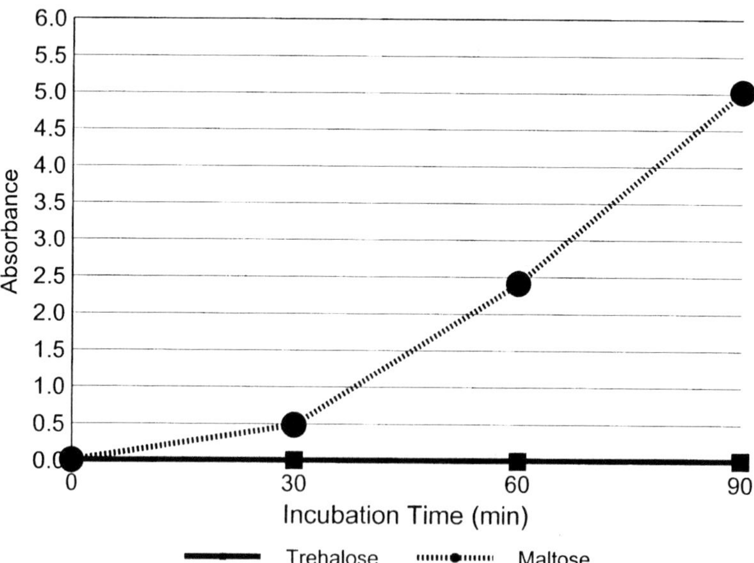

Figure 5 The change in absorbance at 480 nm of 10% solutions of trehalose or maltose containing 5% (w/w) of a polypeptone solution. Solutions were incubated at 120°C for 90 minutes.

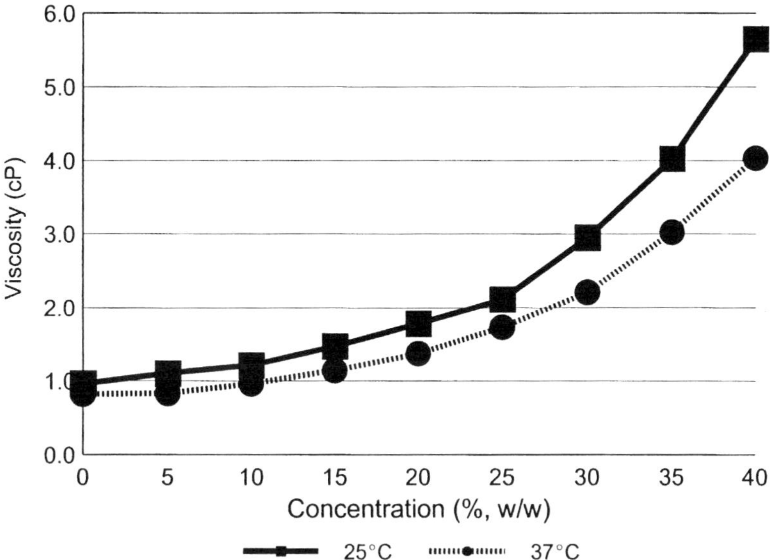

Figure 6 The viscosity of trehalose solutions from 5 to 40% (w/w) at 25 and 37°C measured in centipoise (cP).

from 5–40% (w/w) at 25 and 37°C. Even at 40%, the viscosity did not rise above 5.7 cP.

VIII. NUTRITIONAL PROFILE

Trehalose produced by HBC is the only product currently available in commercial quantities to the food industry. A nutritional analysis of several lots of this product showed that it is composed of more than 90% carbohydrate. The remaining components consisted of water at about 9.7% (dihydrate), trace quantities of lipid with an average content of 0.025%, and low concentrations of protein (0.007–0.008%) and ash (0.002–0.010%). Sodium concentrations ranged from nondetectable to 0.065%, with residual starch being undetectable. These values compared favorably with those of other saccharides listed in the Food Chemicals Codex.

Energy values based on the preceding data calculate to 3.62 kcal/g. This value correlates well with the analytical results, 3.46 kcal/g (HBC, unpublished data, 1997).

IX. SHELF-LIFE AND TRANSPORT

To ensure that its product continued to meet food grade specifications over an extended shelf-life, HBC conducted analyses on commercial grade samples of trehalose stored for up to 24 months. Twenty kilograms of trehalose was stored at 25°C in three-layer Kraft paper bags in which one of the layers was polyethylene. One hundred-gram samples were taken from the same bag at 0, 1, 3, 6, 9, 12, 18, and 24 months and tested for eight variables. The results confirmed that the material maintained integrity over the storage period, with little variation in the product.

Since becoming commercially available in Japan in November of 1995 until the end of 1998, a cumulative total of more than 15,000 metric tons has been sold. It is anticipated that in excess of 15,000 metric tons will be sold during 1999 in Japan. There has not been a recall of trehalose because of degradation of the product.

No data on the stability of trehalose in specific products is yet available. However, because trehalose is chemically, thermally, and pH stable, it suggests that trehalose may aid in the stabilization of the products in which it is used.

Transport of trehalose is not restricted in Japan. It is not anticipated that any such restrictions will be placed on trehalose by any government agencies.

X. COST AND AVAILABILITY

Trehalose is currently being sold in Japan for approximately 300 Yen (¥) per kilogram ($2.50/kg @ 120 ¥ to $1.00 USD). Prices are slightly higher in Korea and Taiwan. To the authors' knowledge, HBC is the only commercially available source of a purified trehalose product. Once regulatory approval is obtained in North America and the European Union, it will be commercially available through a marketing partner(s). Other groups are undoubtedly investigating alternative methods for the production of trehalose, but none are known to be commercially available at this time.

XI. REGULATORY STATUS

Quadrant Holdings (Cambridge, United Kingdom) received approval for trehalose to be used as a novel food from the Ministry of Agriculture, Fisheries and Food in the United Kingdom. The approval, granted in 1991, is for the use of trehalose as a cryoprotectant for freeze-dried foods. Use levels were limited to 5% for each formulation. Trehalose produced by HBC was approved as a food additive in Japan in 1995 and as a food ingredient in Korea and Taiwan in 1998. Hayashibara International Inc. self-affirmed trehalose as GRAS in May of 2000, and received a letter of no objection from the U.S. FDA in October 2000. In the United States the use of Hayashibara trehalose is limited only by current Good Manufacturing Practices. JECFA reviewed the safety profile of trehalose in June of 2000, and assigned the Acceptable Daily Intake as ''not specified.'' Currently regulatory approval is being sought in Europe.

XII. APPLICATIONS

Numerous published and unpublished studies have shown that relatively low concentrations of trehalose in food formulations can reduce sweetness; stabilize protein matrices, flavors, colors and fatty acids; reduce starch retrogradation; maintain the texture of coatings; and prevent weeping (25, 27, 29, 30, 47, 48, HBC, unpublished data, 1997).

A. Protein Stabilization

Trehalose appears more effective in stabilizing proteins against damage caused by drying or freezing than other sugars tested (28–30). Trehalose has also been shown to help maintain delicate protein structures after thawing and to stabilize disulfide bonds, thereby inhibiting the formation of odors and off-flavors.

HBC found that a 5% addition of trehalose to an egg white preparation, subsequently frozen for 5 days and then thawed resulted in almost no protein denaturation compared with a control preparation (Fig. 7). Protein denaturation was measured by the change in turbidity before and after freezing. Relative denaturation of preparations containing other sugars ranged from 14–58% (HBC, unpublished data, 1997).

In a second example, pulverized carrots were mixed with 10% (w/w) of various sugars. Samples were dried for 64 hours at 40°C and stored for 7 days. The superoxide dismutase (SOD)–like activity was measured (Fig. 8). Trehalose appeared to maintain three times as much SOD-like activity as sucrose (HBC, unpublished data, 1997). Studies using seven other vegetables dried with and without trehalose showed a similar SOD-like protective activity. Preservation of enzymes under various stress conditions has been reported by others (29, 41).

The effect of trehalose on protein stability was also demonstrated in a coffee/milk drink. A 3% addition of trehalose (25% of added sugar) reduced the coagulation of casein during sterilization at 121°C for 5 minutes and suppressed color, taste, and pH changes during subsequent storage (Hayashibara, unpublished data, 1997).

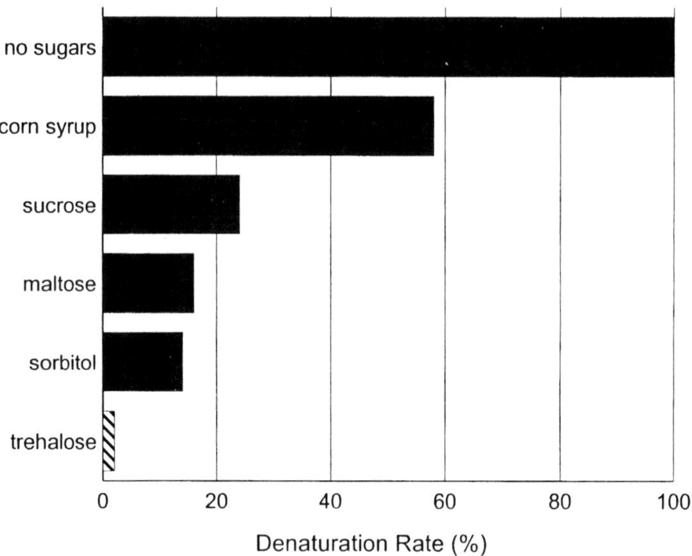

Figure 7 Egg white (95 g) was mixed with 5 g of various sugars and stored for 5 days at −20°C. The samples were thawed and the turbidity was assayed. Egg white without added sugar served as control and was given the relative denaturation value of 100%.

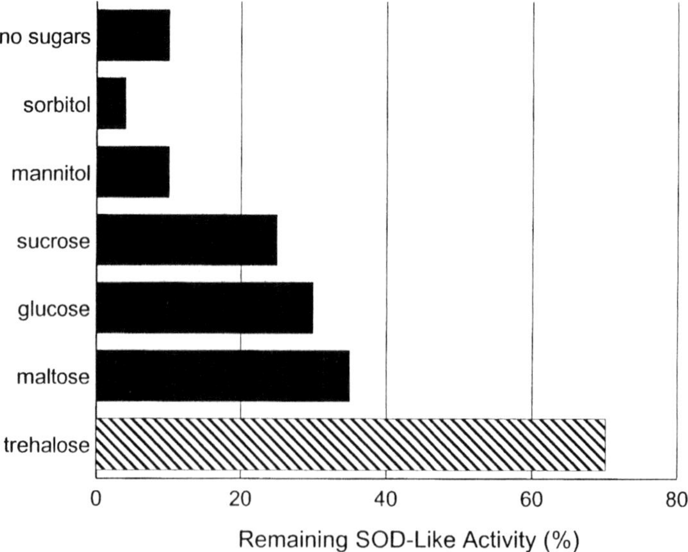

Figure 8 Pulverized carrots were mixed with various sugars and dried at 40°C for 64 hours. The samples were rehydrated and the SOD-like activity was measured. The results presented are the percent of SOD-like activity obtained after compared with the activity before the samples were dried.

B. Stabilization of Starch

Trehalose is currently being used in Japan to retard starch retrogradation in such products as Udon noodles (0.2% of flour), clam chowder (0.4% of product), and traditional Japanese confectioneries (10–50% of sugars). Although the mechanism is not yet fully understood, applications research has shown that trehalose can be effective in stabilizing starch. Basic tests were performed to assess the ability of trehalose to inhibit starch retrogradation. A 1% cornstarch solution was mixed with 6% of various sugars. The mixture was gelatinized and cooled. The turbidity of the solution was tested before and after 12 hours of storage. The percent change was regarded as the amount of retrogradation of the starch solution (Fig. 9) (HBC, unpublished data, 1998). A similar experiment was performed using equal volumes of a 2% starch solution mixed with a 12% sugar solution. After gelatinization, the solutions were stored at 4°C for 12 hours. Results were similar as those obtained in the first experiment (Fig. 9). The percentage of retrogradation with trehalose was 13%, whereas that for sucrose was 50% (HBC, unpublished data, 1998).

Applications of this effect were tested in several Japanese food systems. The hardness of bread made with and without trehalose was measured on day 0

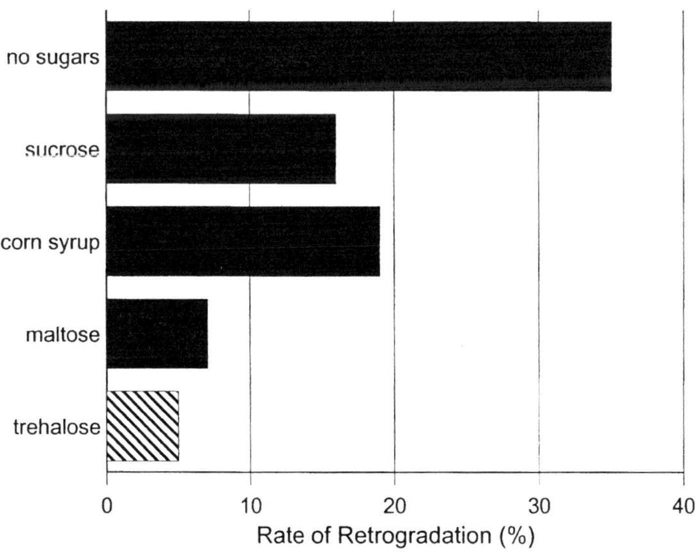

Figure 9 A 1% corn starch solution was mixed with various sugars (6%). The samples were then gelatinized, cooled to 4°C and stored for 12 hours. The percent change in the turbidity before and after storage is reported as the amount of starch retrogradation.

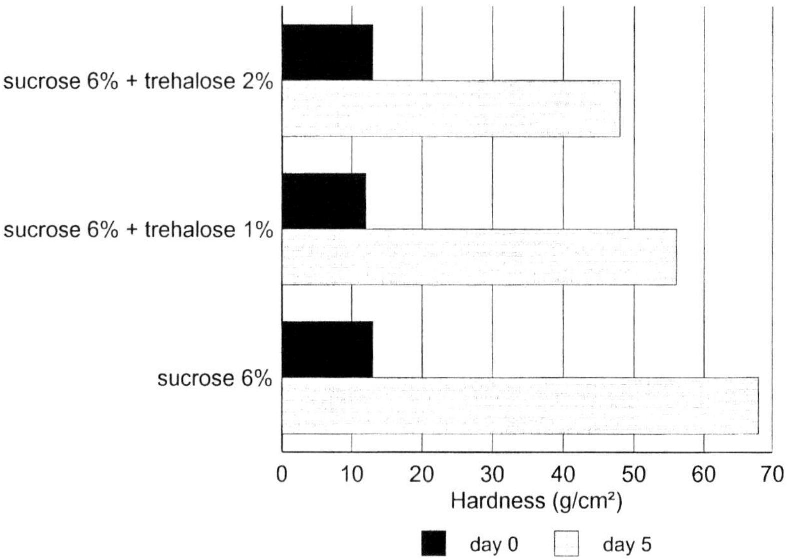

Figure 10 Bread was baked using 6% sucrose and 1%, 2%, or no trehalose. The hardness of the bread was assayed after cooling the bread (4°C) and again after 5 days of storage at 4°C.

and after 5 days storage at 4°C (Fig. 10). The addition of 2% trehalose to the mix provided a reduction in hardness after 5 days of about 32%. The effects of trehalose in frozen sponge cakes and dinner rolls were also investigated. Hardness measurements on thawed sponge cakes showed that trehalose (1.7%) reduced starch retrogradation and produced a product that was 28% softer than controls after thawing. A 1% trehalose addition to dinner rolls containing sucrose resulted in less heat shock when the rolls were thawed using an electric range than when sucrose was used alone. In addition, trehalose-produced rolls maintained a softer texture for up to 16 hours after thawing (Fig. 11) (HBC, unpublished data, 1997).

C. Prevention of Fat Decomposition

Trehalose's ability to protect membrane phospholipid layers subjected to heat stress or freeze/thaw cycles has been examined (28, 44, 45). Results showed that trehalose was more effective in maintaining the integrity of vesicle membranes than standard protectants such as glycerol or dimethylsulfoxide. The ability of trehalose to stabilize free fatty acids has also been studied (HBC, unpublished data, 1998). One hundred grams of four different fatty acid solutions were combined with 1 ml of 5% solutions of sucrose, sorbitol, or trehalose. The mixtures were heated for 1 hour at 100°C. Fatty acid concentrations before and after heat-

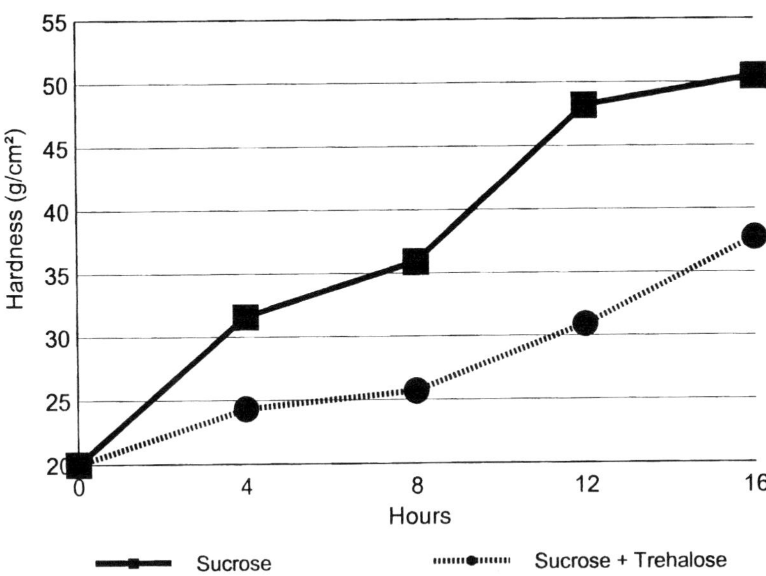

Figure 11 Dinner rolls with or without 1% trehalose (dry weight) were baked and frozen at −18°C for 7 days. The rolls were thawed in an electric range and the hardness was measured every 4 hours for 16 hours.

ing were measured by gas chromatography. Results showed that trehalose markedly reduced the decomposition of the fatty acids tested compared with control samples or those with sucrose or sorbitol (Fig. 12).

D. Texture, Flavor, and Color Stabilization

Colaço and Roser studied the use of trehalose to preserve the fresh qualities of a number of foods (29). They maintained that because trehalose was not particularly sweet, it would not significantly change the flavor of foods to which it was added. The authors blended fresh eggs with trehalose and dried the mixture at 30–50°C. An odorless, yellow-orange powder that could be stored at room temperature was produced. When this powder was rehydrated, the product was reported indistinguishable from an equivalent fresh egg mixture.

The ability of trehalose to preserve the fresh aroma and texture of herbs, fresh fruit slices, and purees was also studied (29). Fruit purees dried with trehalose resulted in shelf-stable powders with little aroma. On rehydration, these powders recovered the texture and smell of fresh fruit. Control samples dried without trehalose were more difficult to reconstitute, and a taste panel described the

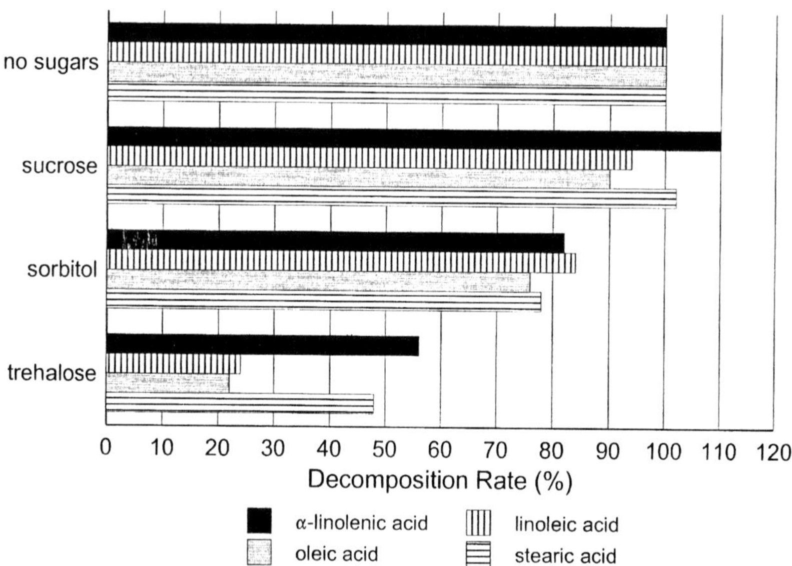

Figure 12 One of four fatty acid solutions (100 g) was mixed with 1 ml of a 5% solution of sucrose, sorbitol, or trehalose or without adding a sugar solution (control). The solutions were heated for 1 hour at 100°C, and the amount of remaining fatty acids was compared with preincubation measurements. The results are recorded as percent decomposition of the control sample. Fatty acids were assayed using gas chromatography.

aroma and flavor as denatured or "cooked." The authors used gas chromatography to quantitate the preservation of fruit volatiles from bananas, which had been vacuum-dried at 37–40°C with or without 10% (w/w) trehalose. After several months storage at room temperature, the banana purees dried without trehalose exhibited a loss of flavor volatiles, whereas those dried with trehalose showed a minimal loss. It was concluded that the purees dried with trehalose retained most of the volatiles typical of fresh fruit (29).

Similar results were noticed for fresh herbs dried with trehalose. Control products were reported to have lost flavor, aroma, and color during storage, whereas rehydrated herbs dried with trehalose maintained their fresh color and organoleptic properties (29).

Spinach was heated at 90°C for 30 seconds to inactivate enzymes. The spinach was shredded and 10% (w/w) of various sugars were added. The spinach/sugar preparations were dried at 55°C for 16 hours and heated in an oven at 100°C for 3 hours to simulate maximum commercial heat stress (HBC, unpublished data, 1998). Color measurements using an L*a*b* colorimeter were taken after 3 hours

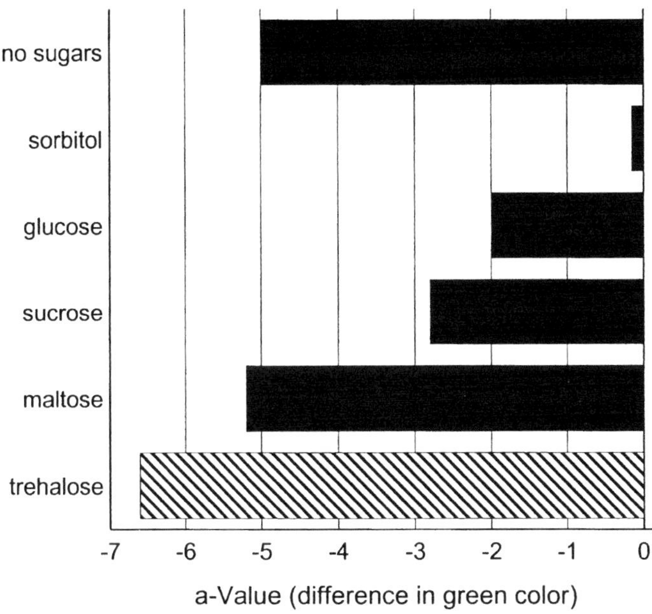

Figure 13 Various sugars (10% w/w) were mixed with fresh spinach. Samples were dried at 55°C for 16 hours, after which they were heated in a microwave at 100°C for 3 hours. Color measurements were taken after heating and reported as a-values.

in the oven. Figure 13 shows the differences in green color (a-value) compared with a control sample where no sugar was added. In this study trehalose provided the most beneficial effect for color preservation.

Ten grams of fish paste was mixed with 5 ml of 5% solutions of various sugars. The prepared fish paste was boiled for 15 minutes. Percent release of trimethylamines (a primary component of fishy odors) was measured for the various sugar-containing preparations and a control sample (Fig. 14). A relative release value of 100% trimethylamine was used as the control when the fish paste was boiled in the absence of any of the sugars. In comparison, fish paste boiled after mixing with trehalose solution released 40%, whereas the fish paste using sucrose released approximately 110% (HBC, unpublished data, 1998). In a similar experiment, trehalose or sorbitol was combined with fish paste at 10 and 20% to test the release of aldehyde, ethylmercaptan, and trimethylamine (fish-related smells) after boiling for 15 minutes. Ten percent trehalose reduced the release of these chemicals compared with sorbitol by 3.1-, 2.9-, and 4.7-fold and in samples containing 20% by 2.8-, 2.4-, and 5.2-fold, respectively.

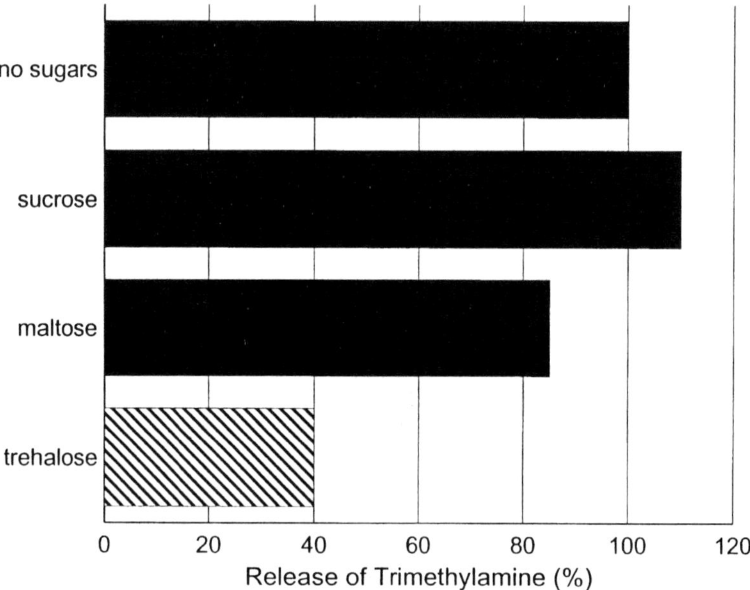

Figure 14 Ten grams of fish paste was mixed with 5 ml of a 5% solution of sucrose, maltose, or trehalose. A fourth sample without the addition of a sugar was used as a control. The samples were boiled for 15 minutes, and the relative release of trimethylamine was measured, with the control representing 100%.

In food formulations, trehalose has been shown to preserve the color and texture of cold or frozen foods. A 4% gelatin raspberry mousse was prepared using sucrose or with a partial substitution of sucrose by trehalose. The mousse was frozen (−18°C) for 1 week and then thawed at 4°C. The addition of trehalose helped maintain 97.9% of the hardness and integrity of the dessert compared with the mousse containing only sucrose (81%) (HBC, unpublished data, 1997). We have not been able to duplicate this data in the United States.

Japanese commercial food processors have used trehalose in a variety of food categories since 1995. They report that trehalose masks bitterness and enhances flavor in beverages, reduces sweetness, retards starch retrogradation, increases moisture-holding capacity in bakery products, prevents Maillard reactions in candies and light-colored soups, enhances reconstitution of dried noodles, and prevents hygroscopicity in jellies and in various toppings (HBC, 1997, unpublished data). Table 6 lists the multifunctional benefits of trehalose reported for foods and provides the approximate amounts required for each effect. These concentrations are based on estimated values used in Japanese commercial products discussed previously and on laboratory trials conducted by HBC.

Table 6 Technical Effects of Trehalose

Food category	Technical effect	Approximate trehalose addition
Bakery products	Moisture retention Shelf-life extension Crumb softener Reduced sweetness Reduced hygroscopicity	2% flour
Frozen bakery products	Protein preservation freeze-thaw stabilization Shelf-life extension Crumb softener	13–18%
Frozen desserts	Freeze-thaw stabilization Texture stabilization	13–18%
Dairy-based foods and toppings	Texture stabilization Texture stabilization Flavor profile improvement	2–12.5%
Dried, frozen, or processed fruits and vegetables	Color stabilization Texture stabilization Flavor profile improvement Masks bitterness	5% of carrier solution
Beverages	Flavor profile improvement Color stabilization Reduced sweetness pH stabilization Masks bitterness	0.4% of product to 50% of sugars
Jellies and gelatins	Moisture retention Reduced sweetness Reduced hygroscopicity Color stabilization Flavor profile improvement	15–30% of sugars
Confectionery	Reduced syneresis Shelf-life extension Moisture retention Reduced sweetness Reduced hygroscopicity Texture improvement Flavor profile improvement	5–40% of product 5–80% of sugars
Meats/fish/eggs	Protein preservation Moisture retention Texture improvement Masks cooking odors	2–10%

XIII. CARIOGENICITY

The cause of dental caries and the method for intraoral plaque–pH telemetry have been well documented and discussed in other chapters of this book (Chapters 17, 19). Trehalose was compressed into a lozenge with mint flavor. The change in pH during consumption of the mint was studied under a standard protocol in which a plaque-covered pH sensor is integrated into a removable, mandibular, restorative device (49, 50). The pH did not drop below the critical value of 5.7 in any of the four subjects tested. With regard to nonfermentability, the mints would, therefore, qualify for the ‘‘toothfriendly’’ claim.

A similar study was performed in Japan where trehalose was incorporated into a chocolate candy (HBC, unpublished data, 1997). Four subjects dissolved the chocolate (5.1 g) in their mouths, and the plaque pH was measured by an indwelling electrode over a 30-minute period. A 10% sucrose solution was used as a positive control. The pH did not drop below 5.7 in any of the four subjects consuming the chocolate. This suggests that trehalose taken under these conditions does not promote dental caries.

In vitro studies on the fermentability of trehalose using *Streptococcus mutans* have shown that trehalose can be fermented (British Sugar, unpublished data, 1998); however, the fermentation rate was lower than that of sucrose. In vivo plaque-pH assays using the two different trehalose-containing products indicate that the time during which trehalose is in contact with dental plaque is insufficient to result in critical plaque acidification. Furthermore, the amount of acid formed during the period of trehalose exposure was either too small to reduce the pH below 5.7 or may have been neutralized sufficiently by increased saliva production during consumption of the mint or chocolate.

XIV. METABOLISM

Humans have long consumed trehalose in various foods, primarily young mushrooms and baker's yeast (5, S Miyake, unpublished data, 1997). At present, trehalose is being used by the Japanese food industry at a rate of more than 1,000 metric tons per month. This will likely increase on an international scale as more functional applications are found and additional regulatory approvals are granted. The mechanism by which dietary trehalose is metabolized in mammals has been studied and appears straightforward. Trehalose is not assimilated intact into the body and has not been detected in blood. Like other disaccharides, it is hydrolyzed on the brush border of the intestinal enterocytes. The enzyme that hydrolyzes trehalose into its two glucose units is trehalase (1, 6). Trehalase is tightly bound to the surface of the external side of the membrane microvilli in the small intestine (51, 52). It is highly specific for trehalose, appears to have the highest concentration in the proximal and middle jejunum, and declines toward the distal ileum (53). Trehalase hydrolyzes trehalose in close proximity to the enterocytes where the two glucose molecules are transported into the body by a well-known active transport system (54). Glucose uptake in the small intestine is efficient, and studies have shown that each 30 cm of jejunum (10% of the total length of the small intestine) can absorb 20 g of glucose per hour (55). In addition, during the first hour after digestion, the stomach releases only about 50 g of glucose into the duodenum (56). Disaccharides are known to retard gastric emptying, and although trehalose has not been specifically tested, it is likely that it would have a similar effect (57). Thus all ingested trehalose is hydrolyzed to glucose and absorbed in the small intestine.

Only a few studies have examined the level of trehalase activity in the gut. These have not always used the same endpoints or experimental conditions, so it is not possible to make direct comparisons. It can be inferred from these studies that, except in a few specific instances, the concentration of trehalase in the small intestine of humans is sufficient to handle substantial amounts (at least 50 g in a single ingestion) of trehalose.

Gudmand-Høyer et al. noted that trehalase and lactase activities in the gut are similar (58). In the experience of the authors, a lactase activity of 6.0 IU/g protein or less results in malaborption of lactose, whereas activities from 6.0 to 8.0 IU/g protein were considered intermediate and may or may not result in malabsorption. The authors suggested that it may be appropriate to use a similar standard for trehalase. They referred to a study of trehalase activity of intestinal biopsies from 248 Danish patients in which the lowest level of activity in the group was 8.3 IU/g. Furthermore, no trehalase deficiencies were identified in more than 500 biopsies specimens of Danish subjects (58).

Welsh described intestinal trehalase activity in 123 Caucasian subjects ranging in ages from 1 month to 93 years from the southwestern United States (59). No significant differences in trehalase activity for all age groups or sex-based differences were found. The lowest recorded values were 7 and 8 IU/g in two infants 0 to 2 years of age ($n = 70$). No statistically significantly differences in trehalase activity were found for any age group or sex. Importantly, trehalase activity did not appear to wane with age.

In a study of 100 consecutive normal biopsy samples from adults (72 men, 28 women), two subjects with low trehalase activity (2.7 and 1.5 IU/g) were identified (60). It was suggested by these authors that a trehalase value < 5 IU/g might result in intolerance of trehalose ingestion, which is similar to the conclusion of Gudmand-Høyer et al. (58). In a second study, Bergoz tested 16 control subjects for their ability to assimilate trehalose (4). Each subject drank 50 g of glucose in water, followed within 2 days by a similar preparation of trehalose. Blood glucose values were assayed after ingesting glucose or trehalose, and the ratios of glucose absorbed were calculated. All control subjects tolerated both test solutions and assimilated the glucose hydrolyzed from trehalose (ratio = 0.70, range, 0.31–1.42); however, the time to peak blood glucose concentrations was slower after trehalose ingestion. A third study by Bergoz et al. reported on 50 hospitalized control subjects (61). Patients were given glucose and trehalose oral tolerance tests as described previously. Control subjects had a median absorption ratio of 0.70 (range, 0.31–1.52) and tolerated the trehalose exposure. These values were essentially the same as in the second study.

Twenty patients in Czechoslovakia with no bowel symptoms or disease were examined for trehalase activity (62). None of these subjects were considered to be trehalase deficient (58).

Murray et al. reviewed several reports on trehalase activity in various groups of patients, and presented data on 369 subjects from the U.K. with healthy intestinal histology (62a). Of the 369 subjects, only 1 was shown to be slightly deficient in trehalase activity. The authors used a different standard to classify trehalase deficiency, and although not identical to the other methods it appears that the data can be compared.

In summary, no indication of trehalase deficiency was reported in 500 biopsies specimens in a Danish population (58). Welsh's group studied 123 subjects from the Southwestern United States. The lowest value of trehalase activity was 7 IU/g in an infant(s) less than 1 year of age; however, there were no statistical differences in means between age groups. In addition, no adults had low enzymatic levels (59). Bergoz et al. assayed biopsies specimens from 100 controls and showed substantial trehalase values in 98 samples (60). Bergoz found no abnormalities in trehalose absorption in 16 control patients (4). Fifty hospitalized patients showed normal trehalose absorption (61). Twenty patients sampled in Czechoslovakia did not display a trehalase deficiency, and only one of the 369 was found to have low activity in the U.K. (62, 62a). Taken together, it appears that when trehalase activity assays or trehalose absorption tests are performed on hundreds of control subjects, only a few individuals can be identified with low trehalase activity. Importantly, this percentage ($< 1\%$) appears substantially less than for lactase and possibly other disaccharidase deficiencies (63).

Trehalase is also found in human peripheral blood, urine, kidney and liver, and urine (64, 65). The serum concentrations of trehalase appear to be fairly constant in an individual, although they can vary widely within the population. Significant sex or age differences have not been reported. The presence of trehalase in these areas of the body do not appear to be related to the absorption of trehalose from the gut.

XV. MALABSORPTION AND INTOLERANCE

Malabsorption and/or intolerance to trehalose has been reported (58–63). The clinical symptoms and the pathophysiology appear to be identical to those seen in other disaccharide malabsorption syndromes and are therefore self-limiting

(58, 63). Few specific studies have been performed to identify the prevalence of trehalose intolerance or have specifically related intestinal trehalase activity with trehalose intolerance. It is believed that trehalose intolerance is substantially less frequent than lactose intolerance with similar self-limiting consequences (63).

As reported previously, 100 consecutive normal biopsies specimens were evaluated for trehalase activity (60). Two subjects had low trehalase activity. One subject with the lowest trehalase activity was tested for trehalose malabsorption and intolerance. No elevation in blood glucose was observed after consumption of the trehalose; intestinal symptoms developed within the first hour. The authors suggested that trehalase values less than 5 IU/g would result in symptoms. Bergoz also identified a 71-year-old woman who reported abdominal pain when she ingested mushrooms (4). The patient and 16 control subjects were tested for glucose and trehalose absorption and tolerance by consuming 50 g of both sugars in water. The woman experienced intestinal distress starting about 20 minutes after ingesting trehalose, but not glucose. The mean absorption ratio of trehalose to glucose was 0.02 for the patient compared with 0.7 for controls.

A 24-year-old man complained of intestinal symptoms after eating mushrooms (62). The patient's father, uncle, and two cousins also reported experiencing similar symptoms when they ate mushrooms. Jejunal biopsies were taken from immediate family members and 20 control patients (62). Trehalose and glucose tolerance was tested with 50 g of each disaccharide. Trehalose ingestion caused intestinal symptoms in the father and son; no trehalase activity was detected in their biopsy tissue.

Murray et al. reported that only one of 369 subjects with normal histology was deficient in trehalase, however, the subject was not tested for malabsorption or intolerance (62a). Gudmand-Høyer et al. studied intestinal biopsies from 97 adult (50 women, 47 men) residents of Greenland (58). Eight subjects had trehalase activity values of less than 6 IU/g protein and another six had values less than 8 IU/g. Three subjects with low trehalase activity (< 6 IU/g) were given 50 g of trehalose dissolved in water. No glucose was assimilated into the blood. Lactase deficiency is found in approximately 60% of Greenlanders, and sucrase deficiency is not uncommon. The percentage of trehalase deficiency in a fairly closed genetic group and in a single family suggests a hereditary basis, as is seen in other disaccharide deficiencies (58, 60).

XVI. AT-RISK POPULATIONS

It appears from studies that the consumption of all disaccharides, including trehalose, can be of concern in patients with intestinal malabsorption disorders (61, 62a, 66, 67). Patients with juvenile diabetes were tested for trehalase activity. It does not appear that this condition results in a significant depression of trehalase activity, although the mean enzymatic activity of these patients in one study was lower than controls (68).

Cerda and coworkers reported that in non-insulin-dependent diabetic patients with chronic pancreatic insufficiency, there was approximately a twofold increase in trehalase and other disaccharidases (69). Patients with chronic renal failure appear to have reduced levels of trehalase, but the reduction appears to be less than for other disaccharides (70). The more common condition of lactase deficiency (lactose intolerance) has not been shown to be correlated with a deficiency of trehalase in the general population (71). From this information, it appears that the consumption of trehalose under various conditions presents no more risk than any other disaccharide in a normal diet.

XVII. SAFETY STUDIES

Trehalose is considered a natural sugar for which humans have evolved and maintained an intestinal disaccharidase. Because trehalose is now only a minor component of the human diet, little interest has been shown in studying possible effects of higher consumption. In the early 1990s, Quadrant Holdings of Cambridge, England, began investigating the possible use of trehalose for the health and food industry and commissioned seven toxicological studies. With the advent of the enzymatic production method of HBC and the subsequent reduction in cost, several additional safety studies were initiated. At present, none of the studies summarized in the following have been published; however, a number of them have been submitted for publication in 2000.

A. Mouse Micronucleus, Chromosome Aberration, and Bacterial Mutation Assays

All studies were performed by standard methods using trehalose (5, HBC, unpublished data, 1997, 1997, 1995). Peak dosing concentrations for each assay were 5000 mg/kg, 5000 μg/ml and 5000 μg/plate, respectively. On the basis of the criteria established by the protocol, it was concluded that trehalose did not show a positive response (negative treatment effect) in any of the test systems.

B. Single Dose Oral Toxicity

Four studies were performed on mice, rats, and beagle dogs. The first study was performed on rats using trehalose produced by HBC (HBC, unpublished data, 1995). The remaining three studies were performed using pharmaceutical grade trehalose produced by Pfanstiehl Laboratories, Inc. (Quadrant Holdings, unpublished data, 1994). The results are provided in Table 7.

Table 7 Four Acute Toxicological Studies Performed on Mice, Rats, and Beagle Dogs Using Two Difference Preparations of Trehalose

Study	Dose	Conclusions
Trehalose crystals: acute oral toxicity to the rat	Oral gavage 16 g/kg	The acute lethal dose is greater than 16 g/kg body weight. Piloerection in all rats within 5 min of dosing lasting through day 1. Occurred at later intervals, but stopped by day 4.
An acute toxicity study of trehalose in the albino mouse	Intravenous (IV) 1 g/kg Oral gavage (OG) 5 g/kg	No treatment-related mortality, body weight changes, or systemic toxicity was evident.
An acute toxicity study of trehalose in the albino rat	Intravenous (IV) 1 g/kg Oral gavage (OG) 5 g/kg	No acute signs of toxicity with either IV or OG dose.
An acute toxicity study of trehalose in the beagle dog	Intravenous (IV) 1 g/kg 6 day washout Oral capsules 5 g/kg	No acute signs of toxicity with either IV or OG dose.

C. Acute Eye Irritation Study

Six male New Zealand White rabbits received a single 0.1-ml dose of a 10% trehalose solution in the right eye (Quadrant Holdings, unpublished data, 1994). The left eye was used as a control. Eyes were examined at 1, 24, 48, and 72 hours after application. No evidence of irritation was observed.

D. 7-Day Continuous Infusion Study

Three groups of five male Sprague-Dawley rats were dosed by continuous infusion with a 10% solution of trehalose delivered at a rate of 400 mg/kg/hr, a 5% solution of dextrose at 200 mg/kg/hr, or a 0.9% NaCl solution at 4 ml/kg/hr (Quadrant Holdings, unpublished data, 1997). The solutions were administered for 7 consecutive days through tail vein catheters. No clinical signs, changes in body weights, or treatment-related findings at necropsy were observed. The only untoward effect reported was osmotic nephrosis in the animals given trehalose. This is thought to result from a lack of trehalase in the kidney of rats.

E. 14-day Toxicity Studies

A 14-day toxicity study in CD-1 albino mice with pharmaceutical grade trehalose was performed (Quadrant Holdings, unpublished data, 1994). Doses of 1 g/kg/day by intravenous (IV) injection, 2.5 g/kg/day by subcutaneous injection, or 5 g/kg/day by oral gavage were administered for 14 consecutive days. No gross or microscopic lesions attributable to trehalose were observed, and trehalose appeared to be innocuous to the mice treated in these studies. In addition, there were no effects on body weight or food consumption.

Beagle dogs were given doses of 1 g/kg/day by IV injection, 0.25 g/kg/day subcutaneous injection, or 5 g/kg/day orally for 14 consecutive days (Quadrant Holdings, unpublished data, 1994). No gross or microscopic lesions attributible to trehalose were observed, indicating that trehalose appeared to be nontoxic. The only condition observed was self-limiting loose stools in dogs given trehalose orally. The loose stools are thought to be caused by osmotic pressure bringing water into the gut, similar to that seen in human and other animals with lactase or other disaccharidase deficiencies.

F. Thirteen-week Toxicity Study

A subchronic 13-week study was conducted to assess the effects of up to 50,000 ppm of trehalose in mice (HBC, unpublished data, 1997). Trehalose was well tolerated and there was no evidence of toxicity. There was a slight reduction in food consumption, with a concomitant reduction in weight gain in male mice. It is believed that palatability was the cause of reduced food intake. Slight increases in plasma glucose and reduction in plasma bilirubin and potassium were observed. The results indicate a no-toxic-effect-level of 50,000 ppm.

G. Embryotoxicity, Teratogenicity, and Two-Generation Reproductive Studies

Trehalose was fed in the diet to mated female rats from 0 to 21 days (HBC, unpublished data, 1998) and to female rabbits from 0 to 29 days of gestation (HBC, unpublished data, 1998). The concentration of trehalose was 2.5, 5, and 10% of the diet for both species, which equated to an intake of 1.4–2.0, 2.8–3.9, and 5.5–7.8 g/kg/day for rats, and 0.21–0.77, 0.48–1.34, and 1.04–2.82 g/kg/day for rabbits, respectively. Analysis of the results from both studies indicated that trehalose did not induce maternal or developmental toxicity, even at the highest concentration.

Two generations of male and female rats were fed trehalose in the diet at concentrations of 2.5, 5, and 10% (HBC, unpublished data, 1999). The consumption by males of the two generations ranged from 1.24–2.90, 2.38–5.65, and

4.89–12.43 g/kg/day, respectively. The range of consumption by female rats (including premating, during gestation, and during lactation) was 1.06–5.46, 2.22–11.49, and 4.40–23.2 g/kg/day, respectively. No treatment effects were observed on any variables of adult males and females, offspring, or reproduction.

XVIII. CONCLUSIONS

Trehalose is a relatively new entry to the food industry, but has had a long history as a natural part of the human diet. Although trehalose consumption in particular foods is not a large part of our modern diet, the total amount of trehalose consumed can be substantial when the contribution of these trehalose-containing foods is considered as a whole.

The ability to digest trehalose depends on the presence of the enzyme trehalase. From available information, it appears that only a relatively small percentage of people in a western population lack this enzyme. Intolerance to trehalose appears to be far less common than intolerance to lactose, and both are believed to have a genetic basis.

More than 30,000 metric tons of trehalose added to food products will be consumed in the Japanese market by the end of 1999 without any known reports of intolerance. In addition, symptoms of intolerance to trehalose are identical to those observed in individuals with intolerances to other disaccharides. Safety studies have demonstrated that there are no consistent untoward effects associated with the consumption of trehalose.

From information presented in this chapter, it would appear that trehalose has functional properties that may be of great interest to the food, cosmetic, and pharmaceutical industries. Heretofore, the cost of trehalose precluded its use in all but the highest value products. With the advent of the new enzymatic production process invented by HBC, trehalose is now priced where its use in cost-sensitive applications can by justified.

Trehalose is approved as a food additive in Japan and as a food ingredient in Taiwan and Korea. It is approved as a novel food for use in the preservation of freeze-dried products in the U.K. Recently JECFA approved trehalose with no specified ADI, and the U.S. FDA had no objection to the GRAS notification submitted by HBC. Regulatory approval in Europe is being sought.

REFERENCES

1. GG Birch. Trehalose. In: ML Wolfrom, RS Tyson, eds. Advances in Carbohydrate Chemistry, vol. 18. New York: Academic Press, 1963, pp 201–225.
2. J Leibowitz. A new source of trehalose. Nature 152:414, 1943.

3. J Leibowitz. Isolation of trehalose from desert manna. J Biochem 38:205–206, 1944.
4. R Bergoz. Trehalose malabsorption causing intolerance to mushrooms. Gastroenterol 60:909–912, 1971.
5. T Sugimoto. Production of trehalose by enzymatic starch saccharification and its applications. Shokuhin Kogyo 38:34–39, 1995.
6. AD Elbein. The metabolism of α,α-trehalose. In: RS Tipson, D Horton, eds. Advances in Carbohydrate Chemistry and Biochemistry, vol. 30. New York: Academic Press, 1974, pp 227–256.
7. P Van Dijick, D Colavizza, P Smet, JM Thevelein. Differential importance of trehalose in stress resistence in fermenting and nonfermenting *Saccharomyces cerevisiae* cells. Appl Environmental Micro 61:109–115, 1995.
8. B Nicolaus, A Gambacorta, AL Basso, R Riccio, M De Rosa, WD Grant. Trehalose in archaebacteria. System Appl Microbiol 10:215–217, 1988.
9. S Leslie, E Israeli, B Lighthart, JH Crowe, LM Crowe. Trehalose and sucrose protect both membranes and proteins in intact bacteria during drying. Appl Environ Microbiol 61:3592–3597, 1995.
10. K Aso, T Watanabe, A Hayasaka. Studies on the MIRIN. II. Fractionation and determination of sugars in Mirin (sweet saké) carbon column chromatography. Tohoku J Agric Res 13:251–256, 1962.
11. GM Brown, DC Rohrer, B Berking, CA Beevers, RO Gould, R Simpson. The crystal structure of α,α-trehalose dihydrate from three independent x-ray determinations. Acta Cryst B28:3145–3158, 1972.
12. T Taga, M Senma, K Osaki. The crystal and molecular structure of trehalose dihydrate. Acta Cryst B28:3258–3263, 1972.
13. S Budavari. The Merck Index, 11th ed. New Jersey: Merck and Co, 1989, p 1508.
14. GG Birch. A method of obtaining crystalline anhydrous α,α-trehalose. J Chem Soc 3:3489–3490, 1965.
15. A Steiner, CF Cori. The preparation and determination of trehalose in yeast. Science 82:422–423, 1935.
16. EM Koch, FC Koch. The presence of trehalose in yeast. Science 61:570–572, 1925.
17. LC Stewart, NK Richtmyer, CS Hudson. The preparation of trehalose from Yeast. J Natl Inst Health 72:2059–2061, 1950.
18. T Suzuki, T Katsunobu, S Kinoshita. The extracellular accumulation of trehalose and glucose by bacteria grown on n-alkanes. Agric Biol Chem 33:190–195, 1969.
19. S Murao, H Nagano, S Ogura, T Nishino. Enzymatic synthesis of trehalose from maltose. Agric Biol Chem 49:2113–2118, 1985.
20. I Schick. Papers from Proceedings of the 2nd International Symposium-Biochemical Engineering-Stuttgart. 1991, pp 126–129.
21. T Nishimoto, M Nakano, S Ikegami, H Chaen, S Fukuda, T Sugimoto, M Kurimoto, Y Tsujisaka. Existence of a novel enzyme converting maltose into trehalose. Biosci Biotech Biochem 59:2189–2190, 1995.
22. T Nakada, K Maruta, H Mitzuzumi, K Tsukaki, M Kubota, H Chaen, T Sugimoto, M Kurimoto, Y Tsujisaka. Putrification and properties of a novel enzyme, maltooligosyl trehalose synthase, from *Arthrobacter* sp. *Q36*. Biosci Biotech Biochem 59: 2210–2214, 1995.

23. T Nakada, K Maruta, H Mitzuzumi, M Kubota, H Chaen, T Sugimoto, M Kurimoto, Y Tsujisaka. Putrification and characterization of a novel enzyme, maltooligosyl trehalose trehalohydrolase, from *Arthrobacter* sp. Q36. Biosci Biotech Biochem 59: 2215–2218, 1995.
24. K Tsusaki, T Nishimoto, T Nakada, M Kubota, H Chaen, S Fukuda, T Sugimoto, M Kurimoto. Cloning and sequencing of trehalose synthase gene from *Thermus aquaticus* ATCC33923. Int J Biochem Biophys 1334:28–32, 1997.
25. B Roser. Trehalose, a new approach to premium dried foods. Trends Food Sci Technol 2:166–169, 1991.
26. H Kawai, M Sakurai, Y Inoue, R Chûjô, S Kobayashi. Hydration of oligosaccharide: anomalous hydration ability of trehalose. Cryobiol 29:599–606, 1992.
27. B Roser. Trehalose drying: a novel replacement for freeze drying. Biopharm 4:47–53, 1991.
28. JH Crowe, JF Carpenter, LM Crowe, TJ Anchordoguy. Are freezing and dehydration similar stress vectors? A comparison of modes of interaction of stabilizing solutes with biomolecules. Cryobiol 27:219–231, 1990.
29. CALS Colaço, B Roser. Trehalose, a multifunctional additive for food preservation. In: M Mathlouthi, ed. Food Packaging and Preservation. Maryland: Aspen, 1994, pp 123–140.
30. AS Rudolph, JH Crowe. Membrane stabilization during freezing: The role of two natural cryoprotectants, trehalose and proline. Cryobiol 22:367–377, 1985.
31. HR Bolin, RJ Steele. Nonenzymatic browning in dried apples during storage. J Food Sci 52:1654–1657, 1987.
32. MO Portmann, GG Birch. Sweet taste and solution properties of α,α-trehalose. J Sci Food Agri 69:275–281, 1995.
33. MC Donnamaria, EI Howard, JR Grigera. Interaction of water with α,α trehalose in solution: Molecular dynamics simulation approach. J Chem Soc Faraday Trans 90:2731–2735, 1994.
34. JL Green, CA Angell. Phase relations and vitrification in saccharide-water solutions and the trehalose anomaly. J Phys Chem 93:2880–2882, 1989.
35. R Karmus, M Karel. The effect of glass transition on Maillard browning in food models. In: TP Labuza, GA Reieccius, eds. Maillard Reactions in Chemistry, Food, Health. London: Royal Society Chemistry, 1993, pp 182–187.
36. CALS Colaço, S Sen, M Thangavelu, S Pinder, B Roser. Extraordinary stability of enzymes dried in trehalose: Simplified molecular biology. Bio Tech 10:1007–1010, 1992.
37. CK Lee, GG Birch. Structural functions of taste in the sugar series: Binding characteristics of disaccharides. J Sci Food Agri 26:1513–1521, 1975.
38. S Shamil, GG Birch, M Mathlouthi, MN Clifford. Apparent molar volumes and tastes of molecules with more than one sapophore. Chem Senses 12:397–409, 1987.
39. M Mathlouthi. Relationships between the structure and the properties of carbohydrates in aqueous solution: Solute-solvent interactions and the sweet taste of D-fructose, D-glucose and sucrose in solution. Food Chem 13:1–13, 1984.
40. M Mathlouthi, C Bressan, MO Portmann. Role of water structure in the sweet taste chemoreception. In: M Mathlouthi, ed. Sweet-Taste Chemoreception. London: Elsevier Applied Science, 1993, pp 141–174.

41. LM Crowe, DS Reid, JH Crowe. Is trehalose special for preserving dry biomaterials? Biophys J 71:2087–2093, 1996.
42. SP Ding, J Fan, JL Green, Q Lu, E Sanchez, CA Angell. Vitrification of trehalose by water loss from its crystalline dihydrate. J Thermal Anal 47:1391–1405, 1996.
43. BJ Aldous, AD Auffret, F Frank. The crystallization of hydrates from amorphous carbohydrates. Cryo-Letters 16:181–186, 1995.
44. JH Crowe, LM Crowe, JF Carpenter, AS Rudolph, CA Wistrom, BJ Spargo, TJ Anchordoguy. Interactions of sugars with membranes. Biochem Biophys Acta 947: 367–384, 1998.
45. LM Crowe, JH Crowe, A Rudolph, C Womersley, L Appel. Preservation of freeze-dries liposomes by trehalose. Arch Biochem Biophys 242:240–247, 1985.
46. F Nazzaro, PA Malanga, F Immacolata, V Zappia, G Donsi, G Ferrari, R Nigro, M De Rosa. Trehalose as innovative agent for food osmo-drying processes. Food, Science and Technology Congress, Como, Italy. September, 1997.
47. B Roser, CALS Colaço. A sweeter way to fresher food. New Scientist 106(15 May): 25–28, 1993.
48. GA MacDonald, T Lanier. Carbohydrates and cryoprotectants for meats and surimi. Food Technol 3:150–159, 1991.
49. T Imfeld. Identification of low caries risk dietary components. In: HM Myers, ed. Monographs in Oral Science, vol. 11. Basel: Karger, 1983, p 117.
50. T Imfeld, HR Muhlemann. Cariogenicity and acidogenicity of food, confectionery and beverages. Pharm Ther Dent 3:53, 1978.
51. G Galand. Brush border membrane sucrase-isomaltase maltase-glucoamylase and trehalase in mammals. Comp Biochem Physiol 94B:1–11, 1989.
52. D Maestracci. Enzymatic solubilization of the human intestinal brush border membrane enzymes. Biochim Biophys Acta 433:469–481, 1976.
53. NG Asp, E Gudman-Høyer, B Andersen, NO Berg, A Dahlqvist. Distribution of disaccharides, alkaline phosphatase, and some intracellular enzymes along the human small intestine. Scand J Gastroenterol 10:647–651, 1975.
54. WJ Ravich, TM Bayless. Carbohydrate absorption and malabsorption. Clin Gastroenterol 12:335–356, 1983.
55. CD Holdsworth, AM Dawson. The Absorption of monosaccharides in man. Clin Sci 27:371–379, 1964.
56. CD Holdsworth, GM Besser. Influence of gastric emptying-rate and of insulin response on oral glucose tolerance in thyroid disease. Lancet 2:700–703, 1968.
57. E Elias, GJ Gibson, LF Greenwood, JN Hunt, JH Tripp. The slowing of gastric emptying by monosaccharides and disaccharides in test meals. J Physiol 194:317–326, 1968.
58. E Gudmand-Høyer, HJ Fenger, H Skovbjerg, P Kern-Hansen, PR Madsen. Trehalase deficiency in Greenland. Scand J Gastroenterol 23:775–778, 1988.
59. JD Welsh, JR Poley, M Bhatia, DE Stevenson. Intestinal disaccharide activities in relation to age, race and mucosal damage. Gastroenterol 75:847–855, 1978.
60. R Bergoz, MC Vallotton, E Loizeau. Trehalose Deficiency. Ann Nutr Metabol 26: 291–295, 1982.
61. R Bergoz, JP Bolte, KH Meyer zum Bueschenfelde. Trehalose tolerance test. Scand J Gastroent 8:657–663, 1973.

62. J Mardžarovová-Nohejlova. Trehalase deficiency in a family. Gastroenterol 65:130–133, 1973.
62a. IA Murray, K Coupland, JA Smith, IS Ansell, RG Long. Intestinal trehalase activity in a UK population: establishing a normal range and the effects of disease. Br J Nutri 83:241–245, 2000.
63. A Dahlqvist. Enzyme deficiency and malabsorption of carbohydrates. In: H Sipple, ed. Sugars in Nutrition. New York: Academic Press, 1974, pp 187–217.
64. A Stork, E Fabian, B Kozakova, J Fabianova. Trehalase activity in diabetes mellitus and in cirrhosis of the liver. In: J Horejsi, ed. Enzymology and its Clinical Use. Univerzita Kavlova: Praha, 1977, pp 37–40.
65. GG Birch. Mushroom sugar. In: GG Birch, LF Green, LG Plaskett, eds. Health and Food. New York: Elsevier Science, 1972, pp 49–53.
66. U Redeck, HC Dominick. Trehalose-load-test in gastroenterology. Monatsschr Kinderheilkd 131:19–22, 1983.
67. NO Berg, A Dahlqvist, T Lindberg, A Nordén. Correlation between morphological alterations and enzyme activities in the mucosa of the small intestine. Scand J Gastroenterol 8:703–712, 1973.
68. H Ruppin, W Domschke, S Domschke, M Classen. Intestinale disaccharideasen bei juvenilem diabetes mellitus. Klin Wochr 52:568–570, 1974.
69. JJ Cerda, H Preiser, RK Crane. Brush border enzymes and malabsorption: Elevated disaccharides in chronic pancreatic insufficiency with diabetes mellitus. Gastroenterol 62:841, 1972.
70. T Dennenberg, T Lindberg, NO Berg, A Dahlqvist. Morphology, dipeptidases and disaccharidases of small intestinal mucosa in chronic renal failure. Acta Med Scand 195:465–470, 1974.
71. NG Asp, NO Berg, A Dahlqvist, J Jussila, H Salmi. The activity of three different small-intestinal β-galactosidases in adults with and without lactase deficiency. Scand J Gastroenterol 6:755–762, 1971.

24
Mixed Sweetener Functionality

Abraham I. Bakal
ABIC International Consultants, Inc., Fairfield, New Jersey

Previous chapters have dealt with the properties and functionality of several currently approved sweeteners or compounds under development or consideration. The requirements for the ideal sweetener or sugar substitute, as usually defined (1) are to:

1. Have the taste and functional properties of sugar
2. Have low-caloric density on a sweetness equivalency basis
3. Be physiologically inert
4. Be nontoxic
5. Be noncariogenic
6. Compete economically with other sweeteners

An analysis of the organoleptic or functional properties of each single sweetener clearly shows that none of the currently known sugar substitutes comes close to the taste and functional properties of sucrose. Most exhibit one or more of these differences:

1. Taste properties such as sweetness lag, undesirable and lingering aftertaste, narrow taste profile, or bitterness. For example, saccharin is generally reported to have a bitter aftertaste (2); stevioside, to have a menthol aftertaste (3); and aspartame, to have a delayed sweetness (4).
2. Lack of bulking properties.
3. Stability problems during processing and storage. For example, aspartame loses its sweetness in aqueous solutions and is not stable at high temperatures (5); thaumatin reacts with tannins and loses its sweetness (6).

4. Competitive prices. For example, saccharin and cyclamate reportedly cost, on a sweetness equivalency basis, less than sucrose and other nutritive sweeteners. On the other hand, in the United States, the new sweeteners cost significantly more (7) than saccharin.

The food industry has partially overcome the bulking limitations in some selected applications. With the expansion of polydextrose use and the availability of polyols such as isomalt, maltitol, and sorbitol, other reduced-calorie products based on combinations of polydextrose and/or on polyols and intense sweeteners have been introduced in the U.S. market. Fibers and fibers and polydextrose have also recently found expanded use as bulking agents. However, this problem is universal for all intense sweeteners; thus, it will not be discussed in this chapter. Except for the bulk issue, by far the most important limitations to consumer acceptability are taste properties and cost.

The use of more than one sweetener provides the food technologist with a tool for overcoming these taste limitations. The advantages of combining sweeteners are many; some of the aims and goals are to:

1. Formulate products that closely imitate the taste and stability of their sugar-sweetened counterparts

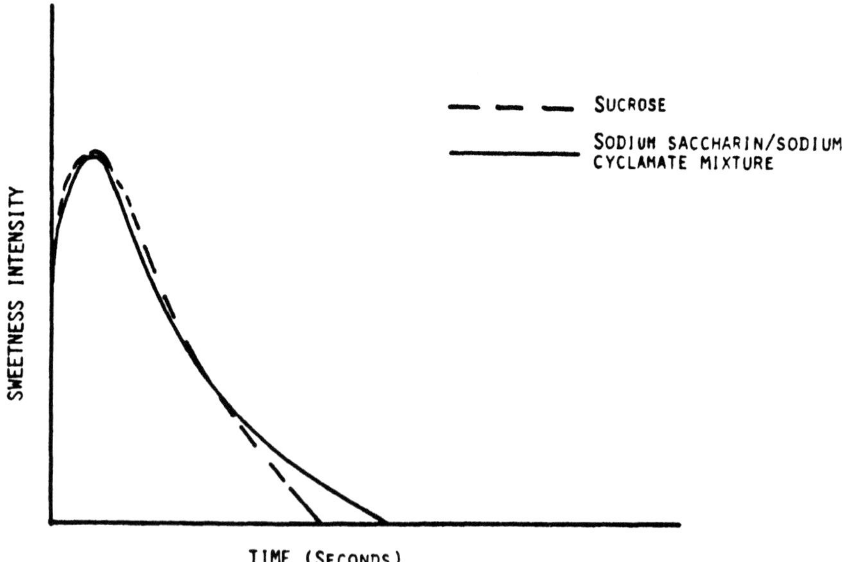

Figure 1 Sweetness profile of aqueous solutions of sucrose versus a sodium saccharin/sodium cyclamate combination (1:10).

2. Create totally new taste experiences by using sweeteners in the same manner the food industry uses flavors (8)
3. Meet cost restrains

Two interesting observations made by Paul (9) in 1921 led to the subsequent practice of combining sweeteners. Paul, as cited by Mitchell and Pearson (10), observed that the relative sweetness of an intense sweetener decreases with increasing concentration and that when combining two sweeteners, each sweetener will at least contribute the relative sweetness on the sweetness concentration curve. In other words, less of each sweetener is required when two or more sweeteners are used to achieve the same final sweetness achieved by the use of a single sweetener. These observations formed the basis for extensive work on the effects and benefits of using a combination of sweeteners.

Indeed, the use of combined sweeteners in food applications has been and is currently being practiced by the food industry. Before the cyclamate ban in the United States, the food industry used a combination of saccharin and cyclamate in the formulation of products. This mixture showed synergistic properties and improved taste profile as shown in Fig. 1. For comparison, the taste profile of saccharin against sucrose is shown in Fig. 2. With the approval of aspartame and acesulfame-K and their expanded use worldwide, several intense sweetener

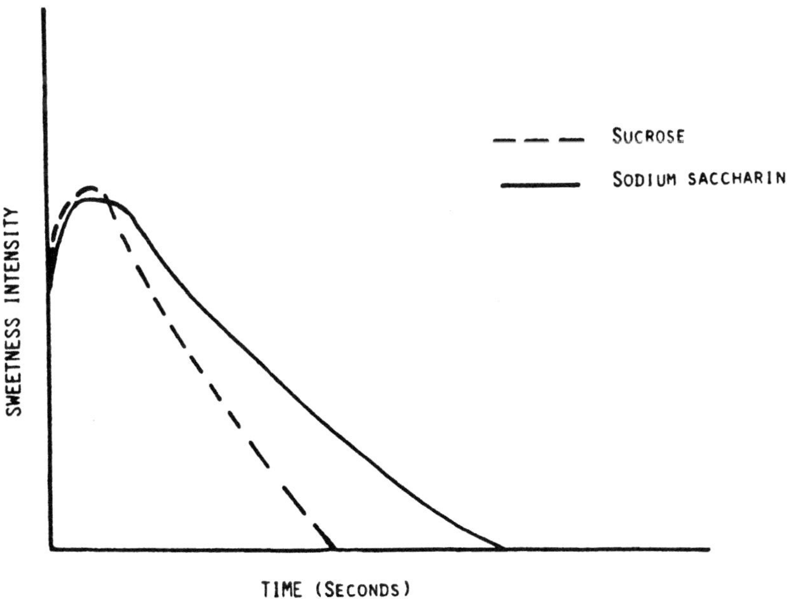

Figure 2 Sweetness profile of aqueous solutions of sucrose versus sodium saccharin.

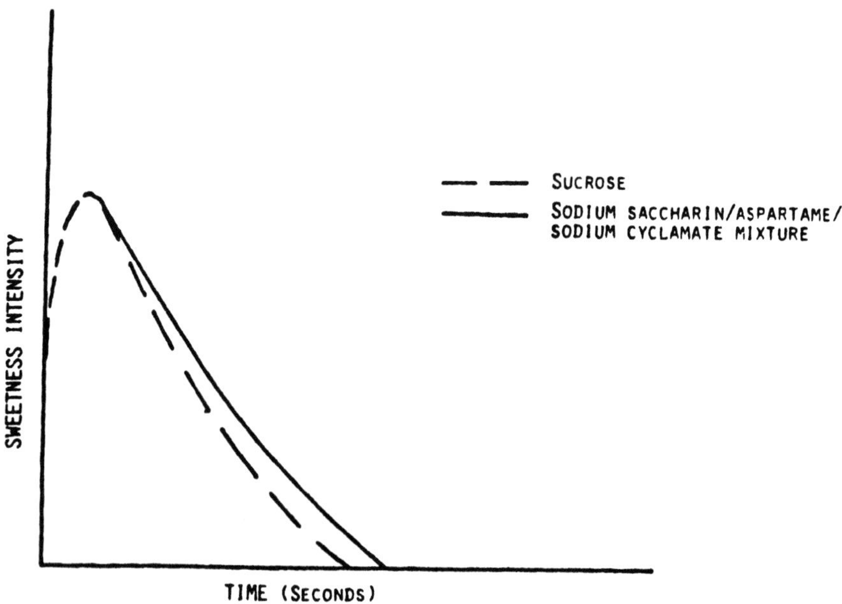

Figure 3 Sweetness profile of aqueous solutions of sucrose versus a sodium saccharin/aspartame/sodium cyclamate combination (1:5:8).

mixtures are being used in many foods and beverages. Fig. 3 shows the taste profile of an aqueous solution sweetened with a saccharin/aspartame/cyclamate mixture in the ratio of 1:5:8, respectively. These data clearly demonstrate that combined sweeteners result in an improved sweetness profile compared with each of the single sweeteners. In addition to the sweetness profile, the quality of the taste is significantly improved, as shown in organoleptic tests conducted in our laboratories. Verdi and Hood (11) provide a good summary of some of the advantages of intense sweetener blends. Applications of combined sweeteners in various products are discussed in the remaining section of this chapter.

I. APPLICATIONS

A. Tabletop

Tabletop sweeteners are a major product category used by consumers for a variety of applications. They are commonly available as powders packed in bulk or sachets, as tablets, or as liquids. The effective amount of sweetener or sweetener combinations required to provide the sweetness equivalency of two teaspoons

Table 1 Typical Effective Amounts of Sweeteners or Sweetener Combinations in Tabletop Applications

Sweetener	Amount (mg)[a]
Sodium saccharin	30–40
Sodium cyclamate	150–200
Aspartame	35–45
Acesulfame-K	50–60
Sucralose	12–15
Sodium cyclamate/sodium saccharin	80/8
Aspartame/sodium saccharin	5/15
Acesulfame-K/aspartame	30/3
Aspartame/sodium saccharin/sodium cyclamate	10/4/30

[a] Effective amount required to provide the sweetness equivalency of two teaspoons (10 g) of sugar to one cup (240 ml) of coffee.
Source: Refs. 1 and 12.

(10 g) of sugar to 1 cup (240 ml) of coffee is given in Table 1. This table clearly illustrates the reduction in the level of sweetness ingested when a combination of sweeteners is used. When a mixture of saccharin/cyclamate is used, the amount of saccharin is reduced by a factor of 5 (i.e., 40 mg alone to 8 mg in combination), and the amount of cyclamate by a factor of 2.5 (i.e., 200 mg alone to 80 mg in combination) compared with the use of any one single sweetener. Similarly, in the case of an aspartame/saccharin mixture, the level of aspartame may be reduced by a factor of 10, and saccharin by a factor of 2.

Several patents and articles describe the synergism among aspartame, saccharin, cyclamate, acesulfame-K, and other sweeteners. Scott described the use of aspartame in combination with saccharin, cyclamate, or both to provide beverages with improved taste and consumer acceptability (12, 13). In a 1973 patent, Scott described the use of aspartame with saccharin in a ratio range of 15:1 to 1:15 (13). A patent issued to General Foods discloses a sweetening composition containing aspartame, saccharin, and cyclamate (14). The inventors state that the sweetness is intense and lacks the lingering or bitter aftertaste associated with these sweeteners singly. A patent issued to E.R. Squibb proposes the use of "dipeptides" and saccharin, with the dipeptides masking the bitter aftertaste of saccharin (15–17). The saccharin to dipeptide ratio is 48:1.

Taste panels conducted in our laboratories on tabletop preparations confirm the superiority of a combination of aspartame with saccharin, cyclamate, or acesulfame-K over any of these sweeteners singly. Panelists judged the taste quality of coffee sweetened with a combination of sweeteners closely resembling the taste quality of sugar-sweetened coffee. Studies conducted in our laboratories suggest that aspartame/saccharin and aspartame/cyclamate combinations have improved stability, even in hot aqueous solutions, and also exhibit superior taste profiles. The superior taste may be associated with flavor-enhancing properties of aspartame.

From 20–30% of tabletop sweeteners are sold in bulk in the United States, a significant portion of which is used in cooking and baking. Consumers who cook with artificial sweeteners are usually those who must restrict their sugar intake for medical reasons. Saccharin, aspartame, acesulfame-K, and sucralose are approved for sale in bulk in the United States.

The liquid tabletop sweetener market represents approximately 10% of the U.S. sales. Aspartame-based products may not be feasible for this application because of their limited stability in aqueous solutions.

In Europe, most tabletop products are formulated with a combination of sweeteners. The commercial products include traditional combinations of saccharin and cyclamate and a combination of acesulfame-K and aspartame.

In summary, tabletop formulations using a single sweetener meet the current needs of the consuming public; however, each preparation has one or more serious defects, such as taste or stability. The consensus of experts and consumers is that tabletop preparations combining two or more sweeteners are superior in taste and stability and more closely imitate the sweetness of sucrose.

B. Carbonated Beverages

Low-calorie soft drinks represent a significant segment of the market. Current market share of low-calorie carbonated beverages reportedly exceeds 25% of the total market. Until the approval of aspartame for use in carbonated beverages in the United States, only saccharin was available. The level of saccharin used to sweeten one fluid ounce varies between 8 and 11mg; the actual level depends on the soft drink flavor and the product brand. To achieve similar sweetness levels with aspartame as the single-source sweetener, significantly greater amounts of aspartame are required. Table 2 summarizes the levels of saccharin and aspartame used in selected soft drinks and syrups, showing that, on the average, the amount of aspartame used is about 1.5 times the amount of sodium saccharin when each of the sweeteners is used alone.

Once regulatory approval was granted to aspartame, it became the most widely used sweetener in soft drinks.

Aspartame-based beverages lose sweetness as a function of storage time, temperature, and pH. Data submitted by G.D. Searle to the Food and Drug Administration (FDA) (18) indicate that about 50% of the initial aspartame remains

Table 2 Typical Concentrations of Saccharin and Aspartame in Selected Soft Drinks and Syrups

Flavor	Sodium saccharin[a] (mg/100 ml)	Aspartame[b] (mg/100 ml)
Soft drinks		
Cola	31–42	57.7
Orange	37–38	92.6
Lemon-lime	26–42	50.1
Root beer	27–37	60.5
Syrups		
Cola	—	347.6
Orange	—	401.0
Lemon-lime	—	234.2
Root beer	—	355.2

[a] From Ref. 12.
[b] From Ref. 14.

in cola beverages stored at 30°C for 24 weeks (Fig. 4). In cola syrups (pH 2.4), 75% of the initial aspartame remains after 2 weeks storage at 30°C (Fig. 5).

Combination sweeteners have been used by the soft drink industry for many years. Superior products were available when cyclamate and saccharin were both approved by the FDA. Most products used 10:1 cyclamate/saccharin combinations because this ratio provided acceptable products. The ban of cyclamate in the United States resulted in significant deterioration in the taste profile of soft drinks that did use saccharin as the single sweetener. The use of aspartame in combination with saccharin is described by G.D. Searle (19). Cola beverages sweetened with aspartame and saccharin at a ratio of 1:1 and stored at 20°C show significantly better sweetness stability than cola beverages sweetened with aspartame alone and stored under the same conditions. The performance of this combination of sweeteners is exceptional. The pH of the aspartame/saccharin product is lower than the pH of the aspartame-sweetened product (pH 2.8 versus 3.05, respectively). The data are shown in Fig. 6.

After the approval of aspartame in the United States, beverage manufacturers announced their plans to introduce soft drinks containing a combination of aspartame and saccharin. One cola beverage marketed at that time contained approximately 18 mg of saccharin and 8 mg of aspartame per 100 ml.

Marketing and other considerations have led major soft drink manufacturers to convert to 100% aspartame in the United States and in the United Kingdom

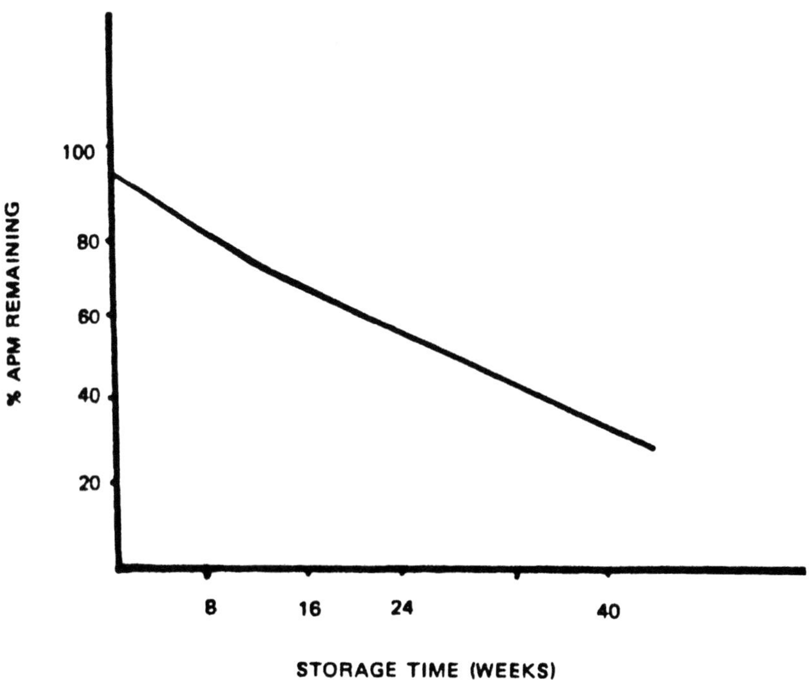

Figure 4 Aspartame (APM) stability in carbonated cola beverages stored at 30°C. (From Ref. 17.)

in bottled and canned drinks. However, in other markets, products containing two or more intense sweeteners were introduced. For example, a major soft drink manufacturer markets products that contain saccharin, aspartame, and acesulfame-K.

Experiments conducted in our laboratory indicate that the addition of very small amounts of aspartame, in accordance with the specifications of the Lavia and Hill patent (15) improve the acceptance of saccharin-sweetened soft drinks. The cola beverages evaluated were sweetened with 33 mg saccharin per 100 ml. The addition of 0.7 mg aspartame to 100 ml of the saccharin-sweetened beverage (0.0007% concentration) resulted in a significant consumer preference for the beverage. It is clear that the use of aspartame at this concentration falls below the sweetness threshold of aspartame, which, according to Beck, is in the range of 0.007–0.001% (4).

The expiration of the U.S. patent on aspartame and the approval of acesulfame-K for use in soft drinks resulted in marketing changes and renewed interest in using a combination of sweeteners. The most important of these devel-

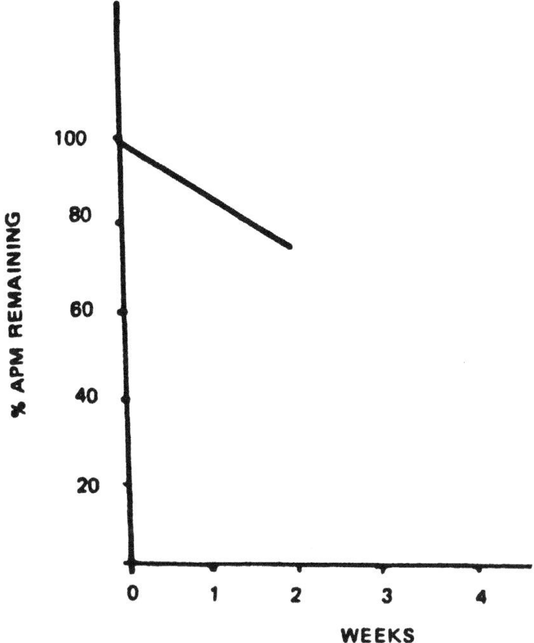

Figure 5 Aspartame (APM) stability in cola syrup at 30°C. (From Ref. 17.)

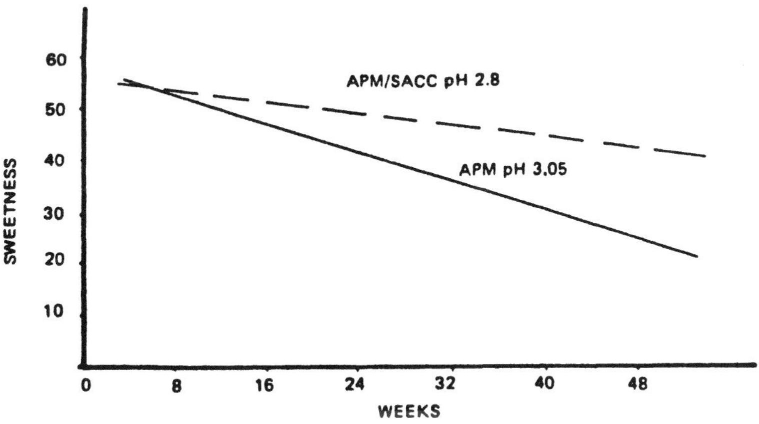

Figure 6 Average scores for sweetness of cola carbonated beverages stored at 20°C over a period of 52 weeks. (From Ref. 17.)

opments is the introduction of a new brand of cola beverage, which is sweetened with a combination of acesulfame-K and aspartame. This product contains 70 mg of aspartame and 22.7 mg of acesulfame-K per 8 oz. Another cola-based beverage uses a combination of saccharin and aspartame. This product contains 63 mg of sodium saccharin and 19 mg of aspartame per 8 oz. For comparison, the 100% sweetened cola contains 125 mg per 8 oz of aspartame.

Acesulfame-K and aspartame show a sweetness enhancement of about 35% when used in combination. In addition to sweetness enhancement, aspartame broadens the taste profile of acesulfame-K and brings the taste of the mixture closer to the sweetness profile of sucrose.

Aspartame and cyclamate combinations may also prove highly beneficial if cyclamates are approved in soft drinks. Acesulfame-K and cyclamate is another combination that is of interest because it yields excellent taste quality and exceptional storage stability.

The recent approval of sucralose in the United States has not yet resulted in a major introduction of soft drinks based on a combination of sucralose with other sweeteners. However, soft drinks containing sucralose as the single sweetener are available in Canada and the United States.

Experiments conducted in our laboratories indicate that neohesperidin dihydrochalcone (NHDC), when used as a single sweetener, is inadequate for the preparation of soft drinks. However, when used in combination with saccharin, NHDC has a synergistic effect and gives improved taste perceptions. This finding is in accordance with observations described in the patent literature (20).

Stevioside was also recently evaluated as a single sweetener in cola beverages. Results were discouraging. The beverages had a licorice-like taste and were judged unacceptable. Fructose and stevioside are successfully combined in Japan to produce reduced-calorie soft drinks (50% reduction). These products enjoy good consumer acceptance in Japan (21).

Moskowitz addressed the issue of sweetness optimization in cola-flavored beverages using combination sweeteners (22) and presented a quantitative model for developing products acceptable to the consumer. Hoppe discussed the effect of various mixtures of sucrose, saccharin, and cyclamate on sweetness perception in aqueous solutions (23). Van Tornout et al. (24) evaluated the taste characteristics of mixtures of fructose with saccharin, aspartame, and acesulfame-K in soft drinks. Their data indicate that combinations of small amounts of fructose with these intense sweeteners result in soft drinks that cannot be distinguished from sucrose-sweetened beverages. All these data indicate the benefits to the consumer that can be derived from the use of combination sweeteners in soft drinks.

C. Dry Mixes

This category encompasses a variety of food products that are sold in dry form and are reconstituted by the consumer before use. They include beverage mixes,

Table 3 Typical Concentrations of Saccharin and Aspartame in Selected Dry-Mix Products

Product	Sodium saccharin[a]	Aspartame[b]
Beverage mix	27–34	40–55[b]
Gelatin dessert	21–28	19–44
Puddings	20–25	50–60

[a] Concentration in mg/100 g of reconstituted product.
[b] Beck (5) suggests using 50–65 mg/100 g.

presweetened cereals, puddings, and desserts. In the United States, until the approval of aspartame, saccharin was the only intense sweetener used in this application. The approval of aspartame resulted in significant proliferation of these products in the market positioned as reduced-calorie and sugar-free. These aspartame-sweetened products have enjoyed wide consumer acceptance. To date, no dry mix products are available in the United States using combinations of saccharin and aspartame, although in the past, cyclamate/saccharin combinations were used. Table 3 provides a comparison of typical concentrations of saccharin and aspartame in selected products. The table clearly shows that, in these applications, the effective aspartame level is about twice the effective sodium saccharin level. The approval of acesulfame-K for use in these products and the expiration of the U.S. patent on aspartame resulted in the introduction of several products sweetened with a combination of aspartame and acesulfame-K.

A combination of saccharin/aspartame, with a concentration of 15–20 mg sodium saccharin and 6–10 mg aspartame, in 100 g of finished product yields a highly acceptable product and is significantly less expensive. In addition, such a combination represents a major reduction in the consumption of each of the sweeteners: almost a 50% reduction in sodium saccharin and about a 75% reduction in aspartame. Most importantly, results of consumer tests conducted in our laboratory clearly indicate the superiority of the products sweetened with a combination of the two sweeteners. Similarly, combinations of acesulfame-K and aspartame yield products with a taste profile much more closely resembling that of the sucrose-sweetened products than those sweetened with either intense sweetener singly.

D. Chewing Gums

Sugarless chewing gums represent a significant segment of the chewing gum market. These gums typically contain sorbitol, mannitol, and/or xylitol as the bulking agents to replace sugar. However, the sweetness intensity of sorbitol and mannitol is about half that of sucrose. These gums receive lower consumer acceptance because of lack of sweetness and flavor; therefore, intense sweeteners have been and are being used to improve the taste qualities of these products.

Saccharin is used in sugarless gums at concentrations between 0.1 and 0.2%. A typical chewing gum stick, weighing approximately 3 g, contains from 3 to 6 mg of saccharin. Several patents indicate the use of insoluble saccharin to produce chewing gums with long-lasting flavor (25, 26). Taste tests comparing the sweetness quality of chewing gums with and without saccharin establish the superiority of the saccharin-sweetened products. Saccharin taste does not seem to be a limiting factor in this application because of the low saccharin concentration.

Figure 7 illustrates the sweetness-duration properties of chewing gums containing saccharin acid compared with chewing gums containing the same concentration of sodium saccharin. Figure 8, which shows the overall quality rating of the same chewing gums on a scale of 0 (dislike extremely) to 4 (neither like nor dislike) to 8 (like extremely) clearly shows consumer preference for the chewing gum containing the insoluble saccharin. Because of its relatively low sweetness intensity, cyclamate is not used as a single sweetener by the chewing gum industry. Before the U.S. ban on cyclamates, cyclamate/saccharin combinations were commonly used at ratios between 2:1 and 10:1. Chewing gums using this combination of sweeteners were acceptable and had taste profiles similar to sugar-based products. FDA regulations allow the use of aspartame in chewing gum, both as a sweetener and flavor enhancer (18). The literature describes the use of aspartame in chewing gum at levels of 0.3% or higher (27). These gums are reported to have longer lasting flavors.

One of the problems encountered by chewing gum manufacturers in their use of aspartame is the relative instability of aspartame in chewing gums. A patent issued to General Foods describes a method for stabilizing aspartame-containing gums by removing the calcium carbonate filler from the gum base formulation and replacing it with talc (28). This procedure apparently alters the pH of the gum base and results in a slower rate of aspartame decomposition.

A method investigated in the industry is the application of aspartame only on the surface of the gum through dusting or glazing (29). This technique, although technically successful, presents consumer acceptance problems. Recently, encapsulation techniques have been investigated and are reportedly used by some chewing gum manufacturers to overcome sweetness loss and to extend the sweetness and flavor of both aspartame- and saccharin-sweetened products.

Beck (5) describes the use of a combination of aspartame and saccharin in chewing gum. The ratio used was 2:1, aspartame to saccharin. This gum was preferred to gum containing any of the single sweeteners. Furthermore, although no stability data are available on this combination of sweeteners, it is believed that stability will be significantly improved. Several products containing aspartame and mixtures of aspartame and saccharin were evaluated. Figure 9 provides a sweetness profile of chewing gums containing a mixture of saccharin and aspar-

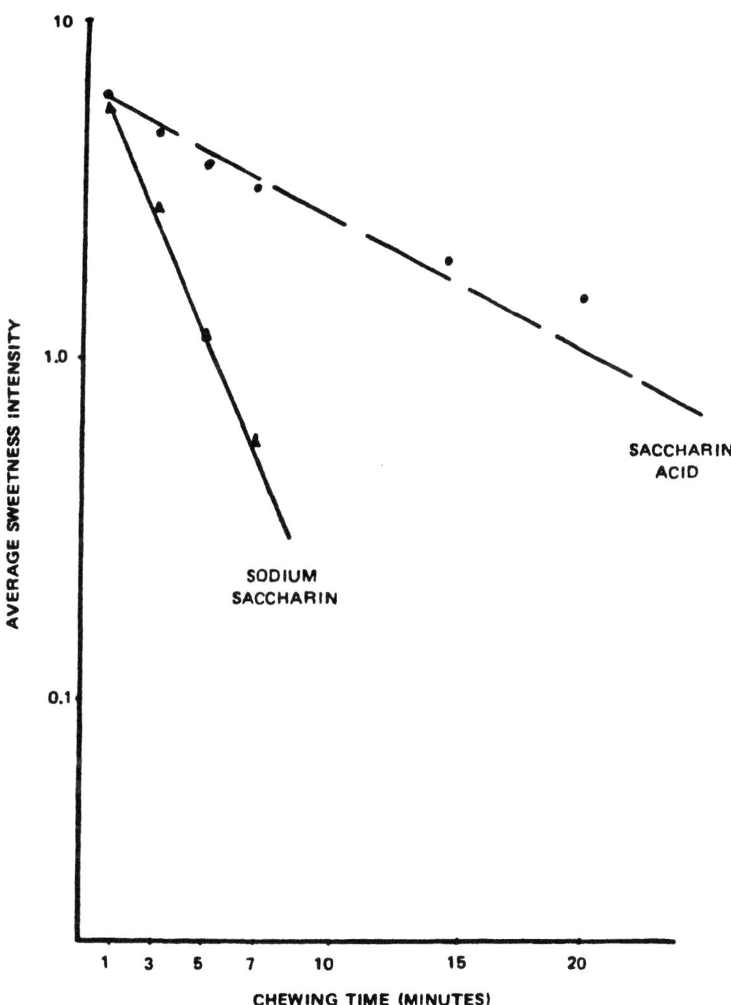

Figure 7 Sweetness intensity of chewing gums with sodium saccharin and saccharin acid.

tame. The data clearly show that this product exhibits longer lasting sweetness properties.

The approval of acesulfame-K in the United States resulted in the introduction of chewing gums sweetened with this sweetener alone and with a combination of acesulfame-K and aspartame. Such products are also available in Europe.

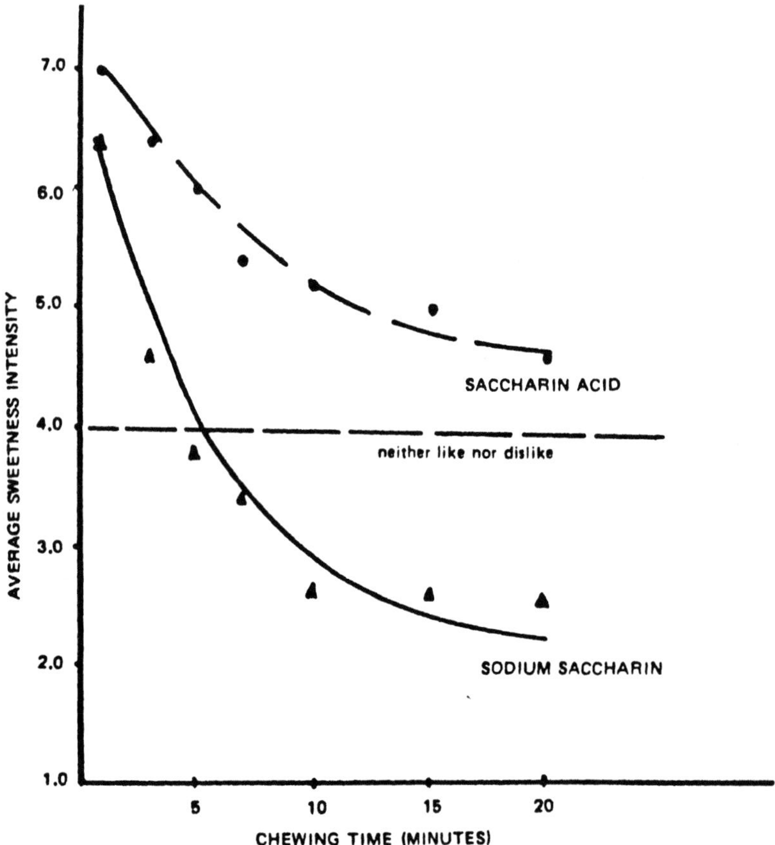

Figure 8 Overall acceptability rating of chewing gums with sodium saccharin and saccharin acid.

Preliminary experiments conducted in our laboratory indicate that the use of a combination of sweeteners in chewing gums presents many advantages to the manufacturer and the consumer. Major U.S. chewing gum manufacturers have recently introduced acesulfame-K/aspartame-sweetened products.

Several other sweeteners have been investigated. Experiments conducted with NHDC indicate that this sweetener is good in chewing gum applications. It is stable and provides an acceptable sweetness profile and flavor-enhancing properties (30). Stevioside is currently used in sugarless gums in Japan. Although the products lack the sensory qualities of saccharin- or aspartame-sweetened

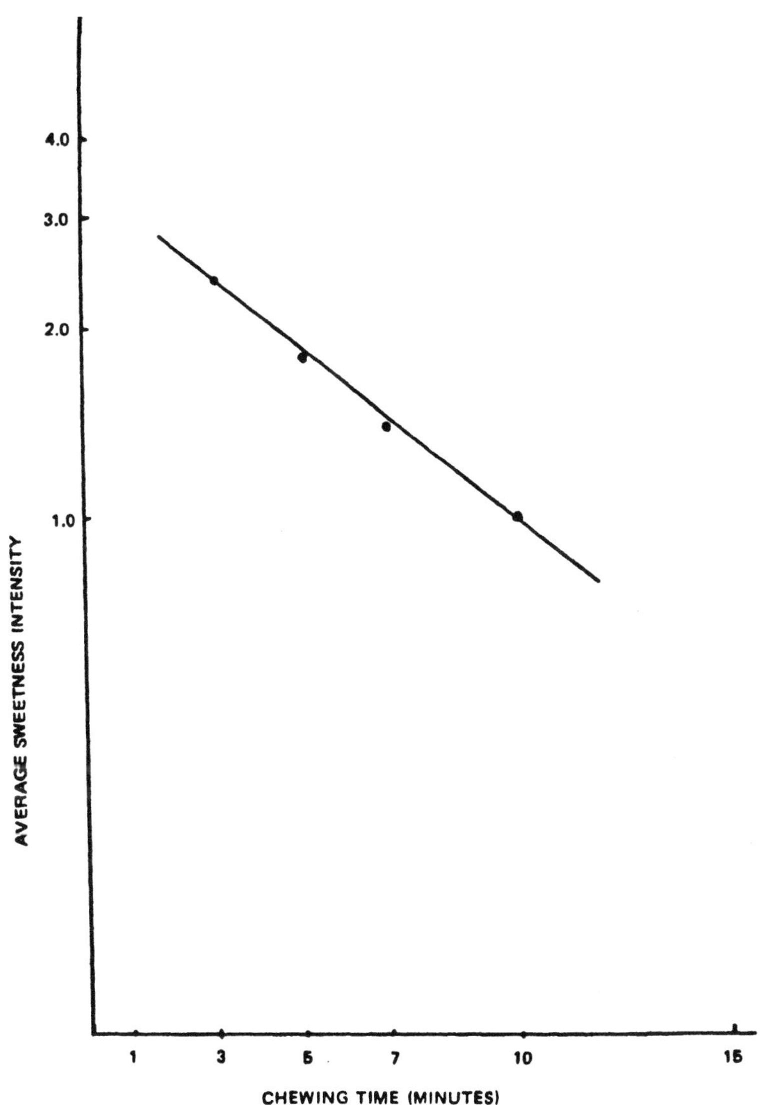

Figure 9 Sweetness intensity of chewing gums containing a mixture of saccharin and aspartame.

products, they enjoy high consumer acceptance in Japan (Maruzen Fine Chemicals, personal communication, 1982).

E. Processed Foods

This application category includes a variety of food products. Heat-processed foods, which include baked goods and canned dried foods, represent a variety of low-calorie products. Refrigerated and frozen products also use low-calorie sweeteners.

Saccharin is extensively used in baked products, especially by diabetics. It is estimated that approximately 40% of the bulk tabletop sweetener products marketed in the United States and Europe are used in baked goods (M. Eisenstadt, personal communication, 1999). Recipes that use saccharin tabletop sweeteners are available (31). Saccharin's heat stability makes it a good functional sweetener for this application.

The availability of a number of bulking agents such as polydextrose and the polyols (e.g., sorbitol, isomalt, and maltitol) have made it possible to introduce several reduced-calorie and sugar-free cookies, cakes, and cake mixes. In the United States, these products use acesulfame-K or a combination of acesulfame-K and aspartame as the intense sweetener.

The instability of aspartame at high temperatures limits its use in this application. However, recipes are available for preparing baked products to which aspartame is added after baking and cooling (32). Aspartame added to baked goods at low levels improves their aroma (33). Lim et al. (34) reported on the use of combinations of intense sweeteners in cookies.

The use of aspartame in frozen desserts and other liquid products in the United States has proliferated since approval in these applications was granted by the FDA. Beck has reported that aspartame-sweetened frozen desserts were preferred by a taste panel to ice milk (4, 5). Aspartame also performs well in frozen beverages (35), no-bake cheesecake, yogurt, and other products. No-sugar-added ice cream and frozen desserts have become available in the United States. These products generally use polyols and polydextrose and a combination of aspartame and acesulfame-K.

Cyclamate/saccharin combinations, when available, were extensively used in baked goods because of the heat stability of this combination. Cyclamate/saccharin ratios range from 2:1 to 10:1 in most uses, with 10:1 being the preferred ratio when organoleptic considerations were taken into account. Similarly, this combination was used in low-calorie frozen desserts, salad dressings, jams, and jellies. Blends of acesulfame-K and aspartame are used in Europe in this category of product applications.

II. CONCLUSIONS

The search for the ideal noncaloric sugar substitute will continue. The ideal sweetener must not only meet the taste and functional properties of sugar but can also undoubtedly create opportunities for new products that are not yet possible with currently available ingredients. On the basis of available evidence, the likelihood of identifying a single compound capable of meeting all requirements of all food applications is extremely small. Thus, the use of multiple sweeteners, specifically designed to meet desired taste and functional characteristics in a specific food application, is the most logical and promising route. Market developments in the last few years clearly suggest that a combination of sweeteners is being well received by consumers.

REFERENCES

1. AI Bakal. Functionality of combined sweeteners in several food applications. Chem Inc September 19, 1983.
2. A Salant. Non-nutritive sweeteners. In: TE Furia, ed. Handbook of Food Additives, 2nd ed. Cleveland: CRC Press, 1972, pp 523–585.
3. GE Inglett. Natural and synthetic sweeteners. Hortscience 5:139–141, 1970.
4. CI Beck. Sweetness, character, and applications of aspartic acid-based sweeteners. In: GE Inglett, ed. ACS Sweetener Symposium, Westport, CT: AVI Publishing, 1974, pp 164–181.
5. CI Beck. Application potential for aspartame in low calorie and dietetic food. In: BK Dwivedi, ed. Low Calorie and Special Dietary Foods. West Palm Beach, FL: CRC Press, 1978, pp 59–114.
6. JD Higginbotham. Talin, a New Natural Sweetener from Tate and Lyle. Reading: Tate and Lyle, 1980.
7. Personal communications with sweetener suppliers, 1999.
8. AB Porter. Effectiveness of multiple sweeteners and other ingredients in food formulation. Chem Ind September 19, 1983.
9. T Paul. Chem Zeit 45:38–39, 1921.
10. ML Mitchell, RL Pearson. Saccharin. In: Nabors, Gelardi, eds. Alternative Sweeteners, 2nd Edition, New York: Marcel Dekker, Inc., 1991, pp 132.
11. RJ Verdi, LL Hood. Food Tech 47:94–102, 1993.
12. D Scott (G. D. Searle). Use of dipeptides in beverages, German Patent 2,002,499, September 20, 1979.
13. D Scott (G. D. Searle). Saccharin-dipeptide sweetening compositions, British Patent 1,256,995, 1971.
14. TP Finucase (General Foods). Sweetening compositions containing aspartame and other adjuncts. Canadian Patent 1,043,158, November 28, 1978.

15. AF Lavia, JA Hill (E. R. Squibb). Sweeteners with masked saccharin aftertaste. French Patent 2,087,843, February 4, 1972.
16. AF Lavia, JA Hill (E. R. Squibb). Sweetening compositions, U.S. Patent 4,001,455.
17. JA Hill, AF Lavia (E. R. Squibb). Artificial sweeteners, U.S. Patent 3,695,898, 1978.
18. U.S. Food and Drug Administration. Aspartame, 21CFR172, 804, March 19, 1982.
19. G.D. Searle and Company. Aspartame in carbonated beverages, Vol. 1, U.S. Food Additive Petition No. 2A3661, August 13, 1982.
20. K Ishii, et al. (Takeda Chemical Industries). Sweetening composition, U.S. Patent 3,653,923, April 4, 1972.
21. H Fujita, T Edahiro. Safety and utilisation stevia sweetener. Shokuhin Kogyo 22(20): 66–72, 1979.
22. HR Moskowitz, et al. Sweetness and acceptance optimisation in cola flavored beverages using combinations of artificial sweeteners—a psycho-physical approach. J Food Quality 2:17–26, 1978.
23. K Hoppe. Taste interaction of citric acid with sucrose and synthetic sweeteners. Nahrung 25:11–14, 1981.
24. P Van Tornout, et al. Sweetness evaluation of mixtures of fructose with saccharin, aspartame or acesulfame-K. J Food Science 50:469–472, 1985.
25. AI Bakal, et al. (Life Savers). Long-lasting flavored chewing gum including sweetener dispersed in ester gums and method. U.S. Patent 4,087,557, May 2, 1978.
26. DAM Mackay, et al. (Life Savers). Long-lasting flavored chewing gum, U.S. Patent 4,085,227, April 18, 1978.
27. BJ Bahoshy, et al. (General Foods). Chewing gums of longer lasting sweetness and flavor. U.S. Patent 3,943,258, March 9, 1976.
28. RE Klose, BJ Bahoshy (General Foods). Chewing gums of improved sweetness retention. U.S. Patent 4,246,286, January 20, 1981.
29. T Cea, M Glass (Warner-Lambert). Aspartame sweetened chewing gum of improved sweetness stability. S. African Patent ZA 80 06,065, September 30, 1981.
30. EB Westall, et al. (Nutrilite Products and Warner-Lambert). Preservation and stability of flavors in chewing gum and confectioneries, German Patent Offen. 2,155,321, May 10, 1972; British Patent 1,310,329; U.S. Patent 3,821,417.
31. GL Becker. Cooking for the Health of It. Elmsford, NY: Benjamin 1981.
32. B Gibbons. Slim Gourmet Sweets and Treats. New York: Harper & Row, 1982.
33. I Chibata, et al. (Tanabe Seiyaku). Aspartyl amide sweetening agents, U.S. Patent 3,971,822, July 27, 1976.
34. Lim, Hyesook, et al. Sensory studies of high, potency multiple sweetener systems for shortbread cookies with and without polydextrose. J Food Science 54:624–628, 1989.
35. TP Finucane (General Foods). Methyl ester of L-aspartyl-L-phenylalanine in storage-stable beverages. Canadian Patent 1,050,812, March 20, 1979.

25
Aspartame-Acesulfame: Twinsweet

John C. Fry
Connect Consulting, Horsham, Sussex, United Kingdom

Annet C. Hoek
Holland Sweetener Company, Geleen, The Netherlands

I. INTRODUCTION

Aspartame-acesulfame is the first commercially viable member of a group of compounds called sweetener-sweetener salts. These salts owe their existence to the fact that some intense sweeteners form positively charged ions in solution, whereas others are negative. It is thus theoretically possible to combine two oppositely charged sweeteners to create a compound in which each molecule contains both ''parent'' sweeteners. To look at this another way, many currently permitted sweeteners are sold as their metal salts, for example, sodium cyclamate, calcium saccharin, acesulfame potassium. In a sweetener-sweetener salt, the positively charged metal ion—sodium, calcium, or potassium—is replaced by another sweetener, which itself carries a positive charge.

Aspartame was probably the first realistic candidate for this role of positively charged sweetener, although there have since been others, such as alitame. Despite the availability of aspartame for decades, the practical difficulties of preparing sweetener-sweetener salts seem to have defeated most researchers. The patent literature records only one attempt at laboratory synthesis (1) via a route involving dissolution of an unstable form of the negatively charged sweetener in a toxic organic solvent, a procedure that produces only small quantities of poor crystals. Realistic commercial manufacture would require a synthesis that could be carried out in an aqueous medium and that satisfied both food industry demands for purity and commercial requirements for economic yield. This was achieved in 1995, when Fry and Van Soolingen (2) invented such a process and used it to produce a range of intense sweetener salts, including the first, unique crystals of aspartame-acesulfame. Patent applications on the Fry-Van Soolingen method have been filed widely, including in Europe (2) and the United States (3). Also the subject of patents is the aspartame-acesulfame compound itself. From all the possible sweetener-sweetener salts, this latter has become the leading candidate for full commercial production, and its introduction is being carried out by Holland Sweetener Company under their trademark Twinsweet™.

The reasons for marketing this particular salt are clear. First, there are the well-known advantages of blending aspartame with acesulfame-K. For example, in the case of liquid products, these two sweeteners together offer greater sweetness stability and longer shelf-life compared with aspartame alone. In addition, aspartame and acesulfame-K exhibit quantitative synergy, which means that, when used jointly, they are a more potent sweetener than would have been expected based on their properties used independently. Furthermore, the quality of sweetness is improved by blending the two. The best features of the sweetness profiles of each come to the fore when they are combined, and a favored blend to achieve this is 60:40 by weight of aspartame and acesulfame-K, respectively. This ratio happens to be equimolar (equal numbers of molecules of each) and is exactly the ratio in which the sweeteners appear in the aspartame-acesulfame salt.

These quantitative and qualitative synergies between this pair of sweeteners have fostered an enormous growth in their joint use, as well as a general acceptance that blended sweetener systems can often offer the consumer a better taste than single sweeteners alone, however "sugarlike" the latter are claimed to be. Yet mechanical blends of aspartame and acesulfame-K are not without technological problems (4, 5). There are issues of dissolution time, hygroscopicity, and the homogeneity of powder mixes, all of which bear on the ease of use of physical mixtures of these sweeteners and the quality of consumer products made with them. In the creation of the novel crystalline form of Twinsweet, where aspartame and acesulfame are combined at the molecular level, these issues have been largely resolved.

The way aspartame and acesulfame are combined in the salt gives rise to yet more advantages. For example, the molecular arrangement in such that, in the solid, access to the free amino group of the aspartyl moiety is hindered. The availability of this group is critical to the (in)stability of aspartame when used conventionally as a separate sweetener in certain low-moisture applications, such as sugar-free confectionery, especially chewing gum. Where these products include flavors high in aldehyde content, there is a risk that aspartame is degraded through reaction with the flavor. This can shorten the shelf-life unacceptably because there is simultaneous loss of both flavor and sweetness. The hindered structure of the solid aspartame-acesulfame salt, however, is less susceptible to aldehyde attack, and the salt can be used successfully to create products of acceptable shelf-life.

It will be apparent, however, that the advantages of aspartame-acesulfame are not confined to its physicochemical properties as a solid. Because it provides two sweetening components in each molecule of the crystalline material, the salt represents a saving on the number of raw materials to be purchased, stored, and handled. In addition, the salt is a more concentrated source of sweetness than the blend because it is free of potassium and lower in moisture content. As a result, the sweetener-sweetener salt provides 11% more sweetness on a weight-for-weight basis than the corresponding equimolar blend, which is a modest but real advantage to those handling large quantities of low-calorie sweeteners.

Finally, aspartame-acesulfame poses no new toxicological issues. The sweetener-sweetener salt dissociates immediately on solution in water. In doing so it releases the same sweetener molecules that would be present were a mere mixture of aspartame and acesulfame-K to have been used. Consequently, the consumer is exposed only to known, permitted sweeteners.

For all the above reasons, aspartame-acesulfame was the sweetener-sweetener salt preferred for commercial development, and these and other attributes of Twinsweet are discussed further below.

II. PRODUCTION

In the patented process (2), aspartame-acesulfame is made by combining aspartame and acesulfame-K in an aqueous acid solution. The trans-salification reaction is depicted in Fig. 1. The sweetener-sweetener salt is subsequently crystallized, separated, washed, and dried.

All the components used are commercially available and of food grade. No unusual forms of the sweeteners and no organic solvents are used. No additional purification of the salt is necessary. This is because, as is apparent from Figure 1, the preparation of the salt is, in effect, also a recrystallization of the starting materials. As might be expected of such a process, the resultant aspartame-acesulfame has a higher degree of purity than even the food-grade raw materials. Moreover, the process introduces no new impurities, and so neatly combines synthesis with purification.

III. PHYSICAL AND OTHER DATA

Table 1 lists the main characteristics of aspartame-acesulfame.

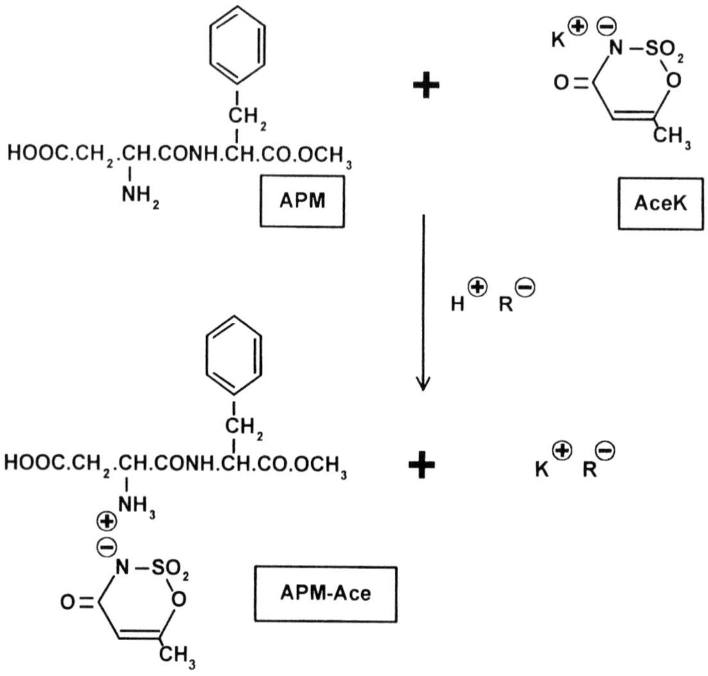

$HOOC.CH_2.CH(NH_2).CONH.CH(CH_2C_6H_5).CO.OCH_3$ **APM** + $K^{\oplus}$ acesulfame$^{\ominus}$ **AceK**

$\xrightarrow{H^{\oplus} R^{\ominus}}$

APM-Ace + $K^{\oplus} R^{\ominus}$

Figure 1 Reaction scheme for the preparation of aspartame-acesulfame (*APM-Ace*) from aspartame (*APM*) and acesulfame-K (*AceK*). (From Ref. 4.)

IV. RELATIVE SWEETNESS

As described previously, each molecule of aspartame-acesulfame contains one molecule of aspartame and one of acesulfame, and these are released immediately when the salt is dissolved. From these observations, it is to be expected that the salt would exhibit the same sweetness as an equimolar blend of the two parent sweeteners, aspartame and acesulfame-K. This is so in practice. Taste panel tests (4, 6), using solutions both in water and in a model soft drink base (a citrate buffer of pH value 3.2), show exactly the same sweetness for the salt and an appropriate equimolar blend of aspartame and acesulfame-K suitably adjusted for the weight caused by the presence of extra potassium ions and moisture in the blend.

Figure 2 shows the concentration dependence of the sweetness of Twinsweet in these two solvents. The acid model system was included because, in practice, most intense sweeteners are used in acid foods, and beverages and taste panel data obtained for such a system are more closely representative of most

Table 1 Physical and Other Data

Appearance	White, odorless, crystalline powder
Taste	Clean sweet taste, with rapid onset and no lingering sweetness or off-taste
Chemical formula	$C_{18}H_{23}O_9N_3S$
Molecular weight	457.56
Loss on drying	Not more than 0.5%
Assay (on dried basis)	Not less than 63.0% and not more than 66.0% of aspartame, not less than 34.0% and not more than 37.0% of acesulfame calculated as acid form
Melting point	Decomposes before melting
Solubility	Temperature (°C) — Solubility (% weight in water)
	10 — 1.82
	21 — 2.75
	40 — 5.53
	75 — 48.1
pH of solution	2–3 (0.3% by weight in water, room temperature)
Tapped bulk density	650–750 kg/m^3

Source: Holland Sweetener Company, Geleen, The Netherlands.

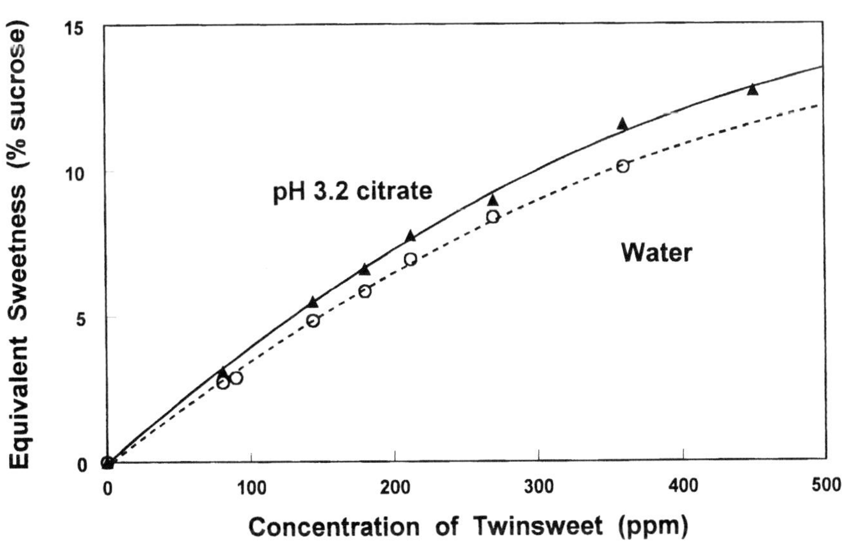

Figure 2 Equivalent sweetness of aspartame-acesulfame (Twinsweet) in water and in citrate buffer of pH 3.2.

real applications than tests conducted with water. Figure 2 indicates a relative sweetness figure for Twinsweet of about 350 times as sweet as sucrose in water and 400 times as sweet in pH 3.2 citrate at 4% sucrose equivalence. The higher relative sweetness in the citrate buffer is consistent with observations of the behavior of aspartame, which also tastes sweeter in acid solution than in water.

Across the full range of concentration, the relative sweetness of Twinsweet is 11% higher than could be obtained from the same weight of an equimolar mixture of aspartame with acesulfame-K. This is because the sweetener-sweetener salt contains only active sweeteners and no potassium. The latter accounts for 19.4% of acesulfame-K by weight but contributes no sweetness. In a similar vein, Twinsweet has a very low moisture content, whereas the aspartame component of a physical blend can contain up to 4.5% moisture (strictly "loss on drying") while remaining within international specification limits. The overall saving on nonsweet components, namely potassium and moisture, means that the aspartame-acesulfame salt is a more effective sweetener on a weight-for-weight

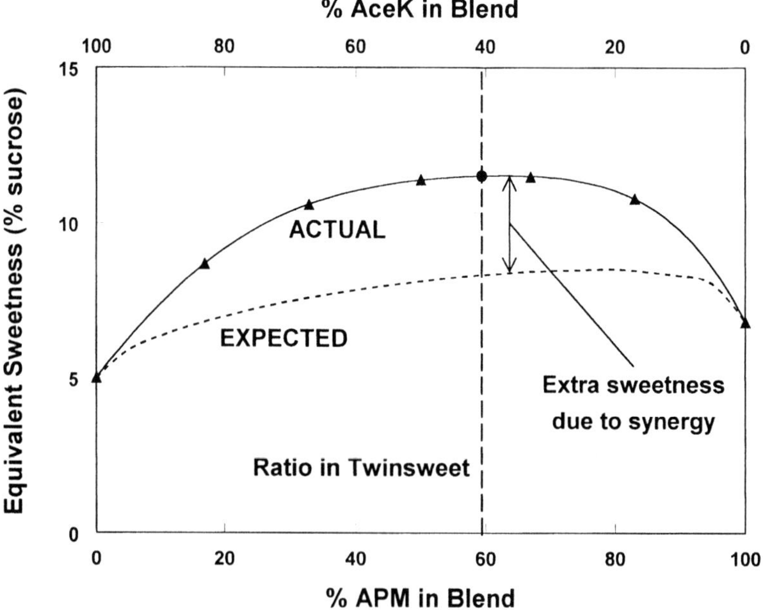

Figure 3 Synergy in a blend of aspartame (*APM*) with acesulfame-K (*AceK*) as a function of blend ratio, also showing the fixed equimolar ratio of aspartame-acesulfame (Twinsweet). Solvent = pH 3.2 citrate buffer, blend concentration = 400 ppm at all ratios, aspartame-acesulfame concentration = 360 ppm. (From Ref. 6.)

basis than a blend. This higher relative sweetness of Twinsweet is in addition to the synergy between aspartame and acesulfame, which is considered later.

Because the salt provides both aspartame and acesulfame, its relative sweetness is enhanced by the quantitative synergy between these two. That is to say, the salt is significantly sweeter than would have been predicted by a simple summation of the characteristics of the individual sweeteners tasted alone. This is illustrated in Fig. 3, which shows how the sweetness of an aspartame: acesulfame-K blend (400 ppm total sweeteners in pH 3.2 citrate) varies with the ratio of the two sweeteners. Two curves are contrasted, namely an ''expected'' curve, which has been calculated from the behavior of the sweeteners when tasted in isolation from each other (dotted line), and the actual results recorded by a taste panel (solid line). The actual sweetness is substantially greater because of the synergy between the two sweeteners.

Also shown in Figure 3 is the effective ratio of the sweeteners as provided by Twinsweet. It will be appreciated that, because the salt is an ionic compound, this ratio is fixed and dictated by the molecular weights of aspartame and acesulfame. The equimolar ratio of the two sweeteners combined in Twinsweet translates to a conventional blend ratio of 60:40 aspartame:acesulfame-K by weight. As can be seen, 60:40 is at or near the peak for quantitative synergy between aspartame and acesulfame-K, and the salt thus provides the maximum quantitative synergy available. In the case of the system illustrated (360 ppm Twinsweet), this synergy boosts the ''expected'' sweetness by 40%.

V. TECHNICAL QUALITIES

Twinsweet represents an advance over mechanical blends of aspartame with acesulfame-K in three main areas. The aspartame-acesulfame salt dissolves more rapidly than the blend, is much less hygroscopic, and exhibits a higher stability than aspartame in certain aggressive environments. These are dealt with individually in the following.

A. Improved Dissolution Rate

Consumers take for granted that powder products, such as desserts, toppings, and beverage mixes, can be reconstituted almost instantaneously, and that tabletop sweeteners dissolve immediately. Achieving such performance in sugar-free and diet products containing intense sweeteners poses difficulties for the formulation technologist. In particular, aspartame is relatively slow to dissolve, especially in cold systems such as might be involved in reconstituting a cold dessert or drink with refrigerated milk or cold water or in sweetening ice tea. The speed of dissolution can be improved by reducing the particle size because this exposes a greater

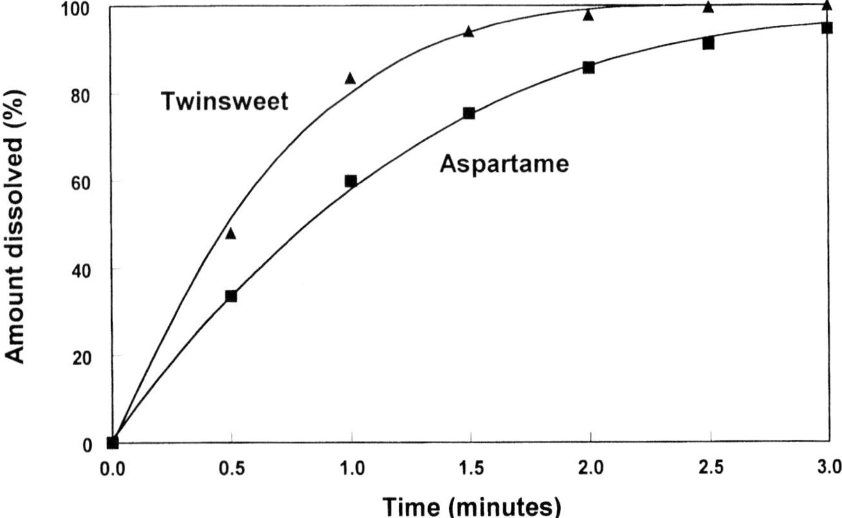

Figure 4 Dissolution profile of aspartame-acesulfame (Twinsweet) compared with aspartame. Solvent = stirred water at 10°C, particle size range of each sweetener = 100–250 μm.

surface area to the solvent. However, as explained further in Section VI, Applications, it is generally desirable not to have very fine fractions in powder products because these contribute to dust and poor flow. Figure 4 contrasts the dissolution time of aspartame with aspartame-acesulfame in cold water for powders of matched particle size. The more rapid dissolution of the salt is evident. In particular, the salt achieves the critical range of 80–95% dissolved in about half the time of aspartame.

B. Absence of Hygroscopicity

Hygroscopicity—the tendency for materials to take up moisture from their surroundings—is an important property of food ingredients. In particular, with powder mixes, hygroscopic materials can draw moisture from the atmosphere during manufacture or from other ingredients with which they are blended. The moisture taken up can change the powder flow characteristics. Among other problems, this can lead to the self-agglomeration or clumping of one or more ingredients, causing further difficulties, depending at which stage it happens. For example, clumped ingredients may be hard for the mixer to disperse, or they may segregate during subsequent handling to produce inhomogeneous mixtures. Agglomerates may even be visible and give rise to consumer complaint. Furthermore, on recon-

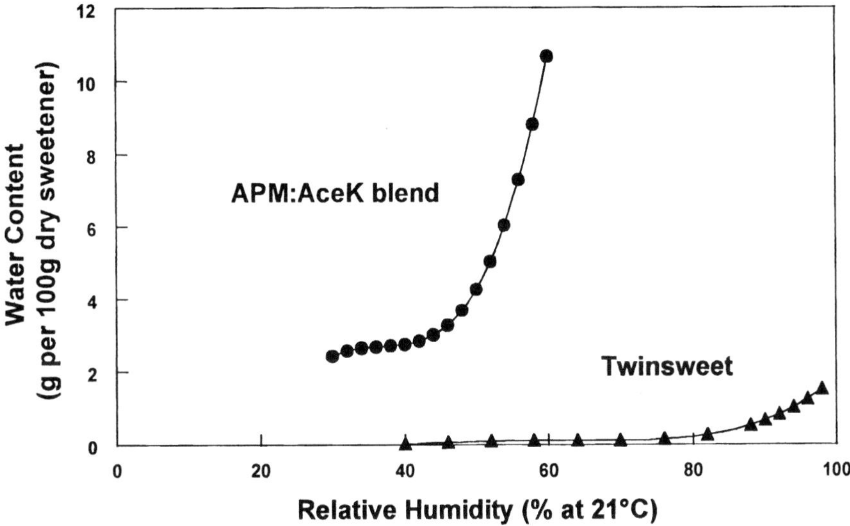

Figure 5 Hygroscopicity. Moisture uptake of aspartame-acesulfame (Twinsweet) is negligible compared with an equimolar physical mixture of aspartame with acesulfame-K (*APM:Acek blend*). (From Ref. 8.)

stitution of the powder, agglomerates may act as if they were single, large particles that dissolve relatively slowly. In addition, powders containing excess moisture are more likely to cake on storage, and they can lose all ability to flow.

Twinsweet has virtually zero hygroscopicity and is remarkably immune to moisture uptake, even when exposed to very high relative humidities. This advantageous behavior of the aspartame-acesulfame salt is illustrated in Fig. 5. The latter shows the moisture taken up by an equimolar mixture of aspartame with acesulfame-K at various relative humidities (RH), a trend that rises steeply above 45% RH at room temperature. Twinsweet takes up little moisture, even in the region of 95% RH. Thus, in circumstances in which moisture uptake is likely to create difficulty, the salt clearly outperforms the blend. This advantage is not confined to easing issues of powder mixing and flow but also simplifies packaging and storage requirements, both for the salt being handled as a bulk ingredient and for mixtures made with it.

C. Stability

Aspartame-acesulfame is stable as a dry solid. The salt shows no breakdown in either of the aspartame or acesulfame moieties on prolonged storage, and periods as long as a year at abuse temperatures of 60°C have failed to show any change in composition. Indeed, there is a suggestion that the salt might be more stable to abuse than aspartame. The reason for this may reside in the fact that, in the salt, the amino group of aspartame is blocked by the presence of acesulfame (Fig. 1) and is thus hindered from taking part in the self-cyclization to diketopiperazine (DKP) and subsequent breakdown of DKP to aspartylphenylalanine (AP). In practice, any blocking effect would be enhanced by the nonhygroscopic nature of the salt, as previously mentioned.

However, cyclization in the presence of water is not the only hazard to aspartame stability. As a dipeptide, aspartame exhibits properties also found in other peptides and in proteins, including an ability to take part in Maillard-type reactions. In some low-moisture applications, such as chewing gum, aspartame can be lost through reaction with the aldehyde groups of certain flavors. This reaction also depends on the accessibility of the sweetener's free amino group and, it is surmised, this may be the reason why Twinsweet is so effective in extending the shelf-life of such products. This is discussed further in Section VI, Applications.

Although aspartame-acesulfame exhibits advantageously high stability as a dry solid, it should be clear that it will have no particular benefit once dissolved. At the moment of solution the salt releases only aspartame and acesulfame, and these then behave in exactly the same way as if they had been released from a physical mixture of aspartame with acesulfame-K. Consequently, in solution, the stability characteristics of the aspartame contributed by Twinsweet are no different from those of aspartame from any other source. Use of the aspartame-acesulfame salt has no consequences, beneficial or deleterious, for the subsequent stability of the aspartame once the salt has been dissolved.

VI. APPLICATIONS

Aspartame-acesulfame can be used wherever both aspartame and acesulfame-K are used jointly and in most applications in which these sweeteners might be used singly. Thus, the salt is suitable for a wide range of products, including beverages, dairy products, tabletop sweeteners, confectionery, and pharmaceutical preparations. The salt's fixed composition means that it always delivers the equivalent of a 60:40 weight ratio of aspartame to acesulfame-K, but at 11% higher sweetness. As previously mentioned, it is the absence of potassium ion that principally makes Twinsweet a more concentrated source of sweetness than a blend, and this is exemplified in Table 2, which gives guidelines for the amounts of Twinsweet for various applications. Broadly, the concentration of Twinsweet required in any product is that which supplies the same number of molecules of aspartame and acesulfame as would be derived from a 60:40 mechanical blend of the two sweeteners.

Table 2 Guidelines for Twinsweet Concentrations in Various Products

Product	Twinsweet concentration ready to consume (ppm)
Beverages	
Carbonated lemon-lime	270
Hot cocoa mix	240
Cold chocolate mix	190
Instant lemon tea	200
Instant lemon drink	220
Desserts/dairy	
Instant pudding mix	380
Gelatin mix	435
Confectionery	
Chewing gum	2700
Hard candy	1000
Chocolate	800
Tabletop sweeteners	
Tablets (1 tablet = 1 tsp of sugar)	11 mg/tablet

Source: Holland Sweetener Company, Geleen, The Netherlands.

However, although suited for use in any product in which the parent sweeteners appear, the salt's particular properties mean that it is especially beneficial for dry or low-moisture materials. Specifically, there are substantial advantages to be gained by using the salt in powder mix products such as instant beverages for cold reconstitution, mixes for hot cocoa–based drinks, instant desserts, toppings, tabletop sweeteners and sugar substitutes, and pharmaceutical powder preparations. In addition, aspartame-acesulfame has a number of benefits in sugar-free confectionery, including chewing gum and hard candy, to which can be added medicated confectionery and chewable tablets. These are discussed in the following paragraphs.

A. Powder Mixes

These products are typically sold as a convenient package of premixed powder suitable for ''instant'' reconstitution by the consumer. Consumption usually follows shortly after reconstitution. Indeed consumption may be immediate, as in the case of certain pharmaceutical preparations that are stirred into water and swallowed directly. Another example is a sugar substitute added to a beverage, stirred briefly, and then drunk. Key elements in successful powder mix products are rapid dissolution, essential to meet consumer expectations of an instant product, and the homogeneity of the mix, which is essential to deliver reproducible product performance. In addition, there are factors that affect the manufacturer, such as ease of mixing of the ingredients and low dust content.

Unfortunately, aspartame and acesulfame-K crystallize in different forms. Aspartame has needlelike crystals, whereas those of acesulfame-K are more cubic. This means it is technologically difficult to create and maintain mixtures of these sweeteners that are homogeneous and stay so throughout the manufacturing and retail chains. There is evidence that this difficulty influences the quality of powder products. Not only is there a tendency for the two conventional sweeteners to separate from each other, they can also redistribute themselves unevenly with respect to the other components of the mixture. For example, Hoek et al. (5) have shown that, in segregation tests of aspartame and acesulfame-K contained in a typical instant beverage powder, the sweeteners can separate to a degree that could be perceptible to consumers as differences in sweetness. In the aspartame-acesulfame salt, however, the two sweeteners are combined at the molecular level and cannot be separated from each other until the moment they are dissolved. Moreover, the rapid dissolution of the salt (see Section V) means there is greater freedom to choose a particle size range that gives a mechanically stable, homogeneous mix. This is an advantage in comparison with aspartame, which dissolves relatively slowly, a factor that may drive the powder mix manufacturer to use very finely ground material to increase dissolution speed. Such fine powders bring other difficulties, however. These include a tendency to cohere, which means that finely-milled aspartame does not flow easily and can clump. Perversely, such clumps can act as very large particles and take an extended time to dissolve. The tendency to form clumps is made worse by the uptake of moisture, a process encouraged by the sweetener's hygroscopicity. In contrast, the complete absence of significant hygroscopicity in the salt has already been noted.

Comparative trials have been made of Twinsweet with an equisweet, equimolar physical blend of aspartame with acesulfame-K in a typical instant beverage mix (5). Under standardized mixing conditions, and in both tumbling and convective mixers, the sweetener-sweetener salt produced a more homogeneous mix, with coefficients of variation for the concentration of both sweetener components half those found with the blend. The mixture based on the salt was also much more resistant to segregation. In tests, columns of powder were deliberately vibrated to induce segregation and subsequently sectioned horizontally and analyzed. The spread of sweetness after this test is shown in Fig. 6, from which it can be seen that the sweetness of the product with Twinsweet is much more narrowly distributed than that using the blend. More particularly, the variation in concentration of the blend is likely to be perceptible to a proportion of consumers, whereas that of the aspartame-acesulfame salt is unlikely to be so detected.

Naturally, although Twinsweet may be used in a powder mix solely because of its advantageous powder flow properties, its rapid dissolution is also of obvious

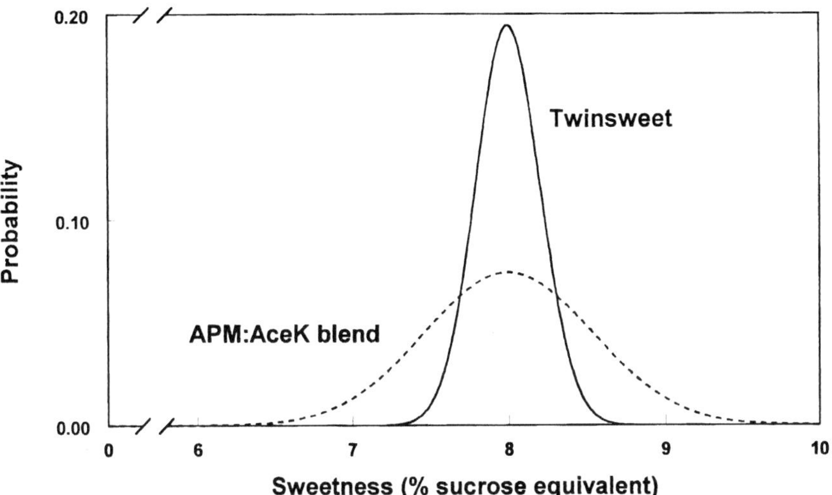

Figure 6 Segregation in a powder mix. Aspartame-acesulfame (Twinsweet) produces a narrower distribution of sweetness than an equisweet, equimolar physical mixture of aspartame with acesulfame-K (*APM:AceK blend*) after forced vibration of an instant beverage powder mix. (From Ref. 5.)

direct benefit. The salt dissolves in about half the time required for aspartame of matched particle size (see Section V), and this means better products for impatient consumers, especially where dissolution is required in a cold solvent, such as chilled milk or iced tea. In addition, when Twinsweet dissolves, it always releases aspartame and acesulfame in an exactly balanced ratio, and this ratio is constant throughout the dissolution process. This is not so when a physical blend of the intense sweeteners is used because the individual sweeteners dissolve at different rates. This means that, until both sweeteners are fully dissolved—a process that can take several minutes in the cold—there is a mismatch in the blend ratio and the taste is not as the product designers intended.

B. Chewing Gum

The aspartame-acesulfame salt has exciting advantages in chewing gum. Fry et al. (7) have demonstrated long-lasting sweetness, a noticeable boost to sweetness after some minutes of chewing, as well as improvements to stability and shelf-life. Long-lasting sweetness is a key feature of chewing gum because sweetness and flavor perception are intimately related. A gum that is no longer sweet is also perceived as having reduced flavor. In general, ordinary chewing gum gives an immediate sweetness that declines quite rapidly after the first 1 or 2 minutes.

Encapsulation of part of the sweetener can lengthen this time but, typically, even encapsulated sweeteners are exhausted after about 10 minutes. Twinsweet improves on this. In comparative trials, gum sweetened with aspartame-acesulfame was significantly sweeter at the end of 15 minutes chewing than gums made with other sweeteners and was most preferred overall (8). This was achieved by direct incorporation of Twinsweet during gum manufacture. The salt was not encapsulated, although some of the sweeteners with which it was compared were coated to extend their sweetness release. Direct addition of the salt thus represents a useful simplification of the alternative, complex process of encapsulation, and the directly added aspartame-acesulfame produces longer sweetness.

However, not only does aspartame-acesulfame extend gum sweetness overall, it also provides a remarkable boost to sweetness after some minutes chewing. Hoek and Fry (8) describe this second peak of sweetness as occurring after 5 to 8 minutes chewing. Again, the effect is achieved with Twinsweet as the sole sweetener and without the use of encapsulated material. The intensity and timing of the second peak depend on the type of gum base, sweetener concentration, product formulation, and the gum manufacturing process. These offer the product developer considerable scope to tailor a sweetness release profile.

As well as profound effects on the sweetness delivery of chewing gum, aspartame-acesulfame can be used to extend shelf-life (7, 8). Gum is a concentrated, low-moisture system and is a good medium for unwanted reactions between flavor compounds and aspartame. Especially aldehyde-rich cinnamon and cherry flavor gums host these reactions, which reduce both flavor impact and sweetness. Twinsweet is much more resistant to attack by aldehydes than is aspartame (see Section V), and this is manifest in substantial improvements to storage stability and acceptability. Figure 7 compares the sweetness during storage of two cinnamon-flavored gums, identical except that one was made with Twinsweet whereas the other contained an equimolar blend of aspartame and acesulfame-K. After 32 weeks storage, the gum sweetened with the blend was deemed barely acceptable by a taste panel after 2 minutes of chewing, and unpleasant after 15 minutes. The same gum with Twinsweet was found to be good at 2 minutes and still acceptable at 15, results that were maintained after a further 20 weeks storage. The dramatic impact of the aspartame-acesulfame salt was underlined by the fact that, after 32 weeks in storage, the gum with the salt contained three times as much aspartame as that sweetened with the blend.

C. Hard Candy

Sugar-free candy, including medicated confectionery, is an area of increasing demand, and relies on the use of polyols as bulk substitutes for the sucrose and glucose syrup used in conventional high-boiled sweets. Often, the lower sweetness of the polyols needs a boost with a modest amount of a low-calorie sweet-

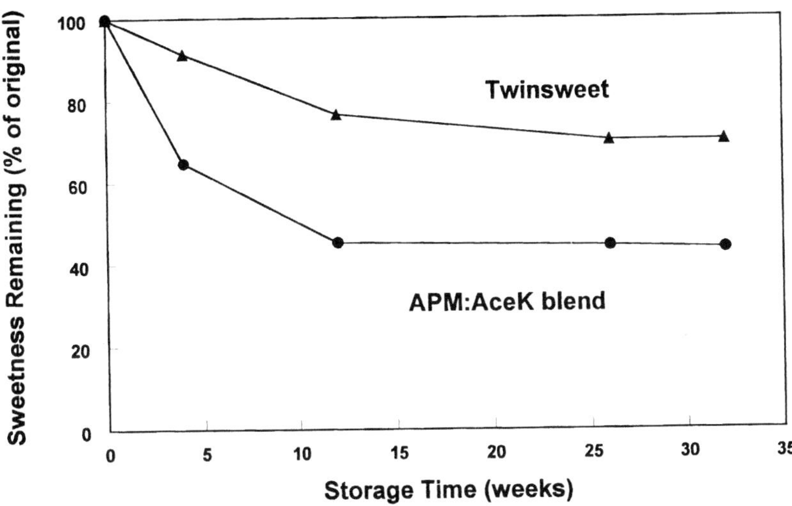

Figure 7 Shelf-life of cinnamon-flavored chewing gum. Aspartame-acesulfame (Twinsweet) retains sweetness on storage longer than an equisweet, equimolar physical mixture of aspartame with acesulfame-K (APM:AceK blend). (From Ref. 8.)

ener. Such an addition is not always easy, particularly where aspartame is involved. Although the addition of aspartame to fruit-flavored candy is rather straightforward, nonacid flavors such as mint can create a problem because sufficient aspartame cannot always be dispersed homogeneously throughout the hot candy mass (9). This difficulty vanishes when Twinsweet is used because it disperses directly in the hot mass to give products with a fuller, more sugarlike sweetness and better flavor impact than achieved with an equisweet blend of sweeteners.

D. Chewable Tablets and Tabletop Sweetener Tablets

A number of the advantages of aspartame-acesulfame already cited combine to make the salt well suited to tabletting processes. The qualities of Twinsweet that contribute to mechanically stable, homogeneous powder mixes are directly relevant to the mixing of powdered ingredients before tabletting. The absence of hygroscopicity means that the salt flows reliably and will not change its flow characteristics on exposure to moist air. At the same time, the relative chemical stability of crystalline Twinsweet leads to a long shelf-life for tablets by minimizing any degradative reactions with flavors or excipients. Finally, rapid dissolution of the salt assists in giving an immediate release of sweetness when the tablet comes to be used.

In common with hard candy, the use of aspartame-acesulfame in a variety of low-moisture products, tablets, and the like is the subject of widespread patent applications (9).

VII. TOXICOLOGY

The toxicology of aspartame-acesulfame is relatively straightforward. The synthesis of the salt is, in effect, a recrystallization and a purification of the raw material sweeteners. Those raw materials are already food grade, and the synthesis of Twinsweet simply results in further reduction of any trace impurities. No new impurities are introduced and, as a result, there are no toxicological issues associated with the manufacture. Furthermore, the salt dissociates immediately on solution to release only known, widely permitted sweetener molecules, namely aspartame and acesulfame. Once in solution then, aspartame-acesulfame salt behaves the same as a mixture of aspartame and acesulfame-K from which the functionless potassium has been removed. Accordingly, human dietary exposure is not to Twinsweet itself, which does not exist in solution, but to the permitted sweeteners from which it is derived, and the toxicological fate of aspartame-acesulfame is the same as that of those existing, permitted sweeteners.

There remains the question of the amount of aspartame-acesulfame used and whether the quantities likely to be used will affect dietary intakes of either aspartame or acesulfame. This is also easily resolved. The salt is used to provide the same amounts of aspartame and acesulfame as would have been present had these "parent" sweeteners been used separately. At the same time, no new uses have been proposed for the salt. It owes its position to the unique way it overcomes technological problems and offers consumer benefits in existing applications. Accordingly, use of aspartame-acesulfame will not affect the human exposure data and predictions on which regulatory approval for aspartame and acesulfame has been based. In short, from a toxicological point of view, there is no difference between Twinsweet and a physical blend of aspartame and acesulfame-K, and the use of aspartame-acesulfame introduces no new toxicological issues.

VIII. REGULATORY STATUS

Aspartame-acesulfame is only applicable to products in which both aspartame and acesulfame-K are permitted to be used jointly. There are countries where quantitative limits are applied to one or both of these sweeteners. In such lands, the use of the salt must conform to the limits in terms of the amounts of aspartame and/or acesulfame released by the salt when it dissolves. Because the salt is a fixed, equimolar ratio, this is a simple calculation.

At the time of writing, both the U.S. Food and Drug Administration (FDA) and the Canadian Health Protection Branch regard the marketing of Twinsweet as being covered by the existing regulations on aspartame and acesulfame-K. Products containing the salt are required to declare this in their ingredient list as "aspartame-acesulfame."

In June 2000 the Joint WHO/FAO Expert Committee on Food Additives (JECFA) concluded that the aspartame and acesulfame moieties in Twinsweet are covered by the acceptable daily intake (ADI) values established previously for aspartame and acesulfame-K. In the European Union the Scientific Committee for Food (SCF) also concluded that the use of Twinsweet raises no additional safety considerations. As a result the product is expected to be included in the forthcoming amendment of Directive 94/35/EC on sweeteners for use in foodstuffs. Regulatory clearance is being sought in numerous other countries. This is principally a matter of administrative process, necessitated because the wording of regulations in some countries does not encompass the concept of sweetener-sweetener salts.

IX. CONCLUSION

Aspartame-acesulfame is a sweetener-sweetener salt in which the potassium ion of acesulfame-K has, in effect, been replaced by aspartame. The result is an intense sweetener combination that can be produced commercially as a pure and stable solid that possesses highly advantageous properties. Aspartame-acesulfame has a high relative sweetness, about 350 times as sweet as sucrose in water and 400 times as sweet in pH 3.2 citrate, because it comprises only synergistic, intensely sweet molecules and contains no significant amounts of functionless potassium ions or moisture. It dissolves more rapidly than an equimolar mechanical mix of aspartame and acesulfame-K, yet releases only the same sweetening molecules as this familiar and widely accepted blend. Also in contrast to a mixture of aspartame with acesulfame-K, the salt is remarkably immune to moisture uptake and this, coupled with its rapid dissolution and excellent powder flow characteristics, makes it an ideal sweetener for use in powder mixes of all types. In addition, aspartame-acesulfame is stable in low-moisture products that can offer a challenging environment to aspartame itself through potential reaction with aldehyde-rich flavors. Not only is the salt resistant to these reactions in products such as chewing gum, it is also responsible for a marked extension of the sweetness release of gum, both effects being achieved without the need to encapsulate the sweetener. Other low-moisture products such as confectionery, tabletop sweeteners, sugar substitutes, pharmaceutical powders, and tablets are also likely to benefit from the special attributes of aspartame-acesulfame. Naturally, there is no reason the salt cannot be used in any application permitted by regulation, including liquid beverages and dairy products.

The salt is made from existing, permitted intense sweeteners. Its manufacture even purifies further the food-grade raw materials, and the process introduces no new impurities. The salt is stable on dry storage, including at elevated temperatures, and dissociates immediately when dissolved to provide an equimolar solution of aspartame and acesulfame. Thus, toxicologically, there is no difference between aspartame-acesulfame salt and an equimolar, mechanical mixture of aspartame and acesulfame-K, and use of the salt introduces no new toxicological issues. The salt is regarded in the United States as being covered by current FDA regulations on aspartame and acesulfame-K, and regulatory clearance is being sought in numerous other countries.

Aspartame-acesulfame, its production, and many applications are the subjects of international patents and patent applications by Holland Sweetener Company, which markets the sweetener-sweetener salt under their trademark Twinsweet™.

REFERENCES

1. DA Palomo Coll. Procedimiento para la preparación de nuevas sales fisiológicamente activas o aceptables de sabor dulce. Spanish Patent ES-A-8 604 766, 1986.
2. JC Fry, J Van Soolingen. Sweetener salts. European Patent Application EP 0 768 041, 1997.
3. JC Fry, J Van Soolingen. Sweetener salts. United States Patent US-A-5827562, 1998.
4. JC Fry. Two in one—an innovation in sweeteners. Proceedings of 1996 International Sweeteners Association Conference: Sweeteners. Openings in an Expanding European Market, Brussels, 1996, pp 58–68.
5. AC Hoek, LFW Vleugels, C Groeneveld. Improved powder mix quality with Twinsweet in Low-Calorie Sweeteners: Present and Future, Proceedings of ISA/IUFoST World Conference on Low-Calorie Sweeteners, Barcelona, 1999, Ed. A Conti. Karger, Basel, 1999, pp 133–139.
6. JC Fry. Two in one—an innovation in sweeteners. Proceedings of the 45th Annual Conference of the International Society of Beverage Technologists, Savannah GA, 1998, pp 83–92.
7. JC Fry, AC Hoek, AE Kemper. Chewing gums containing dipeptide sweetener with lengthened and improved flavour. International Patent Application WO-98/02048-A, 1998.
8. A Hoek, J Fry. Aspartame-acesulfame salt: an innovation in sweetness. Proceedings of Food Ingredients Europe, London, 1997, pp 55–57.
9. JC Fry, AC Hoek, LFW Vleugels. Dry foodstuffs containing dipeptide sweetener. International Patent Application WO-98/02050-A, 1998

26
Polydextrose

Helen Mitchell
Danisco Sweeteners, Redhill, Surrey, United Kingdom

Michael H. Auerbach
Danisco Cultor America, Ardsley, New York

Frances K. Moppett
Pfizer, Inc., Groton, Connecticut

I. INTRODUCTION

Polydextrose is an ingredient designed to be the ultimate companion ingredient to high-intensity sweeteners. Polydextrose gives the bulk, texture, mouthfeel, and functional attributes of caloric sweeteners. These are attributes that are often lost in the formulation of calorie-modified products. The key to the performance of polydextrose is its caloric value of 1 calorie per gram. This, in combination with its excellent water solubility, makes it unique in its application in reduced-calorie and low-calorie foods. When used to replace sugars and fats, polydextrose contributes only 25% of the calories of sugars and 11% of the calories of fats.

Unlike the serendipitous discovery of many unique products, polydextrose is the result of a targeted research program. The goal of the program was to fill the need for a reduced-calorie bulking agent for the reduced-calorie foods market. For such food products to be widely accepted, they must be functionally comparable to their fully caloric counterparts. Achieving desirable body, mouthfeel, and texture is critical to achieving success.

Polydextrose received Food and Drug Administration (FDA) approval for use in select food categories in 1981 (1). It was not until the mid-1980s, however, that significant commercial success was realized for this product. Proven successful formulations and a mushrooming of approved high-intensity sweetener applications have resulted in a secure place for polydextrose in reduced-calorie food products. The ingredient is approved in more than 50 countries and is used widely throughout the world.

II. COMMERCIAL PRODUCTION

Polydextrose is a randomly bonded melt condensation polymer of glucose. This unique product is a patented material invented in Pfizer Central Research Laboratories by Dr. Hans Rennhard (2). The patent describes the process for manufacture and applications of a novel carbohydrate substitute.

Polydextrose is prepared commercially by vacuum bulk polycondensation of a molten mixture of food-grade starting materials. The starting materials are glucose, sorbitol, and either citric acid or phosphoric acid in approximately an 89:10; 0.1–1 mixture. The final product of this reaction is a weakly acidic water-soluble polymer that contains minor amounts of bound sorbitol and citric or phosphoric acid.

The theoretical chemical structure of polydextrose is illustrated in Fig. 1. This structure is drawn to represent the types of bonding that can occur during polymerization. The R group may be hydrogen, glucose, or a continuation of the polydextrose polymer. As evidenced by this representative structure, polydextrose is a very complex molecule, being highly branched with varied glucose

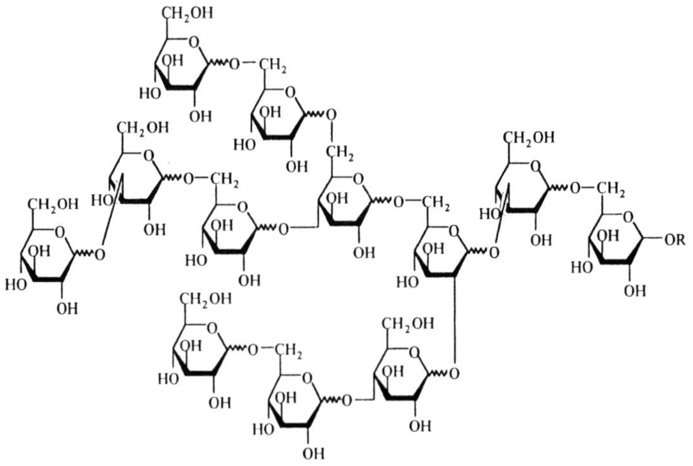

Figure 1 Representative structure for polydextrose. R = H, sorbitol, sorbitol bridge, or more polydextrose.

Table 1 Approximate Molecular Weight Distribution of Polydextrose

Molecular weight range	Percent
162–5,000	88.7
5,000–10,000	10.0
10,000–16,000	1.2
16,000–18,000	0.1

linkages. In fact, all possible glycosidic linkages with the anomeric carbon of glucose are present, including alpha- and beta-1–2, 1–3, 1–4, and 1–6, with some branching; the 1–6 linkage predominates. It is these chemical parameters that result in the water-soluble, reduced-calorie nature of polydextrose.

The typical molecular weight distribution of the polymer is shown in Table 1. The average molecular weight is 2000–2500 with an average degree of polymerization (DP) of 12–15. During the manufacturing process of polydextrose, the size of the polymer is controlled to restrict the formation of large molecular weight molecules. This control prevents the formation of insoluble material. Further discussion on this topic is presented by Allingham (3) and Beereboom (4).

As manufactured, polydextrose conforms to the typical composition described in the product description presented in Table 2. Polydextrose is supplied to the food industry in compliance with Food Chemicals Codex and JECFA compendial specifications.

Table 2 Characteristics of Polydextrose

Compendial designation	Food Chemicals Codex
Appearance	White to light tan powder
Odor	None
Polymer	>90%[a]
Glucose + sorbitol	<6%[a]
Levoglucosan	<4%[a]
Water	<4%[a]
Citric acid (free)	<0.1%
pH (10% w/v solution)	2.5–6.5
Solubility in water (25°C)	80%
Optical rotation	+60
Viscosity (cps, 50% solution)	35

[a] Anhydrous, ash-free basis

Table 3 Polydextrose and Litesse Product Forms

	Polydextrose	Litesse	Litesse II	Litesse Ultra (III)[a]
Taste	Tart Acid	Bland Neutral	Clean Mildly sweet	Very clean Mildly sweet
Color	Cream	Cream	Cream	White
Acidity (mEq/g)	0.1	0.03	0.003	0.002
PH range (10% w/v aqueous)	2.5–3.5	3.0–4.5	3.5–5.0	4.5–6.5
Maillard reaction	Yes	Yes	Yes	No

[a] Litesse® Ultra (III) is a reduced version of Litesse® prepared by catalytic hydrogenation of polydextrose. It does not contain any reducing groups and will not take part in Maillard reactions. Litesse® Ultra (III) produces a very stable, clear, water-white solution and is sugar free.

III. PRODUCT FORMS

Litesse® is the brand name for improved forms of polydextrose. Litesse is produced from polydextrose using additional processing to reduce acidity and bitterness, thereby improving the flavor profile. Litesse II and Litesse Ultra (III) are further refinements. The Litesse family of products is unique in its ability to vary from a bland, neutral powder through a colorless, mildly sweet liquid. Table 3 summarizes the different attributes of the Litesse grades.

IV. PROPERTIES

A. Caloric Content

The key property that gives polydextrose its important role in the formulation of reduced calorie foods is the caloric value of 1 calorie per gram (5). This value is significant when polydextrose is used to replace sucrose, which has a caloric value of 4 calories per gram. The degree of caloric reduction is even greater when polydextrose is used to replace fat, which has 9 calories per gram.

The reason polydextrose has a low caloric value is that it is a large, complex, randomly bonded polymer that is not broken down by mammalian digestive enzymes. The one calorie is a result of the metabolism of microorganisms in the intestinal tract. This subject is discussed more fully in Section VII, Metabolism.

B. Water Solubility

The excellent water solubility of polydextrose differentiates it from the insoluble bulking agents such as cellulosic products. It is soluble to approximately 80%

at 25°C. Polydextrose acts in a similar manner to sucrose in that it results in a clear solution without haze or turbidity. Good mechanical mixing is required to prepare concentrated solutions. The rate of solubility will depend on the efficiency of the mixing equipment as determined by shear speed and rate of addition. Use of moderate heat and slow rate of addition aid in making these solutions. A preblend with another water-soluble ingredient is another technique that will greatly speed the rate of solution of polydextrose.

C. Viscosity

The viscosity that polydextrose contributes to a solution is greater than that contributed by an equal amount of sucrose. Figure 2 illustrates the viscosity of 70% solutions of polydextrose, sucrose, and sorbitol at varying temperatures. In a manner similar to sucrose solutions, the viscosity of polydextrose solutions decreases with an increase in temperature.

The viscosity-enhancing effect of polydextrose plays an important role in food uses such as reduced-calorie dressings and puddings. Figure 3 shows the

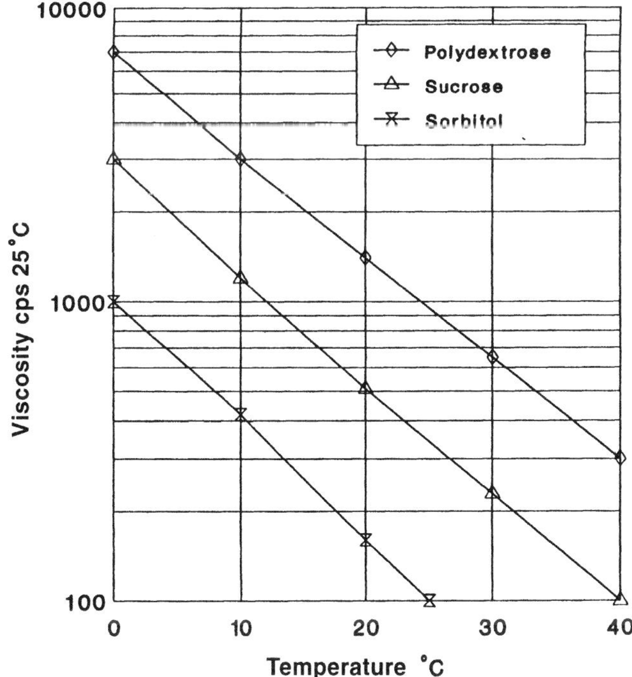

Figure 2 Temperature/viscosity relationships for 70% solutions.

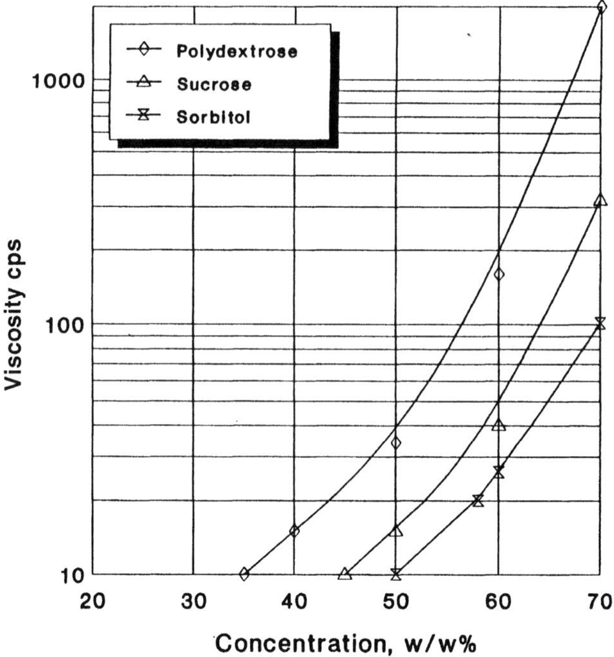

Figure 3 Concentration/viscosity relationship.

viscosity of 20–70% solutions of polydextrose, sucrose, and sorbitol. This illustrates the potential for using polydextrose to replace the viscosity typically provided by high sugar content.

D. Humectancy

Under conditions of high relative humidity, polydextrose is fairly hygroscopic. Figure 4 illustrates the water pickup of polydextrose in storage at 75 and 52% relative humidity. In food products, polydextrose functions as a humectant and can play an important role in product quality by controlling the rate of moisture gain or loss. An example of this property is found in the baked goods area. Polydextrose can retard the loss of moisture, which helps to protect against staling. In this way, polydextrose serves as an important ingredient in extending shelf-life.

Another important characteristic of polydextrose in solution is the effect that it has on water activity (a_w). Water activity is a measure of the availability

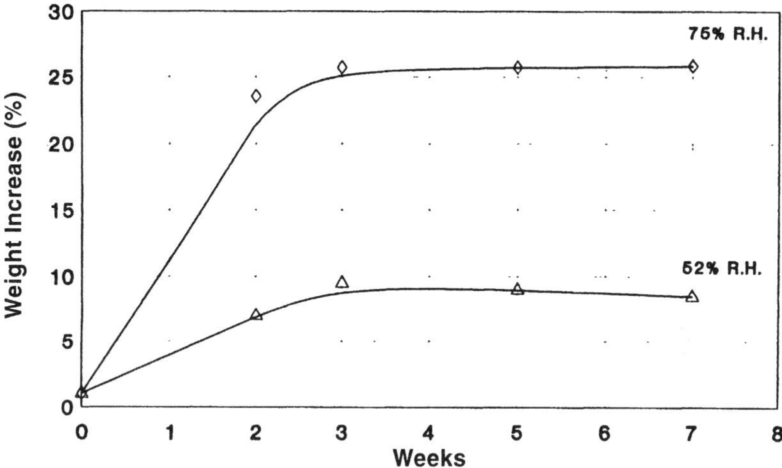

Figure 4 Hygroscopicity of polydextrose at 25°C.

of the water to participate in chemical reactions, physical reactions, and support of microbial growth in a food product.

The water activity of various polydextrose solutions is reviewed in Table 4. At equal concentrations, polydextrose has less effect on water activity than smaller molecular weight products such as sucrose and sorbitol.

E. Taste

Polydextrose is not sweet. It is used to replace the physical functionality of high-calorie ingredients such as sugar, but not the sweetness. In some food products, a less sweet taste may be desirable, and thus at least partial sucrose replacement can be achieved without compensating for the loss in sweetness. In most cases, it is necessary for calorie-modified foods to have the same quality and quantity

Table 4 Water Activity of Various Polydextrose Solutions

	Solution concentration (% w/w)	
Sweetener	50%	60%
Sorbitol	0.90	0.85
Sucrose	>0.95	0.91
Polydextrose	>0.95	0.92

of sweetness as the fully caloric counterpart. In such foods sweetness can be provided by a high-intensity sweetener. It is in these applications that polydextrose is an ideal companion product for high-intensity sweeteners.

F. Noncariogenicity

Studies carried out by Professor Muhleman at the University of Zurich demonstrated that polydextrose has a very low potential for promoting dental caries (6). These tests showed polydextrose (a) to be inert to in vitro systems, (b) not to cause caries in rats, and (c) to pass the human intraproximal plaque pH telemetry test. These three tests are good indicators of the cariogenicity of foods. As a direct result of these studies, the Swiss government allows a ''Safe for Teeth'' labeling associated with polydextrose.

G. Stability

Polydextrose is a very stable ingredient. The hygroscopic nature of the product requires good packaging and reasonable storage under conditions of low humidity to avoid moisture pickup.

Polydextrose solution is also very stable. The high solids level does not allow for the support of microbial growth. The solution can darken after prolonged storage at elevated temperatures, and, therefore, storage under cool temperatures is recommended.

V. FUNCTIONS OF POLYDEXTROSE

The unique combination of properties found in polydextrose results in a variety of functional attributes that are important in the formulation of reduced-calorie, low-calorie, and sugarless products. The principal change that is made in these food products is the removal of sucrose. High-potency sweeteners replace the sweetness lost when sucrose is removed. However, in most food products, a great deal more than sweetness is lost along with the sugar. Sucrose also provides the functional attributes of bulk, mouthfeel, humectancy, viscosity, freezing point depressant, and preservation (7, 8). As illustrated in Table 5, polydextrose can at least partially fill all these roles except for sweetness. In a similar manner, when calorie reduction is achieved by reducing the fat level, there are functional attributes of the fat that polydextrose can replace.

A. Bulk

Polydextrose is most commonly referred to as a ''bulking agent.'' Probably the greatest single challenge when formulating reduced-calorie products is replacing

Table 5 Functional Attributes of Sugar and Polydextrose

	Sugar	Polydextrose
Bulk	Yes	Yes
Mouthfeel/Texture	Yes	Yes
Humectancy	Yes	Yes
Viscosity	Yes	Yes
Freezing point depressant	Yes	Yes
Preservation	Yes	Yes
Sweetness	Yes	No

the bulk of sugar. In liquid products such as beverages, this is not such an important issue. In others, however, it is critical. Probably the most dramatic examples of the importance of a bulking agent are in baked goods and confections, which rely heavily on the bulk of sugar to give the products their character. Any significant reduction in the bulk and body of these products would greatly decrease their acceptability.

Other ingredients that could be used to replace the bulk of carbohydrates and fats include such diverse ingredients as crude dietary fiber, maltodextrin, and sorbitol. However, these products are limited in their usefulness because none of these alternative bulking agents have the combined properties of 1 calorie per gram and water solubility.

B. Mouthfeel/Texture

A pleasant, satisfying mouthfeel and texture is another important contribution of sucrose and fat. Limited success has been achieved using gums to enhance the texture of reduced-calorie foods. At higher levels, a slimy mouthfeel and gelled characteristics become noticeable. The use of polydextrose gives a comparable mouthfeel and general textural eating quality to the fully caloric food product. The product categories that particularly illustrate this function are puddings, frozen desserts, and salad dressings. A rich, creamy mouthfeel is particularly important and expected in these products.

C. Freezing Point Depression

One of the less obvious functions of sugar is that of being a freezing point depressant. This function is very important in making creamy, palatable frozen desserts. If the freezing point of a product is too low, the texture as consumed will be too soft. If the freezing point is too high, an unacceptably hard product results. In conventional ice cream type products, sucrose is the primary freezing point depressant, and alternative ingredients such as polydextrose are typically compared with sucrose.

The effect that an ingredient has on freezing point depression is a function of its molecular weight. Sucrose is a disaccharide. A larger molecule such as polydextrose is somewhat less effective as a freezing point depressant. A smaller molecular weight ingredient such as sorbitol would have a greater effect on freezing point depression than either sucrose or polydextrose. Ideally, a balance can be achieved with several ingredients to match the effect of sucrose.

The comparative effects of polydextrose, sucrose, and sorbitol on freezing point have been reported (9, 10). The freezing point of a 5% solution of these three ingredients was as follows: polydextrose, −.147°C; sucrose, −.298°C; and sorbitol, −.613°C.

D. Preservation/Osmotic Activity

Another function that sucrose serves, particularly at higher use levels, is that of preservation. Sugar at high concentrations reduces water activity (a_w). Under conditions of reduced water activity, there is less water available for the growth of microorganisms and there is also greater osmotic pressure. When the sucrose level is reduced in a recipe, the new product may have shelf-life problems with respect to bacteria, yeast, or mold growth that the original product did not have. There are several ways this problem can be addressed. One approach is to replace sucrose with ingredients that have a similar effect on water activity. Polydextrose can be used to help maintain the soluble solids level while limiting the calories. In this regard, polydextrose can be used to decrease the water activity of a product as described previously. An equal weight of polydextrose will not have as great an effect on a_w as sucrose.

It may be beneficial to make formulation changes in reduced-calorie products to maintain a comparable shelf-life. One technique is to decrease the pH to inhibit microorganisms. Also, the use of chemical preservatives such as sodium benzoate or potassium sorbate may be particularly useful in calorie-modified products.

E. Cryoprotectant

Several applications have been suggested for the cryoprotectant capability of polydextrose. Cryoprotection involves stabilizing foods against the damaging physical effects of freezing. This is a function that is well recognized for many soluble ingredients, particularly sugars. This function has been described in fish products, including surimi (11, 12), and meat products (13).

F. Fiber

Polydextrose conforms to the Japanese definition of a dietary fiber: ''polysaccharides, related polymers and lignins, which are resistant to hydrolysis by the digestive enzymes of man'' (14). The Japanese consumer has a particularly keen interest in supplementing his or her fiber intake as part of a desirable diet. In line with these factors, polydextrose has found a unique market in Japan for use in beverage products fortified with polydextrose as a soluble fiber (15). The fiber properties of polydextrose are reviewed in detail in the literature (16). Polydextrose may also be labeled fiber in Argentina, Egypt, Korea, Poland, and Taiwan.

VI. APPLICATIONS

Polydextrose has been accepted as a food additive by the U.S. Food and Drug Administration (1). The original food additive petition included specific application in eight food categories. These categories are baked goods and mixes; chewing gum; confections and frostings; dressings; frozen dairy desserts; gelatins, puddings, and fillings; hard candy; and soft candy. These food categories are among those described in the Code of Federal Regulations (21 CFR 172.84).

New categories have recently been approved. These include peanut spreads, syrups and toppings, sweet sauces, and fruit spreads.

With the development of new forms of polydextrose and Litesse®, it is now possible to choose a bulking agent that best fits the end-use application (see Table 3 for a summary of polydextrose and Litesse attributes). Selection of the appropriate Litesse form offers greater flexibility in terms of color, taste, and product claims (Litesse Ultra (III) is sugar free).

Polydextrose may be used in the approved food categories to fulfill any of four functions:

1. Bulking agent—to provide bulk or substance to a food
2. Formulation aid—to promote or produce a desired physical state
3. Humectant—to promote retention of moisture
4. Texturizer—to affect the appearance or feel of the food

A. Baked Goods

This food category more specifically includes baked goods and baking mixes (restricted to fruit-, custard-, and pudding-filled pies, cakes, cookies, and similar baked products).

A one-third calorie reduction in these products would require a fat and carbohydrate reduction, which would alter the physical and organoleptic acceptability of the baked good. Therefore, polydextrose plays a particularly important role in the formulation of reduced-calorie baked goods (14, 18). In many baked goods formulations, polydextrose can be used without a high-intensity sweetener.

Recent work reported on the good textural qualities of polydextrose shortbread cookies formulated with high-intensity sweeteners (19). For sweetness, the study evaluated synergistic combinations of aspartame, cyclamate, saccharin, and acesulfame-K in the cookies.

The benefits of polydextrose in cake-type products include volume, tenderness, structure, and eating quality. Typical use levels in such products would be from 7 to 15% by weight.

B. Chewing Gum

Polydextrose can be used to make chewing gum with good shelf-life and elasticity. A high-intensity sweetener would be required to provide sweetness to the gum. Several patents exist for low-calorie gum products that cite the use of polydextrose (20, 21).

C. Confections and Frostings

This area of application represents products in which the functional properties of carbohydrates and fats are critical to the character of the food. Polydextrose is not crystalline, and therefore the crystalline/sugar texture cannot be achieved. The bulk and mouthfeel of taffylike products is more closely matched using polydextrose. Polydextrose use level would typically be in the range of 25%.

D. Salad Dressings

Many pourable salad dressings are surprisingly high in calories, containing up to 60–80 calories per tablespoon. Of their calories, approximately 80–95% come from fat (22). Therefore, making reduced-calorie dressing involves, primarily, reducing the oil level.

Polydextrose finds use in this application by replacing the functionality and mouthfeel that was contributed by the fat. Some dressings do contain a significant level of sugar. Russian, thousand island, and French-style dressings are highest in sugar. These dressings are good candidates for the replacement of sucrose with polydextrose and a high-intensity sweetener. A number of highly palatable reduced-calorie dressings have been formulated with polydextrose and enjoy considerable success in the marketplace.

E. Frozen Dairy Desserts

This area of use typically involves the reduction of both fat and sugar in the formulation of reduced-calorie products (23). Sugar is a multipurpose ingredient providing bulk, mouthfeel, sweetness, and freezing point depression to the final product.

The recipe in Table 6 is for a reduced-calorie, nonfat product.

The firmness of a frozen dessert can be manipulated by altering the freezing point. Freezing point depression is a function of molecular weight of key ingredients, and thus, a balance of polydextrose with a lower molecular weight product such as sorbitol may best match the firmness desired.

Very pleasant-tasting products are possible without the use of a high-intensity sweetener. However, a series of regulatory approvals in 1987 and 1988 extended the use of aspartame to include frozen dairy desserts, and this resulted in a surge of very successful reduced-calorie products (24, 25). Typical use levels of polydextrose in these products range from 7–15%.

Table 6 Nonfat Ice Cream

Ingredient		Percent
Sucrose		11.00
Nonfat milk solids		9.00
Corn syrup solids		7.40
Litesse®		4.00
Dairy-Lo®		3.00
Stabilizer/emulsifier		0.50
Flavor, vanilla, N&A		0.30
Flavor, art. mouthfeel type		0.03
Fat	<0.5%	
Total solids	35%	
Calories	127 calories per 100 g 90 calories per 4 fl. oz, serving at 90% overrun	

Vitamin A should be added at a level ranging from 600–800 IU, depending on the reference ice-cream

Processing Procedure

Dairy-Lo is compatible with conventional ice cream manufacturing equipment and processes.

- Mix liquid ingredients
- Add Dairy-Lo and other dry ingredients with moderate agitation
- Pasteurize at 180°F (83°C) for 25 seconds
- Homogenize
- Cool and age
- Flavor the mix
- Freeze, package
- Harden

Frozen yogurt is a specially product within this category that has become very popular. Frozen yogurt is lower in fat than ice cream, and thus calorie reduction is achieved primarily from sugar replacement. Here again, polydextrose serves the same varied functions of bulk, texture, and freezing point depression.

F. Gelatins, Puddings, and Fillings

A 50% calorie reduction is readily achieved in an instant pudding-type product when polydextrose is used with a high-intensity sweetener (26). Using polydextrose results in a creamier texture with a more uniform dispersion of cocoa and starch. It also functions as a formulation aid and helps prevent lumping and non-uniformity.

G. Hard Candy

A major portion of the sugar in hard candy can be replaced with polydextrose (27). Clarity and good bite are still achievable with a reduced sugar level. A high-intensity sweetener is needed to supply the sweetness expected in hard candy products.

Polydextrose is amorphous and does not crystallize at low temperatures or high concentrations so it can be used to control the crystallization of polyols and sugars and therefore the structure and texture of the final product. This is analogous to conventional sugar confectionery production in which glucose syrups are used to prevent or control sucrose crystallization.

Litesse Ultra (III) is a particularly useful ingredient in the production of sugar-free and reduced-calorie hard candy and offers both product and processing improvements when used at low levels in polyol mixes.

While acknowledging that some polyols work very well in this application, there are still some technical issues associated with their use in hard candy systems. From production scale trials using Litesse Ultra (III) in combination with polyols, the following general conclusions have been drawn.

1. Increasing the proportion of Litesse Ultra (III) increases the molten mass viscosity, improving the handling characteristics of sugar-free candy. Cooling times are reduced and the products may be processed using conventional stamped and depositing technology.
2. Low-level additions of Litesse Ultra (III) reduce the graining of isomalt candies.
3. The stability of Litesse Ultra (III) combination may be equivalent or better than sugar/glucose candy.
4. Clear, transparent products with excellent flavor release are possible.
5. Calorie reductions greater than 50% are possible.

H. Soft Candy

The use of polydextrose as a partial sugar replacement in soft candy serves two key functions: providing bulk and humectancy (28, 29).

I. Other Categories

Sweet sauces, toppings and syrups (30), peanut spreads (31, 32), and fruit spreads (33) are newly approved categories for polydextrose and demonstrate the flexibility of the ingredient across a range of applications where sugar is reduced or replaced.

VII. METABOLISM

The basis for the 1 calorie per gram caloric value of polydextrose lies in the difficulty that mammalian digestive enzymes have in attacking a large and randomly bonded polymer. Extensive feeding studies with man and animals have confirmed the caloric contribution (1, 34–39).

The definitive work on the metabolism of polydextrose studied the disposition of ^{14}C radiolabeled polydextrose in feeding studies. The caloric value of 1 calorie per gram and the testing method were fully endorsed by the U.S. Food and Drug Administration (5).

A portion of the polydextrose molecule is metabolized in the intestinal tract by microorganisms. As by-products of this metabolism, these microorganisms produce volatile fatty acids and carbon dioxide. The volatile fatty acids are absorbed in the large intestine and used as an energy source. Therefore, it is the microbial function that contributes the 1 calorie per gram of ingested polydextrose. This is a phenomenon common to essentially any complex carbohydrate in the diet.

Another area of metabolism that has been closely investigated is the effect of polydextrose on insulin demand and the implications of this on use by diabetic patients. Studies show that the use of polydextrose does not create an insulin demand using glucose tolerance techniques.

As described previously, tests show that polydextrose has a very low cariogenicity potential.

VIII. TOLERANCE

Numerous clinical studies show that polydextrose is well tolerated when consumed in moderation as part of a normal daily diet. This has been confirmed by consumer results in the years since polydextrose has been approved.

When consumed in excessive amounts, polydextrose can have a laxative effect. Studies of adults showed a mean laxative threshold of 90 g/day (40). This compares with a level of 70 g of sorbitol for similar effect.

IX. REGULATORY STATUS

Except where specifically stated otherwise, the comments offered on regulatory issues are geared to U.S. regulations and practices. The reader should also be aware that regulations are constantly being changed and updated.

A. FDA Approval

There were years of extensive research into the safety, properties, and applications of polydextrose before the approval of the food additive petition in 1981 (1, 3) by the U.S. Food and Drug Administration. The regulations allow for the safe use of polydextrose as a multipurpose food additive in specific foods. It is used in accordance with good manufacturing practices as a bulking agent, formulation aid, humectant, and texturizer. Polydextrose is specifically allowed in the food categories listed in Table 7.

There is no maximum established use limit for polydextrose. Good manufacturing practices limit the quantity to the amount necessary to accomplish the intended purpose in the food.

Table 7 Approved Food Categories for Polydextrose

Chewing gum
Confections and frostings
Dressings for salad
Frozen dairy desserts
Fruit spreads
Fruit and water ices
Gelatins, puddings, and fillings
Hard candy
Peanut spreads
Soft candy
Sweet baked goods and mixes
Sweet sauces
Table spreads
Toppings and syrups

B. International

The use of polydextrose in foods has received approval in numerous countries (41). Numerous national and supranational expert groups have assessed polydextrose. Without exception, it was concluded that polydextrose is safe for human use. Both the Joint FAO/WHO Expert Committee on Food Additives (JECFA) and the EU Scientific Committee for Food (SCF) allocated an acceptable daily intake (ADI) of ''not specified'' in 1987 and 1990, respectively.

Polydextrose has been approved as a Miscellaneous Food Additive by the European Union and may be used at *quantum satis* levels. Polydextrose is listed as E1200.

Japan's Ministry of Health and Welfare (MOHW) recognizes polydextrose as a food. Polydextrose also conforms to the generally accepted Japanese definition of dietary fiber. An energy value of 1 kcal/g is accepted.

Polydextrose is approved for food use in confectionery, chewing gum, custard powder/mix, dairy-based desserts, dairy ice mix, dessert mix, frozen/ice confection, reduced/low-fat ice-cream, and yogurt in Australia and New Zealand. An energy value of 1 kcal/g is accepted.

To date polydextrose is approved in more than 50 countries worldwide.

X. LABELING

When polydextrose is used in foods for special dietary purposes such as reduced-calorie foods, it must be labeled in accordance with 21 CFR Part 105. Should a single serving of food contain more than 15 g of polydextrose, the label must read: ''Sensitive individuals may experience a laxative effect from excessive consumption of this product.''

For nutritional labeling, polydextrose should be included in the carbohydrate section of the Nutrition Facts Panel. It may also optionally be included below ''Total Carbohydrate'' under ''Other Carbohydrate.'' The calorie content of 1 kcal/g is used when determining total calories per serving. Use of polydextrose may be taken into consideration when contemplating comparative nutrient content claims such as ''light,'' ''low calorie,'' ''reduced sugar,'' ''sugar-free,'' and/or ''no added sugar'' under 21 CFR 101.60.

''Polydextrose'' is the officially recognized name and should appear as such in the ingredients list.

REFERENCES

1. 21 CFR 172.841, 46 FR 30080 (June 5, 1981), Food Additive Petition No. 9A3441.
2. HH Rennhard (Pfizer). Dietetic foods, U.S. Patent 3,876,794, 1975.

3. RP Allingham. Polydextrose—A new food ingredient: Technical aspects. In: Chemistry of Foods and Beverages: Recent Development. New York: Academic Press, 1982, p 293.
4. JJ Beereboom. Technical aspects of polydextrose, Polydextrose Trade Press Briefing, May 28, 1981.
5. EC Coleman. U.S. Food and Drug Administration, letter dated April 21, 1981, to EF Bouchard.
6. HR Muhlemann. Polydextrose—a low-calorie sugar substitute. Dental tests. Swiss Dent 2(3):29, 1980.
7. RE Smiles. The functional applications of polydextrose. Chemistry of Food and Beverages: Recent Developments. New York: Academic Press, 1982, pp. 305–322.
8. JT Liebrand, RE Smiles, TM Freeman. Functions and applications for polydextrose in foods. Food Technol Aust 37(4), 1985.
9. RJ Baer, KA Baldwin. Freezing points of bulking agents used in manufacture of low-calorie frozen desserts. J Dairy Science 67:2860–2862, 1984.
10. RJ Baer, KA Baldwin. Bulking agents can alter freezing. Dairy Field, Feb. 1985.
11. JW Park, TC Lanier. Combined effects of phosphates and sugar or polyols on stabilization of fish myofibrils. J Food Science 52(6):1509–1513, 1987.
12. JW Park, TC Lanier, DP Green. Cryoprotective effects of sugar, polyols, and/or phosphates on Alaska pollack surimi. J Food Science 53(1):1–3, 1988.
13. JW Parks. Effects of cryoprotectants on properties of beef protein during frozen storage, dissertation, North Carolina State University, 1986.
14. The Foundation for Health and Physical Development, Ministry of Health and Welfare, Japan.
15. M Hamanaka. Polydextrose as a water-soluble food fibre. Food Industry 30(17):73–80, 1987.
16. SAS Craig, JF Holden, JP Troup, MH Auerbach, HI Frier. Polydextrose as soluble fiber: physiological and analytical aspects. Cereal Foods World 43:370, 1998.
17. TM Freeman. Polydextrose for reduced calorie foods. Cereal Foods World 27(10): 515–518, 1982.
18. K Kim, L Hansen, C Setser. Phase transitions of wheat starch-water systems containing polydextrose. J Food Science 51(4):1095–1097, 1986.
19. H Lim, CS Setser, SS Kim. Sensory studies of high potency multiple sweetener systems for shortbread cookies with and without polydextrose. J Food Science 54(3): 625–628, 1989.
20. RE Klose, RE Sjonvall. Low calorie, sugar-free chewing gum containing polydextrose. European Patent 0123742B1, 1987.
21. SR Cherukuri, F Hriscisce, YC Weis. Low calorie chewing gum and method for its preparation, European Patent 0252874A, 1988.
22. FS Goulart. How to read a salad dressing label. Nutr Dietary Consultant, pp 37–38, June 1985.
23. All the satisfaction of ice cream, but its a diet bar. Food Engineering, 56(9):50–51, 1984.
24. DH Goff, WK Jordan. Aspartame and polydextrose in a calorie-reduced frozen dairy dessert. J Food Science 49:306–307, 1984.
25. KC Reddy. Effect of aspartame and polydextrose levels on organoleptic, physio-

chemical and economic values of ice cream and ice milk products. M.S. thesis, Mississippi State University, 1985.
26. Reduced calorie desserts made with aspartame and polydextrose (puddings, gelatins). Food Processing, pp. 21–22, July 1983.
27. KJ Klacik. Continuous production of sugar-free hard candies. Manufacturing Confectioner, pp 61–67, August 1989.
28. JT Liebrand, RE Smiles. Polydextrose for reduced calorie confections. Manufacturing Confectioner, 61(11):35–36, 1981.
29. A Cridland. Developments in dietetic chocolate. Confectionery Manufacturing and Marketing 24(10), 1987.
30. F.A.P. (9A4126). Use of polydextrose in sweet sauces, toppings and syrups. Federal Register 54(49): March, 1989.
31. F.A.P. (7A3998). Use of polydextrose in peanut butter spread, Federal Register 53(16): January 26, 1988.
32. U.S. Patent #4,814,195. Polydextrose in peanut butter.
33. F.A.P. 8A4068. Use of polydextrose in fruit spreads. Federal Register 53(102): May, 1988.
34. SK Figdor, HH Rennhard. Calorie utilization and disposition of [^{14}C] polydextrose in the rat. J Agric Food Chem 29:1181–1189, 1981.
35. JS White, CM Parsons, DH Baker. An in-vitro digestibility assay for prediction of the metabolizable energy of low-calorie dextrose polymeric bulking agents. J Food Science 53(4):1204–1207, 1988.
36. SK Figdor, JR Bianchine. Caloric utilization and disposition of [^{14}C] polydextrose in man. J Agric Food Chemistry 31:389, 1983.
37. L Achour, B Flourie, F Briet, P Pellier, P Marteau, J Rambaud. Gastrointestinal effects and energy value of polydextrose in healthy non obese men. Am J Clin Nutr 59:1362, 1994.
38. GS Ranhotra, JA Gelroth, BK Glaser. Usable energy value of selected bulking agents. J Food Sci 58(5):1176–1178, 1993.
39. NC Juhr, J Franke. A method for estimating the available energy of incompletely digested carbohydrates in rats. J Nutr 122:1425, 1992.
40. Raphan Study (IV). Pfizer communication.
41. Bulking agent moves to international commercialization. Food Processing 47(8), 1986.

27
Other Low-Calorie Ingredients: Fat and Oil Replacers

Ronald C. Deis
SPI Polyols, Inc., New Castle, Delaware

The interest in fats and oils and their presence in foods has gone through a number of developmental stages, each causing a change in priorities at various times in product development cycles—these stages include the elimination of saturated fats, the "fat-free" phase, the fat and calorie reduction phase, and the search for healthy fats from natural and biosynthetic routes. All of this is gradually evolving into a more rational design of better-quality, more healthy products as we learn more about the effects of changes in diet. The 1980s are known for the need to eliminate saturated fats and to increase fiber. The number of companies marketing fiber products surged enormously, then ebbed as the fiber frenzy dwindled. As we entered the 1990s, the message that saturated fats were "bad" progressed into the need to eliminate fats entirely—to be "fat free" was golden. This spawned a race to develop new fat replacers to fill the need—many claiming to be the one ingredient to replace fat in all food systems. As noted by M. Glicksman in 1991, "every food company, ingredient supplier, and biotechnology company is looking for a colorless, odorless liquid that looks, tastes, and functions like oil but has no calories and is less expensive than water"—what Glicksman referred to as the "oily Grail" (1). Some categories, such as dairy and salad dressing, were able to produce reasonably acceptable fat-free or reduced-fat alternatives over time. In other areas, such as baked goods (in which less water was available in the system and any added water caused noticeable differences), the quality of these products was poor, and consumer acceptance has dwindled over time. During this era of product development, the work on fat replacement led to better understanding of ingredient interactions, resulting in a reasonable lowering of fat and calories in foods using—for the most part—conventional food ingredients.

I. FAT IS ESSENTIAL AND FUNCTIONAL

Reports concerning the negative effects of fat and oil consumption have caused many consumers to focus on fat elimination, but this needs to become more balanced to include the body's fat requirements (2). Fats act as important energy sources, especially during growth, or at times when food intake might be restricted. Protein and carbohydrates provide about 4 kcal/g, but fats provide more energy at about 9 kcal/g. Linoleic and linolenic acids are regarded as essential fatty acids that aid in the absorption of vital nutrients, regulation of smooth muscle contraction, regulation of blood pressure, and growth of healthy cells. On the negative side, the U.S. Surgeon General has stated that consumption of high levels of fat is associated with obesity, certain cancers, and possibly gallbladder disease. The Surgeon General also notes that strong evidence exists for a relationship between saturated fat intake, high blood cholesterol, and coronary disease. Rather than viewing all fats as ''bad,'' most nutritionists urge consumers to control the percentage of calories as fat in their diets and to limit levels of saturated fat and polyunsaturates. Current government guidelines state that total fat intake should be no more than 30% of total calories. Saturated fat should make up less than 10% of calories, and monounsaturates should make up 10–15% of calories. Concerted efforts from ingredient suppliers and product developers have reduced the use of saturated fats such as lard, beef tallow, butterfat, coconut oil, and palm oil and increased the use of vegetable oils with higher percentages of polyunsaturates and monounsaturates.

Most of the fat consumed in the United States comes from salad and cooking oils, followed by frying fats and bakery shortenings, then meat, poultry, fish, and dairy products (cheese, butter, margarine). Each of these applications is unique in its requirements for fat functionality (Table 1). In fried foods, oil acts as a heat-transfer medium but also becomes a component of the food. Because of this dual function, the oil must meet a number of requirements—it must have good thermal and oxidative stability, good flavor, good shelf-life, and acceptable cost.

Fats and oils provide important textural qualities to certain foods. Much of this is due to specific melting qualities and crystal structure, and qualities are provided by the ''shortening'' effect of fats, primarily in baked goods. Fats provide baked goods with a characteristic rise, flakiness, tenderness, strength, ''shortness,'' and cell structure that are not apparent in fat-free varieties (4,5). Fats and oils are also essential to lubrication of foods in two ways: as release agents during cooking and as lubricants during chewing, causing a cooling and coating sensation picked up as moistness in baked goods. Fats modify flavor release and affect mouth-feel by providing viscosity and coating effects and also possess their own characteristic flavors (animal fats, olive oil, and peanut oil are classic examples).

Table 1 Functional Properties of Fats in Different Food Categories

Frozen desserts	Fried foods	Meats	Baked goods
Flavor	Flavor	Flavor	Flavor
Viscosity/body	Heat transfer	Mouth-feel	Viscosity/body
Creaminess	Crispness	Juiciness	Richness
Mouthfeel	Aroma	Tenderness	Texture
Opacity	Color	Texture	Shortness
Heat shock	Heat stability	Binding	Tenderness
Stability	Migration	Heat transfer	Flakiness
Overrun			Aeration
Melt			Elasticity
			Leavening
			Lubricity
			Moisture
			Retention
			Shelf-life
			Water activity

Source: Adapted from Ref. 3.

All the functional factors are impacted by the type of fats used. In turn, the type and composition have health consequences. Some of the issues surrounding functionality are better understood by looking at the nature of fat and the composition of the predominant fats and oils available for food use. Fat molecules consist of three fatty acids linked to a glycerol backbone. Native fats and oils are made up of mixtures of a wide range of fatty acids arranged in varying ratios and positions on these triglycerides: This determines the characteristics of a particular fat and creates a wide selection of properties. If a fatty acid is "saturated," it means that the maximum number of attachment sites (four) on the carbon atom are filled by another attached carbon or a hydrogen atom, and only single bonds exist. Unsaturated fatty acids contain one or more double bonds between carbon atoms in the chain. If an unsaturated fatty acid contains one double bond, it is referred to as monounsaturated (oleic, 18:1). If more than one double bond exists, the fatty acid is polyunsaturated (linoleic, 18:2; linolenic, 18:3). If an oil contains a predominance of saturated fatty acids, such as palm oil or coconut oil, it is commonly referred to as a saturated fat. Conversely, if the fat is predominantly unsaturated, it is referred to as an unsaturated fat.

Fats and oils are very complex substances from a number of origins, made up of a number of fatty acids in infinite combinations, and chemically modified in a number of ways. This chemical complexity results in a very complex mix of results in terms of appearance, texture, flavor, mouthfeel, and processing char-

Figure 1 Ingredient support system in fat replacement.

acteristics. Although it was the goal of many initial fat replacers to encompass all categories of food, fat functionality is category-specific—performance in dairy products or salad dressing can be expected to be different than performance in baked goods or snacks. Second, fat functionality is product specific—for example, since the percent of moisture in a cheesecake is far more than that of a cookie, it can be expected that fat replacers will perform differently in these products. Finally, the need to consider adjustment of all other ingredients in the formulation

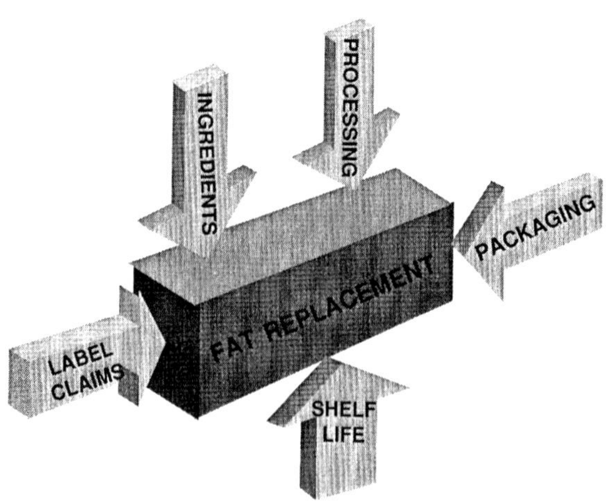

Figure 2 Issues to consider as part of fat replacement.

means that fat replacement is formulation-specific. Fat replacement in any formulation requires attention to more than fat—support ingredients are required to address mouth-feel, texture, aeration or structure, color, flavor, handling characteristics, and shelf stability (Fig. 1). Other factors to consider are cost, regulatory concerns, safety, packaging needs, label claims, and availability (Fig. 2).

II. FAT REPLACERS—AN OVERVIEW

Most fat replacement is accomplished by using water effectively, and air can be entrapped as a texture aid in many products. The first fat replacers were primarily air, water, and emulsifiers. Because early products were unsuccessful and the NLEA eliminated the use of emulsifiers as nonfats, combinations of ingredients in a systems approach found more success. Finally, when the consumer discovered that ''reduced fat'' or ''fat free'' did not necessarily mean less calories, development became more focused toward caloric reduction and fat reduction. The term ''fat replacer'' is a generic term for any bulking agent or ingredient that somehow replaces fat in a system. Fat extenders serve to extend the usefulness of a reduced amount of fat in a food. This could be an emulsifier or something coated with a fat—so that the fat is still a part of the system. A fat substitute actually has the characteristics of fat but is absorbed differently (or not absorbed at all) by the body, resulting in less caloric density. An example of this would be olestra, caprenin, or salatrim. Fat barriers reduce the amount of oil migration into a product (doughnuts, french fries). A number of film-formers such as starches and celluloses could be placed here. Finally, fat mimetics are ingredients that somehow partially imitate fat, usually by binding water. Many of the carbohydrate and protein fat replacers would fall into this category. Another way to look at fat replacers is to place them into general application classes—the simplest classification is to classify them as (a) modified fats or (b) water binders. It is more fair and easier to discuss properties when they are classified by their chemical identity—carbohydrate, protein, or fat. Many fat replacers do function by binding water, but thinking of this in terms of their chemical class helps to explain how they accomplish this and whether more functional ingredients are available. A number of general reviews on fat replacers have been published (6–14) and can be consulted for more specific information. Because virtually every ingredient that participates in water binding or structure setting in foods could be considered a fat replacer, this will be a general overview of those potential ingredients.

A. Carbohydrates

At many ingredient trade shows, about 50% or more of the products promoted as fat replacers have been carbohydrates. Carbohydrates generally mimic fat by binding water, thus providing lubrication, slipperiness, body, and mouth-feel. Carbohydrate adjustments can positively (or negatively) influence shelf-life, freezing characteristics, and mouth-feel by affecting the physical state of the final product (Fig. 2). Processing parameters such as pH, temperature, shear, and compatibility with other ingredients, as well as the rheological character of the carbohydrate, must be considered. Guar, locust bean, and xanthan gums are effective thickeners across a number of food categories (15, 16). Pectins can form soft to hard gels and are widely used in jams, jellies, and tomato-based products. Alginates and carrageenans are commonly used in ice-creams, puddings, fruit gels, and salad dressings. Most gums, depending on form and/or processing conditions, can be used as gelling agents or thickeners.

Gellan gum, approved as a food additive in 1990, is produced by *Sphingomonas elodea* (known earlier as *Pseudomonas elodea*). Gellan exists in two forms—a native, acylated form and a deacylated form. Both forms have a glucose, glucuronic acid, rhamnose backbone (2:1:1), forming a linear tetrasaccharide repeating unit. The acylated from provides elastic gels; the deacylated form provides a more brittle gel. Gellan is compatible with a number of other gums (xanthan, locust bean), starches, and gelatin to manipulate the type of gel, elasticity, and stability and can form strong brittle films exhibiting oil and moisture barrier properties.

Fiber should always be considered for fat replacement and as a replacement for flour and other caloric ingredients (17). "Dietary fiber" as a classification encompasses a wide range of fiber sources that vary in their physical properties. Two subclasses are recognized—soluble and insoluble—which are very different in chemistry and physiological effects. Insoluble dietary fibers, the predominant class, are insoluble in aqueous enzyme solution. About two thirds to three fourths of the dietary fiber in a typical diet is insoluble. Soluble dietary fiber is soluble in an aqueous enzyme system but can be precipitated with 4 parts of ethanol to 1 part of the aqueous mixture. Certain nonabsorbable, nondigestible saccharides are not precipitated in alcohol, and so are not counted as soluble fiber in the current method, even though they contribute to physiological functions and may be beneficial to health. These include insoluble resistant starch, polydextrose, Fibersol-2, fructooligosaccharides, inulin, polyols, and D-tagatose. At this point, these products cannot be claimed in total dietary fiber (TDF), but work is underway to change this.

Cellulose is the most abundant source of insoluble dietary fiber, as well as the most abundant carbohydrate in nature, making up a significant part of the mass of a plant. Cellulose is a linear polymer of beta-1,4-linked D-glucose undigestible by the human gastrointestinal tract (in contrast, the alpha-1,4-linked glucose in starch is highly digestible). Sources are predominantly plant cell walls—foods high in insoluble fiber are whole grains, cereals, seeds, and skins from fruits and vegetables. Cellulose is available in a number of forms—from mechanically disintegrated forms to fermentation-derived to chemically substituted or hydrolyzed versions (Fig. 3). The most common cellulose product found in foods is microcrystalline cellulose (MCC). Many forms are available—MCC as is, MCC with carboxymethylcellulose (CMC) added, or MCC in combination with other gums such as guar and sodium alginate, as well as a range of particle sizes. MCC and its derivatives have been used extensively to replace calories and fat. Other cellulose derivatives such as methylcellulose and hydroxypropyl methyl cellulose (HPMC) have been used for years as multipurpose thickeners because of their ability to hydrate and build viscosity quickly to form clear gels of varying strengths.

Powdered cellulose, at 99+% TDF, is marketed in various fiber lengths to provide a range of water-holding capabilities. Powdered cellulose is derived from wood pulp, treated to remove lignin and other impurities, then milled to a range of fiber lengths from 22 to 120 μm in length. Chemically, it is 90% beta-1,4 glucan plus approximately 10% hemicelluloses. Because it is almost pure TDF, powdered cellulose is considered to be noncaloric. Depending on fiber length, powdered cellulose can retain 3.5–10 times its weight in water (longer fiber lengths are able to retain more water, also increasing the viscosity of the food system).

Insoluble fiber is an important tool for fat reduction and for caloric reduction. Powdered cellulose, cellulose gums, microcrystalline cellulose, and cellulose gels, as well as other plant fibers (oat fiber, soy fiber, wheat fiber, rice flour, hydrolyzed oat flour, etc.), present a wide variety of possibilities in terms of water-binding, viscosity, film-forming, gelling, and pulpiness. Most of these are regarded as "natural," if that is a consideration. When considering these for use, many factors should be considered—water-holding capacity, texture, TDF, caloric density, ingredient legend compatibility, color, and cost to name a few.

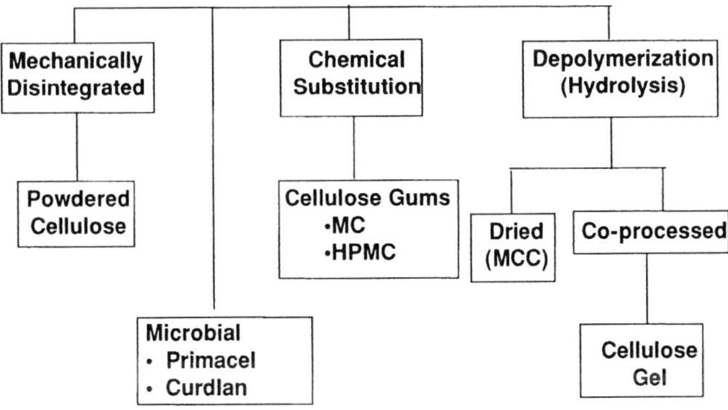

Figure 3 Forms of cellulose commercially available.

The U.S. Department of Agriculture's Agricultural Research Service has had an ongoing program for several years to develop usable products from agricultural by-products such as grain hulls (oat, corn, rice, soybean, peas) and brans (corn, wheat). This has resulted in some new products that have exhibited potential as fat-replacing ingredients, including oatrim, Z-Trim, and Nu-Trim (18). Developed as a fat replacer, oatrim is USDA patented and was licensed to ConAgra, Quaker Oats, and Rhône-Poulenc. Quaker Oats and Rhône-Poulenc joined forces and ConAgra (through Mountain Lake Manufacturing Company) joined with A.E. Staley Co. Oatrim is enzymatically hydrolyzed oat flour containing 5% beta-glucan soluble fiber. Starch in oat flour or bran is hydrolyzed by α-amylase to form a more soluble material (oat β-glucan-amylodextrins) labeled as ''oatrim'' or ''hydrolyzed oat flour.'' Similar to other carbohydrate and protein fat replacers, oatrim can form a gel with water to mimic fat in a number of food applications. From a caloric standpoint, if the gel contains 25% oatrim at 4 kcal/g and the remaining 75% is water at 0 kcal/g, the gel is 1 kcal/g. USDA-conducted studies suggest oatrim might have some hypocholesterolemic benefits, but this product is not currently included in the list of oat products that can carry nutritional labeling to that effect. Oatrim was used in several commercial products, such as fat-free and cholesterol-free milks. On October 9, 1996, USDA announced the development of another product—dubbed ''Z-Trim'' (for zero calories). Whereas oatrim was developed from the inner, starch-containing part of the hull or bran, Z-Trim was developed from the more cellulosic, outer portion. In a process similar to the alkaline/hydrogen peroxide process, which led to a USDA-patented oat fiber also licensed by ConAgra, the hulls of oats were treated in a multistage process to remove the lignin. The resulting cellular fragments were purified, dried, and milled. This dried powder could later be rehydrated to form a gel or be incorporated directly into a food. The USDA applied for a patent in 1995, but this product has not yet been commercialized. Nu-Trim was introduced by USDA in 1998 as a physically modified soluble fiber product with properties similar to oatrim.

Another fiber-containing ingredient that can take the place of fat in foods is resistant starch. Because of its reduced caloric content, it functions mainly as a bulking agent, although it can have other beneficial effects in the finished product. Although the term ''resistant starch'' has only recently become well known, it was coined in the early 1980s, and scientists discussed its dietary effects years before that time. Resistant starches are starches and products of starch degradation that resist enzymatic digestion and act like dietary fiber. Resistant starch is present in many foods—it is naturally found in coarsely ground or chewed cereals, grains, or legumes as a physically inaccessible starch (RS1). It also can be found in bananas, high-amylose starch, and raw potato as naturally resistant or ungelatinized granules (RS2). A third type of resistant starch (RS3) is generated by retrograding starch during food processing. This variety can occur naturally in products such as bread, cereals, and cooked potatoes. Recently identified, another type is classified as RS4. This is representative of starch that has been rendered resistant by chemical modification. Currently, no commercial products of this type exist. As knowledge of TDF developed through the 1980s, resistant starches frequently were discussed, but no significant attempts were made to commercialize them until the 1990s. The first resistant starch released and marketed as such was an RS2 based on a high-amylose corn starch hybrid. The product—Hi-Maize™—was developed by Starch Australasia Ltd. and won the 1995 Australia Institute of Food Science and Technology Industry Innovation Award in Australia. Hi-Maize contained approximately 20–25% total dietary fiber and was introduced into several breads and extruded cereals as a functional fiber in Australia. In 1991, Opta Food Ingredients, Bedford, MA, was granted a U.S. patent for a concentrated, process-tolerant source of RS3-resistant starch in food applications. Independently, National Starch and Chemical Company, Bridgewater, NJ, a major food-starch producer, also was granted a U.S. patent on a common process to produce RS3 starch. The companies agreed to cooperate to commercialize this higher TDF (30%) product, resulting in the 1994 launch of two new products marketed as resistant starches—Crystalean™ (an Opta Food Ingredients product) and Novelose® (a National Starch and Chemical Company product). National Starch continued to expand its Novelose line, offering products with >40% TDF. Resistant starch represents a functional fiber alternative. The RS3-resistant starch, for example, has been shown to significantly improve expansion and eating quality for extruded cereals and snacks SP. The RS2-resistant starch performs well in baked goods because of its small granule size and low water-holding capacity. The RS2 product can be labeled as ''cornstarch,'' and the RS3 as ''maltodextrin,'' both already recognizable terms on many ingredient legends. They compare well to several natural grain sources, contain little fat, are white in color, and neutral in flavor. A 1996 American Institute of Baking study (19) determined that although nearly one third of the resistant RS3 starch consumed is fermented, it produces essentially no energy value.

Two fermentation-derived cellulosic thickeners and stabilizers were introduced in the latter part of the 90s. PrimaCel® (Monsanto) is produced by the microbial fermentation of *Acetobacter xylinum* combined with sucrose and carboxymethylcellulose coagents to promote dispersion. The technology for this food cellulose, also known as ''microfibrous cellulose,'' was originally developed and patented by Weyerhaeuser Company. The product was previously known as Cellulon®. A Generally Recognized As Safe (GRAS) petition was accepted for filing by the U.S. Food and Drug Administration in 1992. These fibers are extremely fine, with a 0.1-μm to 0.2-μm diameter, forming a strong, stable colloidal network. Another fermentation-derived gum has found some success in Japan and was recently given the green light in the United States. On December 16, 1996, the FDA approved curdlan for use as a formulation aid, processing aid, stabilizer, thickener, or texturizer in foods (21 CFR 172.809(b)). Curdlan is a unique polysaccharide with potential uses as a texture modifier and/or gelling agent in processed meats, noodles, surimi-based foods, and processed cooked foods (20). Discovered in 1966 at Osaka University, curdlan (common name) is a polysaccharide produced by *Alcaligenes faecalis* var. myxogenes—it is a linear β-1,3 glucan, insoluble in water, alcohol, and most inorganic solvents, and is indigestible—virtually 100% TDF. Once suspended, curdlan produces a weak low-set gel if heated to 60°C and then cooled to less than 40°C. Gel strength also increases with increased product concentration. If heated to greater than 80°C, a stronger, thermoirreversible gel forms.

Another Asian product, konjac flour, is a centuries-old ingredient obtained by grinding the root of the *Amorphophallus konjac* plant (also known as elephant yam). Konjac is GRAS and has been listed (FCC monograph) as konjac, konjac flour, konjac gum, and konnyaku (21). Konjac's average molecular weight is 200,000 to 2 million daltons (average, 1 million), with short side branches and acetyl groups positioned at C-6 every 6 to 20 sugar units. Konjac is able to form very strong, thermally reversible gels with carrageenan, xanthan gum, and locust bean gum. Adding a base (potassium, sodium, calcium hydroxide, or potassium or sodium carbonate) forms a thermally stable, nonmelting gel. Konjac also forms a heat-stable gel with starch when it is cold-set by raising the pH.

Fibersol®-2 is an indigestible dextrin produced by the acid and enzyme hydrolysis of cornstarch (22). It is claimed as 40–50% soluble fiber and would be stated on an ingredient legend as "maltodextrin." Fibersol-2 contains α-1,4, α-1,6, and β-1,2, β-1,3, and β-1,6 glucosidic bonds. It is soluble at up to 70% dry solids at 20°C and has low viscosity. A similar product with excellent solubility is Benefiber®, a hydrolyzed guar gum (23). Inulin also is not a new dietary component—but only during the last several years has it been heavily marketed in the United States. After starch, inulin is the most abundant nonstructural polysaccharide in nature—being the energy reserve in thousands of plants. As an oligosaccharide, inulin is extremely well known and widely used in Asia. Oligofructose is present naturally in onions, asparagus, leeks, garlic, artichokes, bananas, wheat, rye, and barley. For the purified form, manufacturers generally turn to the more concentrated sources—chicory (>70% inulin on dry solids) and Jerusalem artichoke (also >70%).

Chemically, inulin is a 2 → 1 fructan with the general formula:

Gfn

where G = glucosyl unit, f = fructosyl unit, and n = number of fructosyl units linked n(2). The degree of polymerization (DP) ranges from 2–60. Oligofructose, another product on the U.S. market, contains a mixture of Gfn and independent fructosyl units, with an overall DP of 2–20. In 1992, Zumbro, Inc. (Hayfield, MN) began marketing a Jerusalem artichoke flour based on a 1991 FDA letter verifying that it is recognized as a food. The flour marketed by Zumbro contains 13.2% dietary fiber and 65.7% carbohydrates (78.4% DP-2, 15.2% DP-2 as fructose/fructose and fructose/glucose). Inulin can been used for fat replacement in food products in dry and gel form because a 30–40% solids gel has a fatty feel. The gel strength can be varied to result in a low-calorie fat replacement for specific uses. Several studies indicate the caloric value of inulin is approximately 1.0 to 1.5 kcal/g. Inulin is metabolized preferentially by bifidobacteria in the colon, thereby providing many benefits. However, inulin's uses in foods are still poorly understood. Many Asian and European uses have focused on the health benefits, which are unknown to the U.S. consumer.

In most cultures, plants or plant-derived ingredients (such as flour) have long been used as thickening agents. In plants, starch is a reserve carbohydrate, deposited as granules in the seeds, tubers, or roots. These starch granules differ in size and shape, depending on the plant source (Table 2). Granules of rice starch are small (3–8 μm), polygonal in shape, and tend to aggregate, thereby forming clusters. Cornstarch granules are slightly larger (approximately 15 μm) and round to polygonal. Tapioca granules are even larger (approximately 20 μm), with rounded shapes that are truncated at one end. Wheat starch tends to cluster in several size ranges: normal granules are approximately 18 μm; larger granules average about about 24 μm; and smaller granules average approximately 7 to 8 μm, with round to elliptical shapes. Potato starches are oval and very large, averaging 30 to 50 μm. It is important to note these variations in granule size and shape because they yield distinct differences in viscosity development, stability,

Table 2 Starches—Sources and Modifications

Source
Corn (common, waxy, high amylose)
Rice
Tapioca
Potato
Wheat
Chemical modification
Cross-linked
Substitution
Acid hydrolysis (particle gel)
Physical modification
Pregelatinization (instant)

mouth feel and rate of gelatinization in products. Starch is a carbohydrate polymer, consisting of anhydroglucose units linked together by λ-D-(1 → 4) glucosidic bonds arranged as two major types of polymers, known as amylose and amylopectin. Amylose is primarily linear, containing anywhere from 200 to 2000 anhydroglucose units. Amylopectin is a branched polymer, also connected by λ-1-D (1,4) glucosidic linkages, but with periodic branch points created by β-D-(1,6) glucosidic linkages. Amylopectin is typically much larger than amylose, with molecular weights in the millions. Amylose and amylopectin contain an abundance of hydroxyl groups, creating a highly hydrophillic (or water-loving) polymer that readily absorbs moisture and disperses well in water. Because amylose is linear, it has a tendency to align itself in a parallel nature with other amylose chains, leading to precipitation (in dilute solutions) or retrogradation (in high solids or gels). On the plus side, it also can lead to the formation of strong films, which are extremely useful in certain food applications. The negative side is that amylose can detract from the clearer food products by contributing opacity and also tends to mask delicate flavors (24).

Because of its branching, amylopectin forms clearer gels—often favored in the food industry—that do not form strong films or gels. Retrogradation occurs less readily. In the native form, most starches contain 18–28% amylose, with the remainder as amylopectin. Corn and wheat starches contain approximately 28% amylose, whereas potato, tapioca, and rice varieties are closer to 20%. Two genetic varieties of corn have become popular and well accepted in the industry: ''waxy'' starch, which contains amylopectin with practically no amylose, and high-amylose starches. Of the high amylose starches, two varieties exist: approximately 55% amylose and approximately 70% amylose. Obviously, waxy starches develop weak gels with excellent clarity and are poor film formers. On the other hand, high-amylose starches form very opaque, strong gels and are excellent film formers. Because of the number of hydroxyl groups available on the starch polymer, starches are, very fortunately, easily modified for a wide array of food-industry applications.

During the ''fat-free era,'' a number of starches were developed for the purpose of fat reduction. Examples of this are two potato-based enzymatically converted products from Avebe America, Inc., Princeton, NJ. Paselli® SA-2, and a cleaner flavor version, Paselli Excel. Because of the degree of hydrolysis required to thin these products, these are labeled as ''maltodextrins'' (dextrose equivalent is less than 3). The much blander flavored version was developed to work with delicate flavors. Gels (minimum 18% solids) produced from this product are composed of microparticles 1 to 2 μm in size and are smooth and creamy. Avebe also added an improved taste version of their 6 DE hydrolyzed potato starch, Paselli D-Lite. This was recommended for taste-sensitive dairy applications, such as no-fat vanilla ice-cream, dairy desserts, dairy beverages, or reduced-fat salad dressings. National Starch also designed a series specifically for fat-replacement systems, tagged N-Lite™, labeled as "food starch-modified." One version was recommended specifically for liquid food systems, such as spoonable salad dressings, soups, and microwavable cheese sauces, where it provides lubricity without gelling. A pregelatinized version, LP, was designed to be specific for cold-process liquid applications. Another product designed to function in fat replacement systems was Staley's Instant Stellar®, a modified cornstarch acid hydrolyzed to produce a loose association of crystallites. Xanthan gum aids dispersion and hydration of these crystallites. A "particle gel" network effectively immobilizes water in the formation, creating rheological traits similar to shortening. In addition to their corn-based products, several manufacturers have tapioca starches as part of their product mix, and tapioca and waxy corn are frequently featured as part of the same class or series of product. Tapioca, potato, and rice starches have been recognized to provide flavor advantages over corn, and these corps, plus wheat, are recognized as offering allergenicity advantages because of their low protein levels. Within the processed meats industry, modified corn, wheat, and potato starches compete with proteins as water binders in low-fat meat applications. Using starches as water binders in a fat-replacement system maintains consistent moisture, flavor, and texture throughout product shelf-life.

In a discussion of starches, it should also be noted that syrups produced by enzyme and/or acid hydrolysis of starches from any of these sources can be used to control texture and water activity in food products. High fructose syrups (42% or 55% fructose) are extremely sweet and are low in molecular weight. Corn syrups are available in a wide range of dextrose equivalents, which can be used as a tool for adjusting the level of sweetness (sweetness decreases as the DE is decreased) and also controlling water activity (be aware of the degree of polymerization, or DP, range of the product). Maltodextrins and syrup solids can also be used in this way to manipulate the water activity of a formulation. If calorie reduction is a consideration, monosaccharide and disaccharide sugar alcohols (polyols) should be considered (25, 26). Many of these are available in crystalline or syrup form, with caloric densities ranging from 0.2 to 3.0 kcal/g. These include sorbitol, erythritol, mannitol, isomalt, lactitol, xylitol, and maltitol.

Similar in function to the syrups, a hydrated polydextrose can contribute syruplike qualitites at 1 kcal/g. Polydextrose has been used in ice-creams, confectionery, and jams and jellies for caloric reduction. In high-moisture applications, it can provide a "slippery" mouth-feel that can somewhat mimic fat. In lower moisture applications, polydextrose can provide some hygroscopicity to the final product, which contributes to softening. The original polydextrose product was acidic and slightly bitter because of citrate residual in the product. Newer, less acidic versions are now available for lightly flavored foods.

Glycerine (Fig. 4) is a carbohydrate well known for its use in fruits and candies as a control for drying and graining, as well as softness control (27, 28).

GLYCERINE (GLYCEROL)

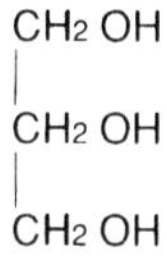

$$\begin{array}{l} CH_2OH \\ | \\ CH_2OH \\ | \\ CH_2OH \end{array}$$

Figure 4 Structure of glycerine.

It is used in jelly candies, fudge, cake icings, cookie fillings, dried fruits, and citrus fruit peels. It prevents oil separation in peanut butter and acts as a softener and humectant in shredded coconut. Glycerine has been used extensively to influence texture and decrease water activity in low-moisture fat-free and fat-reduced products. It is extremely effective, but use is restricted by its sweet, astringent taste. A significant indirect use is the reaction of glycerine with fats and fatty acids to form monoglycerides and diglycerides—emulsifiers important to fat reduction. Beyond the basic emulsifiers, glycerine is used to produce polyglycerol esters, which have become prominent in many of the new entries into the reduced-fat arena. Most of the glycerine now produced is natural—a by-product of production of soaps, fatty acids, and fatty acid esters.

B. Proteins as Fat Replacers

A protein's contribution to fat replacement is determined by the extent of denaturation, which affects flavor and the protein's solubility, gelling properties, and temperature stability (Table 3). Proteins are important as whipping agents, foam and emulsion stabilizers, and dough strengtheners (29). Gelatin and egg albumen have been used extensively in fat-reduced baked goods, frostings, and marshmallows. Soy proteins, egg albumen, wheat gluten, nonfat dry milk, caseinates, and

Table 3 Functionality of Protein-based Fat Replacers

Water-binding
Emulsification
Viscosity
Film formation
Opacity
Whipping and foam stabilization
Binding

whey protein concentrates (WPC) are often used to strengthen fat-reduced pasta, bread, and sweet goods.

Whey proteins and WPC have been used extensively in dairy-based applications (30). Some of these products are based on the ''microparticulated'' concept and consist of extremely small particles within certain size ranges—this is the concept initiated by Simplesse®, which was developed as a whey protein or egg protein–derived fat replacer (31). These particles mimic a fatlike sensation on the tongue. Other products, such as DairyLight® and DairyLo® were based on controlled denaturation to provide a workable viscosity in processing, particularly in ice-creams.

In processed meats, soy protein isolates and concentrates provide high protein quality, meatlike texture and appearance, excellent firmness, reduction in purge (or water loss), and brine retention in injected products. Soy proteins are used for a number of functions, but the top four are emulsification, fat absorption, hydration, and texture enhancement. Because they are concentrated proteins, soy proteins can also be used for film formation, adhesion, cohesion, elasticity, and aeration. Soy flour contains up to 50% protein and also retains carbohydrates, fiber, and fat (unless it is a defatted flour). Soy protein concentrates are made primarily by alcohol extraction of a portion of the carbohydrates from defatted, dehulled soybeans. As awareness of isoflavones has increased, manufacturers have developed other processes to preserve isoflavone content. Most of the fiber is retained in a soy protein concentrate, and these must contain at least 65% protein on a moisture-free basis. The most concentrated soy protein source is isolated soy protein, required to have at least 90% protein on a moisture free basis. These are generally extracted with water from defatted, dehulled soybeans. Because 90% of the product is protein, isolates contain very little fiber or other components.

C. Fat-based Fat Replacers

The fat-based fat replacers have the advantage of a closer relationship chemically with fats, and so physical appearance, thermal stability, and melting points may be a little closer (Table 4). Emulsifiers such as lecithin, monoglycerides and diglycerides, DATEM, SSL, polyglycerol esters, and sucrose esters have been used to extend the available fat in the system, allowing one to increase water content, more fully aerate if needed, improve processing characteristics, stabilize emulsions, and improve shelf-life by complexing with starches and proteins (Table 5).

The concept of structured fats has been around for a number of years. Although medium-chain triglycerides have long been recognized for their nutraceutical potential, their major drawback may be education of the consumer and cost (32). Medium-chain triglycerides (MCT, $C_6 - C_{12}$) are metabolized differently than long-chain triglycerides (LCT, $C_{14} - C_{24}$). LCTs are hydrolyzed, then re-

Table 4 General Categories of Fat-based Fat Replacers

Emulsifiers
Molecular backbones to which fatty acids are attached in such a way that digestion is altered, but functional properties are retained (e.g., Olestra)
Glycerol backbones to which groups with poor digestibility are attached (e.g., Caprenin, Salatrim)

esterified to triglycerides, then imported into chylomicrons, which enter the lymphatic system. MCTs bypass the lymphatic system. They are hydrolyzed to MC fatty acids, which are transported by way of the portal vein directly to the liver, where they are oxidized for energy. They are not likely to be stored in adipose tissue. For enteral and parenteral feeding, their advantage is already known. MCTs provide patients with an energy source similar to glucose, but with twice the caloric value.

The group of fat replacers with the most consumer consciousness at this time are the synthetically structured fats (33, 34). In any discussion of fat substitutes, these hold the most interest because they are designed to look and act like fats, but they contribute fewer calories and less fat. Two approaches to this have been taken: (a) work from a glycerol backbone and attach planned ratios of long-chain (LC) saturated fatty acids with very low caloric density and shorter-chain (SC) fatty acids with slightly lower caloric density than LC fatty acids (caprenin, salatrim); or (b) attach fatty acids to a nonglycerol backbone in such a manner that the molecule is poorly absorbed in the body (olestra). Because the first method results in a triglyceride similar to what could be found in nature, the regulatory route is far shorter: Have it reviewed by an expert panel, file a GRAS petition, and commercialize. The second route is a little more complex. A full food additive petition is required and because of the amount of fat that could

Table 5 Functional Properties of Emulsifiers

Increase water content
Aeration to reduce density
Use fats more efficiently
Improved processing
Stabilize emulsions
Release agent
Starch, protein interactions

potentially be replaced in the diet, approval will be on a category-by-category basis until there is an adequate comfort level with any potential side effects.

Although the "ultimate fat substitute"—heat-stable, fryable—seems to be in the olestra-type class, other developers of fats have chosen a more limited route. Several companies have worked toward synthesizing carefully structured triglycerides with a glycerol backbone, viewing GRAS approval as a much faster route to regulatory acceptance. This is usually done through interesterification, a modification process that results in the rearrangement of the fatty acids of the triglyceride molecule. Through choices of starting materials (different oils or fats), catalysts and/or enzymes, and kinetics, this reaction can be more directed toward a relatively specific end product. This means that the choice of fatty acids involved, as well as their relative ratios, can be limited. Interesterification has been used for some time as a more randomized process to produce plastic fats from animal/vegetable fat blends for use in margarines. The first product commercialized under this grouping was caprenin, a reduced-calorie designer fat consisting of three fatty acids: caprylic (eight carbon atoms, no double bonds), capric (10 carbon atoms, no double bonds), and behenic acid (22 carbons, no double bonds). Behenic acid is only partially absorbed by the body, and the medium-chain fatty acids have lower caloric densities than longer-chain fatty acids, resulting in a total caloric density for caprenin of 5 kcal/g. Caprenin was commercialized as a cocoa butter replacer and was launched in two products. Unfortunately, the product had difficult tempering characteristics and appeared to increase serum cholesterol slightly, resulting in its withdrawal from the market.

As caprenin was being tested, another family of restructured fats was being developed by Nabisco Foods Group, Parsippany, NJ. Salatrim, which is an acronym for short and long acyltriglyceride molecule, is a family of structured triglycerides based on the use of at least one SC fatty acid and at least one LC fatty acid (stearic, C-18). Salatrim triglycerides typically contain one or two stearic acids combined with specific ratios of SC fatty acids (acetic, C-2; propionic, C-3; and butyric, C-4). As with naturally occurring triglycerides, the properties of salatrim are dictated by the fatty acids used, as well as their position of the molecule. The first product, trademarked Benefat 1, was developed to replace cocoa butter in confectionery applications. A GRAS petition was filed in December 1993 and was accepted for filing by FDA in June 1994. Safety studies have shown no effect on serum cholesterol, no effect on absorption of fat-soluble vitamins, and have verified the safety of the molecule. Salatrim is the generic name for this class of molecules, and it is the name used on an ingredient legend. Because of the lower caloric density of stearic acid and the SC fatty acids, salatrim contributes a total of 5 kcal/g. Because the FDA has no regulation for food factors regarding fat reduction (only calories), these claims resulted in some controversy and discussion at FDA. In its 1994 petition, Nabisco claimed that because five-ninths of the fat available in salatrim was used by the body, the product would have a food factor of 5/9—related to both fat and calories.

The ideal fat replacer, or fat substitute, should have a physical appearance, thermal stability, and melting point close to that of the fat being replaced. The "synthetic oil" approach has been to devise a molecular backbone to which fatty acids can be attached such that digestion is altered. Olestra is a mixture of hexa-, hepta-, and octa-fatty acid esters of sucrose (35). As opposed to the glycerol backbone of triglycerides, olestra has a sucrose backbone, to which six to eight long fatty acid chains have been added (70% of the molecules have eight long chains). Olestra is synthesized from sucrose and vegetable oil (cottonseed or soybean), and it has physical properties comparable to conventional fats used in savory snacks and crackers. The complexity of the molecule inhibits the activity of digestive enzymes required to break it down. Therefore, olestra passes through the body undigested, contributing no fat or calories to foods. Olestra is a thermally stable, fryable fat substitute that can substitute for all of the oil in a product, contributing essentially no fat and no calories (36–38). This has been commercialized in several lines of potato and tortilla chip products.

On January 24, 1996, the FDA approved olestra for use "in place of fats and oils in prepackaged ready-to-eat savory (i.e., salty or piquant, but not sweet) snacks. In such foods, the additive may be used in place of fats and oils for frying or baking, in dough conditioners, in sprays, in filling ingredients, or in flavors" (CPR 172.867c). The product must bear an informational statement that says, "This product contains olestra. Olestra may cause abdominal cramps and loose stools. Olestra inhibits the absorption of some vitamins and other nutrients. Vitamins A, D, E, and K have been added." Olestra's approval by the FDA did not come without cost—nearly 30 years, 270 volumes of data, and more than 150 long-term and short-term studies.

The synthetic route to healthy fats is not easy. Other fat substitute projects have either stalled out or have been placed on hold as the olestra project has progressed. Pfizer Food Science Group developed a mixture of fatty acid esters of sorbitol (Sorbestrin), which was reported to be a thermally stable, fryable fat substitute with a caloric content of 1.5 kcal/g. This product is not available commercially, and its use would require a full food additive petition in the United States. Other potential fat substitutes based on the same idea—fatty acid esters of novel backbones—will face the same scrutiny (and expense), so we cannot expect any newcomers to this area in the foreseeable future. Projects (Table 6) reported at ARCO Chemical Company (EPG, or propoxylated glycerol esterified with fatty acids), Frito-Lay Inc. (DDM, or dialkyl dihexadecylmalonate), and Best Foods (TATCA, trialkoxycitrate and trioleyltricarballyates) have not received much press in recent years, but these also would require approval as food additives. Although successes have been achieved recently in what are termed "structured fats," these have not come without considerable time and cost. The

Table 6 Other Acaloric or Low-calorie Fat Substitutes (No FAPs to Date)

EPG (esterified propoxylated glycerol)
ARCO/CPC
Patented 1989
Glycerin reacted with propylene oxide, esterified with fatty acids
DDM (dialkyl dihexabecylmalonate)
Dicarboxylic acid esters of fatty alcohols
Frito-Lay (1986)
TATCA (trialkoxytricarballylate, trialkoxycitrate)
CPC/Best Foods (1985)
Margarines

laborious process of developing an olestra-like product will prevent most ingredient suppliers from exploring that route. New developments will be focused on those categories that will bear the cost.

D. Blending Ingredients

Another strategy that is somewhat old but has resulted in a number of useful products is the use of blends and coprocessed ingredients (Table 7). The number of these products seems to be increasing over the last couple of years as suppliers struggle to come up with more ''user-friendly'' or process-friendly fat-replacement systems (39). Several of these have found their way into fat replacement and can be useful if a cost/benefit exists. If these ingredients are targeted toward one category (such as baked goods or ice-cream), they are definitely worth assessing. If the product claims it ''works in everything,'' one should question what else is needed in the formulation to support such a use. Fat does not have

Table 7 Types of Composites Available

Fruit blends
Syrups/humectants
''Natural''
Flavor/Functional
Stabilizers (e.g., ice-cream, bakery)
Others
N-Flate (National Starch)
K-Blazer(s) (Kraft Food Ingredients)
Yogurtesse (Mid America Dairymen)

the same function in all foods, so it is unlikely that one blend could replace it across all categories. A number of blends have experienced some success, for example: stabilizer systems in ice-creams and baked goods; cellulose gums; many natural blends derived from fruit concentrates and pastes; and a number of specialty ingredients designed to act as fat replàcers. Most of these follow the systems approach, and therefore are combinations of hydrocolloids and fiber, hydrocolloids and emulsifiers, hydrocolloids and opacifiers, etc. The advantage in using blends or coprocessed ingredients depends on the applications: If it works, and the cost is right—why not? In the fast-track world of product development, this could mean the difference between success and failure. Blends also can be a way to better distribute small amounts of small-percentage ingredients on a functional carrier. Examples of this would include emulsifiers on a hydrocolloid carrier or enzyme-emulsifier coprocessed ingredients. Flavors also are often incorporated into these ingredients. These can be flavors that help to supply missing notes that occur through the absence of a highly flavored fat such as butter or may include other flavoring systems. If the supplier will work with the product developer to customize these ingredients, the effort is well worth it. Technical service from the supplier is essential to maximize the ingredient's performance. ''Across the category board'' blends and coprocessed ingredients should be avoided unless the supplier is willing to customize.

III. SUMMARY

Consumers expect more than just fat reduction. They have an increasing awareness that calories are just as important, if not more important, than fat levels, and that other elements of food are important in maintaining overall health. Many of these new ingredients also provide an added bonus—the presence of insoluble or soluble fiber for health benefits and calorie reduction. New fat-reduction strategies can be found in those three words: ''cost,'' ''benefit,'' and ''quality.'' Cost control is a constant consideration in product development, and the ingredients that have made inroads into the market have done so with a cost/benefit of the ingredient in mind. Quality improvement has caused product developers to maintain a watch for new and emerging ingredients that might provide a quality they need for certain formulations. Many say nothing new has emerged from ingredient technology during the last several years. This may seem the case on the surface, because regulatory approval is sometimes difficult, budgets are tight, and product development timelines do not always allow for evaluation of something new. Once developed, several years might transpire before a new ingredient ''breaks through'' in a market. But new products have emerged and new ideas incorporating old products have proven successful. In particular, fiber seems to be making a comeback because of its low calories and nutritional benefits. Many of these ingredients also deliver functional benefits that increase their value to designers trying to formulate lower fat, lower calorie foods.

REFERENCES

1. M Glicksman. Hydrocolloids and the search for the ''oily Grail.'' Food Technol 45(10):94–103, 1991.
2. RC Deis. New age fats and oils. Food Prod Design Nov:27–41, 1996.
3. PA Lucca, BJ Tepper. Fat replacers and the functionality of fat in foods. Trends Ford Sci Techol 5:12–19, 1994.
4. C McWard. Form and function. Baking Snack, Feb:52–54, 1995.
5. A Stockwell. The quest for low fat. Baking Snack May:32–42, 1995.
6. RC Deis. Designing low calorie foods. Food Prod Design Sept:29–44, 1996.
7. J Giese. Fats, oils, and fat replacers. Food Technol 50(4):78–84, 1996.
8. J Giese. Fats and fat replacers: balancing the health benefits. Food Technol 50(9): 76–78, 1996.
9. BF Haumann. The benefits of dietary fats—the other side of the story. INFORM 9(5):366–367, 371, 1998.
10. BG Swanson. Fat replacers: part of a bigger picture. Food Technol 52(3):16, 1998.
11. AM Thayer. Food additives. Chem Eng June 15:26–44, 1992.
12. Anon. Fat reduction in foods. Calorie Control Council, Atlanta, GA, 1996.
13. CC Akoh. Fat replacers. Food Technol 52(3):47–53, 1998.
14. RC Deis. Reducing fat: a cutting edge strategy. Food Prod Design March:33–59, 1997.
15. JD Dziezak. A focus on gums. Food Technol 45(3):115–132, 1991.
16. GR Sanderson. Gums and their use in food systems. Food Technol 50(3):81–84, 1996.
17. L Gorton. Fiber builders. Baking Snack, Feb:46–49, 1995.
18. B. Hardin. Fact sheet: Z-Trim, a calorie-free, high fiber fat replacer. Backgrounder, Z-Trim update conference, Peoria, IL., 1996.
19. GS Ranhotra, JA Gelroth, BK Glaser. Energy value of resistant starch. J Food Sci 61(2):453–455, 1996.
20. Anon. Curdlan—properties and food applications. Tokyo: Takeda Chem. Ind. Ltd., 1994.
21. RJ Tye. Konjac flour: properties and applications. Food Technol 45(3):82–92, 1991.
22. Anon. Fibersol-2. Hyogo, Japan: Matsutani Chem. Ind. Co. Ltd., 1–17, 1999.
23. Anon. The choice is clear—Benefiber. Minneapolis, MN: Sandoz Nutrition. 1995.
24. RC Deis. The new starches. Food Prod Design Feb:34–58, 1998.
25. RC Deis. Low calorie and bulking agents. Food Technol Dec:94, 1993.
26. RC Deis. Adding bulk without adding sucrose. Cereal Foods World 39(2):93–97, 1994.
27. C Croy, K Dotson. Glycerin. INFORM 6(10):1104–1114, 1995.

28. Anon. Glycerine: like wine, it's getting better. INFORM 6(10):1116–1118, 1995.
29. J Giese. Proteins as ingredients: types, functions, applications. Food Technol 48(10): 50–60, 1994.
30. KA Harrigan, WA Breene. Fat substitutes: sucrose esters and Simplesse. Cereal Foods World 34(3):261–267, 1989.
31. GA Corliss. Protein-based fat substitutes in bakery foods. AIB Tech. Bulletin XIV (10):1–8, 1992.
32. A Merolli, J. Lindemann, AJ Del Vecchio. Medium chain lipids: new sources, uses. INFORM 8(6):597–603, 1997.
33. BF Haumann. Structured lipids allow fat tailoring. INFORM 8(10):1004–1011, 1997.
34. CE Stauffer. Fats that cut calories. Baking Snack, May:44–49, 1995.
35. MS Brewer. Olestra and other fat substitutes. FDA Backgrounder BG 95-17, Nov. 13, 1995.
36. RA Yost. Quick-setting sandwich biscuit cream fillings. US Patent #5,374,438, 1994.
37. J Sullivan. Flaky pie shells that maintain strength after filling. US Patent #5,382,440, 1995.
38. J Sullivan. Shortbread having a perceptible cooling sensation. US Patent #5,378,486, 1995.
39. Anon. K-Blazer shortening replacement system for bakery products. Kraft Food Ingredients, Memphis, TN.

Index